MAGNETISM IN CARBON NANOSTRUCTURES

Magnetism in carbon nanostructures is a rapidly expanding field of current materials science. Its progress is driven by the wide range of actual and potential applications for magnetic carbon nanosystems, including transmission elements in spintronics, building blocks of cutting-edge nanobiotechnology, and qubits in quantum computing. These systems also provide novel paradigms for basic phenomena of quantum physics, and are thus of great interest for fundamental research.

This comprehensive survey emphasizes both the fundamental nature of the field, and its ground-breaking nanotechnological applications, providing a one-stop reference for both the principles and the practice of this emerging area. With equal relevance to physics, chemistry, engineering and materials science, senior undergraduate and graduate students in any of these subjects, as well as all those interested in novel nanomaterials, will gain an in-depth understanding of the field from this concise and self-contained volume.

FRANK HAGELBERG is a Professor of Physics at East Tennessee State University and a member of the American Physical Society. His current work focuses on electronic structure methods applied to problems of nanoscience.

MAGNETISM IN CARBON NANOSTRUCTURES

FRANK HAGELBERG
East Tennessee State University

CAMBRIDGE
UNIVERSITY PRESS

CAMBRIDGE
UNIVERSITY PRESS

University Printing House, Cambridge CB2 8BS, United Kingdom

One Liberty Plaza, 20th Floor, New York, NY 10006, USA

477 Williamstown Road, Port Melbourne, VIC 3207, Australia

4843/24, 2nd Floor, Ansari Road, Daryaganj, Delhi – 110002, India

79 Anson Road, #06–04/06, Singapore 079906

Cambridge University Press is part of the University of Cambridge.

It furthers the University's mission by disseminating knowledge in the pursuit of education, learning, and research at the highest international levels of excellence.

www.cambridge.org
Information on this title: www.cambridge.org/9781107069848
DOI: 10.1017/9781107707047

First published 2017

Printed in the United Kingdom by TJ International Ltd. Padstow, Cornwall

A catalogue record for this publication is available from the British Library.

Library of Congress Cataloging-in-Publication Data Names: Hagelberg, Frank.
Title: Magnetism in carbon nanostructures / Frank Hagelberg,
East Tennessee State University.
Description: Cambridge : Cambridge University Press, 2016.
Identifiers: LCCN 2016056468 | ISBN 9781107069848
Subjects: LCSH: Carbon–Magnetic properties. | Nanostructured materials–Magnetic properties. | Magnetic materials.
Classification: LCC QD181.C1 H244 2016 | DDC 620.1/1597–dc23
LC record available at https://lccn.loc.gov/2016056468

ISBN 978-1-107-06984-8 Hardback

We may see how far from unproductive magnetick philosophy is, how agreeable, how helpful, how divine!

William Gilbert, *De Magnete*, 1600

Contents

Preface

Research on magnetic carbon nanostructures is a young and challenging field of materials science. It is in the process of widening our understanding of both items combined in the title of this book, of magnetism, and of carbon nanostructures. Magnetism in Carbon Nanostructures is taken here in the most general sense, which includes intrinsic magnetism as well as magnetism imported into a carbon nanostructure by a magnetic guest species. While effects of the latter type have been extensively studied, the former is the subject of lively current debate. It comprises a broad spectrum of magnetic phenomena not arising from conventional sources of magnetism in materials, i.e. the transition metals of the periodic table, but arising instead from structural irregularities, from dimensional reduction, defects and disorder. Equally persuasive as the promise of novel insight into the nature of magnetism and the basics of carbon nanostructures is the prospect of technological innovation offered by carbon-based magnetic materials. These materials are lighter than their metallic analogs. Also, they are expected to be available at lower fabrication cost and to be more efficient in terms of power consumption. In particular, during the past two decades, a wide range of actual and potential applications of carbon nanostructure magnetism have been identified, extending from fullerene-based magnetic resonance imaging (MRI) contrast agents to zigzag graphene nanoribbon (zGNR) transmission elements in nano-spintronics.

This text aims at a synoptic presentation of these advances, highlighting theory, experiment, and nanotechnological perspectives. Its main objective is to unite what is currently scattered over numerous articles in journals and conference proceedings, or found in individual chapters of multi-author monographs. Beyond these items, a great number of very instructive survey articles, illuminating various facets of carbon nanostructure magnetism, have been published. Likewise, excellent books on magnetism in condensed-matter systems or on carbon nanostructures (most notably on graphene, the topic of several extensive monographs that have appeared during the past few years) are available to readers, many of them covering some of the topics presented in this work. Yet, as far as I see, a comprehensive treatment of the subject, addressing the principles and the practice of carbon nanostructure magnetism in a single volume, is still lacking in this multitude of materials.

Therefore, I hope to satisfy with this publication a growing need of all those interested in novel nano-materials, be it as researchers or developers, as teachers or students. It is intended for a readership at the advanced undergraduate, as well as the graduate and the research level. Familiarity with quantum mechanics and thermodynamics at the level of the regular undergraduate curriculum in physics or chemistry is assumed. Due to the great variety of magnetic carbon nanostructures and, by the same token, the interdisciplinary nature of the field, a wide spectrum of different methods is used to study them. Naturally, this creates barriers that hinder communication between the practitioners of this field. For instance, an experimental chemist or materials scientist, specializing in magnetic metallofullerenes, and a theoretical physicist who explores magnetotransport through graphene, may be interested in each other's areas of expertise, but find it difficult to understand each other's language. Acknowledging this situation, I tried to make this book as self-contained as possible.

In accordance with this goal, I provide condensed introductions into the two principal concepts of this text, magnetism and carbon nanostructures. This material is covered in parts I and II of this book. The remaining three parts, dealing with intrinsic magnetism in carbon nanostructures, with magnetic transport phenomena, and with composite systems, contain the core material of this monograph.

Surveying a quickly expanding field of high current interest confronts the author with the task of hitting a rapidly moving target, a challenge compounded by the great diversity of the contributing science communities. Approaching the topic from various angles, those of physics, chemistry, and materials science, I attempted to provide a viable representation of the state of the art in the area of carbon nanostructure magnetism, but I did not strive to give maximum exposure to the most recent achievements in the field. While it was among my objectives to offer a snapshot of the current state of the field, I aimed still more at an organic presentation that develops, wherever possible, the tenets of carbon nanostructure magnetism from basic principles. Very recent proposals are included, but in some cases, the reader is referred to the relevant articles rather than taken through the often intricate details of what is not sufficiently assessed at this juncture. While I have aimed at a comprehensive survey that contains experimental and nanotechnological aspects, I am well aware that my own perspective, that of a computational physicist, shines through every page of the text.

Most chapters include some exercises which are intended to strengthen the reader's grasp of pertinent concepts, and also to bridge some gaps left in the main text. To avoid major digressions I have appended special-topic sections to some chapters. These supplements add background materials that are meant to aid understanding, or to highlight a theme of interest which nevertheless would block the flow of the narration if treated within the respective chapter. Some sections with additional content, intended to provide deeper or more detailed understanding, were not included in this volume to keep it at a manageable size. These portions are posted on my website: faculty.etsu.edu/HAGELBER/comp_chem.html

Those who have guided my efforts by sharing their knowledge on the topics of this book with me, or giving me insight into their own research, or supported me through words of criticism, or encouragement, or both, are too numerous to be listed here. My whole-hearted thanks go to all of them. I want, however, to acknowledge explictly those colleagues who worked with me on problems related to the subject of the book, namely Anahita Ayasoufi, Tandabany C. Dinadayalane, Oleksij V. Khavryuchenko, Danuta Lesczcynska, Jerzy Lesczcynski, Gilles H. Peslherbe, and, in the very first place, Jianhua Wu who first kindled my interest in carbon nanostructure magnetism and then became my tireless ally in exploring various facets of this field. I thank David Dixon, Jordan Gillenwater, Leo Memolo, and Dianna Pilkenton for immensely efficient help with preparing the figures reproduced in this volume. My gratitude goes to Karyn Bailey, Tessa de Roo, and Roísín Munnelly at Cambridge University Press for invaluable advice and assistance.

Authors tend to acknowledge their families for patiently enduring the great time demands associated with writing a book, and the author's reduced availability during that time. Far from merely tolerating the venture, my family was a constant source of support for me, inspiring in periods of progress, heartening in moments of despair. And so, my very special thanks go to my wife Elizabeth, and my children Ellen, Julia and Philip.

1

Magnetic Carbon Nanostructures?

The label *carbon nanostructure magnetism* joins two notions that do not seem to go together easily, *carbon* and *magnetism*. The main carbon allotropes, after all, are known to be non-magnetic. This appears to be true not only about the well-known solid phases diamond and graphite, but also with respect to the nanoscopic phases manufactured first in the eighties of the last century, and later: fullerenes and carbon nanotubes, single- and few-layer graphene. While intrinsic magnetism is the rule in the d- and f-blocks of the periodic table, it is extremely unusual in the second period, containing light elements with p electrons in their valence shells. Magnetic derivatives from carbon-based nanostructures, such as metallofullerenes with finite magnetic moments in their ground state have been known for decades, but the magnetism of these composites is inherited from elements different from carbon, such as lanthanide atoms with high spins localized in their 4f shells.

By the beginning of this century, however, sightings of intrinsic magnetism in carbon complexes became increasingly frequent and made headlines, not rarely heralded with adjectives like *surprising* [1], *unexpected* [2] or *exotic* [3]. While in the meantime, the initial surprise about carbon magnetism has somewhat worn off, astonishing discoveries continue to be made in this field, such as the first experimental demonstration of spin transport in graphene at the micrometer scale [4] or the first detection of strong spin-orbit coupling in carbon nanotubes [5].

On the other hand, magnetism in carbon nanostructures also continues to be a topic in tension. Foremost, the proposal of intrinsic carbon magnetism due to net magnetic moments at edges or vacant sites in carbon networks is charged with controvery. Proponents point not only at a large body of theoretical and computational work, predicting these effects (for an overview, see [6]), but also at numerous experiments that appear to confirm these predictions. Detractors call attention to the great difficulty of reliably separating signatures of intrinsic magnetism from artifacts due to small admixtures of magnetic impurities [3] and refer to the rather marginal, if not vanishing, net effects yielded by some recent experimental examinations of carbon magnetism [7].

In view of the ongoing debate, it would be premature to state that magnetism in carbon nanostructures is a firmly established discipline within condensed-matter physics or nanoscience. On some of the fundamental tenets of the field, after all, a widely accepted consensus is still lacking. Does this situation suggest that it is too early for a monograph on this subject?

In defense of the present undertaking, I point out that a wide variety of very diverse phenomena fits under the heading of magnetism in carbon nanostructures, some of them fairly traditional and well understood as the result of intense collaborative work between theorists and experimentalists, extending through more than two decades. These may be exemplified by fullerenes with embedded magnetic moieties. Further, the more recent and more disputed zones of the field contain much that touches on essential features of graphene and its derivatives. Therefore, it is of direct relevance to one of the most rapidly advancing branches of current materials science, the exploration of graphene as the first two-dimensional nanomaterial ever fabricated, with an abundance of record-setting electronic and mechanical properties. In this context, one may mention numerous theoretical and computational studies on spin effects in structurally modified or dimensionally reduced graphene where edges, vacancies, or topological defects alter the electronic band structure of the pristine material.

Perhaps more importantly, the union of carbon and magnetism promises more and more highly rewarding technological applications. Controlled fabrication of magnetic materials based on carbon appears to be beneficial in view of their lower production costs when compared with their metal counterparts, their physical and chemical stability, their ease of processing, and their biocompatibility. In all of these rubrics, magnetic carbon compounds are expected to surpass traditional (e.g. metallic) carriers of magnetism by a substantial margin. Further, as suggested by both theoretical and experimental evidence, graphene and carbon nanotubes, and species derived from these prototypes, are of major interest as media for spin transport, and therefore as elements of spintronics networks.

At the same time, the more traditional research into magnetically functionalized carbon nanostructures – hybrids between pure carbon units and magnetic metal species – continues to progress at a rapid pace. In these cases, carbon components provide templates for magnetic units and thus are instrumental for designing composites with tunable magnetic properties. Forming cages of strong cohesion, various types of carbon nanomaterials are used to protect and stabilize encapsulated magnetic subunits.

This book is intended as an introduction to the science of magnetic carbon nanostructures, presenting a comprehensive view of these materials that extends from the basic principles of their architecture to current perspectives for their future use in technology. The text is guided by a threefold motivation, corresponding to three main reasons for the great present interest in carbon-based nanomaterials with finite magnetic moments. These claim relevance as (1) novel materials with a broad range of possible applications, (2) carriers of spin currents and (3) paradigms for novel concepts in nanoscience and condensed-matter physics. A few comments on each of these items:

1. Research on carbon magnetism has identified a wide variety of novel nanosystems, many of them with highly uncommon, intriguing properties. On the most elementary level, strategies have been described to induce ground-state magnetism in carbon nanostructures with more than zero dimensions by dimensional reduction. Further, these structures have been shown to adopt magnetism upon controlled introduction of defects into their lattices, such as adatoms, impurities or vacancies. Of much higher structural complexity are three-dimensional nanoporous carbon networks that display high-temperature ferromagnetism, as exemplified by nanofoam or activated carbon fibers. The prototype of purely organic magnets based on carbon nanostructures is arguably C_{60}-TDAE, a compound consisting of fullerene in combination with an organic molecule. On the other hand, an ample diversity of magnetic metallofullerenes or metallonanotubes have been characterized, species that owe their magnetism to the presence of magnetic metal components. Besides their intrinsic interest for materials science, they are notable for their high relevance to medical therapy and diagnosis. Be it as safe vessels for toxic contrast agents, be it as highly efficient drug delivery systems, these units hold great promise to be developed into powerful instruments of nanobiotechnology.
2. Graphene appears highly suitable as a medium for spin transport. In the first place, this feature is related to the low weight of carbon, which implies weak spin-orbit as well as hyperfine coupling. Both of these interactions induce spin relaxation and decoherence, thus limiting the lifetime of propagating spins. Further, effective strategies for manipulating the spin transport properties of various carbon nanostructures have been outlined. This includes implementation of Rashba interactions, which provide the mechanism underlying the spin transistor concept. Also, for zigzag graphene nanoribbons and zigzag carbon nanotubes, it has been shown that half-metallicity, and thus spin-selective conductance, can be induced by applying external electric fields. These and other effects specific to carbon nanostructures make them appear extremely useful as carrier materials for spin currents in spintronics circuits.
3. Investigation of magnetic interactions in carbon nanostructures has given rise to the discovery of novel phenomena of basic significance for condensed-matter

physics. These are rooted in the *massless Fermion* nature of the low energy states of ideal graphene. Thus, the pioneering work on the quantum spin Hall effect in graphene by Kane and Mele [8] initialized the now-rapidly advancing research on *topological insulators* as a novel class of condensed-matter systems. Likewise, studies on the magnetotransport in graphene led to a basic understanding of phenomena related to *weak antilocalization* and *positive magnetoresistance*. Predictions based on the Dirac rather than the Schrödinger equation, and thus basic tenets of relativistic quantum electrodynamics, can be examined by analyzing graphene [9].

Chapters 2 to 6 are intended to lay a fundament for the treatment of more specific subjects in the later parts of the book. In more detail, Chapter 2 deals with basic concepts of atomic and condensed-matter magnetism, as far as it is of major relevance for this monograph. Chapter 3 provides information about methods that have been used with great success in the study of magnetic carbon nanostructures, including both computational and experimental approaches. Chapters 4 to 6 are meant to acquaint readers with the main protagonists of this text, where particular attention is paid to their elementary magnetic features. Graphene, carbon nanotubes and fullerenes are introduced by presenting their geometric and electronic, as well as basic magnetic properties. The treatment of magnetism in this segment of the text is mostly confined to the interaction between carbon nanostructures and external magnetic fields. Readers well informed about magnetic phenomena in condensed matter systems, and with customary methods to analyze them, may want to start at Chapter 4; those conversant with carbon nanostructures may skip over Chapters 4 to 6.

The essential content of this book is assembled in Chapters 7 to 12, where carbon nanostructures are discussed in terms of intrinsic magnetism, spin transport, charge transport in the presence of a magnetic field and magnetism resulting from interactions with magnetic as well as non-magnetic impurities. Emphasis is placed on graphene and carbon nanotubes, along with secondary structures derived from these systems. Since neither species is intrinsically magnetic, both of them acquire magnetism as their characteristics are manipulated. This can proceed by combining them with foreign elements, a strategy that is outlined in Chapter 12. An alternative route of adopting magnetic features is dimensional reduction. Thus, graphene assumes a ground state that combines ferromagnetic with antiferromagnetic features when the two-dimensional graphene lattice is trimmed down to one- or two-dimensional substructures, termed *zigzag graphene nanoribbons*. An analogous observation is made for carbon nanotubes of the zigzag type when their dimension drops from one to zero. These ideas are outlined in Chapter 7, where the emergence of magnetic phases from edge structures, as well as geometric irregularities such as vacancies in the graphene network or topological defects, is considered. Chapter 8 focuses on spin-orbit coupling. Due to the symmetries of the hexagonal lattice, this effect is predicted to be of second order, and extremely small in graphene. *Rashba interaction*, however, blending the spin-orbit with the Stark effect, makes it possible to tune the former by use of an external electric field. Carbon nanotubes display significantly stronger spin-orbit coupling than graphene, which is a consequence of their curvature and, ultimately, the different mirror symmetry operations valid for the two nanostructures.

Chapters 9 and 10 focus on spin transport phenomena. The present attention paid to this topic and the recent achievements in predicting and detecting spin currents in carbon-based materials are guided mainly by practical interest, namely the potential use of these materials for applications in spintronics. Thus, the small size of the spin-orbit coupling in graphene gives rise to the expectation of long spin relaxation times in this medium, which is essential for controlled operations involving spin currents. Techniques have been identified for how to generate and observe pure spin currents in graphene, and the experimental proof of principle for the effectiveness of these techniques has been given.

Further, carbon nanotubes are of great relevance as nanoelectronic transmission devices. Their enormous mechanical strength and flexibility in conjunction with their supreme electrical properties, including current densities that exceed those of regular metals by a factor of about a thousand, make them highly promising as elements of nanocircuitry. Importantly, the basic geometric and topological features of the tubes, namely their diameter and their chirality, allow for sensitive tuning of their electronic characteristics. Here we consider nanotubes as potential components of spintronics circuits, and especially their use as quantum dots with various spin-related properties that facilitate effective manipulation of their conductance, such as the Pauli blockade effect. These items are presented in Chapter 10, while Chapter 9 summarizes some essential tenets of spintronics, including a basic description of spin circuit elements, among them spin valves, transistors and filters, and also touches on mechanisms of generating, controlling and observing spin currents.

Chapter 11 turns to charge transport through graphene and carbon nanotubes in the presence of a magnetic field. As a magnetic field may be used as a tool to adjust the gap between the valence and the conduction band of graphene-based nanosystems, it can exert a dramatic effect on their conductivity. This is manifested by the magnetoresistance of graphene as well as carbon nanotubes, i.e. the capacity of the external field to either enhance or reduce the electric resistance of the considered nanosystem. The key for understanding magnetoresistance lies in the phenomenon of *weak localization*. This mechanism turns out to cause opposite effects in carbon nanotubes and in graphene, as the former systems tend to display negative magnetoresistance while positive magnetoresistance prevails in the latter. The root for this difference is found in the pseudospinor nature of electronic states close to the Fermi energy of graphene. Another experimentally accessible consequence of this feature is the emergence of an *anomalous quantum Hall effect* in graphene. In this context, we will trace the origin of conductivity quantization in graphene and motivate the peculiar structure of the conductivity levels. The chapter closes with an extension of these arguments to spin transport, providing a survey of the *quantum spin Hall effect* in graphene.

The remainder of this monograph deals with composite and hybrid systems, consisting of carbon nanomaterials in combination with foreign units (Chapters 12 and 14) and extended structures based on carbon allotropes (Chapter 13). A survey of magnetic metallofullerenes is followed by an account on fullerenes enclosing group V atoms (N or P). We motivate the relevance of the hyperfine interaction between the nucleus and the electronic shell of the encapsulated atom for potential implementations of fullerene-based qubits, units of quantum information. We further highlight extended fullerene structures intercalated with alkali atoms (*fullerides*) or organic molecules, most prominently tetrakis-dimethylamino-ethylene (TDAE). The uniting viewpoint in these cases is electron transfer from the guest species to the fullerene moiety, giving rise to a magnetic ground state. The particular interest of the compound TDAE-C_{60} lies in the trait that, here, the combination of non-magnetic components gives rise to a magnetic whole. Our presentation of graphene in conjunction with impurities focuses on the magnetic properties of the widely studied hydrogenated and fluorinated phases of graphene. We further comment on carbon nanotubes with added magnetic metal atoms or nanoparticles and on designs that involve these aggregates as spintronics transmission elements.

Chapter 13 treats carbon networks assembled from the structural motives of the primary carbon allotropes. Examples are provided by nano-onions (fullerenes or fullerene-analogous shells nested into each other) or by graphite and diamond shrunk to nanoscopic size. *Carbon nanofoam*, in contrast, has been understood as a carbon allotrope in its own right. It is composed of elements with open or hyperbolic geometry and thus the converse of fullerenes, which are formed from closed, elliptic clusters. Like the other networks discussed in this chapter, carbon nanofoam is characterized by surprising magnetic features that turn out to be correlated with its geometric structure.

The final chapter of this volume outlines the potential use of various magnetic carbon nanostructures for clinical purposes. These materials have turned out to be of major interest for both diagnosis and therapy. Examples for the former type of application are provided by carbon-based contrast agents for magnetic resonance imaging (MRI), such as metallofullerenes that enclose lanthanide atoms or clusters with high magnetic moments. This concept promises to unite high contrast efficiency with the safety afforded by locking the highly toxic metal component into the stable enclosure of the fullerene shell. Of equal relevance is the therapeutic usage of magnetic carbon nanosystems. Taking advantage of their response to magnetic stimuli, one employs external magnetic fields to guide them toward a specific site within the organism, which is the idea underlying the concept of *magnetic delivery*. Magnetic elements can also be used in the service of thermotherapy, which exploits the sensitivity of cancerous cells to elevated temperatures. Thus it is possible to induce hysteresis losses in magnetically loaded carbon-based nanoparticles, turning them into local sources of heat that counteract the cancerous growth. The very auspicious and very diverse biomedical strategies involving magnetic carbon nanostructures have spurred extensive research with the aim of developing them into clinical tools and, by the same token, enhancing their biocompatibility.

As might be inferred from this brief summary, the topics discussed in this book touch on most of the defining features of carbon nanostructures, attesting to the fundamental nature of carbon magnetism. This field is fundamental, as it builds on the basic physics and chemistry of carbon nanomaterials and, most importantly, keeps raising essential yet unanswered questions about these materials.

Part I

Theories and Methods

2

Basic Notions of Magnetism

This introductory chapter contains some elementary information about magnetism that will be frequently referenced in the later sections of this book. A summary of atomic magnetism (Section 2.1) is followed by a brief outline of magnetic interactions between atoms (Section 2.2) with emphasis on direct and indirect exchange processes. This segment is followed by a discussion of magnetism in condensed matter systems. In particular, the basic magnetic phases are specified in Section 2.3, which introduces the major patterns of magnetic ordering: ferromagnetism and antiferromagnetism, paramagnetism and superparamagnetism, and related schemes.

2.1 Atomic Magnetism

We consider an atom with N electrons and define a reference Hamiltonian $\hat{H}_0$ by stipulating [10]

$$\hat{H}_0 = \sum_i^N \hat{H}_i, \tag{2.1}$$

with $\hat{H}_i$ as a single-particle Hamiltonian for the ith electron, defined as

$$\hat{H}_i = -\frac{\hbar^2}{2m_e}\Delta + \hat{V}_i = \frac{1}{2m_e}\hat{\mathbf{p}}_i^2 + \hat{V}_i = \hat{H}_{kin,i} + \hat{V}_i. \tag{2.2}$$

In this expression, V_i stands for the potential energy of the ith electron and m_e for the electron rest mass. In the presence of an external magnetic field **B**, the single-electron kinetic energy operator $\hat{H}_{kin,i}$ is modified in the following way:

$$\hat{H}_{kin,i} = \frac{1}{2m_e}(\hat{\mathbf{p}}_i - e\mathbf{A})^2, \tag{2.3}$$

where $\mathbf{A}$ is the vector potential that generates $\mathbf{B}$ according to

$$\mathbf{B} = \nabla \times \mathbf{A}. \tag{2.4}$$

The magnetic field is assumed to be homogeneous within the atom. We adopt the *Coulomb gauge*,

$$\nabla \cdot \mathbf{A}(\mathbf{x}) = 0, \tag{2.5}$$

and conclude from Eq. (2.4) that

$$\mathbf{A}(\mathbf{x}) = \frac{1}{2}\mathbf{B} \times \mathbf{x}. \tag{2.6}$$

The Coulomb gauge implies that $[\mathbf{A}, \hat{\mathbf{p}}] = 0$. Combining this relation with Eq. (2.5), we write the single-electron Hamiltonian as

$$\hat{H}_i = \frac{1}{2m_e}\hat{\mathbf{p}}_i^2 + \hat{V}_i + \frac{e}{m_e}\mathbf{A}_i \cdot \hat{\mathbf{p}}_i + \frac{e^2}{8m_e}(\mathbf{B} \times \mathbf{x}_i)^2. \tag{2.7}$$

The external magnetic field thus manifests itself by two additional terms in the single-electron Hamiltonian. In the following, we will inquire about the physical meaning of these two terms and identify them as the sources of *atomic paramagnetism* and *atomic diamagnetism*. Further clarification of the first term is achieved by observing that

$$\begin{aligned} \mathbf{A} \cdot \hat{\mathbf{p}} &= \frac{1}{2}(\mathbf{B} \times \mathbf{x}) \cdot \hat{\mathbf{p}} \\ &= \frac{1}{2}(\mathbf{x} \times \hat{\mathbf{p}}) \cdot \mathbf{B} \\ &= \frac{1}{2}\hat{\mathbf{L}} \cdot \mathbf{B}. \end{aligned} \tag{2.8}$$

In consequence, the single-electron Hamiltonian becomes

$$\hat{H}_i = \frac{1}{2m_e}\hat{\mathbf{p}}_i^2 + V_i + \mu_B \mathbf{L}_i \cdot \mathbf{B} + \frac{e^2}{8m_e}(\mathbf{B} \times \mathbf{x}_i)^2. \tag{2.9}$$

We have introduced here the *Bohr magneton* $\mu_B = \frac{e\hbar}{2m_e} = 5.79\ 10^{-5}$ eV/T. Including, as a purely quantum mechanical phenomenon, the interaction of the magnetic field with the spin of the electron, $\mathbf{S}$, we write the Hamiltonian for the atom in an external magnetic field as[1]

$$\hat{H} = \sum_i \hat{H}_i = \sum_i \frac{1}{2m_e}\hat{\mathbf{p}}_i^2 + V_i + \mu_B(\hat{\mathbf{L}}_i + g_S\hat{\mathbf{S}}_i) \cdot \mathbf{B} + \frac{e^2}{8m_e}(\mathbf{B} \times \mathbf{x}_i)^2. \tag{2.10}$$

[1] For ease of notation, $\hat{\mathbf{L}}$ and $\hat{\mathbf{S}}$ are here defined as dimensionless operators: $\hat{\mathbf{L}} = -i\mathbf{x} \times \nabla_{\mathbf{x}}$, $\hat{\mathbf{S}} = \frac{1}{2}\hat{\boldsymbol{\sigma}}$, where $\hat{\boldsymbol{\sigma}}$ denotes the vector of the three Pauli matrices.

This expression contains, as a dimensionless gyromagnetic ratio, the *spin g factor* of the electron g_S, where $g_S \approx 2$. For a sufficiently small external magnetic field, we may treat the two operators that depend on $\mathbf{B}$ as small corrections of $\hat{H}_0$ and combine them into a perturbation term $\hat{H}_1$, such that

$$\hat{H} = \hat{H}_0 + \hat{H}_1 = \hat{H}_0 + \hat{H}_1^{para} + \hat{H}_1^{dia}, \tag{2.11}$$

distinguishing the paramagnetic (or *Zeeman*) from the diamagnetic Hamiltonian. In the following two subsections, we will subject both terms to close inquiry and discuss their physical content.

2.1.1 Paramagnetism

Focusing on the paramagnetic contribution $\hat{H}_1^{para}$ to the atomic Hamiltonian Eq. (2.11), we sum over the individual orbital angular momenta $\hat{\mathbf{L}}_i$ to obtain the total angular momentum $\hat{\mathbf{L}}$, and proceed likewise with the individual spins $\hat{\mathbf{S}}_i$:

$$\begin{aligned} \hat{\mathbf{L}} &= \sum_i \hat{\mathbf{L}}_i \\ \hat{\mathbf{S}} &= \sum_i \hat{\mathbf{S}}_i . \end{aligned} \tag{2.12}$$

Further, we define the *total angular momentum* $\hat{\mathbf{J}}$ as the sum of $\hat{\mathbf{L}}$ and $\hat{\mathbf{S}}$, corresponding to *L-S* or *Russell–Saunders coupling*, a scheme prevalent in light atoms. In terms of quantum mechanics, this summation amounts to transforming from a basis $|L, S, M_L, M_S\rangle$, i.e. a set of eigenfunctions with L, S, M_L and M_S as good quantum numbers, to a basis $|L, S, J, M_J\rangle$. This is accomplished by use of the *Clebsch–Gordan coefficients* $C_{M_L,M_S;J,M_J}$ [11]:

$$|L, S, J, M_J\rangle = \sum_{M_L,M_S} C_{M_L,M_S;J,M_J} |L, S, M_L, M_S\rangle. \tag{2.13}$$

This transformation changes $\hat{H}_1^{para}$ from $\mu_B(\hat{\mathbf{L}} + g_S\hat{\mathbf{S}}) \cdot \mathbf{B}$ to

$$\hat{H}_1^{para} = \mu_B g_J \hat{\mathbf{J}} \cdot \mathbf{B}. \tag{2.14}$$

This relation contains the g factor for the Zeeman interaction involving the total angular momentum, g_J. We point out that Eq. (2.14) may be rephrased by introducing the operator of the *total magnetic moment*, $\hat{\boldsymbol{\mu}}_F$:

$$\hat{H}_1^{para} = -\hat{\boldsymbol{\mu}}_F \cdot \mathbf{B}, \tag{2.15}$$

and

$$\hat{\boldsymbol{\mu}}_F = -\mu_B g_J \hat{\mathbf{J}}. \tag{2.16}$$

To derive an explicit expression for the composite g factor g_J, one may evaluate $\hat{H}_1^{para}$ in the $|L, S, J, M_J\rangle$ basis, in which the operators $\hat{\mathbf{J}}$ and $\hat{\boldsymbol{\mu}}_F$ are diagonal [12]. It then follows from Eq. (2.16):

$$\begin{aligned}\hat{\boldsymbol{\mu}}_F \cdot \hat{\mathbf{J}} &= -\mu_B(\hat{\mathbf{L}} \cdot \hat{\mathbf{J}} + g_S \hat{\mathbf{S}} \cdot \hat{\mathbf{J}}) \\ &= -\mu_B g_J \hat{\mathbf{J}} \cdot \hat{\mathbf{J}}.\end{aligned} \tag{2.17}$$

At the same time, we have

$$\hat{\mathbf{S}} \cdot \hat{\mathbf{J}} = \frac{1}{2}(\hat{\mathbf{J}} \cdot \hat{\mathbf{J}} + \hat{\mathbf{S}} \cdot \hat{\mathbf{S}} - \hat{\mathbf{L}} \cdot \hat{\mathbf{L}}) \tag{2.18}$$

and

$$\hat{\mathbf{L}} \cdot \hat{\mathbf{J}} = \frac{1}{2}(\hat{\mathbf{J}} \cdot \hat{\mathbf{J}} + \hat{\mathbf{L}} \cdot \hat{\mathbf{L}} - \hat{\mathbf{S}} \cdot \hat{\mathbf{S}}). \tag{2.19}$$

Comparing Eq. (2.17) with Eqs. (2.18) and (2.19), we conclude that

$$g_J = \frac{J(J+1) + L(L+1) - S(S+1)}{2J(J+1)} + g_S \frac{J(J+1) + S(S+1) - L(L+1)}{2J(J+1)}. \tag{2.20}$$

With $g_S = 2$,

$$g_J = 1 + \frac{J(J+1) - L(L+1) + S(S+1)}{2J(J+1)}. \tag{2.21}$$

In the single-electron case, as realized in hydrogen-like atoms, we have $S = \frac{1}{2}$, and therefore

$$-J - \frac{1}{2} \leq M_J \leq J + \frac{1}{2} \tag{2.22}$$

Likewise, the condition $S = \frac{1}{2}$ is satisfied for atoms whose electronic shells can be decomposed into a closed core and a single electron, an architecture exemplified by alkali atoms. Thus, the electronic configuration of a Na atom may be described as $[(1s)^2(2s)^2(2p)^6]3s$, distinguishing an Ne-like atomic core, filled with with ten electrons, from a single-valence electron. Completely filled atomic shells, as well as subshells, contribute neither to the total orbital angular momentum nor to the total spin of the atom.

Inquiring about the energy eigenvalues associated with the paramagnetic Hamiltonian (2.14), we find for the eigenstate $|L, S, J, M_J\rangle$,

$$E_{JM_J} = g_J M_J \mu_B B. \tag{2.23}$$

This result makes it possible to determine the partition function generated by $\hat{H}_1^{para}$. We find

$$Z = \sum_{M_J=-J}^{J} e^{\frac{g_J M_J \mu_B B}{k_B T}} = \sum_{M_J=-J}^{J} e^{x M_J}, \tag{2.24}$$

with $x \equiv \frac{g_J \mu_B B}{k_B T}$, and k_B as the Boltzmann constant. The partition function may be used to evaluate the *magnetization* M of an atomic ensemble with *magnetic moment density* n_S. We let the direction of the magnetic field define the quantization axis of the system and write the magnetization along that axis as

$$M = n_S g_J \mu_B \langle M_J \rangle, \tag{2.25}$$

with $\langle M_J \rangle$ as thermal average of the magnetic quantum numbers M_J. This average may be written as

$$\begin{aligned} \langle M_J \rangle &= \frac{\sum_{M_J} M_J e^{x M_J}}{\sum e^{x M_J}} \\ &= \frac{1}{Z}\frac{dZ}{dx} = \frac{d \ln Z}{dx}. \end{aligned} \tag{2.26}$$

Recognizing that Z is described by a geometric series, we find

$$Z = e^{-Jx} \frac{1 - e^{(2J+1)x}}{1 - e^x}, \tag{2.27}$$

and ultimately

$$Z = \frac{\sinh([2J+1]\frac{x}{2})}{\sinh(\frac{x}{2})}. \tag{2.28}$$

Inserting this expression into Eq. (2.26), we obtain the magnetization M as a function of x and J, namely

$$M = \frac{1}{2} n_s g_J \mu_B [(2J+1)\coth(\frac{2J+1}{2}x) - \coth(\frac{x}{2})]. \tag{2.29}$$

The *maximum magnetization*, M_{max}, is given by[2]

$$M_{max} = n_s g_J \mu_B J. \tag{2.30}$$

The *relative magnetization*, i.e. the ratio between M and M_{max}, is known as the *Brillouin function*, $B_J(y)$, where y is defined as Jx. By Eqs. (2.29) and (2.38), the Brillouin function may be written in the following way:

$$B_J(y) \equiv \frac{M}{M_{max}} = \frac{2J+1}{2J}\coth(\frac{2J+1}{2J}y) - \frac{1}{2J}\coth(\frac{y}{2J}). \tag{2.31}$$

For sufficiently small y, this simplifies to

$$B_J(y) = \frac{(J+1)y}{3J} + O(y^3). \tag{2.32}$$

[2] A word on units: the magnetization is frequently indicated in the Gaussian system. In this system, it is measured in emu/cm^3, which translates into 1.1 $10^{20} \mu_B$/cm^3. The *mass magnetization* is defined in units of emu/gram, or 1.1 $10^{20} \mu_B$/gram.

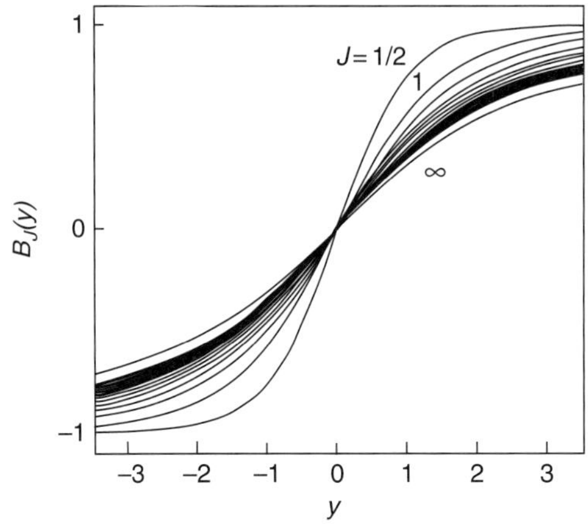

Figure 2.1 Representation of the Brillouin function (Eq. (2.31)), describing the relative magnetization of a paramagnetic sample between the extremes of $J = \frac{1}{2}$ and $J = \infty$.

Exercise 2.1 Show that relation (2.32) describes the Brillouin function $B_J(y)$, as given by Eq. (2.31), in the limit of small y.

This holds, by the definition of the parameter y, for sufficiently small magnetic field strength, or for sufficiently high temperature, or for both. Figure (2.1) provides a sketch of Brillouin functions for various choices of J between the limiting cases of $J = \frac{1}{2}$ and $J = \infty$.

The Brillouin function dictates the *magnetic susceptibility* χ of a paramagnetic substance. The factor χ denotes one of the key properties of magnetic materials, as it provides as measure for the magnetization induced in the material by an external magnetic field **H**[3]:

$$\mathbf{M} = \chi \mathbf{H}. \tag{2.33}$$

This relation implicitly introduces the field **H** that, in isotropic substances, is simply proportional to the field **B**, such that

$$\mathbf{B} = \mu_r \mu_0 \mathbf{H}, \tag{2.34}$$

with μ_r as the *relative magnetic permeability* and μ_0 as the *magnetic permeability of the vacuum* ($\mu_0 = 4\pi 10^{-7} \frac{Vs}{Am}$). From Eqs. (2.29) and (2.30–2.33), we infer the following expression for the paramagnetic case:

[3] We point out that, in most crystals, the magnetic susceptibility is a tensor, such that $M_j = \sum_i \chi_{ji} H_i$, with $i, j = x, y, z$. Further, the magnetic susceptibility may be defined as a quantity that relates the magnetization **M** to the magnetic field **B**. In this case, it carries a dimension (in SI units A/Tm).

$$\chi = \frac{M}{H} = \frac{\mu_0 \mu_r M_{max} B_J}{B}. \tag{2.35}$$

Using for $B_J(y)$ the expansion (2.32), the paramagnetic susceptibility adopts the compact form

$$\chi = \frac{n_s \mu_0 \mu_r}{3k_B T} g_J^2 \mu_B^2 J(J+1). \tag{2.36}$$

Within the limits of the validity of Eq. (2.36), we thus obtain *Curie's law* for paramagnets, stating the inverse proportionality between the susceptibility and the temperature,

$$\chi = \frac{c_{Curie}}{T}, \tag{2.37}$$

and $c_{Curie} = n_s \mu_0 \mu_r g_J^2 \mu_B^2 J(J+1)/3k_B$. For paramagnetic media, χ is a positive quantity, implying that the magnetization induced in the material by the external field enhances the effect of the field. This is in contrast to the diamagnetic response we characterize in the following subsection.

2.1.2 Diamagnetism

The diamagnetic Hamiltonian, $\hat{H}_1^{dia}$, is readily evaluated under the premise made in the preceding subsection, that of an homogeneous magnetic field. This field dictates the quantization axis which, by conventional choice, coincides with the z direction:

$$\mathbf{B} = (0, 0, B). \tag{2.38}$$

This yields for the **B**-dependent factor in the diamagnetic term of single-electron Hamiltonian (2.7):

$$(\mathbf{B} \times \mathbf{x}_i)^2 = B^2(x_i^2 + y_i^2). \tag{2.39}$$

As implied by this relation, the effect of the diamagnetic correction to the ground state energy of an atom is easy to assess in cases of spherical symmetry, which is realized for atoms with completely filled shells. In these cases, it holds that

$$\langle x_i^2 \rangle = \langle y_i^2 \rangle = \frac{1}{3}\langle r_i^2 \rangle. \tag{2.40}$$

Consequently, the energy of the ground state Ψ_0 is increased by an amount ΔE_{dia}, where

$$\Delta E_{dia} = \frac{e^2 B^2}{12 m_e} \sum_i \langle \Psi_0 | r_i^2 | \Psi_0 \rangle. \tag{2.41}$$

To specify the magnetization induced by external magnetic fields in diamagnetic materials, one might invert the relation between energy, the magnetic field, and the magnetic moment. We find

$$\mu = -\frac{\partial E}{\partial B}, \tag{2.42}$$

which holds at zero temperature. Applying this prescription to determine the magnetization of a diamagnetic substance, one obtains

$$M = -\frac{N}{V}\frac{\partial \Delta E_{dia}}{\partial B}, \tag{2.43}$$

where N is the number of atoms, and V the volume they occupy. From this approach,

$$M = -\frac{N}{V}\frac{e^2 B}{6m_e}\sum_i \langle \Psi_0 | r_i^2 | \Psi_0 \rangle. \tag{2.44}$$

From the definition of χ as $\frac{M}{H}$ (see Eq 2.33), we derive

$$\chi = -\frac{N}{V}\frac{e^2 \mu_0}{6m_e}\sum_i \langle \Psi_0 | r_i^2 | \Psi_0 \rangle, \tag{2.45}$$

where $\mu_r \approx 1.0$ has been assumed. In contrast to the paramagnetic case (see Eq. 2.35), the diamagnetic susceptibility turns out to be a negative quantity, implying that the permeated material screens the external magnetic field. Further, from the form of formula (2.45), the outermost electrons of the respective atoms contribute most to the magnetic response. The temperature dependence of χ is negligible in the diamagnetic case.

2.1.3 Spin-Orbit Coupling

So far, we have discussed the behavior of atoms exposed to a magnetic field. In what follows, we will focus on a magnetic interaction within the atom, namely *spin-orbit coupling*, giving rise to the *fine structure* of energy levels in the atomic shell and thus to the emergence of the total angular momentum $\mathbf{J}$, corresponding to J as a good quantum number. As a relativistic effect, this reconfiguration can be derived from the Dirac equation [13]. In a classical frame, the phenomenon may be associated with the appearance of a magnetic field at the site of an electron due to a nuclear current density $\mathbf{j}$ as one transforms into the rest system of the electron, as described by the Maxwell equation,

$$\nabla \times \mathbf{B} = \mu_0 \mathbf{j}. \tag{2.46}$$

Explicitly, the magnetic field felt by the shell electron due to the nuclear current is given by

$$\mathbf{B} = -\frac{\mathbf{v} \times \mathbf{E}}{2c^2}. \tag{2.47}$$

This formula includes the *Thomas correction* [14], which accounts for the fact that the electron's rest frame is a noninertial system and thus includes the effect of the electron's acceleration. Substituting the quantum-mechanical momentum operator and eliminating the electric field $\mathbf{E}$ at the electron's site in favor of the electric potential V, we obtain:

$$\mathbf{B} = \frac{1}{2m_e c^2}(\hat{\mathbf{p}} \times \nabla V). \tag{2.48}$$

The energy involved in the interaction between the electron's spin and the field given by (2.48) yields the Hamiltonian of the *spin-orbit coupling*, $\hat{H}_{SO}$,

$$\hat{H}_{SO}(r) = -\frac{\hbar^2}{2m_e^2 c^2}\frac{1}{r}\frac{dV}{dr}\hat{\mathbf{S}} \cdot \hat{\mathbf{L}}, \tag{2.49}$$

where $\nabla V = \frac{1}{r}\frac{dV}{dr}\mathbf{r}$ has been used, with $\mathbf{r}$ as a radial vector, and the angular momentum operator has been introduced. The ultimate form of the spin-orbit Hamiltonian is

$$\hat{H}_{SO}(r) = \xi_{nL}(r)\hat{\mathbf{S}} \cdot \hat{\mathbf{L}}. \tag{2.50}$$

Integrating the function ξ_{nl} with respect to the radial coordinate results in the *spin-orbit parameter*,

$$\xi_{nl} = \langle \xi_{nL}(r) \rangle = \int_0^\infty R_{nL}(r)\xi_{nL}(r)R_{nL}(r)r^2 dr, \tag{2.51}$$

with R_{nl} as the radial wave function of the electronic state with quantum numbers n and L. This expression scales with the fourth power of the atomic number, Z^4. The magnitude of the spin-orbit parameter, dictating the strength of the spin-orbit interaction, is low by the standard of the exchange interaction (see page 21). For the 3d transition metal, for instance, it amounts to 10–1000 meV. In the $|L, S, J, M_J\rangle$ basis, the expectation value of $\hat{H}_{SO}$ is readily found to be

$$\langle L, S, J, M_J|\hat{H}_{SO}|L, S, J, M_J\rangle = \frac{\xi_{nL}(r)}{2}(J(J+1) - L(L+1) - S(S+1)). \tag{2.52}$$

In the general case, the Russell-Saunders scheme underlying the notion of the total angular momentum J applies to multi-electron coupling, in the sense that the Coulomb and the exchange interaction give rise to a total spin as the sum of the spins of the contributing electrons ($\mathbf{S} = \sum_i \mathbf{S}_i$). An analogous statement holds for the total orbital angular momentum ($\mathbf{L} = \sum_i \mathbf{L}_i$). For completely filled atomic shells, the total spin and the total orbital angular momentum vanish, and so does the total angular momentum. For partially filled atomic shells, S, L and J are seen to obey *Hund's rules* [15]. These three rules may be stated as follows [10]:

1. The quantum number S adopts the maximum value compatible with the Pauli exclusion principle. This requirement leads to

$$S = \frac{1}{2}\{(2l+1) - |2l+1-P|\}, \tag{2.53}$$

where $2l+1$ is the multiplicity of the shell in question and P is the number of electrons occupying it.

2. The quantum number L adopts the maximum value compatible with the Pauli exclusion principle. In particular, it holds

$$L = S|2l+1-P|. \tag{2.54}$$

3. The quantum number J is found from

$$J = |L - S| \tag{2.55}$$

if $P \leq 2l+1$, i.e. the shell is maximally half-filled, and

$$J = L + S \tag{2.56}$$

if $P \geq 2l+1$, i.e. the shell is minimally half-filled.

While there is no general proof for these rules, they have been confirmed experimentally or by computational modeling of many individual systems. Further, their contents are intuitively accessible. For instance, consider the application of the first rule to a two-electron system: the condition of maximum spin implies a symmetric spin function. One then concludes from the Fermionic constraint on the wave function as whole that the space function is antisymmetric, favoring greater distances between the electrons over closer distances and thereby reducing the repulsive electrostatic energy of the system.

2.1.4 Magnetic Hyperfine Coupling

The term *magnetic hyperfine coupling* refers to the mutual influence between the magnetic moments of electrons and nuclei. The dominant contribution to this interaction is the *Fermi contact potential energy*, as described by the Hamiltonian $\hat{H}_{HF}$:

$$\hat{H}_{HF} = \frac{8\pi}{3}\frac{\mu_0}{4\pi} g_S \mu_B g_N \mu_N \hat{\mathbf{S}} \cdot \hat{\mathbf{I}} \delta(\mathbf{x} - \mathbf{X}), \tag{2.57}$$

with μ_0 as the permeability of the vacuum and g_S as the g factor of the free electron (=2.0023). The symbols $\mathbf{X}$, $\hat{\mathbf{I}}$ and μ_N/g_N stand for the position, the spin operator and the magnetic moment/the g factor of the nucleus, respectively, while $\hat{\mathbf{S}}$ denotes the electronic spin operator. Electrons in an s-state $\psi_s(\mathbf{x})$ that are exposed to the Fermi contact interaction experience an effective nuclear magnetic field of the form

$$\mathbf{B}_{nuc} = \frac{2\mu_0}{3}\frac{g_S}{g} g_N \mu_N \, \hat{\mathbf{I}} |\psi_s(\mathbf{X})|^2, \tag{2.58}$$

involving the ratio of the free and the effective g factor of the electron. Often, a more compact expression for the Fermi contact coupling than Eq.(2.57) is encountered, namely

$$\hat{H}_{HF}(\mathbf{X}) = \mu_B g \, \hat{\mathbf{S}} \cdot \mathbf{B}_{nuc} = A\hat{\mathbf{S}} \cdot \hat{\mathbf{I}}, \tag{2.59}$$

with A as the *hyperfine coupling constant* $\frac{2\mu_0}{3} g_S \mu_B g_N \mu_N |\psi_s(\mathbf{X})|^2$. Due to the small ratio between the nuclear and the electronic magnetic moments, the splitting induced by the hyperfine structure is about three orders of magnitude lower than the fine-structure splitting.

2.2 Magnetic Coupling between Several Centers

This section focuses on elementary mechanisms of magnetic interaction. As the most basic one among these, one may consider the interaction between two magnetic dipoles $\boldsymbol{\mu}_1$ and $\boldsymbol{\mu}_2$ whose separation from each other is described by a vector $\mathbf{R}$. The energy due to *magnetic dipole interaction* is found from the expression

$$E = \frac{\mu_0}{4\pi R^3}(\boldsymbol{\mu}_1 \cdot \boldsymbol{\mu}_2 - \frac{3}{R^2}(\boldsymbol{\mu}_1 \cdot \mathbf{R})(\boldsymbol{\mu}_2 \cdot \mathbf{R})). \tag{2.60}$$

For typical situations involving magnetic moments in the order of μ_B and distances of about $1\mathring{A}$, this formula yields energies that correspond to temperatures around one Kelvin. Thus, dipolar magnetic interaction is usually not the mechanism responsible for magnetic ordering, as most magnetic materials display substantially higher transition temperatures.

2.2.1 Direct Exchange

Exchange interactions arise between identical particles as a consequence of Coulomb interactions constrained by the Pauli exclusion principle. Markedly stronger than dipolar interactions, they are the physical cause of magnetic ordering phenomena, such as ferromagnetism, antiferromagnetism or ferrimagnetism. The origin of these interactions is most easily discussed assuming an ideal situation that involves two electrons, placed at positions $\mathbf{x}_1$ and $\mathbf{x}_2$, described by single particle wave functions ϕ_a and ϕ_b, and governed by a Hamiltonian $\hat{H}$. This arrangement is compatible with four different spin functions, partitioned into one singlet

$$\chi_S = |0, 0\rangle = \frac{1}{\sqrt{2}}(|\uparrow_1\downarrow_2\rangle - |\uparrow_2\downarrow_1\rangle) \tag{2.61}$$

and three triplet states

$$\begin{aligned} \chi_{T,1} &= |1, 1\rangle = |\uparrow_1\uparrow_2\rangle \\ \chi_{T,0} &= |1, 0\rangle = \frac{1}{\sqrt{2}}(|\uparrow_1\downarrow_2\rangle + |\uparrow_2\downarrow_1\rangle) \\ \chi_{T,-1} &= |1, -1\rangle = |\downarrow_1\downarrow_2\rangle. \end{aligned} \tag{2.62}$$

The spin states χ_S and $\chi_{T,M}$, with $|M| = 0, 1$, are eigenfunctions of the total spin operator $\hat{\mathbf{S}}$ for the two-particle system, such that

$$\hat{S}^2\chi_{S,T} = (\hat{S}_1 + \hat{S}_2)^2\chi_{S,T} = S(S+1)\chi_{S,T}, \tag{2.63}$$

and S = 0 (1) in the singlet (triplet) case. The overall wave function, composed of a spin and a spatial part, has to satisfy the constraint of antisymmetry with respect to particle exchange. This condition is met by the trial functions ψ_S and ψ_T:

$$\begin{aligned} \psi_S &= \frac{1}{\sqrt{2}}\{\phi_a(\mathrm{x}_1)\phi_a(\mathrm{x}_2) + \phi_a(\mathrm{x}_2)\phi_a(\mathrm{x}_1)\}\chi_S \\ \psi_T &= \frac{1}{\sqrt{2}}\{\phi_a(\mathrm{x}_1)\phi_a(\mathrm{x}_2) - \phi_a(\mathrm{x}_2)\phi_a(\mathrm{x}_1)\}\chi_T. \end{aligned} \tag{2.64}$$

The triplet alternative ψ_T, where the space function is antisymmetric, and the spin function symmetric with respect to particle exchange, may be realized by any one of the three choices (2.62) that are distinguished by their magnetic quantum numbers. Once the two-particle wave functions $\psi_{S,T}$ are found, their energies are obtained from

$$\begin{aligned} E_S &= \langle\psi_S|\hat{H}|\psi_S\rangle \\ E_T &= \langle\psi_T|\hat{H}|\psi_T\rangle. \end{aligned} \tag{2.65}$$

The decomposition

$$\hat{S}^2 = \hat{S}_1^2 + \hat{S}_2^2 + 2\hat{\mathbf{S}}_1 \cdot \hat{\mathbf{S}}_2, \tag{2.66}$$

implies that

$$\begin{aligned} \hat{\mathbf{S}}_1 \cdot \hat{\mathbf{S}}_2 &= -\frac{3}{4}, \text{ if } S = 0, \text{ and} \\ \hat{\mathbf{S}}_1 \cdot \hat{\mathbf{S}}_2 &= +\frac{1}{4}, \text{ if } S = 1. \end{aligned} \tag{2.67}$$

This leads to an *effective Hamiltonian* for the two-electron system. It is partitioned into a scalar and a spin-dependent contribution, according to

$$\hat{H} = \frac{1}{4}(E_S + 3E_T) - (E_S - E_T)\hat{\mathbf{S}}_1 \cdot \hat{\mathbf{S}}_2. \tag{2.68}$$

From Eqs. (2.64) and (2.65), the second term of $\hat{H}$ is proportional to the exchange integral, such that

$$J_{ex} \equiv \frac{E_S - E_T}{2} = \int d\mathbf{x}_1 d\mathbf{x}_2 \phi_a(\mathbf{x}_1)^* \phi_b(\mathbf{x}_2)^* \hat{H} \phi_a(\mathbf{x}_2)\phi_b(\mathbf{x}_1), \tag{2.69}$$

defining the *exchange constant* J_{ex}. This constant is positive if the triplet state is preferred ($E_S > E_T$) and negative if the singlet state is more stable ($E_T > E_S$). The former condition is usually satisfied if the electrons are both in the shell of the same atom, as an antisymmetric spatial (symmetric spin) wave function minimizes the Coulomb repulsion between the electrons. If the electrons are attached to neighboring atoms, the singlet alternative is usually realized, since this configuration is favorable for bond formation between the two adjacent atoms. Generalizing the spin contribution to the Hamiltonian (2.68), one obtains the *Heisenberg Hamiltonian*

$$\hat{H}_H = -\sum_{i,j>i} \tilde{J}_{ij} \hat{\mathbf{S}}_i \cdot \hat{\mathbf{S}}_j, \tag{2.70}$$

which may be used to model arrays of spins. A simple choice for the coefficients $\tilde{J}_{ij}$ is given by $\tilde{J}_{ij} = \tilde{J} = 2J_{ex}$ if the atoms labelled by the indices i, j are adjacent to each other, and $\tilde{J}_{ij} = 0$ otherwise. As will be discussed later in further detail (see Section 3.1.2), the condition $\tilde{J} > 0$ is conducive to a uniform orientation shared by all spins of the array. If $\tilde{J} < 0$, in contrast, opposite orientation between adjacent spins will be favored.

Direct exchange, as described in this subsection, is usually not sufficient to capture the magnetic features of extended quantum systems. Even in homogeneous metals, this mechanism is often not the principal cause of the observed long-range magnetic order. This may be exemplified by the rare earth elements. Their atomic magnetism is determined by their open 4f shells. Since, however, the 4f electrons are strongly localized, spreading only over a small fraction of the interatomic distance in rare earth metals, the direct exchange coupling between 4f shells of neighboring atoms is too small to establish long-range order. For most magnetic materials, the effect of competing *indirect exchange* processes must be included for adequate explanation of their magnetic structure. In these cases, the magnetic order does not result from direct electron wave function overlap, but rather from the coupling between two spin-polarized electrons with an intermediary agent, giving rise to an effective interaction between the two spins. In the following subsection, we consider different schemes of indirect exchange, as typically realized in metals and in isolators.

2.2.2 Indirect Exchange

For metals, the most important mechanism of indirect exchange consists in the mutual influence of two valence electron spins on each other due to their coupling with conduction electrons. This case is known as *RKKY interaction*. We further characterize briefly *superexchange* and *double exchange* processes, which are chiefly operative in isolators.

RKKY

The Ruderman–Kittel–Kasuya–Yosida (RKKY) theory [16, 17, 18] describes the indirect coupling between two magnetic moments located in a crystal. This coupling is mediated by the conduction electrons of the crystal. The *RKKY mechanism* explains, among other phenomena, essential features of the giant magnetoresistance (GMR) effect, most notably the variation of the magnetic phase with the length of a non-magnetic spacer material placed between two ferromagnetic layers [19]. RKKY theory is applicable to voids, but also to other defects, such as impurities or adatoms.

Assuming that the interaction between the spin $\mathbf{S}_a$ located at the position $\mathbf{x}_a$ and the spins $\mathbf{S}_i$ carried by a conduction electron with space coordinates $\mathbf{x}_i$ can be represented by a contact term, we write

$$\hat{H}_{int} = -\tilde{J}_{RKKY} \sum_i \hat{\mathbf{S}}_a \cdot \hat{\mathbf{S}}_i \delta(\mathbf{x}_i), \tag{2.71}$$

with J as exchange constant and $\mathbf{x} = \mathbf{x}_a - \mathbf{x}_i$. This expression may be rephrased by introducing an effective magnetic field $\mathbf{B}_{eff}$,

$$\hat{H}_{int} = -g\mu_B \sum_i \hat{\mathbf{B}}_{eff} \cdot \hat{\mathbf{S}}_i, \tag{2.72}$$

where $\mathbf{B}_{eff}$ is given by

$$\hat{\mathbf{B}}_{eff} = -\frac{\tilde{J}_{RKKY}}{g\mu_B} \hat{\mathbf{S}}_a \delta(\mathbf{x}). \tag{2.73}$$

We define the *spin susceptibility* or *Pauli susceptibility* χ_P as a non-local response function that, in conjunction with the magnetic field, determines the magnetization $\mathbf{M}_P$ induced by the spin $\mathbf{S}_a$, according to

$$\hat{\mathbf{M}}_P(\mathbf{x}) = \int d\mathbf{x}' \chi_P(\mathbf{x} - \mathbf{x}') \hat{\mathbf{B}}_{eff}(\mathbf{x}') = -\frac{\tilde{J}_{RKKY}}{g\mu_B} \chi_P(\mathbf{x}) \hat{\mathbf{S}}_a. \tag{2.74}$$

Using the relation

$$\hat{\mathbf{M}}_P(\mathbf{x}) = -g\mu_B \hat{\mathbf{s}}(\mathbf{x}), \tag{2.75}$$

which associates the magnetization with an induced spin density distribution $\hat{\mathbf{s}}(\mathbf{x})$, we arrive at a formula for this distribution, namely

$$\hat{\mathbf{s}}(\mathbf{x}) = \frac{\tilde{J}_{RKKY}}{(g\mu_B)^2}\chi_P(\mathbf{x})\hat{\mathbf{S}}_a. \tag{2.76}$$

A spin $\mathbf{S}_b$ that is located at $\mathbf{x}$ will be affected by the spin density $\mathbf{s}$. Casting the effective interaction between the localized spin and the conduction spin density into the form (2.71), we arrive at a relation that is symmetric in the two valence electron spins $\mathbf{S}_a$ and $\mathbf{S}_b$:

$$\hat{H}_{RKKY} = -\frac{\tilde{J}^2_{RKKY}}{(g\mu_B)^2}\chi_P(\mathbf{x})\hat{\mathbf{S}}_a \cdot \hat{\mathbf{S}}_b. \tag{2.77}$$

For a free-electron system, the magnetic susceptibility has been evaluated in momentum space [20]. Expressing χ as a function of the difference vector $\mathbf{q} = \mathbf{k} - \mathbf{k}'$, where both $\mathbf{k}$ and $\mathbf{k}'$ lie within the first Brillouin zone, one finds

$$\hat{H}_{RKKY} = -\frac{\tilde{J}^2_{RKKY}}{(g\mu_B)^2 V}\sum_{\mathbf{q}}\chi_P(\mathbf{q})e^{i\mathbf{q}\cdot\mathbf{x}}\hat{\mathbf{S}}_a \cdot \hat{\mathbf{S}}_b, \tag{2.78}$$

with V as reference volume. RKKY interactions result in an oscillatory profile of the magnetic moments as function of the distance from the magnetic impurity, with alternating zones of ferromagnetic and antiferromagnetic coupling. This effect tends to increase the magnetic moment at the impurity site. In contrast to this behavior, *Kondo screening* leads to a reduction of this magnetic moment [10, 21]. This phenomenon is rooted in the *Kondo effect* which consists in the formation of a many-body spin singlet state, resulting from the exchange interaction between the localized spin of the impurity and conduction electron spins at sufficiently low temperature. For $T \gg T_K$, where T_K stands for the *Kondo temperature*, the impurity spin and delocalized spins with opposite orientation only undergo a weak antiferromagnetic coupling. For $T \ll T_K$, however, this coupling is strong enough to surround the impurity spin with a cloud of compensating spins, thus generating a nonmagnetic state. This mechanism augments the resistivity of the material by a spin-dependent part, leading to a resistivity minimum in the Kondo temperature regime.

Superexchange and Double Exchange

Superexchange is an important mechanism of indirect exchange in insulators such as ionic solids, where magnetic metal atoms are usually located too far away from each other to allow for direct interaction. In those cases, an effective increase of the

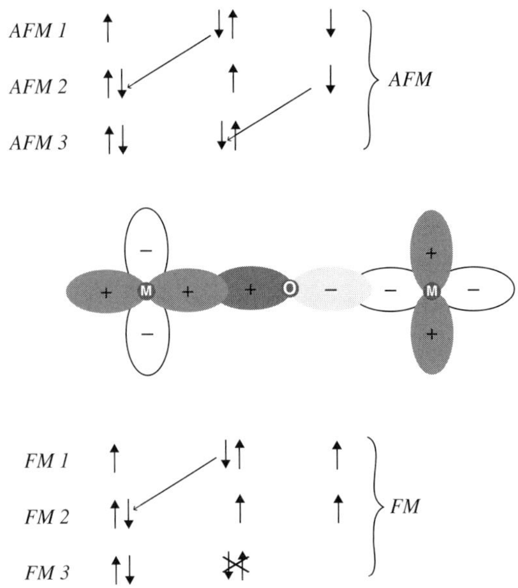

Figure 2.2 Scheme of the superexchange mechanism, as exemplified by a four-electron model system involving two spin-polarized $3d$ electrons coupled by two intermediate $2p$ electrons. Ferromagnetic (FM) and antiferromagnetic (AFM) configurations are compared. A low-energy excitation (AFM3) that arises from the transfer of a $3d$ electron into the half-occupied $2p$ orbital is allowed for the AFM system but forbidden for the FM alternative (FM3). Likewise, emptying the $2p$ reservoir by transfer of both intermediate electrons into the $3d$ states may proceed for the AFM, but not the FM system.

exchange interaction range may still establish a magnetically ordered ground state. Indirect exchange proceeds here typically between localized atoms with finite magnetic moments that are separated by a nonmagnetic intermediate. The paradigm for this process is given by transition metal oxides, such as MnO or NiO, involving two $3d$ transition metal atoms and an O^{2-} anion. The coupling mechanism involves here two $3d$ orbitals in contact with an oxygen $2p$ orbital, as illustrated in Figure 2.2 by the example of a simple four-electron system. The antiferromagnetic ground state, $AFM1$, is here compared with the ferromagnetic alternative, $FM1$. Further, excited states with low transition energies are indicated, arising from electron transfer, which may proceed into or out of the intermediate oxygen $2p$ orbital. Comparing the two competing spin configurations, one realizes that some low-energy excited states are allowed in the antiferromagnetic case but, as a consequence of the Pauli exclusion principle, blocked in the ferromagnetic case, favoring the former over the latter. This feature can be associated with the greater spread of the electrons attached to the intermediate in the antiferromagnetic as opposed to the ferromagnetic situation. In terms of the electron wave function, enhanced spatial extension involves wave length increase and thus momentum reduction, resulting in kinetic energy lowering.

While superexchange often gives rise to an antiferromagnetic ground state, situations are known where ferromagnetism results from this mechanism. This is the case when the components shown in Figure 2.2 form a right angle, i.e. the configuration metal atom - O atom - metal atom involves a ninety-degree bond. In this case, the two $3d$ orbitals may interact with different $2p$ orbitals of the O^{2-} atom, e.g. a p_x and a p_y orbital. Electron transfer from the latter to the metal atoms leaves the p_x, p_y orbitals of the O atom with one electron each. These two tend to form a triplet. Re-donation of the transferred electrons to the oxygen shell will then result in a ferromagnetic coordination between the two $3d$ electrons of the metal component. This interaction is markedly weaker than the more common antiferromagnetic superexchange.

A related indirect exchange process, *double exchange*, correlates ferromagnetic coupling with electron mobility. This mechanism is operative in compounds where the magnetic metal component exists in different states of valency, as exemplified by the ferrimagnetic material magnetite, Fe_3O_4. The iron atom locates here at tetrahedral and octahedral sites. While it adopts a threefold positive charge (Fe^{3+}) at the former, it is seen to be in an equal mixture of Fe^{2+} and Fe^{3+} at the latter site, where the Fe cations of different oxidation states couple ferromagnetically. Other examples for the site-dependent valency of magnetic metal atoms are provided by various manganese-containing compounds, such as the perovskite-type system $La_{1-x}Sr_xMnO_3$, with $0.16 < x < 0.50$ [22, 23]. Figure 2.3 illustrates the elementary mechanism of double exchange by the example of a doped Mn perovskite, containing both Mn^{3+} and Mn^{4+} ions, bridged by an O^{2-} anion. If, as assumed in the figure, O atoms form an octahedron around the Mn centers, the 3d orbital energy of the latter will be split into an e_g and a t_{2g} sublevel. In the configuration shown, *double exchange*, involving the passage of an O^{2-} electron into the e_g subshell of Mn^{4+}, followed by electron transfer from the e_g sublevel of Mn^{3+} into the vacated state of the O atom, requires transfer of a spin-up electron in the first of the two steps, thus forbidding a spin flip to occur during the effective transition between the e_g states of the two Mn centers. It is important to realize that this transition would not proceed if the two t_{2g} substates would display opposite spin orientations, as Hund's first rule (see page 20) would be violated in this modified case. Therefore, double exchange enforces ferromagnetic coordination between the two indirectly interacting metal atoms.

In the following, we survey various types of magnetic long-range order that emerge from the magnetic interactions described in this section.

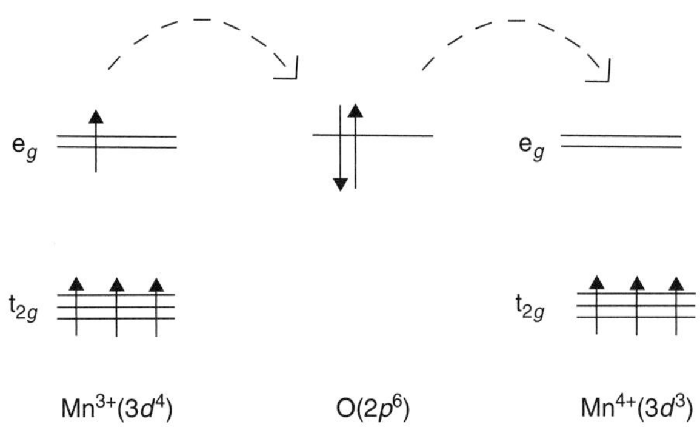

Figure 2.3 Double exchange between two Mn cations of different valency, with an O^{2-} anion as intermediate. An octahedral ligand field, giving rise to a splitting of the Mn(3d) orbital into an e_g and a t_{2g} sublevel, is assumed. The ferromagnetic coordination between the two Mn ions implies that the oxygen electrons can delocalize by transferring a spin-up electron to the e_g subsystem of Mn^{4+}, with a subsequent transfer of the Mn^{3+} e_g electron replenishing the shell of the oxygen anion. Note that this process requires ferromagnetic alignment between the two Mn components, as the configuration of maximum spin multiplicity in the Mn(3d) orbital is strongly favored, in accordance with Hund's first rule. Only two electrons of the O($2p^6$) orbital are included in the sketch.

2.3 Magnetic Order

Introducing magnetic order, we will mostly focus on the major prototypes *ferromagnetism* and *antiferromagnetism*, as they are of chief importance for this monograph. We will further mention *ferrimagnetism* as well as *helically ordered structures* and *spin glasses*.

2.3.1 Ferromagnetism

A ferromagnet presents an example for spontaneous collective order, associated with the uniform alignment of all magnetic moments into a shared direction, without any need for an inducing external field. In the most elementary approach, the mechanism that establishes ferromagnetic order is direct exchange as outlined in the previous section. If, for the sake of simplicity, the orbital angular momentum is assumed to vanish, such that $J = S$, the magnetic Hamiltonian for a ferromagnetic system exposed to an external magnetic field can be derived from (2.70) and may be written as

$$\hat{H}_H = -\sum_{\langle i>j \rangle} \tilde{J}_{ij}\hat{\mathbf{S}}_i \cdot \hat{\mathbf{S}}_j + g\mu_B \sum_i \hat{\mathbf{S}}_i \cdot \mathbf{B}. \tag{2.79}$$

The indices *i,j* refer here to adjacent sites. A useful construct for further simplification of this expression is the *effective molecular field* [24], involving the treatment of the first term in expression (2.79) in analogy to the second, i.e the Zeeman Hamiltonian. This succeeds by introducing the mean field

$$\mathbf{B}_{MF} = -\frac{1}{g\mu_B} z\tilde{J}\langle \hat{\mathbf{S}}_j \rangle \tag{2.80}$$

such that the Hamiltonian (2.79) is approximated by the compact term

$$\hat{H}_H = g\mu_B \sum_i \hat{\mathbf{S}}_i \cdot (\mathbf{B} + \mathbf{B}_{MF}). \tag{2.81}$$

Here, $\tilde{J}_{ij}$ has been equated with $\tilde{J}$, and z stands for the average number of nearest sites. Since the mean field $\mathbf{B}_{MF}$ has the effect of aligning the spins of the system, it is plausibly assumed to be proportional to its magnetization:

$$\mathbf{B}_{MF} = \lambda \mathbf{M}. \tag{2.82}$$

The mean-field formalism encoded by the relations (2.80) to (2.82) has the virtue that it readily extends the quantum-mechanical model for the magnetization of paramagnets, as sketched in Section 2.1, to the ferromagnetic case. To make this model applicable to ferromagnets requires replacing the field $\mathbf{B}$ with the generalized field $\mathbf{B} + \lambda\mathbf{M}$ in the independent variable y of the Brillouin function (see Eq. (2.31)), such that

$$y = \frac{g_J \mu_B J (B + \lambda M)}{k_B T}. \tag{2.83}$$

The difficulty associated with this substitution is that it makes the Brillouin function dependent on the magnetization of the system and thus the very quantity that, according to Eq. (2.31), it is supposed to describe. However, the definition of the variable y in Eq. (2.83) implies that in the assumed case $B = 0$, the relative magnetization is also given by

$$\frac{M}{M_{max}} = y \frac{k_B T}{g_J \mu_B J \lambda M_{max}}. \tag{2.84}$$

Comparing this relation with $\frac{M}{M_{max}} = B_J(y)$, one obtains the solutions for the relative magnetization at a given temperature as the point of coincidence between the two descriptions of $\frac{M}{M_{max}}$. This evaluation can be carried out graphically, as demonstrated in Figure 2.4, where the Brillouin-function parameter J (see Figure 2.1) has been chosen as $J = \frac{1}{2}$, in keeping with the premise $L = 0$.

We note that beyond a certain critical temperature, labeled T_C for *Curie temperature*, Eqs. (2.31) and (2.84) do not have any simultaneous solutions, excepting the place $y = M/M_{max} = 0$. The Curie temperature may be determined by use of the condition that the derivative of the Brillouin function and the function defined by

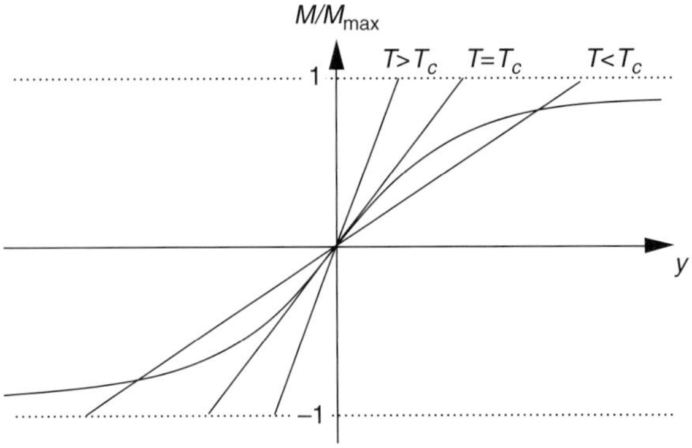

Figure 2.4 Determining the relative magnetization of a ferromagnetic substance by a graphical procedure. The curved line represents the Brillouin function in the case $J = \frac{1}{2}$. The solid lines refer to the relative magnetization by the approach (2.84), with the temperature as a parameter. The relative magnetization is plotted with respect to the variable y (see Eq. (2.83)) under the constraint of vanishing external field ($B = 0$).

Eq. (2.84) coincide at $y = 0$ when $T = T_C$. Equating the two derivatives at the origin, one obtains:

$$T_C = \frac{(J+1)g_J\mu_B\lambda M_{max}}{3k_B}. \tag{2.85}$$

This result is readily confirmed as the approximation to the Brillouin function for small values of y is adopted, see Eq. (2.32). As $M_{max} = n_s g_J \mu_B J$, the formula for T_C may be cast in a more compact form by introducing the quantity $\mu_{eff} = g_J\mu_B\sqrt{J(J+1)}$. This yields

$$T_C = \frac{n_s\lambda\mu_{eff}^2}{3k_B}. \tag{2.86}$$

The order of magnitude of the molecular field B_{MF} may be estimated by considering its maximum value, λM_{max}:

$$B_{MF}^{max} = \frac{3k_B T_C}{g_J\mu_B(J+1)}. \tag{2.87}$$

For typical values of T_C and J, such as 1000 K and 1/2, respectively, one obtains a field in the order of 1500 T, documenting the great strength of the direct exchange interaction that causes ferromagnetic coupling. In the regime $T \approx T_C$, the magnetization of the sample is small, in accordance with an inducing field **B** of low magnitude. This observation may be used to deduce the characteristic behavior of the magnetic susceptibility of ferromagnetic systems around the critical temperature. As the variable y is small in this region (see Eq. (2.83)), the approximation (2.32) can be used for the Brillouin function, resulting in

$$\frac{M}{M_{max}} \approx \frac{(J+1)g_J\mu_B}{3k_BT}(B+\lambda M). \tag{2.88}$$

By use of the critical temperature as given by Eq. (2.85), this relation may be recast in the following way,

$$\frac{M}{M_{max}}(1-\frac{T_C}{T}) \approx \frac{T_C}{T}\frac{B}{\lambda M_{max}}, \tag{2.89}$$

yielding the temperature dependence of the ferromagnetic susceptibility, namely

$$\chi = \lim_{B\to 0}\frac{\mu_0 M}{B} = \frac{c}{T-T_C}, \tag{2.90}$$

with $c \equiv \frac{\mu_0 T_C}{\lambda}$ as the *Curie–Weiss constant*. One concludes from the form of the *Curie–Weiss Law*, Eq.(2.90), that the derivative of the susceptibility and, by extension, of the magnetization is not continuous at $T = T_C$. This implies that the mean field model adopted here describes the change of magnetic order at the Curie temperature as a *second-order phase transition*. This conclusion is based on the definition of the order of a phase transition[4] as well as the fact that the magnetization can be represented as a first derivative of the free energy. This follows directly from the thermodynamic relation between the free energy F and the entropy S in the presence of a magnetic field $\mathbf{B}$:

$$dF = -SdT - \mathbf{M}\cdot d\mathbf{B}. \tag{2.91}$$

One concludes that

$$\mathbf{M} = -(\nabla_{\mathbf{B}}F)_T. \tag{2.92}$$

As a defining characteristic of ferromagnetic materials, they trace out a *hysteresis loop* when cooled below the Curie temperature and exposed to an alternating external magnetic field. As illustrated in Figure 2.5, the magnetization cannot be described as a single-valued function of the inducing field, but it depends on the previous magnetic state of the sample and thus retains a memory of its *magnetization history*. This behavior is rooted in the magnetic composition of a macroscopic ferromagnet. Specifically, ferromagnetic materials are composed of domains which may be understood as magnetic grains whose local magnetization is saturated. Consequently, vanishing overall magnetization of the sample occurs when all partial magnetizations of the individual domains annihilate each other. The non-linear dashed curve in the figure shows the development of the magnetization **M** when

[4] The order of a phase transition is understood as the order of the lowest free-energy differential that exhibits a discontinuity at the transition point.

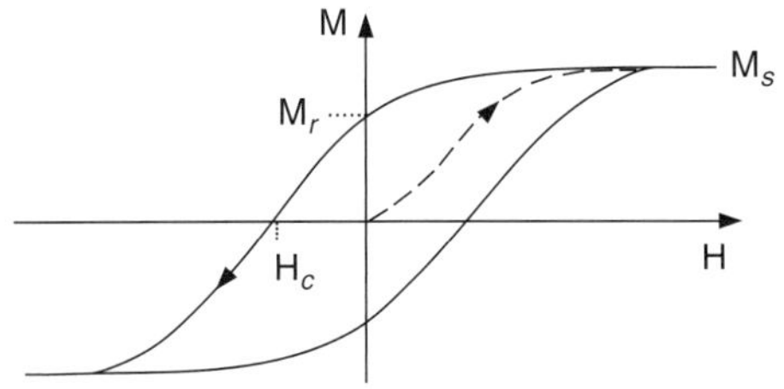

Figure 2.5 Hysteresis loop for a ferromagnet, demonstrating that a memory of the magnetization history is retained by the material. As the external field is reduced from its maximum to zero, the sample preserves a sizable fraction of the saturation value of its magnetization, $\mathbf{M}_s$. The corresponding value is known as the remanent magnetization, $\mathbf{M}_r$. Another point of strategic importance for defining the hysteresis loop is the magnetic field $\mathbf{H}_c$ required to reduce the magnetization to zero, the *coercivity*.

its initial value is zero. In the domain model, saturation of $\mathbf{M}$ is associated with the maximum achievable alignment of all domain magnetizations along the spatial direction dictated by the external field.

The figure clarifies the three distinctive quantities that specify the hysteresis loop for a given ferromagnet. These are the magnetization at the saturation point, $\mathbf{M}_s$, the *remanent magnetization*, $\mathbf{M}_r$, defined as the magnetization retained by the sample when the external field returns from its maximum value to zero, and lastly the *coercivity*, the magnitude of the external field needed to drive the magnetization to zero after the sample had been saturated.

The isotropic interaction, as described by the Heisenberg Hamiltonian (2.70), gives rise to the mean field in ferromagnets (see Eq.(2.80)). It does not account for the existence of *easy axes* along which the magnetization is directed preferentially. The iron lattice, for example, has cubic symmetry, and the easy axes of iron coincide with the cube edges. Magnetic materials tend to be *anisotropic* in the sense that there is a marked directional dependence of the external field needed to achieve magnetic saturation. Along the easy axes, the minimal field strength needed to induce the maximum magnetization, $\mathbf{M}_s$, is lowest, and it is highest along the *hard axes* of the crystal. The *magnetocrystalline anisotropy* is understood as the energy required to turn the direction of magnetization from an easy to a hard axis. The primary cause of this phenomenon is spin-orbit interaction, leading to differences in the orbital overlap between neighboring sites as the spin direction changes in response to a rotation of the external field.

Magnetic nanoparticles can often be characterized as a single magnetic domain. These particles act as one grand magnetic moment, resulting from the concerted effect of all microscopic, i.e. atomic or molecular magnetic moments in the system. This case is referred to as *superparamagnetism*, a phenomenon that emerges below the Curie temperature of the material. In non-spherical particles, usually a direction of preferred magnetization can be identified. In order to reverse its magnetization along this axis, the particle has to overcome an energy barrier. The height of this barrier defines the *Néel relaxation time*, which limits the lifetime of the nanoparticle's magnetization. Compared with regular paramagnets, superparamagnets have markedly higher susceptibilities. Ideally, the coercivity of superparamagnetic substances is zero. Nanoparticles with small coercivity ($\mu_0 H_c \approx 5.0 \cdot 10^{-3}$ T or lower), however, are sometimes termed superparamagnets.

2.3.2 Pauli Paramagnetism versus Spontaneous Symmetry Breaking

Inquiring about the cause of ferromagnetism, we find it to be a manifestation of *spontaneous symmetry breaking* where the symmetry involved is that between the spin-up and spin-down subsets of a material. This process is to be distinguished from spin symmetry breaking that is induced by a disturbance, such as an external magnetic field. In what follows, we first refer to *Pauli paramagnetism* and subsequently contrast the mechanism that establishes that phenomenon with the one that gives rise to ferromagnetism.

Pauli Paramagnetism

The essential features of a paramagnetic metal can be clarified by considering a gas of free electrons within a *Fermi sphere* with radius $|\mathbf{k}| \leq k_F$. If the volume of the sample is L^3, the three-dimensional *density of states* (DOS) as a function of the wave number, $g(k)$, is found from the relation

$$g(k)dk = 2\frac{1}{V}\frac{1}{(2\pi/L)^3}4\pi k^2 dk = \frac{k^2}{\pi^2}dk. \tag{2.93}$$

The factor 2 accounts here for the spin multiplicity. The quantity $g(k)$ indicates the number of states per volume and wave number interval. Relation (2.93) is based on the fact that each quantum state of the free electon system inhabits a phase space cell of size $h^3 = (2\pi\hbar)^3$. We assume a free electron system at zero temperature, such that all states are filled up to the *Fermi energy* E_F, corresponding to a *Fermi momentum* $p_F = \hbar k_F$. Therefore,

$$E_F = \frac{\hbar^2 k_F^2}{2m_e^*}, \tag{2.94}$$

with m_e^* as *effective electron mass*. The number of electrons (N) per volume (V) may be derived from integrating over the density of states $g(k)$ up to the Fermi wave number, according to

$$n \equiv \frac{N}{V} = \int_0^{k_F} g(k)dk = \frac{k_F^3}{3\pi^2}. \tag{2.95}$$

We introduce the density of states as a function of energy by defining

$$g(E) \equiv \frac{dn(E)}{dE}. \tag{2.96}$$

A useful integral representation of $g(E)$ is given by

$$g(E) = 2\int \frac{d^dk}{(2\pi)^d}\delta(E - E(\mathbf{k})), \tag{2.97}$$

where $d = 1, 2, 3$ stands for the dimensionality of the considered system, and the integration extends over the first Brillouin zone. Inverting equation (2.96), we find

$$n = \int_0^{E_F} g(E)dE = \frac{1}{3\pi^2}(\frac{2m_e^*}{\hbar^2})^{\frac{3}{2}}E_F^{\frac{3}{2}}, \tag{2.98}$$

where Eq. (2.94) has been used. Eq. (2.96) in conjunction with Eq. (2.98) yields

$$g(E_F) = \frac{dn(E)}{dE}|_{E=E_F} = \frac{3n}{2E_F}. \tag{2.99}$$

This result makes it possible to approximate the susceptibility of the free electron gas. Exposing this gas to a small external magnetic field **B** splits the degeneracy between the two spin orientations, lowering the energy of electrons with field-parallel magnetic moments by an amount $\mu_B B$, and adding this amount of energy to the electrons with the opposite magnetic moment orientation. Equilibrium, associated with a common Fermi energy for both spin orientations, is attained as a fraction of electrons undergoes spin flip, aligning their spins with the field. Defining the subset with field-parallel spins as the spin-up moiety, we estimate the volume density $n\uparrow$ of the electrons that augment the spin-up reservoir by

$$n\uparrow = \frac{1}{2}g(E_F)\mu_B B = -n\downarrow. \tag{2.100}$$

This assessement is based on the fact that the Pauli exclusion principle forbids electrons with $E < E_F$ to alter their spin. The resulting difference between the up and down spin populations gives rise to a finite magnetization, namely

$$M = \mu_B(n\uparrow - n\downarrow) = g(E_F)\mu_B^2 B. \tag{2.101}$$

Employing this result in combination with Eq.(2.99), we find the Pauli susceptibility χ_P at $T = 0$:

$$\chi_P = \frac{M}{H} \approx \frac{\mu_0 M}{B} = \mu_0\mu_B^2 g(E_F) = \frac{3n\mu_0\mu_B^2}{2E_F}. \tag{2.102}$$

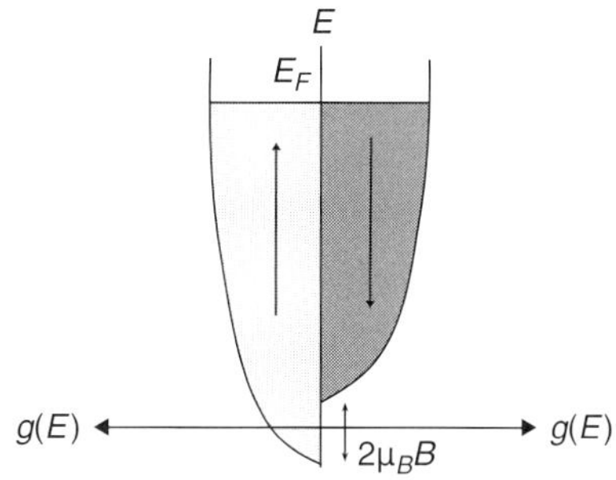

Figure 2.6 Energy of a free electron gas versus the density of states, partitioned with respect to spin orientation. A small magnetic field is applied, resulting in an energy splitting between the spin-up and the spin-down moiety. This effect is amplified in the figure.

Making allowance for $T > 0$ introduces a temperature-dependent correction of this expression, which, for T in the order of room temperature, usually does not exceed the percent range. Figure 2.6 encapsulates the essence of Pauli paramagnetism, illustrating the effect of an external magnetic field to deepen the well of the spin-up moiety while making the spin-down well more shallow. The Fermi energy, however, is the same for both spin orientations.

Spontaneous Symmetry Breaking

With the arguments presented in the preceding subsection, a criterion for the prevalence of ferromagnetic over paramagnetic coupling can be formulated. Such a criterion will identify situations in which spontaneous symmetry breaking, associated with different densities of state for the spin-up and the spin-down moieties of the system at the Fermi level, is energetically preferred over the paramagnetic condition of the same $g(E_F)$ for both subsets. It is of relevance in this context that the fractional magnetic moments of the transition metals[5] imply that their ferromagnetic order is not a mere consequence of the magnetic moments of localized atoms, but also of an itinerant effect, referred to as *band magnetism*. The origin of this spontaneous effect can be most easily studied by inspecting, once more, a free electron gas.

Symmetry breaking between the two spin populations is favored in cases in which the energy expense associated with inducing a difference between them is offset by a simultaneous release of magnetic potential energy. Since the system of reference is a free electron gas, the former component may be understood as kinetic energy increase. More specifically, we assume that spin-down electrons within the

[5] The magnetic moment per atom in Fe, for instance, amounts to about $2.2\mu_B$

energy interval $[E_F - \delta E, E_F]$ undergo spin flip, thus occupying the interval interval $[E_F, E_F + \delta E]$ in the well of the spin-up population. Upon spin reversal, each electron gains an amount of energy δE, and the number of transferred electrons per volume is $g(E_F)\frac{\delta E}{2}$ where the factor $\frac{1}{2}$ is due to the premise that spin flip is restricted to one-half of the electrons. The addition to the system's kinetic energy per volume is therefore given by

$$\Delta E_{kin} = \frac{1}{2} g(E_F)(\delta E)^2 \tag{2.103}$$

The spin flip results in a finite magnetic moment, associated with the magnetization

$$M = \mu_B(n\uparrow - n\downarrow), \tag{2.104}$$

where $n\uparrow$ and $n\downarrow$ are the densities of electrons with up- and down-spin orientation after the transfer, according to

$$\begin{aligned} n\uparrow &= \frac{n}{2} + \frac{1}{2} g(E_F)\delta E \\ n\downarrow &= \frac{n}{2} - \frac{1}{2} g(E_F)\delta E. \end{aligned} \tag{2.105}$$

Combining Eqs.(2.103) and (2.104) with the mean-field theory approach, we obtain the following reduction of potential energy per volume:

$$\Delta E_{pot} = -\int_0^M B_{MF} d\tilde{M} = -\int_0^M \lambda \tilde{M} d\tilde{M} = \frac{1}{2}\mu_B^2 \lambda (n\uparrow - n\downarrow)^2. \tag{2.106}$$

Setting $U \equiv \mu_B^2 \lambda$, we arrive at

$$\Delta E_{pot} = -\frac{1}{2} U g(E_F)^2 \delta E^2, \tag{2.107}$$

yielding an overall energy change of

$$\Delta E = \Delta E_{kin} + \Delta E_{pot} = \frac{1}{2}(1 - U g(E_F)) g(E_F) \delta E^2. \tag{2.108}$$

From this relation, the factor $(1 - Ug(E_F))$ decides if the system experiences energy gain or loss. In particular, if the condition

$$U g(E_F) > 1 \tag{2.109}$$

is satisfied, transition from paramagnetic to ferromagnetic order leads to a decrease of total energy, and the latter state is favored over the former. This is the content of the *Stoner criterion*. The final configuration involves an asymmetry between the densities of states of the spin-up and the spin-down electrons at $E = E_F$, as displayed in Figure 2.7. With reference to this figure, the ultimate economy of symmetry breaking between the two spin reservoirs is thus to lower the energy of

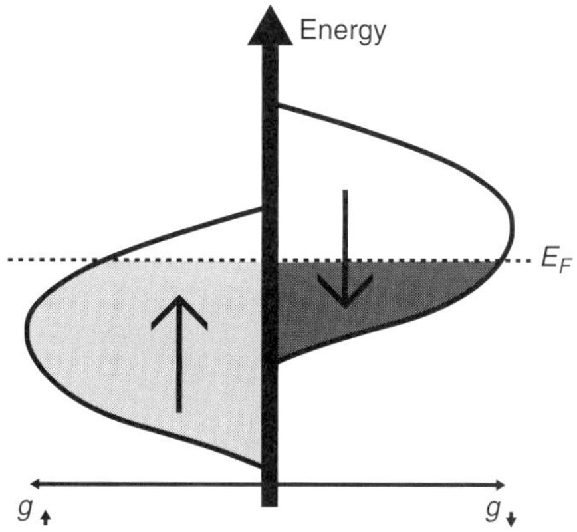

Figure 2.7 Scheme of the density-of-states (DOS) distributions for spin-up and spin-down populations resulting from spontaneous symmetry breaking, as described by relations (2.103)–(2.109). This configuration involves a DOS asymmetry at the Fermi level, $g_\uparrow(E_F) \neq g_\downarrow(E_F)$, reflecting a transition form paramagnetic to ferromagnetic order.

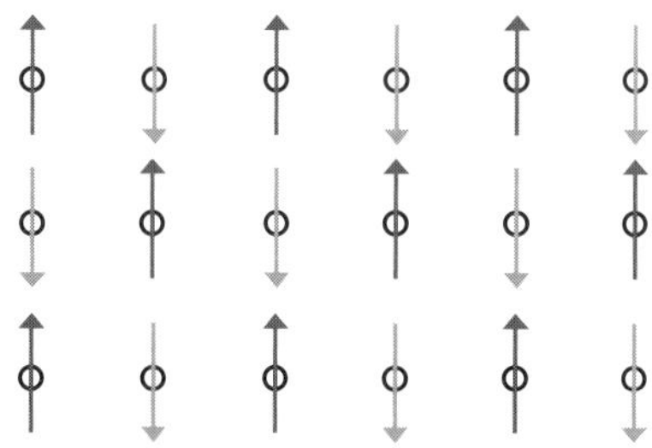

Figure 2.8 The simplest type of antiferromagnetic order: a lattice composed of two interpenetrating sublattices with antiparallel magnetizations.

spin-up states close to the Fermi level while increasing their number at the expense of the spin-down reservoir. By formula (2.108), the preference for this stabilization process increases with $g(E_F)$, i.e. the number of electrons available at the Fermi level for energy lowering.

2.3.3 Antiferromagnetism and Ferrimagnetism

The prototypical realization of antiferromagnetic order consists in two interpenetrating ferromagnetic lattices with opposite spin orientations. For the system displayed in Figure 2.8, the exchange interaction is positive ($J > 0$) within each one of the two sublattices, and negative ($J > 0$) between nearest neighbors.

The *Weiss model of an antiferromagnet* (e.g. [24]) assigns a molecular field to each of the two sublatticers, according to

$$\begin{aligned} \mathbf{B}^A_{MF} &= -|\lambda|\mathbf{M}^B \\ \mathbf{B}^B_{MF} &= -|\lambda|\mathbf{M}^A. \end{aligned} \tag{2.110}$$

As implied by these conditions, the magnetization of sublattice A(B) is proportional to the molecular field of sublattice B(A). The molecular field constant, as defined by Eq. (2.82) is constrained to assume a negative value. No external magnetic field is applied. If the additional stipulation is made that

$$\mathbf{M}^A = \mathbf{M}^B \equiv \mathbf{M}, \tag{2.111}$$

the Brillouin function, as outlined in Section 2.1.1, can be used equally for the magnetization of both sublattices, yielding

$$\frac{M}{M_{max}} = B_J\left(\frac{g_J \mu_B J|\lambda|M)}{k_B T}\right). \tag{2.112}$$

Therefore, the transition temperature formalism, as captured by relations (2.85) and (2.86), carries over to the antiferromagnetic case, with the small caveat that λ is to be replaced with $|\lambda|$. We find

$$T_N = \frac{(J+1)g_J\mu_B|\lambda|M_{max}}{3k_B} = \frac{n_s|\lambda|\mu^2_{eff}}{3k_B}. \tag{2.113}$$

The transition temperature T_N in antiferromagnetic systems is called *Néel temperature*. If $T < T_N$, the partial magnetizations $\mathbf{M}^A$ and $\mathbf{M}^B$, while being equal in magnitude, point in opposite directions, implying zero magnetization for the system as a whole. Once more, the low-field approximation (2.32) to the Brillouin function may be used to derive a closed expression for the magnetic susceptibility, in analogy to steps (2.88)–(2.90). This leads to

$$\chi = \lim_{B\to 0} \frac{\mu_0 M}{B} \propto \frac{1}{T + T_N}. \tag{2.114}$$

Comparison with the parallel results for paramagnets, (2.37), and ferromagnets, (2.90), suggests a universal formula for the temperature dependence of the susceptibility, namely

$$\chi \propto \frac{1}{T - \theta}. \tag{2.115}$$

This relation is valid in the paramagnetic regime, i.e. for ferromagnets and antiferromagnets it holds in the region above their respective transition temperatures. The *Weiss temperature* θ vanishes for paramagnetic materials. For ferromagnets, $\theta = T_C$, and for antiferromagnets, $\theta = -T_N$. The sign of the Weiss temperature, which is positive (negative) for ferromagnetic (antiferromagnetic) systems, thus serves as a fingerprint for the magnetic nature of the considered material. For real antiferromagnets, such as FeO, the parameter θ has been found to deviate very markedly from the negative Néel temperature. This is a consequence of the premise made here that the molecular field of each sublattice depends solely on the magnetization of the other one. A more adequate model is given by the extended approach

$$\begin{aligned}\mathbf{B}^A_{MF} &= c_{AA}\mathbf{M}^A + c_{AB}\mathbf{M}^B \\ \mathbf{B}^B_{MF} &= c_{BA}\mathbf{M}^A + c_{BB}\mathbf{M}^B.\end{aligned} \tag{2.116}$$

From the assumed symmetry of the two sublatties, the two coupling coefficients in this equation system are equal ($c_{AB} = c_{BA}$), and so are the two intra-sublattice coefficients ($c_{AA} = c_{BB}$).

Using further the constraint $\mathbf{M}_A = -\mathbf{M}_B$, which holds for vanishing external field, we readily find from Eq. (2.113) that $T_N = n_s(c_{AA} - c_{AB})\mu^2_{eff}/3k_B$, with $c_{AA} - c_{AB}$ playing the role of the parameter $|\lambda|$. Be $c \equiv n_s\mu^2_{eff}/3k_B$. Further, making admission for an external magnetic field of small magnitude and using the relation (2.37) between the magnetic field and the magnetization in the paramagnetic regime, we write

$$\begin{aligned}\mathbf{M}^A &= \frac{c}{T}(\mathbf{B} + c_{AA}\mathbf{M}^A + c_{AB}\mathbf{M}^B) \\ \mathbf{M}^B &= \frac{c}{T}(\mathbf{B} + c_{AB}\mathbf{M}^A + c_{AA}\mathbf{M}^B).\end{aligned} \tag{2.117}$$

Exercise 2.2 From the system of equations (2.117), obtain the susceptibility $\chi = \lim_{B\to 0}\mu_0\frac{M_A+M_B}{B}$. In particular, show that χ is given by

$$\chi = \mu_0\frac{2c}{T - c(c_{AA} + c_{AB})}. \tag{2.118}$$

By fitting the susceptibility derived in Exercise 2.2 to the Curie–Weiss form (2.115) ($\chi \propto \frac{1}{T-\theta}$), we arrive at

$$\theta = c(c_{AA} + c_{AB}). \tag{2.119}$$

Thus,

$$\frac{T_N}{\theta} = \frac{c_{AA} - c_{AB}}{c_{AA} + c_{AB}}. \tag{2.120}$$

This clarifies that the relation $T_N = -\theta$ is only satisfied in the special case $c_{AA} = 0$, i.e. in the limit of vanishing intra-sublattice contributions to the molecular field that determines the antiferromagnetic configuration.

Relaxation of the condition that the two sublattices have magnetizations of equal magnitudes leads from an antiferromagnetic to a *ferrimagnetic system*. A well-known example for ferrimagnets is provided by *ferrites*, compounds with the chemical formula MO·Fe_2O_3, where M stands for a divalent cation such as Zn^{2+}, Co^{2+}, Fe^{2+}, Ni^{2+} and others. To balance the negative charge of the O^{2-} moiety, Fe exists here in the Fe^{3+} state of oxidation. In *normal spinels*, the M^{2+} sits at tetrahedral sites, being surrounded by four oxygen ligands, while Fe^{3+} locates at octahedral sites, involving six oxygen neighbors. There are twice as many octahedral as tetrahedral sites. The two different sites are associated with antiparallel magnetic moments of different size. In the special case of magnetite (M = Fe^{2+}), the iron component exists both in the ferric (Fe^{2+}) and the ferrous (Fe^{3+}) form, combining to create an *inverse spinel*, defined by the feature that half of the octahedral sites are populated by M^{2+} while all other sites are occupied by Fe^{3+}. This gives rise to two sublattices with opposing magnetic moments of magnitudes 5.0 μ_B and 4.5 μ_B, associated with Fe^{3+} alone and a Fe^{3+}/Fe^{2+} mixture, respectively.

By their definition, ferrimagnetic materials are described by the aproach (2.116), with $c_{AB} = c_{BA}$, but $c_{AA} \neq c_{BB}$. From this premise, one may conclude (see e.g. [10]) that ferrimagnets exhibit spontaneous magnetization below the Curie temperature T_C and that, for a certain temperature in the regime $T < T_C$, both partial magnetic moments of the ferrimagnetic compound may cancel each other out. This critical temperature of vanishing net magnetization is referred to as the *compensation temperature*, T_{comp}.

2.3.4 Helimagnets and Spin Glasses

The magnetic structure of layered materials arises from an interplay between the dominant coupling constants, corresponding to interaction between adjacent layers and between next-nearest-neighbor layers. This is the case for many rare earth metals with stacked hcp lattice structures where ferromagnetic alignment is found within each layer. The magnetic interaction between nearest and next-nearest layers is here described by exchange constants $\tilde{J}_1$ and $\tilde{J}_2$, respectively. In these cases, a *helical spin ordering* may emerge such that the magnetizations of adjacent layers are rotated with respect to each other by an angle α, as shown for a simple model in Figure 2.9. If only nearest- and next-nearest neighbor interactions are included and the Heisenberg Hamiltonian (2.70) is employed to describe this situation, the energy of the spin helix may be indicated as

$$E = -2NS^2(\tilde{J}_1 \cos\alpha + \tilde{J}_2 \cos 2\alpha), \tag{2.121}$$

with S as spin quantum number and N as the number of atoms per layer.

To obtain quantitative information about the angle α, one minimizes the energy by evaluating $dE/d\alpha = 0$. This yields

$$(\tilde{J}_1 + 4\tilde{J}_2 \cos\alpha)\sin\alpha = 0. \tag{2.122}$$

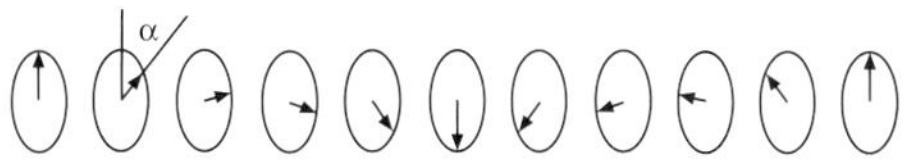

Figure 2.9 Basic illustration of helical order. The magnetizations in adjacent layers are rotated with respect to each other by an angle α.

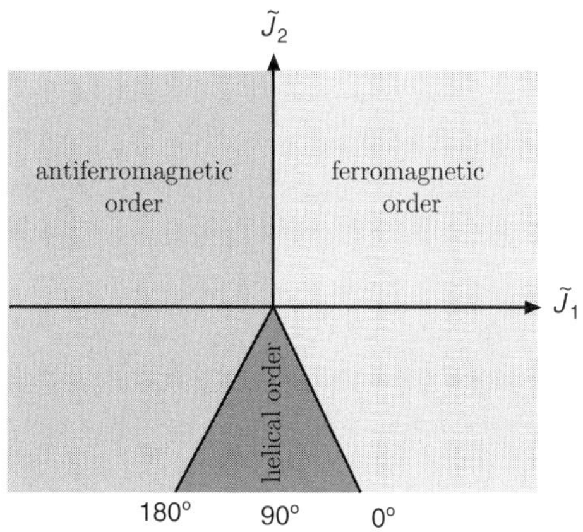

Figure 2.10 The regimes of ferromagnetic, antiferromagnetic and helical order in a phase diagram spanned by the coupling parameters $\tilde{J}_1$ and $\tilde{J}_2$.

The real-valued solutions of this equation correspond to three different magnetic ordering schemes. Ferromagnetic order is generated by the choice $\alpha = 0$, while $\alpha = \pi$ results in antiferromagnetism. An intermediate solution for α is given by

$$\cos\alpha = -\frac{\tilde{J}_1}{4\tilde{J}_2}. \tag{2.123}$$

This equation defines a helimagnetic structure. It may be used to infer a numerical relation between the magnitudes of the two coupling constants: $|\tilde{J}_1| \leq 4|\tilde{J}_2|$. By inserting $\cos\alpha$ as given by (2.123) into the energy equation (2.121) and using the conditions $E_{HM} < E_{FM}$ and $E_{HM} < E_{AFM}$ as stability requirements for a helimagnetic system with E_{HM}, one may conclude [10] that $\tilde{J}_2 < 0$, corresponding to antiferromagnetic interaction between next-neighbor layers. The three magnetic prototypes are summarized in the phase diagram 2.10.

Alloys that consist of highly diluted magnetic solutes in a non-magnetic metallic medium have been shown to exhibit a magnetic phase, although the impurities are distributed randomly and their concentrations are low enough to exclude direct exchange as the source of the magnetic order. These are *spin glasses*, as exemplified by systems of the form $\mathrm{Me}_{100-x}\mathrm{TM}_x$, where Me stands for the metal host (e.g.

Cu or Au) and TM for a transition metal component (e.g. Mn or Fe), and the percentage of the TM phase is smaller than one percent. In view of the large average separation between adjacent magnetic impurity atoms, the only effective mechanism for magnetic coupling between them is RKKY interaction (see Section 2.2.2). Since the sign and the size of the RKKY coupling parameter depend critically on the distance between the interacting spins, the coupling may be ferromagnetic or antiferromagnetic, both at varying strengths. These mixed magnetic interactions give rise to a *freezing* of magnetic moments in a metastable state, characterized by statistically distributed spin orientations.

The signature of a spin glass is a cusp in the susceptibility as a function of temperature, marking the transition from the paramagnetic to the spin glass phase. The transition temperature tends to increase with the solute concentration.

The dilute phase in spin glasses exists in a well-defined concentration range. More specifically, the concentration must be sufficiently high to allow for efficient indirect exchange through the RKKY process. An upper limit on the concentration results from the requirement that the contribution of direct exchange to the magnetic phase should be negligible, and that no clustering of the solute atoms should occur.

3

The Tools: Computational and Experimental Techniques

This chapter combines brief summaries of computational as well as experimental procedures that have been instrumental for the study of nanostructure magnetism. All of the methods sketched here have been successfully applied to magnetic carbon nanostructures. None of them, however, is specific to this type of materials. The models and techniques included in this chapter are pertinent to condensed matter systems or have, in some cases, an even wider scope of usage, including matter at the molecular scale. They thus have the status of standard methods. Procedures more specifically adjusted to exploring magnetism in carbon-based materials, such non-local measurement of spin signals in carbon transmission elements, will be introduced later in their distinct theoretical or experimental contexts.

Computational Methods

In the first part of this chapter we survey a variety of methods that have proved to be efficient in modeling carbon nanostructure magnetism. Procedures based on the *tight-binding* and the *Hubbard model*, as sketched in Section 3.1, have played important roles in clarifying the electronic architecture of graphene and its derivatives. Tight-binding schemes have successfully addressed salient phenomena pertaining to these materials, among them the nature of electronic states at the Fermi level of graphene as massless Dirac Fermions. Lifting the methodological constraints typical for these methods, such as the restriction of interatomic coupling to next-neighbor sites, or the representation of a given lattice site by a single electron, leads to more general theories of electronic structure. These are the topic of Section 3.2, which deals with *ab initio* and *density functional theory* (DFT) methods. In all cases covered in this survey, emphasis is placed on the capacity of a given computational approach to describe magnetic states and to represent magnetic properties.

3.1 The Tight-Binding and the Hubbard Model

The theory of one- or two-dimensional carbon nanostructures, represented respectively by the nanotube and the graphene paradigms, has been largely developed by use of the tight-binding or the Hubbard model. In this section we will introduce some general traits of these approaches, insofar as they are of relevance to this text.

Both models rest on the *tight-binding premise*. According to this condition, the system in its ground state may be described as a lattice populated with atoms in a hydrogen-like configuration. Specifically, the electronic system of the atoms is subdivided into one active electron, localized at the lattice site, and an inert core whose internal structure is not made explicit. We will see in Chapter 4 that this decomposition is, for many purposes, adequate to represent carbon atoms in graphene where the active electron has p_z character. This electron may spread through the lattice by a process of hopping between lattice sites.

Both the tight-binding and the Hubbard model are most conveniently defined in the second-quantization formalism. More specifically, the Hamiltonians that underlie these approaches are expressed in terms of creation (annihiliation) operators $\hat{c}^\dagger_{\sigma,i}$ ($\hat{c}_{\sigma,j}$) that create (destroy) an electron with spin σ at the site i (j). Imposing the Fermionic anticommutation relations

$$\begin{aligned} \{\hat{c}_{\sigma,i}, \hat{c}^\dagger_{\sigma',j}\} &= \delta_{ij}\delta_{\sigma\sigma'} \\ \{\hat{c}^\dagger_{\sigma,i}, \hat{c}^\dagger_{\sigma',j}\} &= 0 \\ \{\hat{c}_{\sigma,i}, \hat{c}_{\sigma',j}\} &= 0 \end{aligned} \tag{3.1}$$

secures the anti-symmetry of the many-electron states on which the operators $\hat{c}^\dagger_{\sigma,i}$ ($\hat{c}_{\sigma,j}$) act. The *vacuum* $|0\rangle$ is understood as the state where no electrons are present. Therefore,

$$\hat{c}_{\sigma,i}|0\rangle = 0. \tag{3.2}$$

The creation operator $\hat{c}^\dagger_{\sigma,i}$, in contrast, generates a localized electron with a definite spin according to

$$\hat{c}^\dagger_{\sigma,i}|0\rangle = \varphi(\mathbf{x} - \mathbf{R}_i)\chi(\sigma), \tag{3.3}$$

where the wave function $\varphi(\mathbf{x} - \mathbf{R}_i)$ may describe an electron attached to an atomic site i, and the spin variable σ can assume the values $\uparrow$ and $\downarrow$. From these ingredients, an elementary Hamiltonian can be built that nevertheless is capable of describing qualitatively a wide range of extended electronic systems. It is the *hopping Hamiltonian* $\hat{H}_{hop}$, defined as

$$\hat{H}_{hop} = -\sum_{i,j,\sigma} t_{ij}\hat{c}^\dagger_{\sigma,i}\hat{c}_{\sigma,j}. \tag{3.4}$$

In the case $i \neq j$, the factors t_{ij} are given by *hopping integrals*. Explicitly,

$$t_{ij} = \int d\mathbf{x}\varphi^*(\mathbf{x} - \mathbf{R}_i)\hat{H}\varphi(\mathbf{x} - \mathbf{R}_j). \tag{3.5}$$

If $i = j$, $t_{ij} = 0$. The quantities t_{ij} are related to the probability that an electron localized at site j undergoes tunneling to site i, corresponding to the destruction of the electron at the former ($\hat{c}_{\sigma,j}$) and its creation at the latter site ($\hat{c}^\dagger_{\sigma,i}$). This process is symmetric between sites i and j, as reflected by the symmetry of the hopping integral: $t_{ij} = t_{ji}$. It is assumed that this transition proceeds without a spin flip. To make the hopping approach practicable, the number of sites connected by transfer processes may be restricted. The *tight-binding approximation* consists in allowing these processes only between next-neighbor sites, as encoded by the *tight-binding Hamiltonian* $\hat{H}_{TB}$,

$$\hat{H}_{TB} = -\sum_{\sigma,\langle i,j\rangle} t_{ij}\hat{c}^\dagger_{\sigma,i}\hat{c}_{\sigma,j}, \tag{3.6}$$

where the symbol $\langle i, j\rangle$ confines i and j to next-neighbor sites. Obviously, the tight-binding model as specified by $\hat{H}_{TB}$ greatly oversimplifies the physical reality of electronic systems. Even if the hopping Hamiltonian (3.4) is chosen rather than the tight-binding Hamiltonian (3.6), a grave deficiency of the approach remains, namely the neglect of any interaction between electrons. This element is incorporated by the *Hubbard model*, which extends the tight-binding theory by admitting interelectronic repulsion. This proceeds by inclusion of a Coulomb term $\hat{H}_C$ of the following form,

$$\hat{H}_C = U\sum_{i=1}^{N} \hat{n}_{\uparrow,i}\hat{n}_{\downarrow,i}, \tag{3.7}$$

with N as the overall number of sites, $\hat{n}_{\sigma,i} = \hat{c}^\dagger_{\sigma,i}\hat{c}_{\sigma,i}$, and U as the Coulomb integral,

$$U = \int d\mathbf{x}_1 d\mathbf{x}_2 |\varphi(\mathbf{x}_1)|^2 \frac{e^2}{4\pi\varepsilon_0 r_{12}} |\varphi(\mathbf{x}_2)|^2, \tag{3.8}$$

with $r_{12} = |\mathbf{x}_1 - \mathbf{x}_2|$. Note that $\hat{n}_{\sigma,i}$ is the *number operator* for the site with index i. Thus, the Coulomb component is defined in such a way that $\langle\hat{H}_C\rangle = 0$ in the half-filled case, which is of relevance for many carbon nanostructures, most prominently for graphene. Equivalently, the only non-zero contributions to $\hat{H}_C$ stem from twofold-filled lattice sites, involving completely occupied electronic states. The full Hubbard Hamiltonian may thus be written in the following way,

$$\hat{H}_{Hub} = -t\sum_{\langle i,j\rangle,\sigma} \hat{c}^\dagger_{\sigma,i}\hat{c}_{\sigma,j} + U\sum_{i=1}^{N} \hat{n}_{\uparrow,i}\hat{n}_{\downarrow,i}, \tag{3.9}$$

where the approximation $t_{ij} = t$, for all i, j has been applied, reducing the hopping integrals to a single parameter. The structure of the Hubbard Hamiltonian may be viewed as an elementary partition into a kinetic energy term, associated with electron hopping between lattice sites, as described by $\hat{H}_{TB}$, and a potential energy contribution, as given by $\hat{H}_C$. The tight-binding premise seems to be in conflict with the presence of an electrostatic potential term, as the Coulomb force is long-range. While this concern is warranted, it is mitigated by the fact that the Coulomb interaction is screened in extended electronic systems. Several generalizations that augment $\hat{H}_{Hub}$ by additional terms not contained in the basic scheme (3.9) have been proposed (e.g. [25]).

The two partial Hamiltonians (3.6) and (3.7) that yield the full Hubbard Hamiltonian (3.9) have complementary ground state solutions, consisting of itinerant wave functions for $\hat{H}_{TB}$ and of localized functions for $\hat{H}_C$. To comment first on the latter, we make reference to a case of major relevance for this text: the *half-filled* configuration in which each site is occupied by one electron only. For this model, the solution of lowest energy may be written in the following way:

$$\Psi_{0,C} = \prod_i \hat{c}^\dagger_{\sigma,i}|0\rangle. \tag{3.10}$$

Under the rule of $\hat{H}_C$, i.e. the condition of small or vanishing hopping matrix elements t_{ij}, the energy cost for twofold occupation of a given site is high, so single occupation prevails. The degeneracy of this maximally localized state is 2^N. With an eye on later utilization of this formalism to describe graphene (see Section 4.1), we apply Eq. (3.10) to the case of a bipartite lattice, assuming that the set of all lattice sites can be subdivided into two subsets, A and B, each of which contain half of them. We find immediately eigenstates for this situation, namely

$$\Psi_{0,hop} = \prod_{i\varepsilon A} \hat{c}^\dagger_{\uparrow,k} \prod_{j\varepsilon B} \hat{c}^\dagger_{\downarrow,k}|0\rangle, \tag{3.11}$$

with

$$E_{A,B} = \sum_{i\varepsilon A} E_A + \sum_{i\varepsilon B} E_B \tag{3.12}$$

as the energy eigenvalue.

The ground state wave function of $\hat{H}_{hop}$ is easily found if the creation/annihilation operators are redefined in a basis consisting of eigenvectors of the matrix $\boldsymbol{t}_{ij}$,

$$\tilde{\hat{c}}_{\sigma,i} = \sum_k a^k_i \hat{c}_{\sigma,k}, \tag{3.13}$$

with $\mathbf{a}^k$, $1 \leq k \leq N$, as an orthonormal set of vectors, given by

$$t_{ij} a^k_j = \varepsilon_k a^k_i. \tag{3.14}$$

Inserting the expansion (3.13) into $\hat{H}_{TB}$ (3.6) and using the eigenvalue equation (3.14) along with the orthonormality of the eigenvectors $\mathbf{a}^k$ and, further, the Fermion anticommutation relation (3.1) yields

$$\hat{H}_{TB} = -\sum_{\sigma,k} \varepsilon_k \hat{\tilde{c}}^{\dagger}_{\sigma,k} \hat{\tilde{c}}_{\sigma,k}(\mathbf{k}) = -\sum_{\sigma,k} \varepsilon_k \hat{\tilde{n}}_{\sigma,k}. \tag{3.15}$$

The wave functions generated by the operation $\hat{\tilde{c}}^{\dagger}_{\sigma,i}|0\rangle$ represent de-localized states, as is made explicit by the choice

$$\hat{\tilde{c}}_{\sigma,\mathbf{k}} = \frac{1}{N} \sum_{\sigma,j} \hat{\tilde{c}}_{\sigma,j} a^k_j e^{-i\mathbf{x}_j\cdot\mathbf{k}}, \tag{3.16}$$

which is the Fourier transform of $\hat{c}_{\sigma,j}$. These operators create *Bloch waves* according to

$$|\sigma,\mathbf{k}\rangle = \hat{\tilde{c}}^{\dagger}_{\sigma,\mathbf{k}}|0\rangle. \tag{3.17}$$

The reader will find some basic information on Bloch waves in the first special-topic section of this chapter. Whenever the kinetic energy component of the problem considered dominates the Coulomb component, such that $U << t$, perturbation theory built on the wave functions $|\sigma,\mathbf{k}\rangle$ is a natural solution strategy. These functions determine the ground state $\Psi_{0,TB}$ of $\hat{H}_{TB}$,

$$\Psi_{0,TB} = \prod_{E(\mathbf{k})<E_F} \hat{\tilde{c}}^{\dagger}_{\uparrow,\mathbf{k}} \hat{\tilde{c}}^{\dagger}_{\downarrow,\mathbf{k}}|0\rangle, \tag{3.18}$$

a state filled with Bloch waves up to the Fermi energy.

3.1.1 Emergence of Ferromagnetism and Antiferromagnetism within the Hubbard Model

In all its apparent simplicity, the Hubbard model is known to capture a wide range of condensed-matter phenomena, among them phase transitions between metallic and insulating states and, of particular importance for the topics of this monograph, the emergence of ferromagnetism and antiferromagnetism. Both effects arise from a rivalry between the kinetic (or hopping) component and the potential (or Coulomb) component of the Hubbard Hamiltonian. Intuitively, the first type of phase transition is based on the notion that the Coulomb interaction causes a reduction of electron mobility at half-filling, since any move of an electron from one lattice site to another involves an increase of the potential energy by the margin U. The size of the evolving energy gap evidently depends on the magnitude of U. A qualitative explanation of ferromagnetism may be given within the Hubbard model by making reference to a non-magnetic system in its ground state. Specifically, we assume that all electronic states are occupied by two electrons with opposite spins up to the Fermi level, with $N_\uparrow = N_\downarrow = N$. The system is thus dominated by the kinetic operator $\hat{H}_{TB}$. In terms of the interelectronic repulsion described by $\hat{H}_C$, the potential energy of such a system is UN^2. This amount can be slightly lowered as admission is made for a small imbalance between the up and down spin reservoirs. If δn is the number of electrons whose spin orientation is changed, the potential energy release is found from

$$\delta\langle\hat{H}_C\rangle = U(N + \delta N)(N - \delta N) - UN^2 = -U(\delta N)^2. \tag{3.19}$$

This transfer requires that a number δN of electrons vacates sites with $E < E_F$ and occupies sites with $E > E_F$. The energy expense associated with this reorganization is obtained from the density of states $g(E)$ at the Fermi level, according to

$$\delta N = g(E)|_{E=E_F}\delta E. \tag{3.20}$$

The overall amount of energy required for this hopping process is

$$\delta\langle\hat{H}_{TB}\rangle = \delta N\delta E = \frac{1}{g(E)}|_{E=E_F}(\delta N)^2. \tag{3.21}$$

Expressions (3.19) and (3.21) combine to yield the overall energy change in the framework of the Hubbard model, resulting in

$$\begin{aligned}\delta\langle\hat{H}_{Hub}\rangle = \delta\langle\hat{H}_{TB}\rangle + \delta\langle\hat{H}_C\rangle &= (-U + \frac{1}{g(E)}|_{E=E_F})(\delta N)^2 \\ &= (-Ug(E)|_{E=E_F} + 1)\frac{(\delta N)^2}{g(E)|_{E=E_F}}.\end{aligned} \tag{3.22}$$

Therefore, $Ug(E)|_{E=E_F} > 1$ provides a criterion for cases in which the ferromagnetic solution is preferred over the non-magnetic alternative, in accordance with the result (2.108) derived in Section 2.3.2, where U is interpreted in terms of the effective molecular field approach.

Reversing the previous scenario, we now inspect a case where the Coulomb component strongly dominates the hopping term. For clarity, we adopt the simplifying frame of a two-sites configuration involving two elementary spin states, up ($\uparrow$) and down ($\downarrow$). With these premises, we define a fourfold degenerate ground state whose energy to leading order is zero. The four functions are partitioned into one singlet,

$$\chi_1 = |0, 0\rangle = \frac{1}{\sqrt{2}}(|\uparrow_1\downarrow_2\rangle - |\uparrow_2\downarrow_1\rangle), \tag{3.23}$$

and three triplet states,

$$\begin{aligned}\chi_2 &= |1,1\rangle = |\uparrow_1\uparrow_2\rangle \\ \chi_3 &= |1,0\rangle = \frac{1}{\sqrt{2}}(|\uparrow_1\downarrow_2\rangle + |\uparrow_2\downarrow_1\rangle) \\ \chi_4 &= |1,-1\rangle = |\downarrow_1\downarrow_2\rangle.\end{aligned} \tag{3.24}$$

Since, by assumption, the electrostatic interaction substantially outweighs the kinetic component, the latter can be treated as a perturbation. The degeneracy of the four states is lifted by the hopping Hamiltonian. Applying $\hat{H}_{TB}$ to any one of the triplet states yields zero. A non-zero result, in contrast, is obtained for the singlet. More specifically,

$$\begin{aligned}\hat{H}_{TB}|\uparrow_1\downarrow_2\rangle &= -t(|\uparrow_1\downarrow_1\rangle + |\downarrow_2\uparrow_2\rangle) \\ \hat{H}_{TB}|\uparrow_2\downarrow_1\rangle &= -t(-|\uparrow_1\downarrow_1\rangle - |\downarrow_2\uparrow_2\rangle),\end{aligned} \tag{3.25}$$

such that

$$\hat{H}_{TB}|0,0\rangle = -2t\frac{1}{\sqrt{2}}(|\uparrow_1\downarrow_1\rangle + |\downarrow_2\uparrow_2\rangle) = -2t\chi_5. \tag{3.26}$$

Thus, $\hat{H}_{TB}$ projects the singlet on a first excited state of the system, χ_5, whose energy eigenvalue is U. This state is orthogonal to the space spanned by $\{\chi_i\}, i = 1, .., 4$. In summary,

$$\langle\chi_i|\hat{H}_{TB}|\chi_i\rangle = 0, \;\; i = 1, .., 4, \tag{3.27}$$

implying that the degeneracy of $\{\chi_i\}, i = 1, .., 4$ is preserved to first-order perturbation theory. In second order, however, the ground state degeneracy is lifted. For the ground state energy correction $\Delta E_{2,0}$, we find

$$\Delta E_{2,0} = \sum_l \frac{\langle\chi_1|\hat{H}_{TB}|\chi_l\rangle\langle\chi_l|\hat{H}_{TB}|\chi_1\rangle}{E_0 - E_l} = \frac{2t\cdot 2t}{0-U} = -\frac{4t^2}{U}. \tag{3.28}$$

Thus, at second order, the result of the perturbation is lowering the energy of the singlet by an amount of $\frac{4t^2}{U}$. This effect can be expressed more concisely as the action of an effective Hamiltonian $\hat{H}_{2,eff}$, such that

$$\hat{H}_{2,eff} = -\frac{4t^2}{U}\hat{P}_{singlet}, \tag{3.29}$$

with $\hat{P}_{singlet}$ as projector on the singlet. This conclusion may be expressed in a way that makes the spins of the interacting centers explicit. Introducing the spin operators

$$\hat{\mathbf{S}}_k \equiv \frac{1}{2}\hat{\boldsymbol{\sigma}}_k, \quad k = x, y, z, \tag{3.30}$$

where the components $\hat{\boldsymbol{\sigma}}_k$ refer to the Pauli matrices, we acknowledge that

$$\hat{\mathbf{S}}_1 \cdot \hat{\mathbf{S}}_2 = \frac{1}{2}[(\hat{\mathbf{S}}_1 + \hat{\mathbf{S}}_2)^2 - \hat{\mathbf{S}}_1^2 - \hat{\mathbf{S}}_2^2] = \frac{1}{4} - \hat{P}_{singlet}. \tag{3.31}$$

This result is based on the observation that $\hat{\mathbf{S}}_1 \cdot \hat{\mathbf{S}}_2 = \frac{1}{2}(0 - \frac{3}{2})\hat{\mathbf{1}}$ for the singlet and $\hat{\mathbf{S}}_1 \cdot \hat{\mathbf{S}}_2 = \frac{1}{2}(2 - \frac{3}{2})\hat{\mathbf{1}}$ for the triplet state. We thus arrive at

$$\hat{H}_{2,eff} = \frac{4t^2}{U}(\hat{\mathbf{S}}_1 \cdot \hat{\mathbf{S}}_2 - \frac{1}{4}) \tag{3.32}$$

Exercise 3.1 Show that $\hat{H}_{2,eff}$ is identical with the effective Hamiltonian (2.68) if $E_T = 0$ and $E_S = \frac{-4t^2}{U}$.

In the light of this derivation, the preference of the Hubbard model for antiferromagnetic ordering is intuitively accessible. A spin flip interaction between adjacent sites can occur as a *second-order process* if the spin orientations of the electrons at these sites are opposite to each other. In this case, a single hop that leads to double occupancy of one of the two sites can take place. A second hop may restore the original state or result in an alternative state with the spins exchanged. Evidently, the first of these two steps cannot happen in the ferromagnetic case, as it would be in conflict with the exclusion principle.

3.1.2 The Heisenberg Hamiltonian

We note that the spin-dependent part of expression (3.32) has the form of a *Heisenberg Hamiltonian* (see Eq. (2.70)):

$$\hat{H}_H = \pm\tilde{J} \sum_{\langle i,j \rangle; i \neq j} \hat{\mathbf{S}}_i \cdot \hat{\mathbf{S}}_j. \tag{3.33}$$

As mentioned in Section 2.2.1, $\hat{H}_H$ applies to magnetic systems in the *insulating magnet* limit, involving electrons localized at distinct lattice sites, and thus excluding itinerant electrons as magnetic agents. While Eq. (3.33) restricts the Heisenberg Hamiltonian to interactions between adjacent sites, this constraint is readily removed by the generalization $\hat{H}_H = \pm\sum_{i,j;i\neq j} \tilde{J}_{ij}\hat{\mathbf{S}}_i \cdot \hat{\mathbf{S}}_j$. The derivation of the perturbation operator $\hat{H}_{2,eff}$ in the previous section provides a justification for the structure of $\hat{H}_H$ in the special case of a half-filled lattice in the low-energy regime, and for $S = \frac{1}{2}$.

Clearly, the sign in front of the exchange coupling constant $\tilde{J}$ decides about the magnetic phase specified by $\hat{H}_H$. A negative coupling constant (the case of $-\tilde{J}$, as $\tilde{J}$ is a non-negative quantity) implies that parallel spins at neighboring sites are favored, while antiparallel spins (corresponding to $+\tilde{J}$) are preferred if the sign is positive. With reference to the former case, defining the *Heisenberg ferromagnet*, we observe that a ground state solution for $-\tilde{J}\sum_{\langle i,j\rangle}\hat{\mathbf{S}}_i\cdot\hat{\mathbf{S}}_j$ is readily obtained by the stipulation

$$\hat{S}_{z,i}|0\rangle = S|0\rangle \tag{3.34}$$

for every lattice site i. In view of the expansion

$$\pm\tilde{J}\sum_{\langle i,j\rangle;i\neq j}\hat{\mathbf{S}}_i\cdot\hat{\mathbf{S}}_j = \pm\tilde{J}\sum_{\langle i,j\rangle;i\neq j}[\hat{S}_{z,i}\hat{S}_{z,j}+\frac{1}{2}(\hat{S}_i^+\hat{S}_j^- + \hat{S}_i^-\hat{S}_j^+)], \tag{3.35}$$

where $\hat{S}_i^{\pm}$ denotes the spin ladder operator acting at site i, $\hat{S}_i^+ = \hat{S}_{x,i} + i\hat{S}_{y,i}$ and $\hat{S}_i^- = \hat{S}_{x,i} - i\hat{S}_{y,i}$, it is obvious that the *Néel state* is not an eigenstate of $\hat{H}_H$. From Eq. (3.35), the Heisenberg Hamiltonian reverses pairs of adjacent spins if they are ordered in an antiferromagnetic configuration. However, $\hat{H}_H$ provides an estimate for an upper limit on the energy of the Néel state [26].

We mention a simplified approach to the description of spin ensembles: the *Ising model* [27, 28] treats the spin in a semi-classical frame, excluding from the Heisenberg Hamiltonian all but one spin component (conventionally chosen as the z component). Further, the operator $\hat{S}_z$ is replaced by a discrete variable that adopts the values +1 or -1. The *Ising Hamiltonian* may thus be written as

$$\hat{H}_{Is} = \pm\tilde{J}\sum_{\langle i,j\rangle}S_{z,i}S_{z,j}, \tag{3.36}$$

and $S_{z,i}, S_{z,j} = -1, +1$. Exact solutions have been identified for the one- and two-dimensional Ising models [28].

While the Heisenberg model, or schemes derived from it, may provide adequate descriptions of magnetic insulators, it does not apply to magnetic metals such as Fe, Co and Ni. For those systems, an approach should be chosen that is capable of representing the band structure of the considered material, as may be achieved the Hubbard model.

In the following, we will consider magnetic systems in the insulating magnet limit, involving electrons localized at distinct lattice sites, and thus excluding itinerant electrons as magnetic agents. We describe electrons by single-particle wave functions φ_i (i =1,2). Generalizing $\hat{H}_C$ as given by expression (3.7), we make admission not only for intra-site but also for inter-site effects, resulting in an extended Hamiltonian $\hat{H}_C$. Including inter-site contributions leads to partitioning $\hat{H}_C$ into a Coulomb and an exchange term,

$$\hat{H}_C = \frac{1}{2}\sum_{i,j,\sigma,\sigma'} \langle i, j|\frac{e^2}{4\pi\varepsilon_0 r_{12}}|i, j\rangle \hat{c}^\dagger_{\sigma,i}\hat{c}^\dagger_{\sigma,i}\hat{c}_{\sigma',j}\hat{c}_{\sigma',j}$$
$$- \langle i, j|\frac{e^2}{4\pi\varepsilon_0 r_{12}}|j, i\rangle \hat{c}^\dagger_{\sigma,i}\hat{c}_{\sigma',i}\hat{c}^\dagger_{\sigma',j}\hat{c}_{\sigma,j}, \tag{3.37}$$

where $\langle i, j|$ denotes $\langle\varphi_i(\mathbf{x}_1)\varphi_j(\mathbf{x}_2)|$, and accordingly for $|i, j\rangle$.

Exercise 3.2 Label the exchange integral as J_{ex} (see Eq.(2.69)) and expand the second term in Eq. (3.37), to arrive at:

$$\sum_{\sigma,\sigma'} J_{ex}\hat{c}^\dagger_{\sigma,i}\hat{c}_{\sigma',i}\hat{c}^\dagger_{\sigma',j}\hat{c}_{\sigma,j} = J_{ex}[\frac{1}{2}(\hat{c}^\dagger_{\uparrow,i}\hat{c}_{\uparrow,i} + \hat{c}^\dagger_{\downarrow,i}\hat{c}_{\downarrow,i})(\hat{c}^\dagger_{\uparrow,j}\hat{c}_{\uparrow,j} + \hat{c}^\dagger_{\downarrow,j}\hat{c}_{\downarrow,j})$$
$$+ \frac{1}{2}(\hat{c}^\dagger_{\uparrow,i}\hat{c}_{\uparrow,i} - \hat{c}^\dagger_{\downarrow,i}\hat{c}_{\downarrow,i})(\hat{c}^\dagger_{\uparrow,j}\hat{c}_{\uparrow,j} - \hat{c}^\dagger_{\downarrow,j}\hat{c}_{\downarrow,j}) \tag{3.38}$$
$$+ \hat{c}^\dagger_{\uparrow,i}\hat{c}_{\downarrow,i}\hat{c}^\dagger_{\downarrow,j}\hat{c}_{\uparrow,j} + \hat{c}^\dagger_{\downarrow,i}\hat{c}_{\uparrow,i}\hat{c}^\dagger_{\uparrow,j}\hat{c}_{\downarrow,j}]$$

We define spin operators $\hat{S}_k$, $k = i, j$, by

$$\hat{\mathbf{S}}_k \equiv \sum_{\sigma,\sigma'} \frac{1}{2}\hat{c}^\dagger_{\sigma,i}\hat{\boldsymbol{\sigma}}_{\sigma,\sigma'}\hat{c}_{\sigma',i}. \tag{3.39}$$

Here, $\hat{\boldsymbol{\sigma}}$ stands for the vector of the Pauli matrices $\boldsymbol{\sigma}_x$, $\boldsymbol{\sigma}_y$ and $\boldsymbol{\sigma}_z$. The sum on the right-hand side of Eq. (3.38) may then be written more succinctly as [26]

$$\sum_{\sigma,\sigma'} \hat{c}^\dagger_{\sigma,i}\hat{c}_{\sigma',i}\hat{c}^\dagger_{\sigma',j}\hat{c}_{\sigma,j} = \frac{\hat{n}_i\hat{n}_j}{2} + 2\hat{S}^z_i\hat{S}^z_j + \hat{S}^+_i\hat{S}^-_j + \hat{S}^-_i\hat{S}^+_j = \frac{\hat{n}_i\hat{n}_j}{2} + 2\mathbf{S}_i\cdot\mathbf{S}_j, \tag{3.40}$$

and $\hat{n}_i \equiv \sum_\sigma \hat{c}^\dagger_{\sigma,i}\hat{c}_{\sigma,i}$. The symbols $\hat{S}^+$ and $\hat{S}^-$ refer to the spin ladder operators $\hat{S}^x + i\hat{S}^y$ and $\hat{S}^x - i\hat{S}^y$, respectively. By comparison with Eq. (3.37), the potential energy Hamiltonian $\hat{H}_C$ thus becomes

$$\hat{H}_C = \sum_{i,j,\sigma,\sigma'} \frac{U}{2}\hat{n}_{\sigma,i}\hat{n}_{\sigma',j} - J_{ex}\frac{\hat{n}_i\hat{n}_j}{4} - J_{ex}\mathbf{S}_i\cdot\mathbf{S}_j. \tag{3.41}$$

Beyond solutions of maximum stability for diverse magnetic phases, the Heisenberg model allows for approximate descriptions of excited spin states, or *spin waves*. Some basics of these magnetic excitations are summarized in a supplemetary section at http://faculty.etsu.edu/HAGELBER/comp_chem.html.

We continue with a brief overview of another class of theoretical methods designed to provide comprehensive models of electronic systems. Operating from first principles, they engender techniques of materials simulation that differ substantially from the procedures reviewed in this section.

3.2 Electronic Structure Theory

Numerous methods for the computational treatment of electronic structures have been applied with great success to carbon nanostructures. This holds in particular, but by far not exclusively, for molecular carbon units or carbon clusters such as fullerenes or axially confined carbon nanotubes. In the following, we give a very condensed survey of the quantum chemical procedures most frequently used to model carbon nanostructures, and for capturing their magnetic properties. We subdivide the considered schemes into *ab initio* and *density functional theory* (DFT) approaches. Here, *ab initio* refers to the methodological constraint that results are generated without any use of empirical or semi-empirical data. This condition restricts the input information for an *ab initio* calculation to the identities and spatial positions of the atomic constituents. Most electronic structure calculations on carbon nanostructures, on the other hand, are based on DFT, where allowance is made for model parameters.

Both types of strategies aim at approximating solutions to the time-independent, non-relativistic Schrödinger equation

$$\hat{H}_{el}\psi = E_{el}\psi, \tag{3.42}$$

with ψ as the electronic wave function of the studied system, $\hat{H}_{el}$ as the Hamiltonian for its electronic degrees of freedom, and E_{el} as the corresponding eigenvalue. Further, both rely on the *Born–Oppenheimer approximation*, which assumes the separability of the nuclear and the electronic coordinates that characterize the system. The intuitive content of this assumption is that the fast electronic motion averages to yield a well-defined *potential energy surface* that determines the comparatively slow nuclear motion. Specifically, the Hamiltion of the system may be partioned into three components:

$$\hat{H} = \hat{T}_N + \hat{T}_{el} + \hat{V}(\mathbf{x}, \mathbf{X}), \tag{3.43}$$

where the operator $\hat{T}_N$ stands for the kinetic energy of the nuclei, $\hat{T}_{el}$ for that of the electrons, and the multiplicative operator $\hat{V}$ for the potential energy of the nuclei and the electrons. If the Born–Oppenheimer approximation is adopted, the electronic Hamiltonian may be represented as the sum of the two latter terms, namely

$$\hat{H}_{el} = \hat{T}_{el} + \hat{V}_{el}(\mathbf{x}, \mathbf{X}). \tag{3.44}$$

The coordinate set $\mathbf{x}(\mathbf{X})$ denotes here the electronic (nuclear) degrees of freedom. The potential energy $\hat{V}_{el}$ contains two components: the interelectronic repulsion $\hat{V}_{ee}$ as well as the attraction between the nuclei and the electrons, $\hat{V}_{eN}$. The electronic wave function $\psi(\mathbf{x}, \mathbf{X})$ depends on the coordinates $\mathbf{x}$ as variables while the nuclear positions $\mathbf{X}$ enter as parameters. For two different choices of these positions, the stationary Schrödinger equation will yield to different solutions $\psi(\mathbf{x}, \mathbf{X})$.

3.2.1 Ab Initio *Approaches*

Electronic structure calculation methods from first principles are often classified into *Hartree–Fock* and *post-Hartree–Fock* techniques. Our goal is here to give an overall idea about these two classes, while providing only few technical details. The reader finds more extensive coverage of the topic in monographs on quantum chemistry or molecular physics (e.g. [30, 32, 33]). In both types of method the electronic Hamiltonian is split into one-particle and two-particle terms,

$$\hat{H}_{el} = \sum_{i=1}^{N} \hat{h}_i + \sum_{i=1}^{N} \sum_{j>i}^{N} \frac{e^2}{4\pi\varepsilon_0} \frac{1}{|\mathbf{x}_i - \mathbf{x}_j|}, \tag{3.45}$$

where the former expression contains the kinetic energy of the electrons as well as their attraction by the nuclei, according to

$$\hat{h}_i = -\frac{\hbar^2}{2m_e} \frac{\partial^2}{\partial \mathbf{x}_i^2} - \sum_{k=1}^{N} \frac{e^2}{4\pi\varepsilon_0} \frac{Z_k}{|\mathbf{x}_i - \mathbf{X}_k|}, \tag{3.46}$$

with m_e as the electron mass, and ε_0 as the vacuum permittivity. The eigenvalue equation

$$\hat{h}\, \Phi_i(\mathbf{x}) = \varepsilon_i \Phi_i(\mathbf{x}) \tag{3.47}$$

is solved by the single-particle functions $\Phi_i(\mathbf{x})$ named *molecular orbitals*. It is often convenient to solve Eq. (3.47) in a basis of *atomic orbitals* ϕ_k, functions that are centered on individual nuclei and describe electronic states in the shell of the kth atom in the described structure. The *linear combinations of atomic orbitals* (LCAO) approach represents any molecular orbital as a sum over atomic orbitals:

$$\Phi_i = \sum_{k=1}^{K} a_{ik} \phi_k. \tag{3.48}$$

In general, the wave functions ϕ_k are not orthogonal. Including spin, we move from a purely spatial molecular orbital to one that depends on the generalized coordinate $\mathbf{x}' = (\mathbf{x}, \omega)$. A *spin orbital* ζ is defined as

$$\zeta(\mathbf{x}') = \Phi(\mathbf{x})\sigma(\omega). \tag{3.49}$$

The symbol ω denotes here a formal coordinate of the spin function σ, to parallel the spatial variable $\mathbf{x}$. The completely antisymmetric product of molecular orbitals,

$$\psi(\mathbf{x}_1', \mathbf{x}_2', \ldots, \mathbf{x}_N') = \frac{1}{\sqrt{N!}} \det[\zeta_{k_1}(\mathbf{x}_1'), \zeta_{k_2}(\mathbf{x}_2'), \ldots, \zeta_{k_N}(\mathbf{x}_N')] \;\; k_1, .., k_N = 1, ., N, \tag{3.50}$$

is referred to as *Slater determinant*.

Gauge-Including Atomic Orbitals

In accordance with the focus of this text on magnetism, we mention the *gauge-including atomic orbital* (GIAO) approach as an important method for representing molecules in magnetic fields. A well-known problem arises when the evaluation of magnetic molecular properties such as the *magnetic shielding tensor*[1] or the *magnetizability* of the molecule is based on the expansion (3.48). These properties are found to display a dependence on the origin of the Cartesian coordinate system employed in the calculation. This is a consequence of the form of the magnetic Hamiltonian (2.7) in conjunction with the finite atomic basis set used in (3.48). Due to the incompleteness of this basis, the cancellation of the origin dependence in the various contributions to a magnetic molecular property is incomplete. A remedy for this situation was proposed first by London [29]. It consists in a gauge transformation of the atomic orbitals, involving the vector potential $\mathbf{A}$:

$$\phi_k = \phi(\mathbf{x} - \mathbf{X}_k) \rightarrow \exp(-\frac{ie}{\hbar}\mathbf{A} \cdot \mathbf{x})\phi(\mathbf{x} - \mathbf{X}_k). \tag{3.51}$$

This GIAO choice of the atomic basis functions eliminates the artificial dependence on the specific coordinate system.

Hartree–Fock Theory

Hartree–Fock theory is based on the *variational principle*. It defines a prescription for obtaining the many-electron determinant ψ_0 that minimizes the expectation value of the total electronic energy E_0, given by

$$E_0 \equiv \langle \psi_0 | \hat{H}_{el} | \psi_0 \rangle. \tag{3.52}$$

Representing the wave function ψ_0 in terms of spin orbitals leads to the *canonical Hartree–Fock equations*:

$$\hat{F}(\mathbf{x}'_1)\zeta_i(\mathbf{x}'_1) = \varepsilon_i \zeta_i(\mathbf{x}'_1). \tag{3.53}$$

The spin orbitals that minimize the electronic energy of the system are here described as eigenstates of the *Fock operator*, which is given by

$$\hat{F} \equiv \hat{h} + \sum_{j=1}^{N} \hat{U}_j - \hat{J}_j. \tag{3.54}$$

The terms $\hat{U}_j$ and $\hat{J}_j$ denote the *Coulomb* and the *exchange operator*, respectively. Both are integral operators, defined by their actions on spin orbitals, according to

[1] The magnetic shielding tensor measures the extent to which the electronic system of the molecule influences the magnetic field experienced by the nuclei.

$$\hat{U}_j(\mathbf{x}_1')\zeta_i(\mathbf{x}_1') = \frac{e^2}{4\pi\varepsilon_0}[\int d\mathbf{x}_2'\zeta_j^*(\mathbf{x}'_2)\frac{1}{r_{12}}\zeta_j(\mathbf{x}'_2)]\zeta_i(\mathbf{x}'_1) \tag{3.55a}$$

and

$$\hat{J}_j(\mathbf{x}_1')\zeta_i(\mathbf{x}_1') = \frac{e^2}{4\pi\varepsilon_0}[\int d\mathbf{x}_2'\zeta_j^*(\mathbf{x}'_2)\frac{1}{r_{12}}\zeta_i(\mathbf{x}'_2)]\zeta_i(\mathbf{x}'_1). \tag{3.55b}$$

The structure of these operators implies that the Hartree–Fock equations are highly nonlinear. The spin orbitals that minimize the electronic energy are derived in traversing a sequence of iterative cycles that continues until a pre-installed convergence criterion is met. The first step of the procedure is the implementation of the Coulomb and exchange operators by selecting a primary set of spin orbitals, and the convergence speed of the iteration process depends critically on the quality of this initial guess. The electronic energy E_0, in most cases of interest the ground state energy of the studied system, depends on the Coulomb and the exchange term in the following way:

$$E_0 = \sum_i \langle\zeta_i|\hat{h}_i|\zeta_i\rangle + \frac{1}{2}\sum_{i,j=1}^{N} U_{ij} - J_{ij}, \tag{3.56}$$

with the matrix elements

$$U_{ij} \equiv \langle\zeta_i\zeta_j|\frac{e^2}{4\pi\varepsilon_0}\frac{1}{r_{12}}|\zeta_i\zeta_j\rangle \tag{3.57}$$

and

$$J_{ij} \equiv \langle\zeta_i\zeta_j|\frac{e^2}{4\pi\varepsilon_0}\frac{1}{r_{12}}|\zeta_j\zeta_i\rangle. \tag{3.58}$$

Hartree–Fock theory realizes an approximation of an N-electron problem by a set of N single-electron problems, thus accomplishing a reduction in complexity that makes a large number of chemical systems accessible to *ab initio* treatment. Expressing the Hartree–Fock equations in the framework of the *LCAO* approach transforms them into a system of matrix equations, namely

$$\mathbf{Fa} = \mathbf{Sa}\boldsymbol{\varepsilon}. \tag{3.59}$$

Here, the coefficient matrix **a** with elements a_{ik} (see Eq. (3.48)) and the orbital energy eigenvalue matrix $\boldsymbol{\varepsilon}$ have been introduced. The latter is diagonal with elements $\varepsilon_{ij} = \varepsilon_i\delta_{ij}$. Further, the *overlap matrix* **S** accounts for the non-orthogonality of the atomic orbitals, containing the elements

$$S_{ij} = \int d\mathbf{x}\ \phi_i(\mathbf{x})^*\phi_j(\mathbf{x}). \tag{3.60}$$

Formula (3.59) embodies the *Roothaan equations*. In principle, the exact solution of the Hartree–Fock equations can be accomplished in case of a complete basis of atomic orbitals.

Magnetic systems (or systems with non-zero spin) may be addressed within the Roothaan formalism through augmentation of the set of spin-paired molecular orbitals by *singly occupied molecular orbitals*, or *SOMOs*. Solutions of lower energy and therefore higher stability may be generated by allowing for differences between arrays of the form $\mathbf{a}^{\alpha}$ ($\boldsymbol{\varepsilon}^{\alpha}$) and $\mathbf{a}^{\beta}$ ($\boldsymbol{\varepsilon}^{\beta}$). This implies a splitting of Eq. (3.59) into two subequations:

$$\mathbf{F}^{\alpha}\mathbf{a}^{\alpha} = \mathbf{S}\mathbf{a}^{\alpha}\boldsymbol{\varepsilon}^{\alpha}, \tag{3.61a}$$

$$\mathbf{F}^{\beta}\mathbf{a}^{\beta} = \mathbf{S}\mathbf{a}^{\beta}\boldsymbol{\varepsilon}^{\beta}. \tag{3.61b}$$

Both subequations are coupled, since the Coulomb component of $\mathbf{F}^{\alpha}$ ($\mathbf{F}^{\beta}$) contains the orbitals with $\beta(\alpha)$ spin orientation. The gain in variational freedom as one goes from Eq. (3.59) to Eqs. (3.61) comes with a sacrifice, since the solution of the *unrestricted Hartree–Fock* (UHF) determinant, as the wave function derived from the solutions of Eq. (3.61) is called, is no longer an eigenfunction of the total electronic spin operator $\hat{S}^2$. This may give rise to unrealistic spin expectation values for the considered system. This deficiency can, in principle, be alleviated by projecting the wave function that results from the variational procedure on spin operator eigenfunctions [30].

Post-Hartree–Fock Methods

Hartree–Fock has opened up an enormous range of chemical systems to computational treatment as it achieves a *mean-field approximation* to the energies and wave functions of many-electron systems, reducing them to many one-electron systems. This highly successful approach, however, has a sensitive drawback: since the Hartree–Fock determinant consists of single-particle states, it fails to include the *correlation between electrons*. Conceptually, we distinguish between a dynamic and a static component of the overall correlation effect.

The dynamic part stems from the approximate treatment of the interelectronic repulsion term by Hartree–Fock theory. Each electron feels the average over the Coulomb and exchange potentials due to all remaining electrons. Moving in an effective one-body instead of the accurate many-body interelectronic potential, two electrons may approach each other too closely, which results in an overestimation of the repulsion between them.

Static correlation typically manifests itself in bond dissociation, which is insufficiently captured by the Hartree–Fock method. This shortcoming is frequently illustrated by the example of H_2 dissociation. The asymptopic state of this process is given by two well-separated spin-one-half hydrogen radicals. The restricted ground state determinant for the stretched H_2 molecule, on the other hand, may be expanded into a covalent and an ionic component, where the latter alternative corresponds to an H^--H^+ or a H^+-H^- scheme. The unrestricted determinant ceases to have spin singlet character in the asymptotic regime but can be described as a hybrid between spin singlet and spin triplet, both of which contribute with equal weights. Thus, neither the restricted nor the unrestricted Hartree–Fock scheme does justice to the physical reality of bond dissociation.

Various cures have been devised to remedy the correlation problem. We mention *configuration interaction* and *many-body perturbation theory* as the principal paradigms that govern the field of post-Hartree–Fock approaches.

The systematic solution of the electron correlation problem consists in replacing the single-determinant approach that underlies the various realizations of the Hartree–Fock method by a multideterminant picture. Thus, the variational search for the optimal wave function now proceeds in a subspace defined by N_C linearly independent determinants, starting with the Hartree–Fock solution ψ_0,

$$\psi = \sum_{k=0}^{N_C} c_k \psi_k, \tag{3.62}$$

where N_C stands for the number of configurations represented by the included determinants ψ_k. This extension of the Hartree–Fock formalism defines the configuration interaction method. From the variational principle, one concludes that the linear combination (3.62) that contains the Hartree–Fock wave function as one of the basis vectors corresponds to a lower energy than that of the Hartree–Fock limit, E_0. This energy decrease is ascribed to the inclusion of electron correlation. Taking into account the full correlation effect is tantamount to solving the Schrödinger equation (3.42), and so we may define the correlation energy E_{co} as the energy difference that bridges the gap between the accurate non-relativistic total energy E_{el} of the stationary molecular system and the Hartree–Fock limit:

$$E_{co} = E_{el} - E_0. \tag{3.63}$$

If the configuration basis $\{\psi_k\}$ is complete, the expansion (3.62) yields the energy eigenvalue E_{el} of the stationary Schrödinger equation, and thus the precise correlation energy. Completeness cannot be achieved in practice. However, a simple construction principle can be outlined whose extension *ad infinitum* generates a complete basis: starting with the Hartree–Fock determinant ψ_0, one replaces an occupied orbital j with an unoccupied orbital s, and thus creates a sequence of substitutions ψ_j^s with coefficients c_j^s. Extending this scheme, one obtains double and triple substitutions, ψ_{jk}^{st}, ψ_{jkl}^{stu} etc. In Eq. (3.62), the Hartree–Fock determinant acts as the *reference configuration*, i.e. the state on which the substitutions are based.

Excited states or states involving non-covalent bonds often require inclusion of multiple reference determinants for adequate representation within the *ab initio* formalism. The latter case is realized by systems containing unpaired electrons with opposite spins (often labeled as *antiferromagnetic solutions* in condensed matter theory), as exemplified by the H_2 molecule at the dissociation limit (see above). In these situations, the generalization of the approach (3.62) from one to several reference determinants (the *multireference configuration interaction* procedure, MRCI [31]) provides an improved model.

Electron correlation effects can often be understood as minor corrections. In this case, $E_{co} \ll E_0$, and a partition may be used that subdivides the Hamiltonian $\hat{H}_{el}$ into a reference Hamiltonian $\hat{H}_0$ as zeroth-order term and a perturbation operator $\hat{H}_{el} - \hat{H}_0 \equiv \hat{V}$ that accommodates electron correlation. If this subdivision is made, the electronic energy can be expressed in the form of perturbation series, namely

$$E_{el} = E_{el}^{(0)} + E_{el}^{(1)} + E_{el}^{(2)} + . \tag{3.64}$$

Closer inspection reveals that the Hartree–Fock limit E_0 is equal to the sum of the first two terms on the right-hand side of this relation, $E_{el}^{(0)} + E_{el}^{(1)}$. By Eq. (3.63), the many-body perturbation scheme yields

$$E_{co} = E_{el}^{(2)} + E_{el}^{(3)} + \dots \tag{3.65}$$

as correlation energy.

Other techniques not discussed in the present context have been successful in describing ground states with high precision.[2] This has been done, for instance, by use of various *Quantum Monte Carlo* methods [35] that implement statistical integration schemes.

Alternative methods of treating electron correlation are provided by density functional theory, as briefly surveyed in the following section.

3.2.2 Density Functional Theory

For atomic or molecular clusters, or for complex molecules, the traditional description of the electronic system in terms of wave functions faces insurmountable obstacles even when confronted with the means of modern computation, as it poses a $3N$-dimensional problem, with N as the number of electrons involved. The single-electron density, in contrast, is defined in three-dimensional space, and thus seems to provide an intuitively as well as mathematically much more accessible tool for characterizing the molecular electrons.

[2] Also, in generating excited states by forcing the respective wave functions to be orthogonal to the ground state.

In presenting the basic tenets of DFT in a condensed overview, we will accentuate its parallels to and deviations from *ab initio* theory. Once more, we employ the Hamiltonian operator $\hat{H}_{el}$, as introduced in Eq. (3.44) to describe a system of N electrons. Writing the operator in a more abstract form as

$$\hat{H}_{el} = -\frac{\hbar^2}{2m}\sum_{i=1}^{N}\frac{\partial^2}{\partial \mathbf{x}_i^2} + \sum_{i=1}^{N}\hat{V}_{ext}(\mathbf{x}_i) + \hat{V}_{int}(\mathbf{x}_1, \mathbf{x}_2, \ldots, \mathbf{x}_N), \tag{3.66}$$

we distinguish an *external potential* $V_{ext} = \sum_{i=1}^{N} V_{ext}(\mathbf{x}_i)$ from a contribution V_{int} due to the mutual repulsion between any two electrons. The external potential accounts for the attraction between the electrons and nuclei, but could also contain any other interaction expressible as a simple partition into additive one-particle terms, such as the influence of an electrostatic field. The ground state expectation values of any operator satisfying this condition can be expressed as functionals of the ground state electron density. This feature, in fact, holds for all ground state variables of a given system and thus also includes the two-particle terms that describe the interelectronic contribution to $\hat{H}_{el}$. DFT exploits this result using the density as a much less redundant, conceptually more simple and computationally more tractable descriptor of a system than its wave function.

Implementations of the idea to replace the wave function as the solution of electronic structure problems with the density reach back to the early approaches of Thomas and Fermi [36, 37]. Here as in later realizations of this concept, the task of identifying the density is reduced to minimizing a total energy functional that can be decomposed into a kinetic energy, an external potential and an interelectronic repulsion component according to

$$E[n(\mathbf{x})] = T[n(\mathbf{x})] + V_{ext}[n(\mathbf{x})] + V_{int}[n(\mathbf{x})], \tag{3.67}$$

where the density is subjected to the normalization constraint $\int d\mathbf{x}\, n(\mathbf{x}) = N$. While the Thomas–Fermi method neglected salient elements of *ab initio* procedures such as, for instance, the exchange interaction, and was merely a qualitative approximation to these, incremental modifications of the density functional methodology improved its accuracy and made it applicable to a wide range of physical systems [38, 39].

The *Hohenberg–Kohn theorems* [40] established the density approach as electron structure theory in its own right, providing an alternative to the wave function based formalism surveyed in the previous section.

The first of the two theorems states that the position space ground state density of an N electron system uniquely determines the wave function of this system. The second theorem extends the variational principle that was used in order to obtain the Hartree–Fock equations (3.53) from wave functions to densities. As much as the ground state wave function ψ_0, the ground state density n_0 minimizes the total energy of the studied system.

It therefore holds,

$$E[n_0] \leq E[n], \qquad n \neq n_0, \tag{3.68}$$

which implies the possibility of computing n_0 by use of a variational procedure. The second theorem of Hohenberg and Kohn lays the groundwork for the practical use of DFT in electronic structure calculations. The pivotal step consists in translating the abstract condition

$$\delta\{E[n(\mathbf{x})] - \lambda(\int d\mathbf{x}\, n(\mathbf{x}) - N)\} = 0, \tag{3.69}$$

where the normalization constraint has been included with λ as Lagrangian multiplier, into a set of equations that yield the ground state density of a physical system. The proposal put forth by Kohn and Sham [41] acquired a high level of prominence, as it yielded a practicable approximation to solving this problem.

As the variation in Eq. (3.69) is to be carried out with respect to n, the equation may be reformulated in terms of a *functional derivative*, with n as the independent quantity:

$$\frac{\delta}{\delta n(\mathbf{x})}\{E[n(\mathbf{x})] - \lambda(\int d\mathbf{x}\, n(\mathbf{x}) - N)\} = 0. \tag{3.70}$$

For the unknown energy functional $E[n]$, Kohn and Sham took the following approach:

$$E[n(\mathbf{x})] = T[n(\mathbf{x})] + \int d\mathbf{x}\, V_{ext}(\mathbf{x})n(\mathbf{x}) + \int d\mathbf{x}\, V_{coul}(\mathbf{x})n(\mathbf{x}) + \tilde{E}_{xc}[n(\mathbf{x})]. \tag{3.71}$$

The expression on the right side contains, besides functionals involving the external potential and the Coulomb potential V_{coul} that governs the mutual repulsion between electrons, two additional functionals: that of the kinetic energy and an *exchange-correlation energy functional* $\tilde{E}_{xc}[n]$ that summarily contains the quantum and many-body effects influencing the total energy of the system. Varying $E[n(\mathbf{x})]$ up to first, or linear, order in the density, we find by combining Eqs. (3.71) and (3.69) that

$$\frac{\delta T[n]}{\delta n} + V_{ext}(\mathbf{x}) + V_{coul}(\mathbf{x}) + \frac{\delta \tilde{E}_{xc}[n]}{\delta n} = \lambda. \tag{3.72}$$

The crucial move made by Kohn and Sham in addressing the challenge posed by this equation was mapping the real system of N interacting electrons onto an auxiliary model system of N non-interacting electrons governed by an effective single-particle potential V^{KS}. Both systems are assumed to have the same ground state density and total energy.

Before commenting on the actual construction of the fictitious system, we point out that this device makes it possible to represent the solution of the stationary Schrödinger equation for the problem at hand as in Hartree–Fock theory, i.e. as a single determinant of molecular orbitals, where the spatial orbitals Φ_i are eigenstates of a single-particle Hamiltonian $\hat{h}$, according to (compare with Eq. (3.47))

$$\hat{h}\Phi_i = E_i\Phi_i, \; i = 1, 2, .., N \tag{3.73}$$

and

$$\hat{h} = -\frac{\hbar^2}{2m}\frac{\partial^2}{\partial \mathbf{x}_i^2} + V^{KS}(\mathbf{x}_i). \tag{3.74}$$

The relations (3.73) are termed *Kohn–Sham equations*. It is important to note that the single determinant, in contrast to its role in *ab initio* theory, has no meaning beyond its auxiliary function of furnishing the ground state density. The electronic density results from the spatial orbitals as

$$n = \sum_{i=1}^{N} |\Phi_i|^2. \tag{3.75}$$

With these premises, the kinetic energy functional is easily determined to be

$$T_s[n] = \sum_{i=1}^{N} \langle\Phi_i| - \frac{\hbar^2}{2m}\frac{\partial^2}{\partial \mathbf{x}_i^2}|\Phi_i\rangle, \tag{3.76}$$

where the subscript s stands for *single particle*.

This functional differs from its counterpart in Eq. (3.72), where interacting electrons are adopted. An expression analogous to Eq. (3.72), however, is obtained by introducing the single-particle potential in the following way:

$$V^{KS}(\mathbf{x}) = \frac{\delta T[n]}{\delta n} - \frac{\delta T_s[n]}{\delta n} + V_{ext}(\mathbf{x}) + V_{coul}(\mathbf{x}) + \frac{\delta \tilde{E}_{xc}[n]}{\delta n}, \tag{3.77}$$

or shorter:

$$V^{KS}(\mathbf{x}) = V_{ext}(\mathbf{x}) + V_{coul}(\mathbf{x}) + V_{xc}(\mathbf{x}), \tag{3.78}$$

based on the definition

$$V_{xc}(\mathbf{x}) \equiv \frac{\delta T[n]}{\delta n} - \frac{\delta T_s[n]}{\delta n} + \frac{\delta \tilde{E}_{xc}[n]}{\delta n} \tag{3.79}$$

of the *exchange-correlation potential*. This completes the assembly of the single-particle potential.

Making reference to spatial orbitals only (see Eqs. (3.73, 3.76)), we have operated thus far under the tacit assumption of a closed-shell system. If this condition is dropped, the formalism has to be generalized in close analogy to the extension that leads from the restricted to the unrestricted Hartree–Fock model, as discussed in the previous section. The density function is then partitioned according to

$$n(\mathbf{x}) = n^{\alpha}(\mathbf{x}) + n^{\beta}(\mathbf{x}) \tag{3.80}$$

to distinguish between the subsets of electrons with α and β spins. The functional of the electronic energy then is modified by going from $E[n(\mathbf{x})]$ to $E[n^{\alpha}(\mathbf{x}), n^{\beta}(\mathbf{x})]$ or, alternatively, to $E[n(\mathbf{x}), m(\mathbf{x})]$. The quantity $m(\mathbf{x})$ $(= n^{\alpha}(\mathbf{x}) - n^{\beta}(\mathbf{x}))$ is here defined as *magnetic* or *spin-polarized density*.

A generalized Kohn–Sham theory that replaces density as basic variable with the *current density* has been shown to widen the range of DFT applications considerably, to include, for instance, interactions with magnetic fields and electromagnetic waves [42, 43].

Exchange-Correlation Potentials

The main challenge for any application of DFT to physical systems is identifying adequate expressions for the exchange correlation potential V_{xc}. The most elementary approximation to this quantity is based on the homogeneous electron gas that allows for exact analytical descriptions of the exchange and correlation energy as a function of the density in the high- and low-density regimes. This defines the *local* solution to the problem of representing V_{xc}, as originally proposed by Kohn and Sham [41]. Later authors refined this approach by suggesting various procedures of interpolating between the known cases of density extremes ([44]–[47]). The *local density approximation* (LDA) treats the exchange-correlation effect as position independent. In the framework of LDA, the energy E_{xc} becomes (see for instance [48])

$$E_{xc}^{LDA}[n] = \int d\mathbf{x}\, n(\mathbf{x})\varepsilon_{xc}[n], \tag{3.81}$$

with $\varepsilon_{xc}[n]$ as exchange-correlation energy per electron. Variation of this expression with respect to the density yields the LDA exchange-correlation potential:

$$V_{xc}^{LDA}(\mathbf{x}) = \varepsilon_{xc}[n(\mathbf{x})] + n(\mathbf{x})\frac{\delta\varepsilon_{xc}[n]}{\delta n}. \tag{3.82}$$

By construction, the LDA technique should be useful in those cases where the density varies slowly in space. Experience, however, has shown it to be a surprisingly powerful method even where this prerequisite is not satisfied, as is the regular situation for atoms, molecules and crystals. This success may be rationalized by the observation that many quantities of major interest, such as the total energy of a system, involve integrals over ε_{xc} that are relatively insensitive to the local features of the exchange-correlation effect. Procedures that operate with extensions of local exchange-correlation potentials to open-shell situations are labeled *local spin density* (LSD) methods.

The next higher step on the ladder of DFT implementations is reached as allowance is made for spatial variation of ε_{xc} with the density. The exchange- correlation energy per electron now depends not only on n itself, but also on its first or higher derivatives with respect to position. Various realizations of this so-called *generalized gradient approximation* (GGA) have been proposed (e.g. [49]–[51]).

We add two remarks on special types of functionals used in DFT. In the context of Hartree–Fock theory, we observe that the diagonal terms of the Coulomb and the exchange integral matrices **U** and **J**, respectively, are identical (compare with Eqs. (3.57 and 3.58)):

$$U_{ii} = J_{ii}. \tag{3.83}$$

As the Hartree–Fock equations depend on the difference of **U** and **J** (as made explicit in Eqs. (3.54) and (3.59)), they are uncompromised by any spurious interaction of an electron with itself. There is no such cancellation in Kohn–Sham theory, since the Coulomb and the exchange-correlation term that appear in formula (3.78) for the single particle potential V^{KS} are of different origin. Specifically, V^{KS} combines the exact Coulomb with an approximate exchange-correlation component, resulting in the appearance of an unphysical *self-interaction effect*. Curing this malady introduces a further potential term into the Kohn–Sham equations (3.73). The explicit form of this contribution, labeled $V_{SIC,i}$, where SIC stands for *self-interaction correction*, changes with the theory adopted to approximate the exchange-correlation potential [45]. This addition imports an orbital dependence into the single-particle Hamiltonian, as indicated by the index i, and thus detracts from the formal simplicity of the original approach.

Mixtures between the DFT and Hartree–Fock schemes are widely used (e.g. [52]). These employ *hybrid functionals*, which combine a single-particle correlation potential with an exchange potential that incorporates both a DFT and an *ab initio* component. These functionals interpolate between the exact exchange expression (3.55b), which is composed of Kohn–Sham orbitals, and one or various DFT representations of the exchange effect. The frequently applied *Becke 3-parameter* approximation, for instance, implements both Hartree–Fock and non-local DFT (specifically GGA, see above) corrections to the LDA (or LSD) approach. In the popular (Becke, three-parameter, Lee-Yang-Parr) potential, two parameters balance the admixtures due to *ab initio* and GGA to the local density exchange, and a third parameter modifies the local density correlation by the correlation functional of Lee, Yang and Parr [53]. Pragmatic as hybrid functionals are, they are also practical, adding a high degree of flexibility to the DFT methodology and alleviating the self-interaction problem.

Experimental Methods

Providing a condensed survey of some of the experimental methods frequently used to determine magnetic properties of carbon nanostructures, we focus first on a spectroscopic technique, namely electron spin resonance (ESR), involving the use of microwaves. The following two sections summarize the fundamentals of non-spectroscopic experimental procedures that have been employed with great success in the study of magnetic phenomena associated with carbon nanostructures. These are vibrating sample magnetometry (VSM), based on classical electrodynamics, and superconducting quantum interference device (SQUID) technology, which exploits a variety of quantum effects for magnetic flux measurements of high accuracy.

Sketches of two frequently used tools for the experimental exploration of magnetic carbon nanostructures are found at http://faculty.etsu.edu/HAGELBER/comp_chem.html. These are x-ray magnetic circular dichroism (XMCD) spectroscopy and magnetic force microscopy (MFM). The former is a further spectroscopic method, operating in a distinctly different energy regime than ESR. The XMCD technique is based on the idea of making the x-ray absorption of a magnetic sample spin dependent by use of circularly polarized incident x-ray radiation. The MFM method is a magnetic variant of atomic force microscopy (AFM): the magnetic texture of a surface is recorded by detecting the interactions between the surface and a magnetized microscope tip.

Additional experimental methods of relevance for this text, namely those pertaining specifically to applications of carbon nanostructures in spintronics, will be introduced in Chapters 9 and 10.

3.3 Electron Spin Resonance

Electron spin resonance (ESR) measurement consists in observing the resonant absorption of microwave radiation by paramagnetic ions or molecules. More specifically, one monitors transitions between spin states in open-shell electronic systems whose degeneracy is lifted by an external magnetic field. The simplest possible case is analogous to that of a free electron, with two levels, $|\frac{1}{2}\rangle$ and $|-\frac{1}{2}\rangle$. The magnetic interaction is then described by the Zeeman Hamiltonian ($g = g_S$)

$$\hat{H}_0 = -\boldsymbol{\mu} \cdot \mathbf{B} = g\mu_B B\hat{S}_z, \tag{3.84}$$

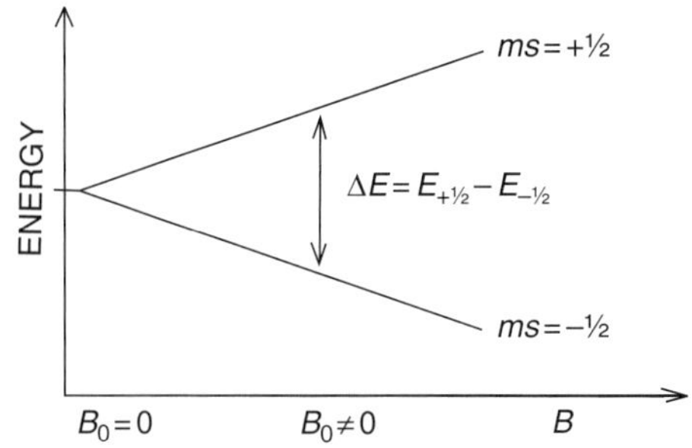

Figure 3.1 Free electron energy levels in a magnetic field, determining the ESR transition frequency between the two spin eigenstates.

where the magnetic field **B** is assumed to point in the z direction. Applying the spin Hamiltonian (3.84) to its two spin eigenstates, one may assign a transition frequency $f_{\pm}$ to each B value of the magnetic field, such that the resonance condition

$$f_{\pm} = \frac{1}{2\pi\hbar} g\mu_B B \tag{3.85}$$

is satisfied, as illustrated in Figure 3.1, which shows the linear dependence of the two alternative electron energies on the magnetic field.

The absorption signals detected in ESR experiments result from the difference between absorbed and emitted photons. As each photon may induce excitation from the lower to the higher level or de-excitation between these two levels, and since the probabilities of these processes increase linearly with the population of the ($|-\frac{1}{2}\rangle$) and the ($|\frac{1}{2}\rangle$) states, respectively, the net absorption is proportional to $N_- - N_+$. The symbols N_- and N_+ signify the numbers of electrons occupying the two levels. At equilibrium they are related by the Boltzmann probability distribution:

$$\frac{N_+}{N_-} = \exp(\frac{E_+ - E_-}{k_B T}) = \exp(\frac{g\mu_B B}{k_B T}). \tag{3.86}$$

Under normal conditions, i.e. if neither the temperature is too low nor the magnetic field is too high, the following approximation applies:

$$\frac{N_+}{N_-} \approx 1 + \frac{g\mu_B B}{k_B T}. \tag{3.87}$$

The further assumption that the two populations do not deviate too much from $N/2$, with N as the overall size of the ensemble, yields for the population difference the compact expression

$$N_+ - N_- \approx \frac{N g\mu_B B}{2k_B T}. \tag{3.88}$$

Thus the strength of the net absorption signal, and therefore the sensitivity of the ESR method, rises with increasing magnetic field and decreasing temperature. Due to the size of the electron's magnetic moment, which exceeds magnetic moments of nuclei by two to three orders of magnitude, ESR allows for lower magnetic fields and higher transition frequencies than NMR, which translates into better spectral resolution for typical applications of ESR as compared with NMR.

Before considering additional spin interactions that modify the absorption behavior of a spin system, we comment on the nature and the significance of the *g factor* in Eqs. (3.84)–(3.88). In particular, it should be noted that the spin quantization axis is not always parallel to the magnetic field, as has been tacitly assumed in Eq. (3.84). The g factor thus has to be generalized into a g tensor that in the simplified case described by Eq. (3.84) is *isotropic*. *Anisotropic* situations are realized when the electron's spin and its orbital angular momentum couple to form the total angular momentum of the electron. Further, local fields, generated by the atomic environment of the electron spin, are sources of anisotropy. This more complex configuration is addressed by rewriting Eq. (3.84) as

$$\hat{H}_0 = \mu_B \begin{pmatrix} B_x & B_y & B_z \end{pmatrix} \begin{pmatrix} g_{xx} & g_{xy} & g_{xz} \\ g_{yx} & g_{yy} & g_{yz} \\ g_{zx} & g_{zy} & g_{zz} \end{pmatrix} \begin{pmatrix} \hat{S'}_x \\ \hat{S'}_y \\ \hat{S'}_z \end{pmatrix}. \tag{3.89}$$

The primes decorating the spin operator components accentuate that the spin operator in Eq. (3.89) is not the same as that in Eq. (3.84), as it acts on spin-orbit corrected states.[3] For a single spin that undergoes only Zeeman interaction with the magnetic field, resonance is found at the frequency $\frac{\mu_B}{2\pi\hbar}(g_{xx}B_x + g_{yy}B_y + g_{zz}B_z)$.

While the extended Hamiltonian H_0 in Eq. (3.89) contains the fine structure of the considered interaction, it leaves out the *hyperfine structure*, which results from the coupling between the electronic and the nuclear spins of the studied system (see Section 2.1.4). Incorporating this effect, we extend $\hat{H}_0$ in the following way:

$$\hat{H}_{mag} = \hat{H}_0 + \hat{H}_{hfs} \tag{3.90}$$

and

$$\hat{H}_{mag} = A\hat{\mathbf{I}}\hat{\mathbf{S}}, \tag{3.91}$$

with $\hat{\mathbf{I}}$ as nuclear spin operator and A as *hyperfine coupling constant*. Assuming an isotropic situation, and introducing the raising and lowering operators associated with $\hat{\mathbf{I}}$ and $\hat{\mathbf{S}}$, we recast the magnetic Hamiltonian into [56]

[3] For further clarification, we consider the z component of both spin operators, which is $\hat{S}_z = \frac{1}{2}(|\frac{1}{2}\rangle\langle\frac{1}{2}| - |-\frac{1}{2}\rangle\langle-\frac{1}{2}|)$ if a pure spin state is considered and $\hat{S}_z = \frac{1}{2}(|\frac{1}{2}'\rangle\langle\frac{1}{2}'| - |-\frac{1}{2}'\rangle\langle-\frac{1}{2}'|)$ in case of states perturbed by spin-orbit interaction [55]. Here, $|\pm\frac{1}{2}'\rangle$ denotes the spin state $|\pm\frac{1}{2}\rangle$ modified to include the spin-orbit effect, for instance by perturbation theory.

$$\hat{H}_{mag} = g\mu_B B\hat{S}_z + A\hat{I}_z\hat{S}_z + \frac{A}{2}(\hat{I}_+\hat{S}_- + \hat{I}_-\hat{S}_+). \tag{3.92}$$

The action of this operator on a spin state with $S = I = \frac{1}{2}$ is readily evaluated. Selecting the substate $|m_S, m_I\rangle = |\frac{1}{2}, -\frac{1}{2}\rangle$, we find

$$\hat{H}_{mag}|\frac{1}{2}, -\frac{1}{2}\rangle = (\frac{1}{2}g\mu_B B - \frac{A}{4})|\frac{1}{2}, -\frac{1}{2}\rangle + \frac{A}{2}|-\frac{1}{2}, \frac{1}{2}\rangle. \tag{3.93}$$

Treating the remaining three states in an analogous manner and diagonalizing the resulting 4×4 Hamiltonian matrix yields the energy eigenvalues for all four spin states:

$$\begin{aligned} E(\frac{1}{2}, \frac{1}{2}) &= \frac{1}{2}g\mu_B B + \frac{A}{4}, \\ E(\frac{1}{2}, -\frac{1}{2}) &= -\frac{A}{4} + \frac{1}{2}g\mu_B BC, \\ E(-\frac{1}{2}, \frac{1}{2}) &= -\frac{A}{4} - \frac{1}{2}g\mu_B BC, \\ E(-\frac{1}{2}, -\frac{1}{2}) &= -\frac{1}{2}g\mu_B B + \frac{A}{4}, \end{aligned} \tag{3.94}$$

with $C = \sqrt{1 + (\frac{A}{g\mu_B B})^2}$. Analyzing these four expressions, one notices that at zero field, the two-spin states arrange into a $|-\frac{1}{2}, \frac{1}{2}\rangle$ singlet at $E = -\frac{3A}{4}$ and a triplet at $E = \frac{A}{4}$. The high-field limit, however, is determined by two distinctly separated bands with $m_s = -\frac{1}{2}$ and $m_s = \frac{1}{2}$, as displayed in Figure 3.2. This limit is defined by $B >> \frac{A}{g\mu_B}$, reducing the list (3.94) to the simple expression

$$E = \pm\frac{1}{2}g\mu_B B \pm \frac{A}{4}. \tag{3.95}$$

The possible transitions between the four substates are constrained by angular momentum conservation. This requirement selects processes with $\Delta m_s = \pm 1$, $\Delta m_I = 0$ (interband transitions), as shown in the figure, or $\Delta m_I = \pm 1$, $\Delta m_s = 0$ (intraband transitions).

3.3.1 The ESR Spectrometer

The great diversity of ESR spectrometers that have been designed since the first introdcution of the ESR method in 1944 by Evgeny Zavoisky [57] may be understood as variations and extensions of the basic setup sketched in Figure 3.3. The *Klystron* generates microwaves that are guided into the *cavity*, which contains the studied sample. Waveguides direct the radiation into the cavity and direct the cavity output to the detector. The sample is exposed to a direct current (DC) magnetic field that varies with time. It is provided by an electromagnet that sweeps through a

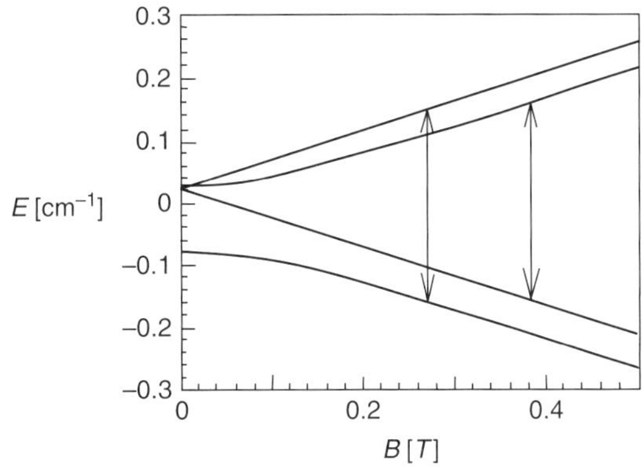

Figure 3.2 Energy levels for an electron spin interaction with a nuclear spin $I = \frac{1}{2}$ as a function of a magnetic field. The arrows indicate the allowed inter-band transitions. From lowest to highest, the following (m_S, m_I) labels apply to the four lines: $(-1/2, 1/2)$, $(-1/2, -1/2)$, $(1/2, -1/2)$, $(1/2, 1/2)$.

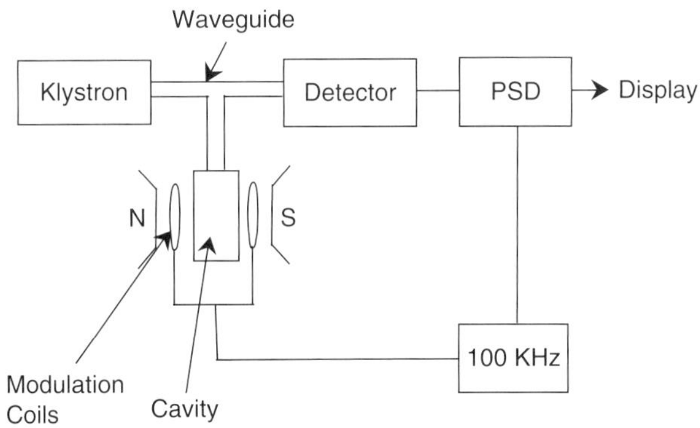

Figure 3.3 Main elements of an ESR spectrometer. The phase-sensitive detector (PSD) is designed to maximize the signal-to-noise ratio.

set range of field strengths. At resonance, the absorption is maximum, such that the transition(s) of interest are found at field values where the cavity output is minimal. Direct observation of the absorption profile, however, is marred by an unfavorable signal-to-noise ratio.

An improved design is accomplished by superposing the DC field with an oscillating magnetic field whose amplitude is less than the breadth of the absorption curve. The overall exterior magnetic field may thus be described as

$$\mathbf{B}(t) = \mathbf{B}_0 + \mathbf{b}_0 t + \mathbf{B}_{mod} \sin(2\pi f_{mod} t), \tag{3.96}$$

where $\mathbf{B}_0$, $\mathbf{b}_0$, $\mathbf{B}_{mod}$, and f_{mod} denote the initial field, the constant sweep rate of the field, the modulation field and its frequency, respectively. Correspondingly, a

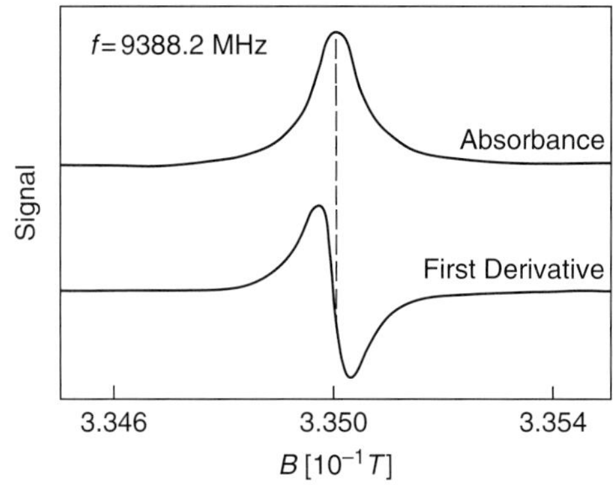

Figure 3.4 The absorption curve and its first time derivative, as generated by small-amplitude modulation of the magnetic field.

modulated signal is detected whose amplitude translates into the change of the absorption within a small interval of magnetic field values rather than the absorption itself at a set value of the field. From Figure 3.4, this differential scanning technique yields the time derivative of the absorption curve. This procedure makes it possible to determine resonance energies with substantially higher precision than direct observation of the absorption process.

ESR spectrometers are available in various regimes of frequency. The most widely used among these is the X band, spreading from 9 to 10 GHz. High-field ESR is performed in the W band that extends from 55 to 100 GHz and requires magnetic field strengths up to about 3.4 T.

3.4 Vibrating Sample Magnetometry

The principle of vibrating sample magnetometry (VSM) makes use of electromagnetic induction to determine the magnetization of a sample and its associated magnetic moment. As shown in Figure 3.5, the sample is attached to a rod and placed between the pole pieces of an electromagnet, which provide a homogeneous external magnetic field $\mathbf{H}_0$ at the location of the sample. The rod carries out a vibration, with 60–80 Hz and 1 mm as typical values for the frequency and the amplitude. This generates an oscillating magnetic field in pickup coils mounted on the poles of the electromagnet, translating into an induced voltage signal.

By Faraday's law, the induced voltage is related to the magnetic flux through

$$V_{ind} = -\frac{\partial \Phi}{\partial t} = -N_w \mathbf{a} \cdot \frac{\partial \mathbf{B}}{\partial t}, \tag{3.97}$$

with N_w as the number of windings in a pickup coil and $\mathbf{a}$ as the vector of its cross-sectional area. Using the relation $\mathbf{B} = \mu_0(\mathbf{H}_0 + \mathbf{M})$, one concludes from the time-independence of the field $\mathbf{H}_0$,

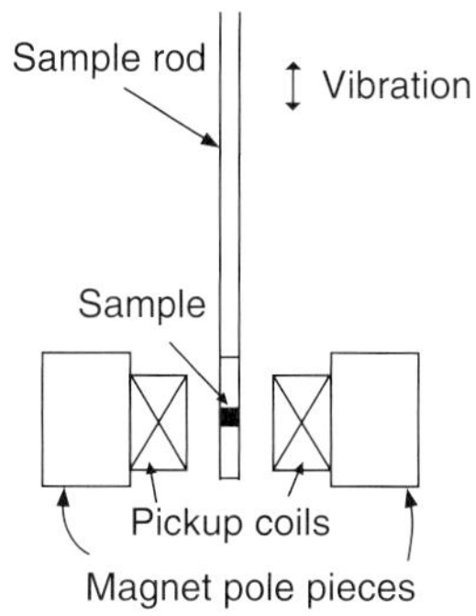

Figure 3.5 Sketch of the central part of a vibrating sample magnetometer. The magnetic sample is set into a vibrating motion and thus induces a voltage in pickup coils attached to the poles of an electromagnet [58].

$$\frac{\partial \mathbf{B}}{\partial t} = \mu_0 \frac{\partial \mathbf{M}}{\partial t}. \tag{3.98}$$

If the vibration of the sample follows a sinusoidal profile, the observed voltage will be proportional to $|\mathbf{M}_0|$, the magnitude of the full magnetization of the sample. As the latter can be analyzed at different values of the external field $\mathbf{H}_0$, vibrating sample magnetometry is an expedient tool for tracing the hysteresis loop of ferromagnetic materials. The VSM technique may be used to measure small magnetic moments on the order of μemu, the equivalent of the magnetic moment of one nanogram of iron [58].

3.5 Superconducting Quantum Interference Devices (SQUIDs)

A superconducting quantum interference device (SQUID) is capable of detecting magnetic flux at the order of one *magnetic flux quantum* in superconductors Φ_0^S, given by[4]

$$\Phi_0^S = \frac{h}{2e} = 2.07 \cdot 10^{-15}\text{Wb}. \tag{3.99}$$

Magnetic flux threaded through a superconducting loop exists in whole multiples of Φ_0^S. Capitalizing on quantum interference effects between two superconductors, SQUID magnetometers, operating with standard resolutions of about a femto-Tesla (fT), may be employed to measure extremely small magnetic fields. Two types of SQUIDs are in use, involving DC and AC current. In the following, we will provide a sketch of the former type, which, at low temperature, is of higher resolution.

The basic setup of a regular DC SQUID is shown in Figure 3.6. It may be described as a parallel arrangement of two *Josephson junctions*. A Josephson

[4] In the following we will refer to the quantity $\Phi_0 = \frac{h}{e}$ as *magnetic flux quantum*. That Φ_0 and Φ_0^S differ by a factor of two is due to the charge of a Cooper pair, namely $-2e$.

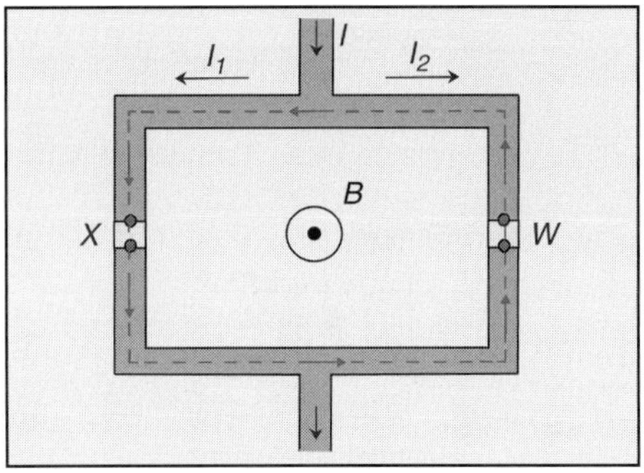

Figure 3.6 Basic design of a DC SQUID. Sketch of a DC SQUID containing two Josepson junctions at the sites marked with X and W. Shown is a superconducting loop, oriented perpendicular to the magnetic field $\mathbf{B}_a$ to be measured [59].

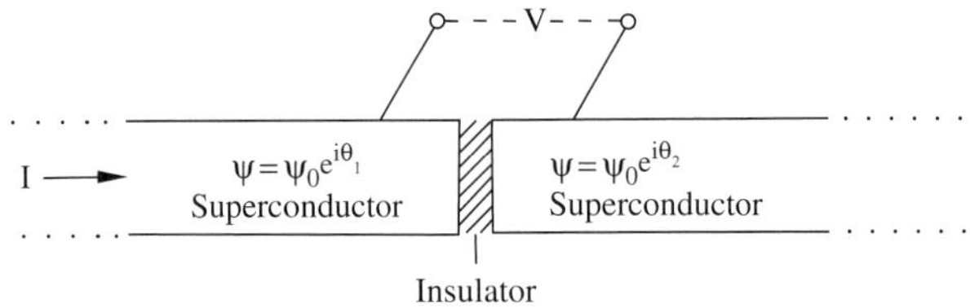

Figure 3.7 Representation of a Josephson junction connecting two superconducting elements. Across the insulating layer between them, a phase change occurs in the Cooper pair wave function ψ [59].

junction consists of two superconductors separated by a narrow insulating layer, designed sufficiently thin to allow for tunneling of Cooper pairs between them.

The interference effect exploited in SQUID measurements is related to the phase coherence of the Cooper pair wave function Ψ in a superconducting medium. This *macroscopic wave function* may be written as

$$\Psi_{CP} = \Psi e^{\frac{i}{\hbar}\mathbf{p}\cdot\mathbf{x}}, \tag{3.100}$$

with $\mathbf{x}$ as the center of mass of the Cooper pair and $\mathbf{p}$ as the net momentum of the pair. We denote the Cooper pair wave function on the left and the right of each insulating layer with Ψ_{CP1} and Ψ_{CP2}, respectively. Specifically, we set [60]

$$\begin{aligned} \Psi_{CP1} &= \sqrt{n_1} e^{i\theta_1} \\ \Psi_{CP2} &= \sqrt{n_2} e^{i\theta_2}. \end{aligned} \tag{3.101}$$

Here, n_i, i=1,2 stands for the probability density of the state Ψ_{CPi}. The interaction between the two sides can be cast in the form of the quantum-mechanical two-state problem. From the premise that the superconducting media on both sides are identical, one then derives two coupled time-dependent Schrödinger equations for the two wave functions:

$$i\hbar\frac{\partial\Psi_{CP1}}{\partial t} = E_1\Psi_{CP1} + \Delta E\Psi_{CP2}$$
$$i\hbar\frac{\partial\Psi_{CP2}}{\partial t} = E_2\Psi_{CP2} + \Delta E\Psi_{CP1}, \tag{3.102}$$

where the energy ΔE reflects the coupling of the two states across the junction. Inserting Eqs. (3.101) into Eqs. (3.102) yields

$$\frac{\partial n_1}{\partial t} = \frac{2\Delta E}{\hbar} n_{CP} \sin(\theta_2 - \theta_1) = -\frac{\partial n_2}{\partial t}. \tag{3.103}$$

Here, $\frac{\partial n_1}{\partial t} = -\frac{\partial n_2}{\partial t}$ has been used in conjunction with $n_1 = n_2 = n_{CP}$, as obtained from the constraint of current conservation, as well as the assumed identity of the two coupled superconductors.

Exercise 3.3 Derive relation (3.103) by inserting the wave functions (3.101) into the coupled Schrödinger equations (3.102), and separating the real and the imaginary parts.

As the left-hand side of Eq. (3.103) is proportional to the current of Cooper pairs, the *supercurrent* I_{CP} from port 1 to port 2, this relation may be cast in the compact form

$$I_{CP} = I_0 \sin(\theta_2 - \theta_1) = I_0 \sin(\Delta\theta), \tag{3.104}$$

where, by Eq. (3.103), the *critical current* I_0 is proportional to the coupling energy ΔE. The supercurrent through the loop shown in Figure 3.7 is the current due to Cooper pairs tunneling through two identical Josephson junctions, a and b, connected in parallel. Therefore, it holds that [61]

$$I_{CP} = I_0 \sin(\Delta\theta_a) + I_0 \sin(\Delta\theta_b) = 2I_0 \cos(\frac{\Delta\theta_b - \Delta\theta_a}{2}) \sin(\frac{\Delta\theta_a + \Delta\theta_b}{2}). \tag{3.105}$$

From the requirement that the Cooper-pair wave function be single-valued, we obtain

$$\Delta\theta_b - \Delta\theta_a = 2\pi M + 2\pi \oint_{C'} \frac{\hat{\mathbf{x}}}{\lambda} \cdot d\mathbf{l}, \tag{3.106}$$

where M is an integer. The circulation integral is performed by integrating along the superconducting loop with respect to the line element $\mathbf{l}$. The integration excludes the regime of the two insulating layers, as indicated by the subscript C'. The unit vector $\hat{\mathbf{x}}$ points in the direction of wave propagation. Acknowledging that the Cooper pair mass is $2m_e$, with m_e as electron mass, one finds that the wave length λ of a Cooper pair is related to the expectation value of its momentum through

$$\langle \mathbf{p} \rangle = 2m_e \langle \mathbf{v} \rangle = \frac{h}{\lambda} \hat{\mathbf{x}}. \tag{3.107}$$

Further, one observes that the *supercurrent density* j_{CP} is given by

$$\mathbf{j}_{CP} = -2en_{CP} \langle \mathbf{v} \rangle. \tag{3.108}$$

Combining Eqs. (3.107) and (3.108), the wavelength becomes

$$\lambda = \frac{hen_{CP}}{m_e |\mathbf{j}_{CP}|}. \tag{3.109}$$

The Cooper pair density n_{CP} is here assumed to be a constant. What is the effect of a magnetic field **B** threaded through the superconducting loop displayed in Figure 3.6? The presence of the field is reflected in the wave function (3.100) by the appearance of an additional phase associated with a generalized Cooper pair momentum $2m_e\mathbf{v} - 2e\mathbf{A}$, with **A** as *magnetic vector potential*. Including the corresponding phase, $-\frac{2e}{\hbar} \oint \mathbf{A} d\mathbf{l}$, and taking into account Eq. (3.109), the difference between the phase shifts across the Josephson junctions, relation (3.106), changes in the following way:

$$\Delta\theta_b - \Delta\theta_a = 2\pi M - 2\pi \frac{m_e}{hen_{CP}} \oint_{C'} \mathbf{j}_{CP} \cdot d\mathbf{l} - \frac{2e}{\hbar} \oint \mathbf{A} d\mathbf{l}. \tag{3.110}$$

To simplify this statement, we rephrase the right-hand side, turning the closed-loop integral over the magnetic vector potential into a surface integral over the magnetic field **B**, and introducing the magnetic flux quantum Φ_0^S,

$$\Delta\theta_b - \Delta\theta_a = 2\pi M - \frac{2\pi}{\Phi_0^S} \oint_{C'} \frac{m_e}{2n_{CP}e^2} \mathbf{j}_{CP} \cdot d\mathbf{l} - \frac{2\pi \Phi_{ext}}{\Phi_0^S}, \tag{3.111}$$

where Φ_{ext} denotes the flux due to an external magnetic field through the surface S enclosed by the loop C ($\Phi_{ext} = \int\int_S \mathbf{B} \cdot d\mathbf{S}$). As the line integrals over the supercurrent density can be taken deep inside the loop in Figure 3.6 where $\mathbf{j}_{CP}$ is exponentially small [61], the second term on the right-hand side can be neglected. The supercurrent Eq. (3.105) then becomes

$$I_{CP} = 2I_{0,CP} \cos(\pi \frac{\Phi_{ext}}{\Phi_0^S}) \sin(\Delta\theta_b + \pi \frac{\Phi_{ext}}{\Phi_0^S}). \tag{3.112}$$

In consequence, the maximum supercurrent I_{CP}^{max} is given by

$$I_{CP}^{max} = 2I_{0,CP} | \cos(\frac{\Phi_{ext}}{\Phi_0^S} \pi)|. \tag{3.113}$$

Thus, the maximum detectable supercurrent exhibits an oscillatory profile as a function of the magnetic flux traversing the superconducting loop, as illustrated in Figure 3.8. This effect can in principle be used to measure external magnetic fields by recording I_{CP}^{max}.

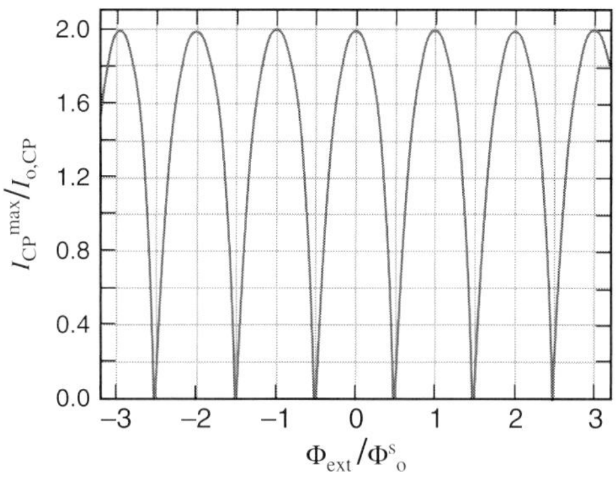

Figure 3.8 The maximum supercurrent, normalized to the critical current $I_{0,CP}$, as a function of the external magnetic flux Φ_{ext} through the superconducting loop, in units of the magnetic flux quantum Φ_0^S [59].

In terms of applying this formalism to determine external magnetic fields, however, one has to be aware that the the flux Φ is not necessarily identical with the external flux Φ_{ext}. More specifically, the relation between the two quantities is

$$\Phi = \Phi_{ext} + \Phi_L, \tag{3.114}$$

where the inductive contribution Φ_L is generated by a potential admixture of the circular current in the loop. Formula (3.113) describes the actual supercurrent maximum in the presence of an external magnetic field to a good approximation in the limiting case of *negligible screening* where the induced magnetic flux is small as compared with the magnetic flux quantum. More realistic situations are discussed in [61]. The ultimate experimental test for magnetism in carbon nanostructures lies in the detection of magnetic fields in these units. Various portions of this monograph, especially Chapters 7 and 8, will make reference to graphene- or fullerene-based magnetic nanostructures that were analyzed by use of recent SQUID technology.

3.6 Special Topic: Bloch States

We consider an electron, described by the wave function $\psi(\mathbf{x})$, in a *Bravais lattice* characterized by the periodic potential $V(\mathbf{x})$. The term Bravais lattice is understood as a periodic arrangement of points generated by discrete translations of the form $\mathbf{R} = n_1\mathbf{a}_1 + n_2\mathbf{a}_2 + n_3\mathbf{a}_3$, with $\mathbf{a}_1$, $\mathbf{a}_2$ and $\mathbf{a}_3$ as the *primitive translation vectors* of the lattice. The wave function ψ satisfies the stationary Schrödinger equation,

$$\hat{H}\psi(\mathbf{x}) = (-\frac{\hbar^2}{2m_e}\sum_{i=1}^{3}\frac{\partial^2}{\partial x_i^2} + V(\mathbf{x}))\psi(\mathbf{x}) = E\psi(\mathbf{x}), \tag{3.115}$$

with $i = x, y, z$. *Bloch's theorem* states that the solution ψ is of the form

$$\psi(\mathbf{x}) = \psi(\mathbf{x})_{\mathbf{k}} = e^{i\mathbf{k}\cdot\mathbf{x}}u_{\mathbf{k}}(\mathbf{x}). \tag{3.116}$$

The expression $u_{\mathbf{k}}(\mathbf{x})$ denotes a function that has the periodicity of the lattice. Therefore,

$$\psi_{\mathbf{k}}(\mathbf{x}+\mathbf{R}) = e^{i\mathbf{k}\cdot\mathbf{R}}\psi_{\mathbf{k}}(\mathbf{x}). \tag{3.117}$$

The solutions $\psi(\mathbf{x})$, as given by Eq. (3.116), are termed *Bloch functions*. One concludes readily from Eq. (3.115) that the function $u_{\mathbf{k}}(\mathbf{x})$ solves the following equation:

$$[\frac{1}{2m_e}(\hat{\mathbf{p}}+\hbar\mathbf{k})^2 + V(\mathbf{x})]u_{\mathbf{k}}(\mathbf{x}) = E_{\mathbf{k}}u_{\mathbf{k}}(\mathbf{x}). \tag{3.118}$$

The periodic nature of the potential $V(\mathbf{x})$ implies the periodicity of its Fourier transform $\tilde{V}(\mathbf{k})$. For clarity, we restrict the following discussion to one dimension. From $V(x) = V(x+na)$, where $n = \pm 0, 1, 2, \ldots$, one concludes that

$$\tilde{V}(k) = \sum_R V_R e^{-ikR}, \tag{3.119}$$

where the summation extends over all translation vectors $R = na$, and $V_R = \int_{-\frac{a}{2}}^{\frac{a}{2}} V(x)e^{ikx}dx$. Correspondingly, $V(x)$ may be expanded into a Fourier series, namely

$$V(x) = \sum_G V_G e^{iGx}, \tag{3.120}$$

where $G = n2\pi/a$. G is the counterpart of R for the *reciprocal lattice*, and the set of wave numbers with $-\pi/a \leq k \leq +\pi/a$ defines the *first Brillouin zone*. Likewise, the solution of the stationary Schrödinger equation (3.115) in one dimension may be written as a Fourier series:

$$\psi(x) = \sum_k c(k)e^{ikx}. \tag{3.121}$$

The values of the discrete variable k are determined by the length of the one-dimensional lattice, L. Specifically, $k = n2\pi/L$. The series $\{c(k)\}$ is obtained by inserting the expansion for $V(x)$, (3.120), and that for $\psi(x)$, (3.121), into Eq. (3.115). This yields

$$\sum_k \frac{\hbar^2}{2m_e}k^2c(k)e^{ikx} + \sum_{k,G} V_G c(k)e^{i(k+G)x} = E\sum_k c(k)e^{ikx}. \tag{3.122}$$

Collecting in this equation the coefficients of e^{ikx}, one finds

$$(\varepsilon_k - E)c(k) + \sum_G V_G c(k-G) = 0, \tag{3.123}$$

with $\varepsilon_k = \hbar^2k^2/2m_e$. As the set of the values adopted by G is infinite, so is the linear system of Eqs. (3.123). In practice, however, useful approximations to $c(k)$

can often be obtained by reducing the system to a few choices for G. The energy eigenvalues that result from Eq. (3.123) for a given k may be labeled E_{nk}, where the index n specifies the *energy band* that contains the eigenvalue. In most cases, the wave number k that enters Eq. (3.123) is chosen from the first Brillouin zone.

Generalizing to three dimensions, one may express the reciprocal lattice translation vector $\mathbf{G}$ in the form

$$\mathbf{G} = n_1\mathbf{b}_1 + n_2\mathbf{b}_2 + n_3\mathbf{b}_3, \tag{3.124}$$

and

$$\begin{aligned} \mathbf{b}_1 &= \frac{2\pi(\mathbf{a}_2 \times \mathbf{a}_3)}{\mathbf{a}_1 \cdot (\mathbf{a}_2 \times \mathbf{a}_3)} \\ \mathbf{b}_2 &= \frac{2\pi(\mathbf{a}_3 \times \mathbf{a}_1)}{\mathbf{a}_1 \cdot (\mathbf{a}_2 \times \mathbf{a}_3)} \\ \mathbf{b}_3 &= \frac{2\pi(\mathbf{a}_1 \times \mathbf{a}_2)}{\mathbf{a}_1 \cdot (\mathbf{a}_2 \times \mathbf{a}_3)}. \end{aligned} \tag{3.125}$$

With reference to the reciprocal lattice spanned by the vectors $\mathbf{b}_i$, $i = 1, 2, 3$, the first Brillouin zone is understood as the set of all points in reciprocal space whose distance to the origin $\mathbf{k} = \mathbf{0}$ is smaller than that to any other lattice point $\mathbf{G} = n_1\mathbf{b}_1 + n_2\mathbf{b}_2 + n_3\mathbf{b}_3$.

The origin of the reciprocal lattice is termed Γ *point*. The periodicity of reciprocal space implies the existence of several *high-symmetry points* within the first Brillouin zone. This may be seen by applying Bloch's theorem to the state $\psi_{\mathbf{k}+\mathbf{G}}$,

$$\begin{aligned} \psi_{\mathbf{k}+\mathbf{G}} &= \exp(i(\mathbf{k}+\mathbf{G}) \cdot \mathbf{x}) u_{\mathbf{k}+\mathbf{G}}(\mathbf{x}) \\ &= \exp(i\mathbf{k} \cdot \mathbf{x}) \exp(i\mathbf{G} \cdot \mathbf{x}) u_{\mathbf{k}+\mathbf{G}}(\mathbf{x}) \\ &= \exp(i\mathbf{k} \cdot \mathbf{x}) \tilde{u}_{\mathbf{k}}(\mathbf{x}), \end{aligned} \tag{3.126}$$

where $\tilde{u}_{\mathbf{k}}$ and $u_{\mathbf{k}+\mathbf{G}}$ have identical periodicities. As any wave vector $\mathbf{k}$ can be manipulated in this manner, it is always possible to reduce the wave function to the first Brillouin zone. Therefore, the eight points $(\pm\frac{1}{2}, \pm\frac{1}{2}, \pm\frac{1}{2})$ can be considered as identical since they can be obtained from each other by adding or subtracting a basis vector of the reciprocal lattice. This gives an example for another high-symmetry point (in FCC lattices termed the *L point*).

Part II

Carbon and Its Nanoscale Allotropes

4
Graphene

In this chapter, as well as the following two, we present the structural protoypes of carbon nanosystems, namely graphene, carbon nanotubes and fullerenes. These represent nanoscopic allotropes of carbon in two, one and zero dimensions. With respect to the dimensionality and also to the chronology of carbon nanostructure research, Chapters 4 to 6 of this text go in reverse order. The reason for this choice is the fundamental significance of graphene for intrinsic magnetism in carbon-based materials. Besides providing general information about graphene, this chapter introduces the dimensionally reduced graphene nanoribbons. The zigzag types of these species have been shown to display magnetism in their ground state and may be understood as the basic units of carbon nanostructure magnetism.

In elemental carbon, four valence electrons occupy the $2s$ and $2p_x/p_y$ orbitals, yielding the occupation scheme $1s^2$, $2s^2$, $2p^2$ for the electronic shell of the carbon atom. The ground state of the carbon atom is, in spectroscopic notation, $^3P_0$, involving a spin triplet in conjunction with a total orbital angular momentum $L = 1$, and a total angular momentum $J = 0$. This configuration confines carbon to two chemical bonds only, as exemplified by the carbene CH_2. Carbon compounds, however, usually involve three or four bonds. This is achieved by promoting a $2s$ electron into a $2p$ orbital, which leads to the formation of hybridized molecular orbitals. We distinguish between sp, sp^2 and sp^3 hybridization. In these cases, the $2s$ and two $2p$ orbitals combine, giving rise to two (sp), three (sp^2) or four (sp^3) mixed s-p orbitals, as represented by linear combinations of atomic orbitals of s and p character. In the regime of carbon-based molecules and polymers, examples for these hybridization schemes are provided by acetylene (the sp case), polyacetylene (sp^2), and methane (sp^3). Among carbon allotropes, the sp^2 and sp^3 pattern are prototypically realized by the graphite and diamond phases of carbon, respectively.

Graphene realizes a perfectly planar form of carbon, which may be understood as a single layer of graphite. Thus, carbon atoms are arranged in a two-dimensional lattice of monatomic thickness, and with hexagonal symmetry. While several layers of graphite display largely graphitic behavior, new features emerge in a perfectly two-dimensional array of carbon atoms. Fabricating such an array was accomplished first by Geim and Novoselov [64] who used a micromechanical cleavage method [65], transferring a single sheet of graphite onto the surface of a substrate. However, as A. Geim remarked in his Nobel Prize lecture *Random Walk to Graphene*, "It is not the observation and isolation of graphene, but its electronic properties that took researchers by surprise" [67].

In what follows, we first introduce the geometric and electronic structure of monolayer (Sections 4.1 and 4.2) and bilayer (Section 4.3) graphene. Section 4.4 discusses the impact of a perpendicular magnetic field on the electronic states of a graphene sheet. The final section of this chapter considers graphene nanoribbons, laying the ground for our treatment of intrinsic graphene magnetism in Chapter 7.

4.1 Geometric Structure of Graphene

The honeycomb network of graphene (see Figure 4.1) is described by a Bravais lattice vector of the form

$$\mathbf{R} = n_1\mathbf{a}_1 + n_2\mathbf{a}_2, \tag{4.1}$$

with primitive vectors $\mathbf{a}_1$ and $\mathbf{a}_2$, such that

$$\begin{aligned} \mathbf{a}_1 &= \frac{a_0}{2}(3, \sqrt{3}) \\ \mathbf{a}_2 &= \frac{a_0}{2}(3, -\sqrt{3}), \end{aligned} \tag{4.2}$$

where $a_0 \approx 1.42\,\AA$ is the lattice parameter. Note that $\mathbf{a}_1$ and $\mathbf{a}_2$ are neither normalized nor orthogonal. From Figure 4.1, the direct lattice of graphene is characterized by two inequivalent sites, A and B, such that the graphene unit cell comprises two atoms. Consequentially, the graphene sheet can be decomposed into two interpenetrating sublattices, one that contains all sites of type A, and the other one consisting of all B sites.

Defining a vector $\mathbf{e}_3$ that is perpendicular on both $\mathbf{a}_1$ and $\mathbf{a}_2$, and implementing the operation

$$\begin{aligned} \mathbf{b}_1 &= \frac{2\pi(\mathbf{a}_2 \times \mathbf{e}_3)}{\mathbf{a}_1 \cdot (\mathbf{a}_2 \times \mathbf{e}_3)} \\ \mathbf{b}_2 &= \frac{2\pi(\mathbf{e}_3 \times \mathbf{a}_1)}{\mathbf{a}_2 \cdot (\mathbf{e}_3 \times \mathbf{a}_1)}, \end{aligned} \tag{4.3}$$

we construct the basis vectors $\mathbf{b}_1$ and $\mathbf{b}_2$ that define the translation vectors of the reciprocal lattice and obtain

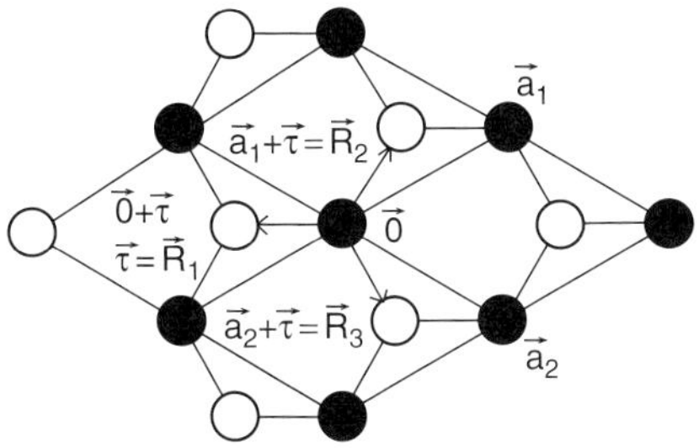

Figure 4.1 Geometric structure of graphene. The basis vectors $\mathbf{a}_1$ and $\mathbf{a}_2$ of the Bravais lattice vector are indicated, along with the lattice parameter and the unit cell and the reference vector $\boldsymbol{\tau} \equiv a_0(-1,0)$ that connects a lattice point $\mathbf{R}_A$ with one of its nearest neighbors.

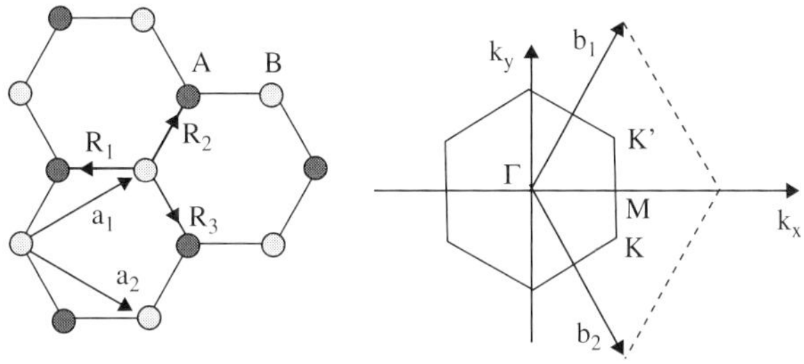

Figure 4.2 The Wigner–Seitz cell as well as the first Brillouin zone of graphene.

$$
\begin{aligned}
\mathbf{b}_1 &= \frac{2\pi}{3a_0}(1, \sqrt{3}) \\
\mathbf{b}_2 &= \frac{2\pi}{3a_0}(1, -\sqrt{3}).
\end{aligned}
\tag{4.4}
$$

From Figure 4.2, the reciprocal lattice is hexagonal, as the direct lattice. The high-symmetry points are indicated. Besides Γ and K, these comprise M points located midway between two adjacent K points.

4.2 Electronic Structure of Graphene

Realizing that graphene is stabilized by sp^2 bonding, and adopting the tight binding approach, one may derive the essential features of the graphene band structure not too far from the Fermi energy E_F in an analytical manner. The starting point of this endeavor is the tight-binding Hamiltonian (see Section 3.1)

$$
\hat{H} = -t \sum_{i,j,\sigma} \hat{a}^{\dagger}_{\sigma,i} \hat{b}_{\sigma,j} + h.c. - t' \sum_{i,j,\sigma} \hat{a}^{\dagger}_{\sigma,i} \hat{a}_{\sigma,j} + \hat{b}^{\dagger}_{\sigma,i} \hat{b}_{\sigma,j} + h.c., \tag{4.5}
$$

with $\hat{a}/\hat{a}^{\dagger}$ $(\hat{b}/\hat{b}^{\dagger})$ as annihilation/creation operators for an electron at site A(B). The first sum extends over all adjacent carbon atom sites; the second over all next-nearest neighbor sites. The factor t stands for the nearest-neighbor electron-hopping integral:

$$t = \int d\mathbf{x}\, \varphi^*(\mathbf{x} - \mathbf{R}_{A_k}) \hat{H} \varphi(\mathbf{x} - \mathbf{R}_{B_l}). \tag{4.6}$$

This integral, also referred to as *nearest-neighbor transfer energy* or *nearest-neighbor energy*, connects atomic wave functions $\varphi^*(\mathbf{x} - \mathbf{R}_{A_k})$ and $\varphi^*(\mathbf{x} - \mathbf{R}_{B_l})$, which are located at adjacent lattice sites of A and B character. The factor t' is defined correspondingly for the case of next-nearest neighbor atoms. Since the sp^2 scheme determines carbon bonding in the graphene sheet, a qualitative difference exists between the carbon s, $p_{x,y}$ (in-plane) atomic orbitals and the out-of-plane orbital p_z. While the former group forms bonds of σ character, the latter builds π bonds, oriented perpendicular to the graphene plane. The energy regime close to the Fermi energy, i.e. in a range of about 1 eV around E_F, has been found to be dominated by occupied bands of the π and unoccupied bands of the π^* type, while the corresponding bands of σ character are shifted to higher or lower energy. This justifies treating the out-of-plane and in-plane orbitals as independent of each other and thus studying the highest occupied and the lowest unoccupied bands by identifying the atomic wave functions φ with the carbon p_z orbitals.

The eigenfunctions of the Hamiltonian (4.5) involve Bloch functions of the form

$$\Phi_A(\mathbf{k}, \mathbf{x}) = \frac{1}{\sqrt{N}} \sum_j^N e^{i\mathbf{k}\mathbf{R}_{A_j}} \varphi(\mathbf{x} - \mathbf{R}_{A_j}), \tag{4.7a}$$

$$\Phi_B(\mathbf{k}, \mathbf{x}) = \frac{1}{\sqrt{N}} \sum_j^N e^{i\mathbf{k}\mathbf{R}_{B_j}} \varphi(\mathbf{x} - \mathbf{R}_{B_j}), \tag{4.7b}$$

where we sum over N unit cells and distinguish between two linear combinations, including the contributions from the A and from the B sites. The overall eigenfunction for the π subsystem of graphene is given by a superposition of these two expressions,

$$\Psi(\mathbf{k}, \mathbf{x}) = A(\mathbf{k})\Phi_A(\mathbf{k}, \mathbf{x}) + B(\mathbf{k})\Phi_B(\mathbf{k}, \mathbf{x}), \tag{4.8}$$

with appropriately chosen coefficients A and B. These coefficients and the energy eigenvalues of the wave function (4.8) may be obtained from solving the stationary Schrödinger equation, which, written in matrix form, reads

$$\begin{pmatrix} H_{AA} & H_{AB} \\ H_{BA} & H_{BB} \end{pmatrix} \begin{pmatrix} A \\ B \end{pmatrix} = E \begin{pmatrix} S_{AA} & S_{AB} \\ S_{BA} & S_{BB} \end{pmatrix} \begin{pmatrix} A \\ B \end{pmatrix}. \tag{4.9}$$

The matrix elements H_{ij} and S_{ij} are defined as

$$H_{ij} = \int d\mathbf{x}\, \Phi_i^* \hat{H} \Phi_j,\; S_{ij} = \int d\mathbf{x}\, \Phi_i^* \Phi_j, \tag{4.10}$$

with $i, j = A, B$.

The elements of the Hamiltonian matrix on the left-hand side of Eq. (4.9) are readily evaluated when the *next-neighbor approximation* (3.6) of the tight-binding model is adopted. Within this approach, electronic interaction is restricted to adjacent lattice sites. This implies that A(B) sites will interact only with the closest B(A) sites, but not with any other A(B) sites. Setting $E_{2p} \equiv \int d\mathbf{x}\, \varphi_i^* \hat{H} \varphi_i$, H_{AA} and H_{AB} simplify to

$$H_{AA} = \frac{1}{N} \sum_j^N \sum_n^N e^{i\mathbf{k}\cdot(\mathbf{R}_{A_n} - \mathbf{R}_{A_j})} E_{2p} \delta_{jn} = E_{2p}, \tag{4.11a}$$

$$H_{AB} = -\frac{t}{N} \sum_j^N \sum_n^3 e^{i\mathbf{k}\cdot\mathbf{R}_n} = -t S(k), \tag{4.11b}$$

where $\mathbf{R}_n$ (n = 1–3) stands for the vectors that connect the A site $\mathbf{R}_{A_j}$ with its three nearest neighbors, and the function $S(k)$ is defined as [68]

$$S(k) = e^{i\mathbf{k}\cdot\mathbf{R}_1} + e^{i\mathbf{k}\cdot\mathbf{R}_2} + e^{i\mathbf{k}\cdot\mathbf{R}_3}. \tag{4.12}$$

With reference to Figure 4.1, the loci of the three nearest neighbors of any lattice site $\mathbf{R}_A$ can be represented as $\mathbf{R}_1 = \boldsymbol{\tau}$, $\mathbf{R}_2 = \boldsymbol{\tau} + \mathbf{a}_1$ and $\mathbf{R}_3 = \boldsymbol{\tau} + \mathbf{a}_2$.

The Hermitian nature of the Hamiltonian matrix demands that $H_{AB} = H_{BA}^*$, which may be derived immediately from Eq. (4.11). Note that $H_{BB} = H_{AA} = E_{2p}$. Correspondingly, the overlap matrix is found from the normalization condition $S_{BB} = S_{AA} = 1$, as well as $S_{AB} = S_{BA}^* = sS(k)$, where s is the *overlap integral*

$$s = \int d\mathbf{x} \varphi^*(\mathbf{x} - \mathbf{R}_A) \varphi(\mathbf{x} - \mathbf{R}_B), \tag{4.13}$$

with $\mathbf{R}_A$ and $\mathbf{R}_B$ as two adjacent sites in the graphene lattice.

To obtain a dispersion relation $E = E(\mathbf{k})$ for the π and π^* subsystems of graphene, we solve the secular equation $\det(\mathbf{H} - E\mathbf{S}) = 0$. This yields

$$E(\mathbf{k})_\pm = \frac{E_{2p} \pm t|S(k)|}{1 \pm s|S(k)|} \tag{4.14}$$

and

$$|S(k)| = \sqrt{1 + 4\cos(\frac{3}{2}k_x a_0)\cos(\frac{\sqrt{3}}{2}k_y a_0) + 4\cos^2(\frac{\sqrt{3}}{2}k_y a_0)}. \tag{4.15}$$

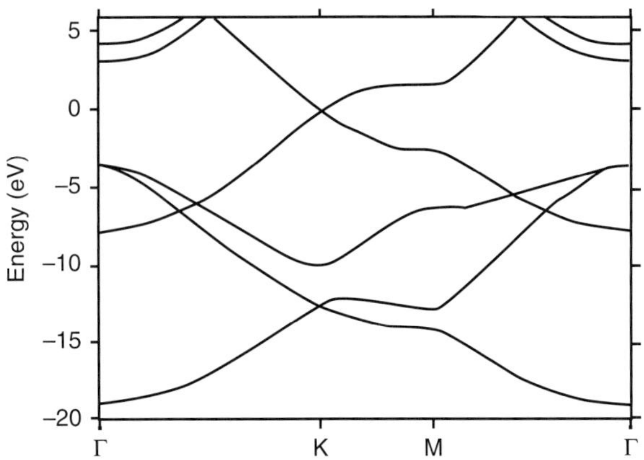

Figure 4.3 The dispersion curves of the graphene π and π^* bands according to formula (4.14), evaluated along the **k**-space triangle **K**-**Γ**-**M**. Reprinted with permission from [71]. Copyright (2008) by the American Physical Society.

The dispersion curves according to Eq. (4.14) are displayed in Figure 4.3, where E_{2p} has been set to zero. At the K-points, $E_+(\mathbf{k})$ and $E_-(\mathbf{k})$ are found to coincide, as both reduce to E_{2p}. In the state of charge neutrality, this energy separates the occupied π from the unoccupied π^* band and is thus to be identified with E_F. The accuracy of the analytical expression depends on the choice of the two governing parameters t and s. Customarily, t is seen to vary between 2.7 eV and 3.3 eV [69]. With t in this range and $s = 0$, corresponding to orthonormal atomic orbitals $\varphi(\mathbf{x}-\mathbf{R}_A)$ and $\varphi(\mathbf{x}-\mathbf{R}_B)$ at adjacent sites A and B, better than qualitative agreement with *ab initio* calculations of $E(\mathbf{k})_\pm$ can be accomplished in the energy range close to E_F [69, 70].

4.2.1 Massless Dirac Fermions

The π and π^* bands become degenerate at the six symmetry points of **K** character within the first Brillouin zone of graphene, as Figure 4.4 illustrates in a three-dimensional representation. This configuration implies that graphene, while lacking a finite band gap, is yet not to be characterized as a regular metal whose Fermi energy lies in the conduction band. Rather, the Fermi energy draws here a sharp demarcation line between the valence and the conduction bands, a case referred to as *semi-metallic*. The distinction between non-equivalent lattice points, made for the real-space lattice of graphene, holds for its reciprocal lattice as well, and we distinguish between two *Dirac points*, labeled **K** and **K**′. In the vicinity of these points, we observe *linear dispersion*, as reflected geometrically by a distinctive double-cone shape of the three-dimensional band structure at these points (for an early artistic rendition of this motif, see the cover image of this monograph).

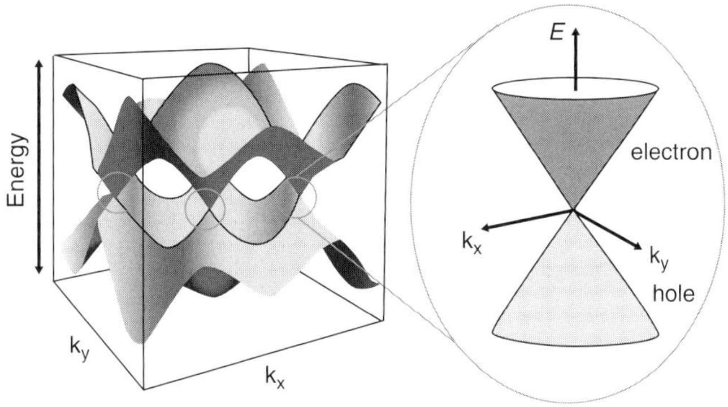

Figure 4.4 The graphene energy landscape of the first Brillouin zone close to the Fermi level E_F. The valence and the conduction band coincide at the six **K**-points (or *Dirac points*). The conical shape of the energy surfaces in the neighborhood of these points reflects the phenomenon of linear dispersion close to E_F.

Close to these so-called *Dirac points*, we may expand the energy $E_\pm(\mathbf{k})$ according to [72]

$$E_\pm(\mathbf{k}) = \pm\hbar v_F|\mathbf{k}-\mathbf{K}| + O[(\frac{|\mathbf{k}-\mathbf{K}|}{K})^2], \tag{4.16}$$

with v_F as the *Fermi velocity*, defined as

$$v_F \equiv \frac{1}{\hbar}\frac{\partial E}{\partial k}, \tag{4.17}$$

as evaluated at the Fermi energy, and with an analogous result for $\mathbf{k} = \mathbf{K}'$. The Fermi velocity thus turns out to be a constant, not depending on energy nor momentum. Confinement to a constant velocity draws a parallel between graphene electrons close to E_F and massless particles. In the following, we will rewrite the Hamiltonian matrix (4.11) in **k**-space to emphasize this analogy.

Identifying the energy E_{2p} with the zero point of the energy scale and assuming that the overlap matrix element S_{AB} vanishes, we write the stationary Schrödinger equation in **k**-space as

$$\mathbf{H}\Psi_\mathbf{k} = \pm E(\mathbf{k})\Psi_\mathbf{k}, \tag{4.18}$$

with $E(\mathbf{k}) = t|S(\mathbf{k})|$. We will identify the positive solution of this equation with electronic excitations, and the negative with hole states.

From the form of the Hamiltonian matrix (4.11), one concludes that low-energy charge carriers close to the corners of the first Brillouin zones (see Figure 4.2) may be understood according to the model of *massless Dirac Fermions*. This feature is made explicit by expanding the function $S(\mathbf{k})$ (4.12) around $\mathbf{k} = \mathbf{K}$ to first order,

$$S(\mathbf{q}) = (1 + e^{i(\mathbf{K}+\mathbf{q})\cdot\mathbf{a}_1} + e^{i(\mathbf{K}+\mathbf{q})\cdot\mathbf{a}_2})e^{i(\mathbf{K}+\mathbf{q})\cdot\tau}, \tag{4.19}$$

with $\mathbf{q} \equiv \mathbf{k} - \mathbf{K}$. From Figure 4.2 in conjunction with formula (4.3), one obtains the coordinates of the **K** and **K'** sites as[1]

$$\mathbf{K} = \frac{2\pi}{3a_0}(1, -\frac{1}{\sqrt{3}}), \mathbf{K}' = \frac{2\pi}{3a_0}(1, \frac{1}{\sqrt{3}}). \tag{4.20}$$

As is readily verified, it holds that $S(\mathbf{K}) = S(\mathbf{K}') = 0$. We insert the **K** coordinates into Eq. (4.19) and approximate

$$S(\mathbf{k}) \approx (1 + (-\frac{1}{2} - i\frac{\sqrt{3}}{2})(1 + i\mathbf{q}\cdot\mathbf{R}_2) + (-\frac{1}{2} + i\frac{\sqrt{3}}{2})(1 + i\mathbf{q}\cdot\mathbf{R}_3)e^{-iK_x a_0}. \tag{4.21}$$

Taking into account the relations $\mathbf{R}_2 + \mathbf{R}_3 = a_0\hat{\mathbf{x}}$ and $\mathbf{R}_2 - \mathbf{R}_3 = \sqrt{3}a_0\hat{\mathbf{y}}$, where $\hat{\mathbf{x}}$ and $\hat{\mathbf{y}}$ denote unit vectors in the horizontal and vertical directions of the graphene lattice as shown in Figure 4.1, we find that the Hamiltonian matrix (4.11) reduces, up to a factor of unit magnitude, to

$$\mathbf{H}_\mathbf{K} = \frac{3}{2}a_0 t\boldsymbol{\eta}\cdot\mathbf{q} = \hbar v_F \boldsymbol{\eta}\cdot\mathbf{q}, \tag{4.22}$$

from Eq. (4.16). Here $\boldsymbol{\eta} = (\eta_x, \eta_y)$ stands for the array containing the $x-$ and $y-$ Pauli matrices. The matrices η_i, with $i = x, y, z$, however, do not operate in a space of spinors, but of *pseudo-spinors*. More specifically, under a parity transformation that interchanges the two graphene sublattices, the array $\boldsymbol{\eta}$ reverses its signs, in contrast to the behavior of $\boldsymbol{\sigma}$, the array of spin operator matrices, under the same operation. While being uniformly represented by the Pauli matrices, $\boldsymbol{\sigma}$ and $\boldsymbol{\eta}$ act in different spaces and differ from each other in terms of their behavior under parity and time-reversal transformations, $\hat{P}$ and $\hat{T}$. Specifically,

$$\begin{aligned} &\hat{P}: \\ &\eta_x \to \eta_x,\ \eta_{y,z} \to -\eta_{y,z} \\ &\sigma_i \to \sigma_i \end{aligned} \tag{4.23}$$

and

$$\begin{aligned} &\hat{T}: \\ &\eta_i \to \eta_i \\ &\sigma_i \to -\sigma_i \end{aligned} \tag{4.24}$$

[1] Note that other representations of **K**, **K'** might be chosen, such as

$$\mathbf{K} = \frac{4\pi}{3a_0}(0, +\frac{1}{\sqrt{3}}), \mathbf{K}' = \frac{4\pi}{3a_0}(0, -\frac{1}{\sqrt{3}}).$$

If this convention is adopted, it holds that $\mathbf{K} = -\mathbf{K}'$.

and $i = x, y, z$. Eq. (4.22) yields an explicit formula for the Fermi velocity, namely

$$v_F = \frac{3}{2\hbar} a_0 t. \tag{4.25}$$

For the Hamiltonian (4.22) one finds the following pair of eigenvectors:[2]

$$\Psi_{\mathbf{K}}^{+} = \frac{1}{\sqrt{2}} \begin{pmatrix} 1 \\ e^{i\phi} \end{pmatrix} e^{i\mathbf{q}\cdot\mathbf{x}}, \text{ and } \Psi_{\mathbf{K}}^{-} = \frac{1}{\sqrt{2}} \begin{pmatrix} 1 \\ -e^{i\phi} \end{pmatrix} e^{i\mathbf{q}\cdot\mathbf{x}}. \tag{4.26}$$

Expanding $S(\mathbf{k})$ around the Brillouin zone corner $\mathbf{K}'$ which is inequivalent to $\mathbf{K}$ leads to the analogous Hamiltonian matrix:

$$\mathbf{H}_{\mathbf{K}'} = \hbar v_F \boldsymbol{\eta}^* \cdot \mathbf{q}, \tag{4.27}$$

The conjugate of $\boldsymbol{\eta}$ is given by $\boldsymbol{\eta}^* = (\eta_x, -\eta_y)$. We note that the eigenfunctions of $\mathbf{H}_{\mathbf{K}'}$, in parallel to those of $\mathbf{H}_{\mathbf{K}}$, are

$$\Psi_{\mathbf{K}'}^{+} = \frac{1}{\sqrt{2}} \begin{pmatrix} 1 \\ e^{-i\phi} \end{pmatrix} e^{i\mathbf{q}\cdot\mathbf{x}}, \text{ and } \Psi_{\mathbf{K}'}^{-} = \frac{1}{\sqrt{2}} \begin{pmatrix} 1 \\ -e^{-i\phi} \end{pmatrix} e^{i\mathbf{q}\cdot\mathbf{x}}. \tag{4.28}$$

Exercise 4.1 Show that $\Psi_{\mathbf{K}}^{+}$ ($\Psi_{\mathbf{K}}^{-}$) is eigenfunction of $\mathbf{H}_{\mathbf{K}}$ (4.22) with eigenvalue $+\hbar v_F|\mathbf{q}|$ ($-\hbar v_F|\mathbf{q}|$), thus describing a conduction (valence) state. Make the analogous statement for the K' point eigenfunctions (4.28).

Note that, by the transformation laws (4.23, 4.24), the low-energy Hamiltonians (4.22, 4.27) are invariant with respect to both space and time reflection, where the latter feature follows from $\hat{T}\mathbf{H}_{\mathbf{K}}\hat{T}^{-1} = \mathbf{H}_{\mathbf{K}'}^{*} = \mathbf{H}_{\mathbf{K}}$, with $\hat{T}$ as the time-reflection operator. For both pairs of eigenfunctions, Eqs. (4.26) and (4.28), the wave function accumulates a *geometric phase* θ [73] as it is transported adiabatically along a closed loop in $\mathbf{k}$ space. We infer this feature directly from the definition of the geometric phase in conjunction with the form of the eigenfunctions. Explicitly,

[2] Frequently, an alternative pair of solutions, namely

$$\Psi_{\mathbf{K}}^{+} = \frac{1}{\sqrt{2}} \begin{pmatrix} e^{-\frac{i}{2}\phi} \\ e^{\frac{i}{2}\phi} \end{pmatrix} e^{i\mathbf{q}\cdot\mathbf{x}}, \text{ and } \Psi_{\mathbf{K}}^{-} = \frac{1}{\sqrt{2}} \begin{pmatrix} e^{-\frac{i}{2}\phi} \\ -e^{\frac{i}{2}\phi} \end{pmatrix} e^{i\mathbf{q}\cdot\mathbf{x}}$$

is encountered in the literature on graphene (e.g. [72]). It is obtained by applying a gauge transformation on the wave functions (4.26), resulting in a double-valued, and thus ambiguous, solution. The single-valued functions (4.26) are thus the physically more adequate choice.

$$\begin{aligned}\theta &= -i \oint d\mathbf{q}\langle\Psi(\mathbf{q})|\nabla_{\mathbf{q}}|\Psi(\mathbf{q})\rangle \\ &= -i \oint dt\langle\Psi(\mathbf{q}(t))|\frac{\partial}{\partial t}|\Psi(\mathbf{q}(t))\rangle \\ &= -i \oint dt\{\frac{1}{2}(1, e^{\mp i\phi})\begin{pmatrix} 0 \\ i\frac{\partial\phi}{\partial t}e^{\pm i\phi}\end{pmatrix} + \frac{i}{\hbar}\frac{\partial\mathbf{q}}{\partial t}\cdot\mathbf{x} + \frac{i}{\hbar}\mathbf{q}\cdot\frac{\partial\mathbf{x}}{\partial t}\}.\end{aligned} \tag{4.29}$$

Since the second and the third term in the integrand vanish when integrated along a closed path, and since $\frac{\partial\phi}{\partial t} = \omega = \frac{2\pi}{T}$, with T as the time needed to traverse the loop, this calculation yields $\theta = \pi$. The solutions thus change their sign as the angle ϕ is incremented by 2π, in direct consequence of the pseudospinor character of the eigenfunctions. We emphasize that this property is related to the structure of the tight-binding Hamiltonian (4.9), accommodating the two inequivalent lattice sites A and B.

Further, the peculiar form of the so-called *valley Hamiltonians* (4.22) and (4.27) implies that the solutions (4.26) and (4.28) are likewise eigenfunctions of the *helicity operator*, defined as $\boldsymbol{\eta}\cdot\mathbf{q}/|\mathbf{q}|$, reflecting the chiral nature of the low-energy electron and hole states in graphene. Specifically, the helicity, understood as the eigenvalue of the helicity operator, is $+1$ (-1) for electrons (holes) in the $\mathbf{K}$, and -1 $(+1)$ for electrons (holes) in the $\mathbf{K}'$ valley.

We observe that the transition from the $\mathbf{k}$-space Hamiltonians $\mathbf{H}_{\mathbf{K}}$ and $\mathbf{H}'_{\mathbf{K}}$ to real-space Hamiltonians can be made by replacing $\hbar\mathbf{q}$ with $-i\hbar\nabla$. For the $\mathbf{k} = \mathbf{K}$ alternative, this yields

$$\mathbf{H} = -i\hbar v_F\boldsymbol{\eta}\cdot\nabla, \tag{4.30}$$

and correspondingly for $\mathbf{k} = \mathbf{K}'$. Thus, in the close vicinity of the Dirac points, the Hamiltonian adopts the form of a *Weyl Hamiltonian*, describing massless particles represented by two-component spinors, as a limiting case of the Dirac Hamiltonian.

We point out that the Fermi energy of graphene does not need to coincide with the Dirac points or the *charge neutrality point*. If $E_F = 0$ and $T = 0$K, no electrons (holes) are present in the conduction band (valence band) and the Dirac points strictly separate the completely filled π valence band from the completely empty π^* conduction band, as reflected by the Fermi-Dirac distribution:

$$f(E) = \frac{1}{1 + e^{\frac{E-E_F}{k_B T}}}. \tag{4.31}$$

Applying a positive or negative gate voltage to the graphene sheet, however, shifts the position of the Fermi energy, as illustrated by Figure 4.5. In this case, E_F in Eq. (4.31) is to be replaced with the chemical potential μ provided by the gate. A positive voltage lifts E_F into the conduction band, which is now partially filled, corresponding to a finite electron carrier density n_i. Conversely, a negative

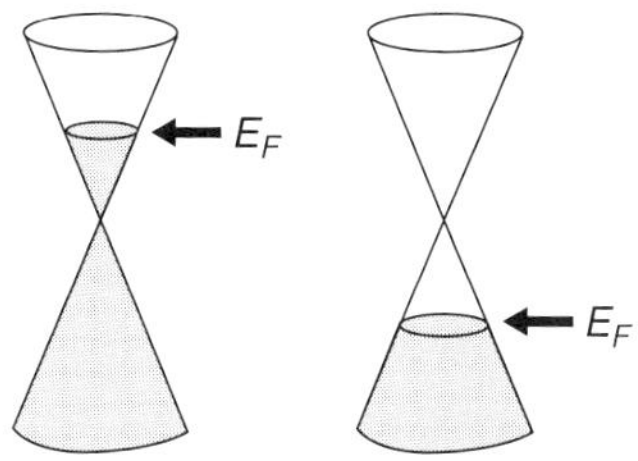

Figure 4.5 Fermi-level pinning in graphene: a positive gate voltage lifts the Fermi energy into the conduction band (left image); a negative gate voltage lowers it into the valence band (right image), and so manipulates the conductivity of the sample. The presence of impurities in graphene also displaces E_F from its position at the charge-neutrality point.

voltage pushes the position of E_F down into the valence band, generating hole conductivity. Also, doping graphene with impurities results in a shift of the Fermi energy position.

In the vicinity of the Dirac points, simple expressions can be found for the graphene density of states, $g(E)$. Employing formula (2.97) for the case $d = 2$, and expanding around $\mathbf{k} = \mathbf{K}$ we once more substitute under the integral the variable $\mathbf{q} = \mathbf{k} - \mathbf{K}$ for $\mathbf{k}$. To first order, we may set $E = E(|q|)$, assuming rotational symmetry about the Dirac point [75]. Within the limits of this approximation, $d^2q = 2\pi q$, and[3]

$$g(E) = 2 \cdot 2 \int \frac{d^2q}{(2\pi)^2} \delta(E - E(\mathbf{q})) = \frac{2}{\pi} \frac{q(E)}{|dE/dq|}. \tag{4.32}$$

The second factor 2 besides the spin multiplicity acknowledges that an identical result is obtained if the calculation is performed with $\mathbf{K}'$ as reference point. From Eq. (4.16), as well as the definition of the Fermi velocity, $g(E)$ is evaluated to yield

$$g(E) = \frac{2}{\pi} \frac{|E|}{\hbar^2 v_F^2}, \tag{4.33}$$

[3] Note that the allowed $\mathbf{k}$ values for the electron gas in a two-dimensional sample with dimension $L_x \times L_y$ are given by

$$k_x = n_x \frac{2\pi}{L_x}$$
$$k_y = n_y \frac{2\pi}{L_y},$$

and $n_x, n_y = 0, \pm 1, \pm 2 \ldots$ It follows that every point in the grid of allowed points (k_x, k_y) occupies an area $\frac{2\pi}{L_x} \times \frac{2\pi}{L_y} = \frac{(2\pi)^2}{A}$ in two-dimensional $\mathbf{k}$ space, with $A = L_x L_y$. The number of grid points per unit area of $\mathbf{k}$ space is thus $\frac{A}{(2\pi)^2}$. An element $dk_x dk_y$ in two-dimensional $\mathbf{k}$ space therefore contributes $\frac{dk_x dk_y}{(2\pi)^2}$ to the electron density.

which is a *low-energy approximation* to the graphene DOS distribution. This expression accentuates the semi-metallic character of graphene. As $E_F = 0$, the DOS vanishes at the Fermi energy. The zero of the DOS draws a demarcation line between the valence and conduction electrons, but instead of an energy gap these two systems are separated by a single point. The approximate analytical expression for the DOS distribution allows us to explore the temperature dependence of the intrinsic conduction electron density,[4] n_i, in graphene. Folding $g(E)$ with the *Fermi–Dirac* distribution, we obtain:

$$n_i(E) = \int_0^{E_{max}} g(E)f(E)dE = \frac{2}{\pi\hbar^2 v_F^2}\int_0^{E_{max}} \frac{E}{1+e^{\frac{E-E_F}{k_B T}}}dE. \tag{4.34}$$

The upper limit E_{max} refers here to the top of the conduction band. For an ideal graphene sheet, E_F may be chosen as 0, and in this case a closed solution exists for the *Fermi–Dirac integral* in Eq. (4.34) if E_{max} is extended to infinity. Due to the exponential dependence of the Fermi–Dirac distribution on the energy difference, the error introduced by this maneuver is small. Thus,

$$n_i(E) = \frac{\pi}{6}(\frac{k_B}{\hbar v_F})^2 T^2. \tag{4.35}$$

As in semiconductors, the conductivity in graphene increases with the temperature. However, an exponential dependence of n_i on the temperature is characteristic for semiconductors, while a parabolic temperature dependence is found for the semi-metal graphene. The expression underscores the great significance of the Fermi velocity for the electronic structure of graphene, since, in the framework of our approximation, this is the only materials parameter that determines the conduction electron density.

4.3 Bilayer Graphene

Beyond isolated graphene sheets, vertically stacked multilayer graphene has attracted the attention of theorists and experimentalists alike [76]. Most studies on these materials have focused on bilayer graphene as the simplest among them [77]. Further, bilayer graphene can claim high practical as well as systematic interest. Exhibiting a gate-tunable band gap, it is of major relevance for nanoelectronics [78, 79], and its low-energy excitations are interpreted in terms of chiral Fermions with *non-zero mass*, as will be clarified in this section.

Bilayer graphene has been fabricated in the laboratory by a variety of methods, among them unzipping of multi-walled carbon nanotubes [83] and plasma etching

[4] *Intrinsic electron density* is a term from semiconductor physics, referring to the electrons of the sample that contribute to its conductivity. The *extrinsic electron density* of a sample, in contrast, is associated with dopant species.

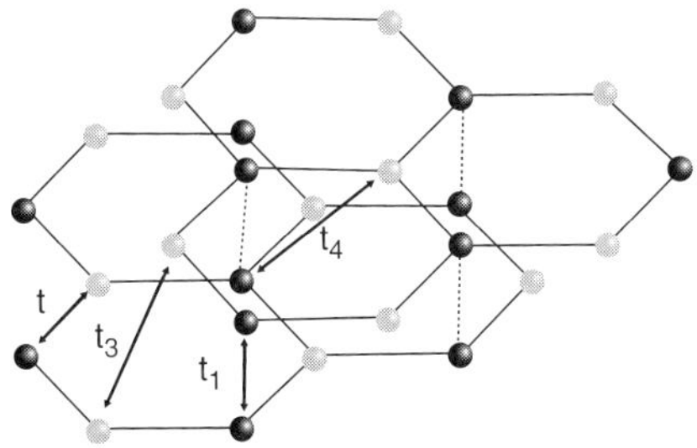

Figure 4.6 Geometric structure of bilayer graphene. The stacking of the two lattices results from rotation by $\pi/3$ of the two layers with respect to each other.

[84]. Figure 4.6 shows the atomic structure of bilayer graphene. As in graphite, the two parallel sheets are rotated with respect to each other by an angle of $\pi/3$, forming a geometry in which the A sublattices of the two sheets lie on top of each other. This is in accord with the *Bernal stacking* observed for graphite.

In what follows, we will apply the tight-binding approach (4.11) to the case of two interacting graphene sheets. In particular, we will show that for bilayer graphene, the tight-binding Hamitonian close to the Dirac points may be understood in terms of chiral Fermions endowed with an effective mass. This Hamiltonian may by written as a direct generalization of the expression in Eq. (4.30) by allowing for hopping channels that connect the adjacent A and B sites within either sheet as well as the A sites of the two sheets. This is reflected by the Hamiltonian matrix in $\mathbf{k}$ space for bilayer graphene, employing the tight-binding approximation with $t' = 0$ and $s = 0$. Extending the single-layer Hamiltonian matrix, we write [75]

$$\mathbf{H}(\mathbf{k}) = \begin{pmatrix} 0 & tS(\mathbf{k}) & t_\perp & 0 \\ tS^*(\mathbf{k}) & 0 & 0 & 0 \\ t_\perp & 0 & 0 & tS^*(\mathbf{k}) \\ 0 & 0 & tS(\mathbf{k}) & 0 \end{pmatrix}, \tag{4.36}$$

where the reference energy E_{2p} has been set equal to zero and the sequence of basis states is chosen as $\{\psi_{A1}, \psi_{B1}, \psi_{A2}, \psi_{B2}\}$, with the indices 1 and 2 labeling the two graphene layers. We have introduced the hopping parameter $t_\perp$ that couples, in the assumed minimal model, two vertically stacked adjacent A sites (see Figure 4.6). The four eigenvalues of the matrix (4.36) may be expressed in compact form as

$$E_j(\mathbf{k}) = \pm\frac{1}{2}t_\perp \pm \sqrt{\frac{1}{4}t_\perp^2 + t^2|S(\mathbf{k})|^2}, \tag{4.37}$$

with $j = 1, \ldots, 4$. Expanding $S(\mathbf{k})$ around $\mathbf{k} = \mathbf{K}$, and setting once more $\mathbf{q} = \mathbf{k} - \mathbf{K}$, we examine the eigenenergies (4.37) in the limit of small $|\mathbf{q}|$ and obtain

$$E_j(\mathbf{k}) \approx \pm\frac{1}{2}t_\perp \pm \frac{1}{2}t_\perp(1 + 2(\frac{3a_0t}{2t_\perp})^2q^2) = \pm\frac{1}{2}t_\perp \pm \frac{1}{2}t_\perp(1 + 2\frac{(\hbar q)^2}{2m^*t_\perp}), \tag{4.38}$$

where we have used Eq. (4.25), and equated $\frac{t_\perp}{2v_F^2}$ with the effective mass m^*. Two of the four bands (4.37) are degenerate at $\mathbf{k} = \mathbf{K}, \mathbf{K}'$, with a parabolic dispersion law at low energy,

$$E_j(\mathbf{k}) \approx \pm\frac{(\hbar q)^2}{2m^*}, \tag{4.39}$$

and $j = 1, 2$. Thus, the charge carriers in bilayer graphene behave like *massive Dirac Fermions*. In analogy to the case of massless Dirac Fermions (see Eq. (4.30)), the interaction between the two bands (4.39) may be described by an effective Hamiltonian of the form

$$\mathbf{H_K} = \frac{\hbar^2}{2m^*}\begin{pmatrix} 0 & (k_x - ik_y)^2 \\ (k_x + ik_y)^2 & 0 \end{pmatrix}. \tag{4.40}$$

Comparing the eigenfunctions of this operator, namely

$$\Psi_{\mathbf{K}}^{+} = \frac{1}{\sqrt{2}}\begin{pmatrix} e^{-i\phi} \\ e^{i\phi} \end{pmatrix} e^{i\mathbf{q}\cdot\mathbf{x}}, \text{ and } \Psi_{\mathbf{K}}^{-} = \frac{1}{\sqrt{2}}\begin{pmatrix} e^{-i\phi} \\ -e^{i\phi} \end{pmatrix} e^{i\mathbf{q}\cdot\mathbf{x}}, \tag{4.41}$$

with those for single-layer graphene (see Eq. (4.26)), we find that there is no double-valued representation about the K point for the former states, in contrast to the latter.

The significance of bilayer graphene in the area of nanoelectronics lies, for the most part, in the prospect of tuning the energy gap of this system by applying a gate voltage V perpendicular to the carbon planes. Doing this modifies the Hamiltonian (4.36) in the following way [85, 79]:

$$\mathbf{H}(\mathbf{k}) = \begin{pmatrix} \frac{V}{2} & tS(\mathbf{k}) & t_\perp & 0 \\ tS^*(\mathbf{k}) & \frac{V}{2} & 0 & 0 \\ t_\perp & 0 & -\frac{V}{2} & tS^*(\mathbf{k}) \\ 0 & 0 & tS(\mathbf{k}) & -\frac{V}{2} \end{pmatrix}. \tag{4.42}$$

Assuming that $\hbar v_F k \ll V \ll |t_\perp|$, one obtains, for electron and hole energies in the regime close to the Dirac points, the following dispersion relation:

$$E(\mathbf{k}) \approx \pm\left(\frac{V}{2} - \frac{V\hbar^2 v_F^2}{t_\perp^2}k^2 + \frac{\hbar^4 v_F^4}{t_\perp^2 V}k^4\right). \tag{4.43}$$

It follows that the band separation at the point $k = 0$ amounts to $\Delta E = V$, opening the possibility of tuning the size of the gap directly by applying a gate voltage. As the weaker interaction between B sublattice sites is included, the energy surfaces around the Dirac points exhibit a trigonal warping, an effect that is likewise obtained if the Hamiltonian for single-layer graphene is extended from linear to quadratic order [75].

4.4 Landau Levels

What effect does a magnetic field **B** applied in a direction perpendicular to the graphene sheet have on the electronic structure of graphene? To answer this question, we combine the Hamiltonian matrices of the K and the K' valley, as given by Eqs. (4.22) and (4.27), respecively, into one expression, and treat the influence of the magnetic field in the minimal coupling approximation. We thus introduce a vector potential **A** and stipulate

$$\hat{\pi}_x \equiv -i\hbar\frac{\partial}{\partial x} - eA_x, \hat{\pi}_y \equiv -i\hbar\frac{\partial}{\partial y} - eA_y. \tag{4.44}$$

The resulting Hamiltonian may then be written as

$$\hat{\mathbf{H}} = v_F \begin{pmatrix} 0 & -(\hat{\pi}_x + i\hat{\pi}_y) & 0 & 0 \\ -(\hat{\pi}_x - i\hat{\pi}_y) & 0 & 0 & 0 \\ 0 & 0 & 0 & \hat{\pi}_x - i\hat{\pi}_y \\ 0 & 0 & \hat{\pi}_x + i\hat{\pi}_y & 0 \end{pmatrix}, \tag{4.45}$$

with eigenfunctions

$$\mathbf{\Psi} = \begin{pmatrix} \psi_{AK'} \\ \psi_{BK'} \\ \psi_{AK} \\ \psi_{BK} \end{pmatrix}. \tag{4.46}$$

To represent the vector potential, we adopt the following gauge: $\mathbf{A} = (-By, 0, 0)$, where B is constant. This potential gives rise to a magnetic field along the z axis. While other choices are possible, this assignment is in accordance with the customary use of the x (y) coordinate to describe a periodic (finite) dimension of graphene-based structures (see, for instance, Section 11.4). For the K-valley components, the stationary Schrödinger equation leads to the following equation system,

$$\begin{aligned} E\psi_{AK} &= v_F(\hat{\pi}_x - i\hat{\pi}_y)\psi_{BK} = v_F\hat{\pi}_-\psi_{BK} \\ E\psi_{BK} &= v_F(\hat{\pi}_x + i\hat{\pi}_y)\psi_{AK} = v_F\hat{\pi}_+\psi_{BK}, \end{aligned} \tag{4.47}$$

with E as energy eigenvalue and $\hat{\pi}_-(\hat{\pi}_+) \equiv \hat{\pi}_x - i\hat{\pi}_y(\hat{\pi}_x + i\hat{\pi}_y)$. It is to be noted that the commutator $[\hat{H}, \hat{p}_x]$ vanishes. Therefore, the operator $\hat{p}_x$ in Eq. (4.47) can be replaced with the momentum eigenvalue $\hbar k_x$. For the B sublattice, we find

$$E^2\psi_{BK} = v_F^2\hat{\pi}_+\hat{\pi}_-\psi_{BK}. \tag{4.48}$$

This compact equation is usually expressed by introducing bosonic creation and annihilation operators $\hat{a}^+$ and $\hat{a}^-$, respectively, defined as

$$\hat{a}^+ \equiv \frac{1}{\sqrt{2\hbar eB}}\hat{\pi}_+$$
$$\hat{a}^- \equiv \frac{1}{\sqrt{2\hbar eB}}\hat{\pi}_-. \tag{4.49}$$

They obey the commutator relation

$$[\hat{a}^-, \hat{a}^+] = 1. \tag{4.50}$$

Exercise 4.2 Prove Eq. (4.50).

Therefore, Eq. (4.48) may be written as

$$\hat{a}^+\hat{a}^-\psi_{BK} = \frac{1}{2\hbar eB}\frac{E^2}{v_F^2}\psi_{BK} = \frac{l_B^2}{2\hbar^2 v_F^2}E^2\psi_{BK}, \tag{4.51}$$

where we have implicitly introduced the *magnetic length* by $l_B \equiv \sqrt{\frac{\hbar}{eB}}$. Eq. (4.51) maps the Landau-level problem on the well-known harmonic oscillator problem. The eigenvalues of the operator $\hat{a}^+\hat{a}^-$ are $N = 0, 1, 2, \ldots$ Thus,

$$\frac{l_B^2}{2\hbar^2 v_F^2}E^2 = N, \tag{4.52}$$

and

$$E_N^\pm = \pm\hbar\omega_D\sqrt{N}, \text{ and } N = 0, 1, 2, \ldots, \tag{4.53}$$

where the *Dirac frequency*, $\omega_D \equiv v_F\sqrt{\frac{2eB}{\hbar}} = \frac{\sqrt{2}v_F}{l_B}$, has been substituted. Landau levels with negative energy lie in the valence band, those of positive energy in the conduction band. For the A sublattice in the K valley, the analogous treatment yields the following spectrum,

$$E_N = \pm\hbar\omega_D\sqrt{N+1}, \tag{4.54}$$

and $N = 0, 1, 2\ldots$ Comparing Eqs. (4.53) and (4.54), we find that only the sublattice B contributes to the level $E = 0$, $(N = 0)$, while all other levels contain admixtures from both sublattices. We summarize Eqs. (4.53) and (4.54) into

$$E_N = \pm\hbar\omega_D\sqrt{N + \frac{1}{2} \pm \frac{1}{2}}, \tag{4.55}$$

with $N = 0, 1, 2, \ldots$ From the equivalent analysis for the K' valley, one obtains a zero level formed by the sublattice A component alone. In contrast to the energies of the harmonic oscillator spectrum, Landau levels are not equidistant, but the separation decreases as the magnitude of the level index N increases, as displayed in

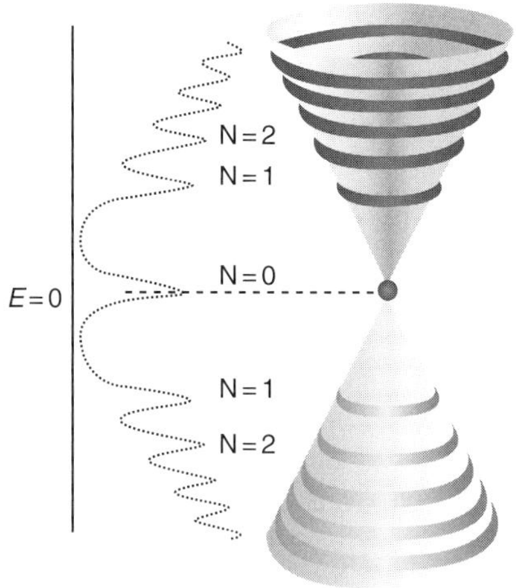

Figure 4.7 Landau levels for massless Dirac Fermions in the K valley, corresponding to electron and hole states [87]. All levels, with the exception of $E = 0$, exhibit twofold pseudo-spin degeneracy with respect to the wave functions ψ_{AK} and ψ_{BK}. The dotted line on the left refers to the density-of-states distribution.

Figure 4.7. We point out that the Landau levels obtained for massless Fermions, as considered here, deviate strongly from those derived in the free-electron case [86], where a harmonic-oscillator spectrum is found, namely

$$E_N = \hbar\omega_c(N + \frac{1}{2}), \tag{4.56}$$

and $\omega_c = \frac{eB}{m}$.

In the following, we will address the curious absence of a zero-point energy in the spectrum (4.55) and interpret it as a manifestation of the geometric phase for the graphene electronic system close to the Dirac points. Specifically, we will view the circular motion of electrons in the graphene plane under the influence of a perpendicular magnetic field from a semiclassical perspective, applying the Bohr–Sommerfeld quantization condition $\oint d\mathbf{x} \cdot \mathbf{p} = 2\pi N\hbar$ to the circular electron orbit. To adjust this rule to the magnetic interaction considered here, and to take into account the effect of the zero-point energy, we extend the Bohr–Sommerfeld formula in two directions [88]:

1. replacing $\mathbf{p}$ with $\mathbf{p} - e\mathbf{A}$, and
2. including an additional phase of π on the right-hand side of the quantization rule to accommodate the Maslov index m = 2 for a cyclotron orbit [89].

This approach leads to

$$\hbar \oint d\mathbf{x} \cdot \mathbf{k} - e \oint d\mathbf{x} \cdot \mathbf{A} = 2\pi (N + \frac{1}{2})\hbar. \tag{4.57}$$

To make Eq. (4.57) applicable to cyclotron motion in graphene, one has to incorporate the geometric phase, which adds an angle π on the left-hand side (see Eq. (4.29)), cancelling the Maslov contribution on the right. The quantization condition now takes the form

$$\hbar \oint d\mathbf{x} \cdot \mathbf{k} - e \oint d\mathbf{x} \cdot \mathbf{A} = 2\pi N\hbar. \tag{4.58}$$

The dynamics of cyclotron motion is given by the Lorentz force:

$$\hbar \frac{\partial \mathbf{k}}{\partial t} = -e \frac{\partial \mathbf{x}}{\partial t} \times \mathbf{B}, \tag{4.59}$$

implying

$$\mathbf{k}(t) - \mathbf{k}(0) = \frac{-eB}{\hbar}(\mathbf{x}(t) - \mathbf{x}(0)) \times \hat{\mathbf{e}}, \tag{4.60}$$

with $\hat{\mathbf{e}}$ as a unit vector that points in the direction of the constant magnetic field $\mathbf{B}$, here the positive z direction. Using relation (4.60) to evaluate the first term on the left-hand side of condition (4.58), we find

$$- e\mathbf{B} \oint d\mathbf{x} \times \mathbf{x} = 2eB\pi R^2 = 2e\Phi, \tag{4.61}$$

with R as the radius of the cyclotron orbit. The second term of the circulation integral, the Bohm–Aharonov component, reduces to $e\Phi$. The quantization condition thus turns into the compact statement

$$\Phi = \frac{Nh}{e} = N\Phi_0. \tag{4.62}$$

This equation in conjunction with Eq. (4.60) establishes[5]

$$\pi k_N^2 = 2\pi N \frac{e}{\hbar} B. \tag{4.63}$$

This, when combined with the dispersion relation for massless Fermions in graphene, $E = v_F \hbar k$, yields the spectrum (4.55). From the preceding derivation, the unusual form of the Landau-level spectrum in graphene is rooted in the presence of a geometric phase for electron orbits.

The energy eigenfunctions that belong to the spectrum (4.54, 4.55) are characterized by three labels: the Landau level index N, the wave number k_x and the

[5] Relation (4.60) implies that the electron trajectory in $\mathbf{k}$ space is its trajectory in real space, rotated by $\pi/2$, and scaled with $-eB/\hbar$ [90]. Using Eq. (4.60) to express the magnetic flux through the cyclotron orbit in terms of the corresponding orbit in $\mathbf{k}$ space and applying the flux quantization condition (4.62) leads to Eq. (4.63).

valley index, K versus K'. Solving the stationary Schrödinger equation for the Hamiltonian (4.45) yields the following solutions:

$$\begin{aligned} \mathbf{\Psi}^{K}_{N,k_x} &= \frac{Z_N}{\sqrt{L}} e^{ik_x x} \begin{pmatrix} 0 \\ 0 \\ sgn(N)(-i)\chi_{|N|-1,k_x} \\ \chi_{|N|,k_x} \end{pmatrix}, \\ \mathbf{\Psi}^{K'}_{N,k_x} &= \frac{Z_N}{\sqrt{L}} e^{ik_x x} \begin{pmatrix} \chi_{|N|,k_x} \\ sgn(N)(-i)\chi_{|N|-1,k_x} \\ 0 \\ 0 \end{pmatrix}. \end{aligned} \tag{4.64}$$

The level index N assumes here negative as well as positive values ($N = \ldots -1, 0, +1, \ldots$), and sgn denotes the sign function. As $sgn(0) = 0$, the expressions for $\mathbf{\Psi}^{K'}_{N,k_x}$ and $\mathbf{\Psi}^{K}_{N,k_x}$ are consistent with the vanishing of one of the two pseudospin components at the zero-energy level. The factor Z_N is 1 for $N = 0$ and $\frac{1}{\sqrt{2}}$ for $N \neq 0$. Box normalization is applied, with $L_x \times L_y = L^2$ as the area of the graphene sheet. The functions $\chi_{|N|,k_x}$ are defined by

$$\chi_{|N|,k_x} \propto e^{\frac{1}{2}\frac{(y-y_0)^2}{l_B^2}} H_N[\frac{(y-y_0)}{l_B}], \tag{4.65}$$

with $y_0 = \frac{\hbar}{eB}k_x = l_B^2 k_x$, and H_N as the *Hermite polynomials*. The zero-energy Landau level is *topologically protected*, i.e. it persists as inhomogeneities are introduced into the magnetic field [9, 64].

Exercise 4.3 Find the counterpart of the solution system (4.64) if the vector potential $\mathbf{A} = (0, Bx, 0)$ is adopted.

Experimental evidence for Landau levels in graphene has been obtained by infrared spectroscopy [91], and by scanning tunneling microscopy (STM) [92], determining the conductivity as a function of the sample bias.

The observation that the Landau spectrum (4.55) does not depend on the wave number k_x implies a *macroscopically large* degeneracy of each Landau level. Specifically, if the size of the graphene sample is $L_x \times L_y$, it holds that

$$k_x = \frac{2\pi}{L_x} N_x, \tag{4.66}$$

with $N_x = 0, \pm 1, \pm 2, \ldots$ Since the center y_0 in Eq. (4.65) has to lie within the graphene sheet ($0 \leq y_0 \leq L_y$), we have

$$|k_x| \leq \frac{L_y}{l_B^2} = \frac{eB}{\hbar} L_y. \tag{4.67}$$

By the relation (4.66), every Landau level has therefore a high degree of degeneracy g beyond that associated with the pseudospin and the valley degrees of freedom. It follows from Eq. (4.66) [9],

$$g = \frac{eB}{\hbar} \frac{L_x L_y}{2\pi} = \frac{\Phi}{\Phi_0}, \tag{4.68}$$

where Φ denotes the overall magnetic flux through the sheet, and Φ_0 the magnetic flux unit $\frac{h}{e}$. The architecture of Landau levels and the corresponding electronic states is the basis for one of the signature phenomena associated with graphene, namely an *anomalous quantum Hall effect*. This feature will be covered in greater detail in Section 11.4.

We close this section with a remark on the magnetization induced in graphene by a perpendicular external magnetic field, as specified by the respective susceptibility (see Eq. (2.35)). The response to the field is strongly diamagnetic for an ideal graphene sample at $T = 0$K, a feature shared by graphene and graphite [80, 81]. Use of linear response theory [82] leads to the expression[6]

$$\chi = -g_v g_S \frac{e^2 v_F^2}{6\pi} \delta(E_F), \tag{4.69}$$

with $g_S = g_v = 2$ as the multiplicities associated with the spin and the valley degrees of freedom. This result is quite extraordinary, as it predicts a singular susceptibility at the Dirac point of pristine graphene. On the other hand, the susceptibility according to Eq. (4.69) drops to zero if the Fermi energy deviates from the zero of energy, i.e. lies below or above the Dirac point. This, however, is the case for doped or gated graphene. This behavior is unique to monolayer graphene as opposed to multilayer graphene or graphite, where pronounced diamagnetism is observed. The diamagnetic profile of graphene is rooted in the band topology of this system, i.e. the crossing of two bands at the Dirac point. Thus, it is among the anomalies of graphene of which the Landau level at $N = 0$ and, associated with this feature, the conductivity profile in the graphene quantum Hall effect provide further examples.

4.5 Graphene Nanoribbons

Decreasing the dimension of graphene by one or two yields *graphene nanoribbons* (GNRs), periodic graphene bands of finite width, or rectangular fragments

[6] In SI units: $[\chi]$ = A/T. Note that the dimension of the magnetization of a two-dimensional system is *magnetic moment/area*, with *Ampére* as the respective SI unit.

of graphene, respectively. Besides their systematic importance in the context of carbon nanostructures, these reduced systems are of major relevance for the nanoelectronic use of these structures. Due to its vanishing band gap, the two-dimensional graphene sheet is not suitable as an element of semiconductor electronics. However, as one imposes quantum confinement on the electronic system by limiting at least one of the two periodic dimensions to finite width, one generates novel materials with non-zero band gaps. Thus, band gaps in excess of 100 meV have been reported for GNRs with widths in the order of 10 nm or narrower [84]. When GNRs were first fabricated in the laboratory by means of lithographic techniques [84, 93], the GNR energy gap ΔE_{gap} as a function of the ribbon width W was found to obey a simple relation [84], namely,

$$\Delta E_{gap} = \frac{\alpha}{W - W^*}, \tag{4.70}$$

where α and W^* are fitting parameters.

Further, GNRs play a crucial role in the field of carbon-based spintronics, as they allow for the controlled design of carbon devices with well-defined magnetism.

The possibility of tuning both the nanoelectronic and the nanomagnetic properties of GNRs is rooted in their unique geometric features. Thus, the structure of one-dimensional GNRs varies between two prototypes that differ from each other by the shape of their graphene edges. These are armchair graphene nanoribbons (aGNRs) and zigzag graphene nanoribbons (zGNRs), as displayed in Figure 4.8. In both of these structural alternatives, electrons are subjected to quantum confinement in one dimension while their motion along the other dimension is unconstrained. Both GNR types exhibit a finite band gap that increases with decreasing ribbon width, but only zGNRs have magnetic ground states corresponding to electronic states localizing at the zigzag edges. In the following, we will give a comparative description of the electronic structure of aGNRs and zGNRs in terms of the Weyl Hamiltonian formalism outlined in Section 4.2.

4.5.1 Zigzag Graphene Nanoribbons

With the unit cell vectors chosen as indicated in Figure 4.8, $\mathbf{a}_1 = a(1, 0)$ and $\mathbf{a}_2 = a(\frac{1}{2}, \frac{\sqrt{3}}{2})$ $(a = \sqrt{3}a_0)$, the two inequivalent corner points of the first Brillouin zone for zGNRs have coordinates $\mathbf{K} = (K, 0)$ and $\mathbf{K}' = (-K, 0)$, where $K = \frac{4\pi}{3a_0}$ [72].

Combining the Bloch function approach with the spinor formalism (4.26), we write the solution for the sublattice A in the following form

$$\Psi_A(\mathbf{x}) = e^{i\mathbf{K}\cdot\mathbf{x}}(\mathbf{k})\psi_A(\mathbf{x}) + e^{i\mathbf{K}'\cdot\mathbf{x}}(\mathbf{k})\psi'_A(\mathbf{x}), \tag{4.71}$$

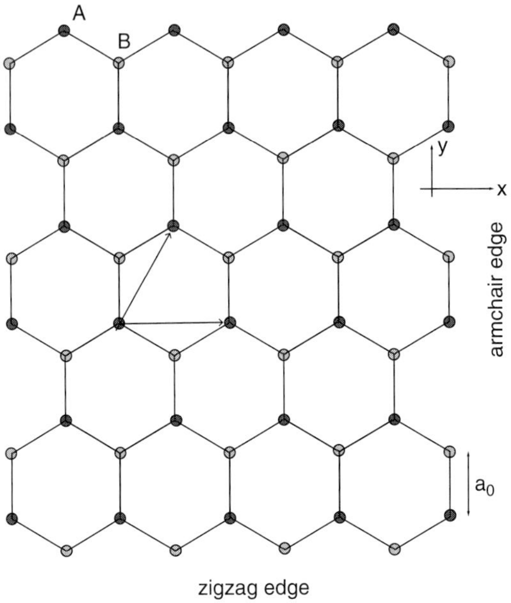

Figure 4.8 The two structural prototypes of graphene nanoribbons (GNRs): zigzag graphene nanoribbons (zGNRs) and armchair graphene nanoribbons (aGNRs), so termed after the patterns that limit carbon chains. Note that in the zero-dimensional case where the periodic dimension is reduced to finite width, zigzag nanoribbons are laterally bounded by armchair edges, and vice versa.

with a corresponding expression for the B sublattice. The x-direction, with reference to Figure 4.8, is chosen as the periodic coordinate. Consequently, the spinor wave function $\boldsymbol{\psi}$ may be represented by

$$\boldsymbol{\psi}(\mathbf{x}) = e^{ik_x x} \begin{pmatrix} \psi_A(y) \\ \psi_B(y) \end{pmatrix}. \tag{4.72}$$

Note that the wave number k_x is given in units of $2\pi/a$, with $a = \sqrt{3}a_0$ as the unit cell length of the zGNR. The expression $\boldsymbol{\psi}$ and its companion spinor $\boldsymbol{\psi}'$ are eigenstates of the Hamiltonians (4.22) and (4.27), respectively. We impose boundary conditions on the two wave function components at the ribbon edges $y = 0$ and $y = L$ by declaring that

$$\Psi_A(y = L) = 0, \text{ and } \Psi_B(y = L) = 0. \tag{4.73}$$

From Eq. (4.71), these conditions are satisfied if

$$\psi_A(L) = \psi_A'(L) = \psi_B(0) = \psi_B'(0) = 0. \tag{4.74}$$

Applying the Weyl Hamiltonian (4.30) to the eigenstate (4.72), we find

$$\begin{pmatrix} 0 & k_x - \frac{\partial}{\partial y} \\ k_x + \frac{\partial}{\partial y} & 0 \end{pmatrix} \begin{pmatrix} \psi_A \\ \psi_B \end{pmatrix} = \varepsilon \begin{pmatrix} \psi_A \\ \psi_B \end{pmatrix}, \tag{4.75}$$

setting $\varepsilon \equiv \frac{E}{\hbar v_F}$. From Eq. (4.75), equations for ψ_A and ψ_B are readily found:

$$(k_x^2 - \frac{\partial^2}{\partial y^2})\psi_B = \varepsilon^2 \psi_B, \tag{4.76a}$$

$$\psi_A = \frac{1}{\varepsilon}(k_x - \frac{\partial}{\partial y})\psi_B. \tag{4.76b}$$

$$\tag{4.76c}$$

Eq. (4.76a) is solved by

$$\psi_B(y) = B_I e^{\zeta y} + B_{II} e^{-\zeta y}, \tag{4.77}$$

with $\zeta^2 \equiv k_x^2 - \varepsilon^2$. The explicit form of $\psi_A(y)$, in turn, is immediately derived from Eq. (4.76b). The boundary conditions for zGNRs (see Eq. (4.74)) imply that $B_I + B_{II} = 0$, and further that

$$e^{-2\zeta L} = \frac{k_x - \zeta}{k_x + \zeta}. \tag{4.78}$$

The parameter ζ results as a real number if k_x is chosen positive. From Eq. (4.77), this choice yields *edge states*. Defining ζ as an imaginary number, i.e. $\zeta = ik_n$, with k_n as a real number, yields, in contrast, a periodic mode within the nanoribbon. From Eq. (4.78), the corresponding wave number is $k_x = k_n/\tan(k_n L)$. A mirror solution for **k**-values close to $\mathbf{K}'$ is generated as k_x is replaced with $-k_x$, as dictated by the coordinates of the two inequivalent corner points **K** and $\mathbf{K}'$ of the first Brillouin zone for zGNRs.

Exercise 4.4 Verify the statement: $k_x > 0$ implies that the parameter ζ is real-valued.

4.5.2 Armchair Graphene Nanoribbons

The electronic structure of aGNRs may be understood with arguments strictly analogous to those used in case of zGNRs. From Figure 4.8 we obtain the aGNR boundary conditions, in parallel to those for zGNRs, as given by Eq. (4.73):

$$\Psi_A(x = 0) = \Psi_B(x = 0) = \Psi_A(x = L) = \Psi_B(x = L) = 0. \tag{4.79}$$

The counterpart of Eq. (4.72) for aZGRs is, once more with reference to Figure 4.8,

$$\boldsymbol{\psi}(\mathbf{x}) = e^{ik_y y} \begin{pmatrix} \psi_A(x) \\ \psi_B(x) \end{pmatrix}. \tag{4.80}$$

Setting up the boundary conditions in keeping with Eqs. (4.74), and solving again equations of the type (4.76) for ψ_A, ψ'_A, ψ_B and ψ'_B [72], we obtain the condition $\sin[(k_n + K)L] = 0$ with $k_n = \frac{n\pi}{L} - \frac{4\pi}{3a_0}$ and energy eigenvalues $\varepsilon = \sqrt{k_y^2 + k_n^2}$. In case of aGNRs, all solutions turn out to be periodic, so no edge-localized states are found for these species.

4.5.3 Chiral Graphene Nanoribbons

The GNR prototypes discussed so far may be considered as special realizations of a more general geometric model, the *chiral GNR* (cGNR). Nanoribbon structures that deviate from the aGNR and zGNR patterns may be defined by use of the *chiral vector* $\mathbf{C}_h$ that connects crystallographically equivalent sites in the graphene mesh,

$$\mathbf{C}_h = n\mathbf{a}_1 + m\mathbf{a}_2, \tag{4.81}$$

with n, m = 0, 1, 2... Figure 4.9 gives the example of a chiral GNR with n = 8 and m = 1. The zigzag and armchair paradigm are characterized by the conditions $n \neq 0, m = 0$ and $m = n$, respectively. Choices of n, m that differ from these fundamental solutions induce nanoribbon edge roughness, as demonstrated by the species shown in Figure 4.9. Alternatively, cGNR structures may be characterized by the *chiral angle* whose value is 5.8° in the structure shown.

cGNRs with well-defined chirality have been fabricated by use of carbon nanotube unzipping [95] or surface-assisted molecular assembling [97] methods. In the latter case, the geometric and catalytic properties of selected metal substrate surfaces are employed to manufacture cGNRs by polymerizing precursor molecules.

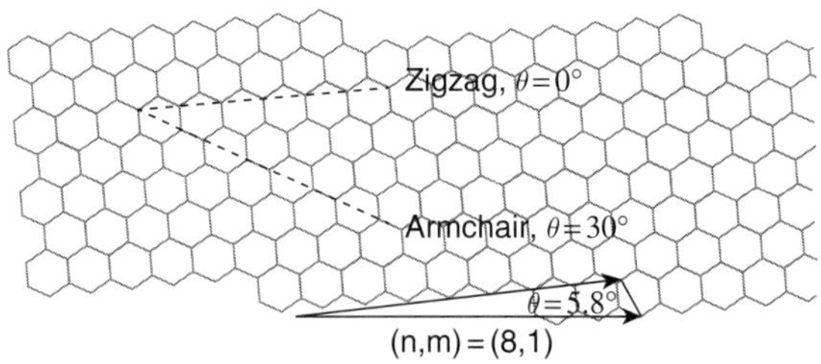

Figure 4.9 Geometric structure of a chiral graphene nanoribbon with n = 8, m = 1. Reprinted with permission from [94]. Copyright (2011) by Nature Publishing Group.

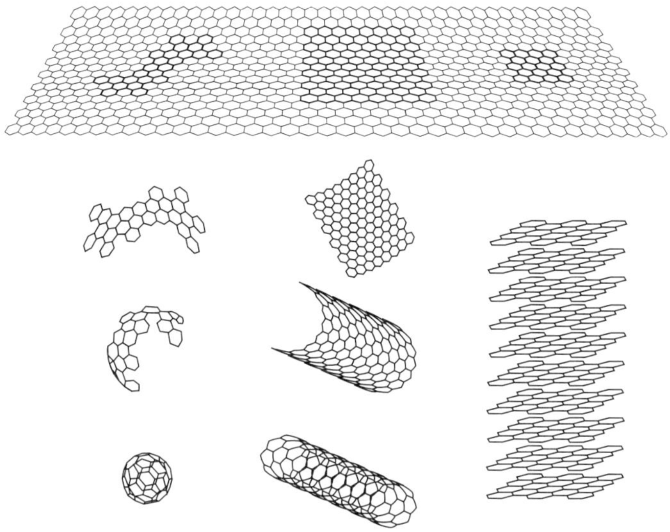

Figure 4.10 The relation between the graphene honeycomb lattice and the single-walled carbon nanotube as well as the fullerene geometry. Reprinted with permission from [102]. Copyright (2007) by Nature Publishing Group.

The derivations of the GNR spectra, as presented above, involve idealized systems with perfectly regular edge structures. This limit, however, cannot be attained with present means of GNR manufacturing. Several authors have addressed the question of how edge roughness affects the electronic properties of GNRs. In simulations of both zGNRs and aGNRs with edge imperfections [98, 99], significantly different behavior of the two GNR protoypes was found. While zGNRs tend to retain their conduction properties in the presence of these structural irregularities, aGNRs may turn into Anderson insulators, and thus undergo a drastic change of their conduction features. Also, non-uniform GNR edges may induce Coulomb blockade phenomena [100] or anomalies in the quantum Hall effect [101].

As will be further outlined in Chapter 7, the existence of electronic edge states for zGNRs, but not for aGNRs, is the reason that the former display ground-state magnetism while the latter do not.

In the following two chapters we will discuss two carbon allotropes whose structures are derived from the planar honeycomb structure of graphene. These are *single-walled carbon nanotubes* (SWCNTs) and *fullerenes*. The first of these species may be described as a GNR rolled up into a seamless cylinder, while the second may be generated from graphene fragments that are molded into spherical shape. The emergence of the SWCNT and the fullerene geometry from the graphene sheet is illustrated in Figure 4.10.

5

Carbon Nanotubes

Carbon nanotubes (CNTs) are classified according to various structural types. While rolling a graphene nanoribbon into a cylinder gives rise to single-walled carbon nanotubes (SWCNTs), nesting several SWCNTs with different radii into each other results in multi-walled carbon nanotubes (MWCNTs). A comparison between these two CNT families is shown in Figures 5.1 and 5.2.

In what follows, we will focus on SWCNTs, detailing their structural, electronic and energetic properties in Sections 5.1 to 5.3. Section 5.4 addresses the behavior of SWCNTs in magnetic fields. We will present two limiting cases involving magnetic fields oriented in the directions parallel and perpendicular to the tube axis and subsequently discuss the magnetic susceptibility of SWCNTs as a function of the angle between the tube axis and the field direction. The final section deals with transport properties of SWCNTs, with emphasis on nanoelectronic circuits that contain SWCNTs as transmission elements and, specifically, as quantum dots.

5.1 Geometric Features of Single-Walled Carbon Nanotubes

The basic structural types of SWCNTs may be understood in analogy to those introduced in the previous subsection on GNRs. Specifically, we distinguish between zigzag, armchair and chiral SWCNTs. The former two types are obtained by rolling zGNRs and aGNRs, respectively, into cylindrical shape. Once more, the chiral alternative is defined by the chiral vector, characterized by a chiral angle different from 30° (armchair tubes) and 0° (zigzag tubes). The relation between the geometric quantities that uniquely determine the SWCNT type are shown in Figure 5.3 and models of the three SWCNT prototypes are displayed in Figure 5.4.

The lattice translation vector $\mathbf{T}$ determines the SWCNT periodicity in axial direction. It may be represented as a linear combination of the basis vectors $\mathbf{a}_1$ and $\mathbf{a}_2$,

$$\mathbf{T} = t_1\mathbf{a}_1 + t_2\mathbf{a}_2, \tag{5.1}$$

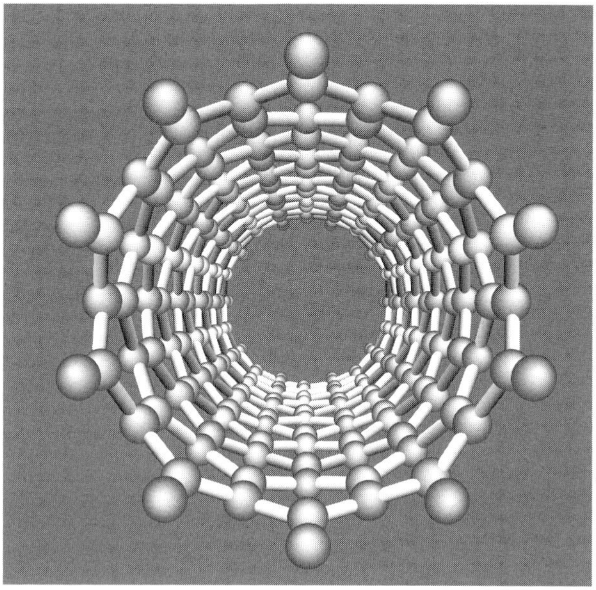

Figure 5.1 Model of a single-walled nanotube (SWNT).

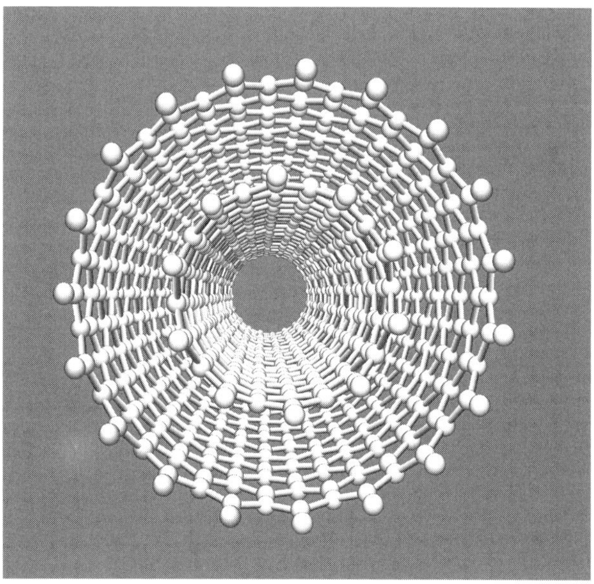

Figure 5.2 Model of a two-layered, multi-walled nanotube. Figures 5.1 and 5.2 were generated using the program *Nanotube modeler* [103].

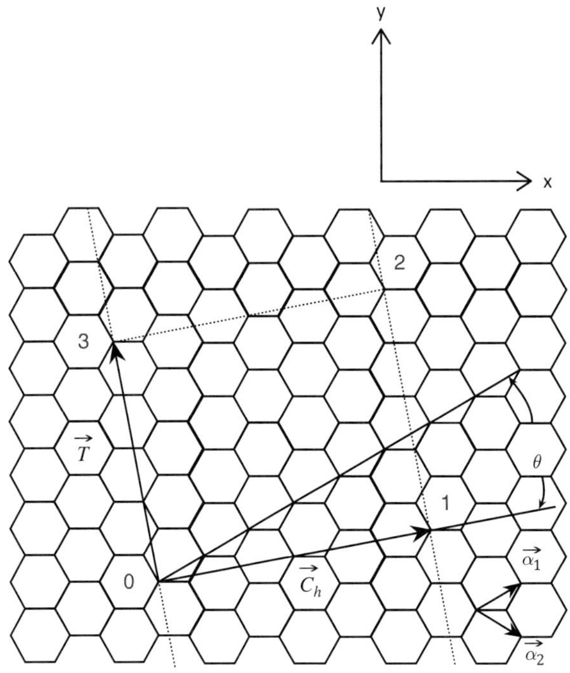

Figure 5.3 The quantities defining the SWCNT structure: the lattice translation vector **T**, the chiral vector $\mathbf{C}_h$, and the chiral angle θ. The unit cell of the tube is given by the rectangle 0123.

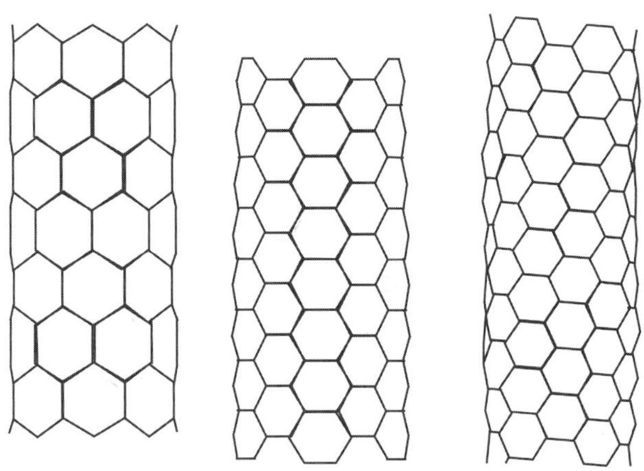

Figure 5.4 The three structural types of single-walled carbon nanotubes (SWCNTs): representations of zigzag, armchair and chiral SWCNTs.

with integer t_1, t_2. The chiral vector is given by

$$\mathbf{C}_h = n\mathbf{a}_1 + m\mathbf{a}_2, \tag{5.2}$$

where n, m are integers, such that $0 \leq |m| \leq n$. This vector specifies the SWCNT type, as it maps a lattice point into another one, such that these two points coincide when the respective graphene fragment is rolled up into a tube. Its length equals the tube circumference. The coefficient pair (n, m) classifies SWCNTs in terms of zigzag, armchair or chiral tubes. With reference to Figure 5.3, the choice $n \neq 0, m = 0$ yields a zigzag tube (zSWCNT), composed of circumferential polyene rings which display the geometric pattern of zGNR edges. Likewise, the condition $n = m$ specifies armchair tubes (aSWCNTs), while all other admitted combinations of n and m determine chiral tubes.

Once a chiral vector has been selected, the chiral angle is found from the following operation,

$$\cos(\theta) = \frac{\mathbf{C}_h \cdot \mathbf{a}_1}{|\mathbf{C}_h||\mathbf{a}_1|}, \tag{5.3}$$

and the tube diameter d_t results from

$$d_t = \frac{|\mathbf{C}_h|}{\pi} = \frac{a_0\sqrt{3}}{\pi}\sqrt{n^2 + m^2 + mn}. \tag{5.4}$$

As the translation vector of the tube points in the axial (or tubular), and the chiral vector in the circumferential (or lateral) direction of the tube, they are orthogonal to each other, such that $\mathbf{T} \cdot \mathbf{C}_h = 0$. This implies

$$t_1(2n + m) + t_2(2m + n) = 0. \tag{5.5}$$

In accordance with the definition of the translation vector, among the solutions for t_1, t_2, those of minimum magnitude are to be selected as the coordinates of $\mathbf{T}$. These are

$$t_1 = \frac{2m + n}{gcd}, \quad t_2 = -\frac{2n + m}{gcd}, \tag{5.6}$$

Exercise 5.1 Show that t_1 and t_2, as given by relation (5.6), are the correct choice for the coordinates of the translation vector $\mathbf{T}$. In the first step, prove statement (5.5).

where gcd is the greatest common divisor of $2n+m$ and $2m+n$, and the stipulation has been made that $t_1 > 0$. From a quick calculation, the surface area of an SWCNT unit cell is

$$A_{unit\ cell} = |\mathbf{C}_h \times \mathbf{T}| = \frac{\sqrt{3}(n^2 + nm + m^2)3a_0^2}{gcd}. \tag{5.7}$$

Combining this result with the geometric finding that the surface area of a single hexagon is described by the product $|\mathbf{a}_1 \times \mathbf{a}_2|$, we obtain an expression for the number of hexagons, N_{cell} in an SWCNT unit cell, or equally, the number of graphene unit cells contained in an SWCNT unit cell:

$$N_{cell} = \frac{|\mathbf{C}_h \times \mathbf{T}|}{|\mathbf{a}_1 \times \mathbf{a}_2|} = \frac{2|\mathbf{C}_h|^2}{3a_0^2 gcd}. \tag{5.8}$$

As there are two carbon atoms in each graphene unit cell, $2N_{cell}$ indicates the number of atoms contained in an SWCNT unit cell. We point out that N_{cell} reduces to the same result for both zigzag and armchair tubes, namely $2n$. In the following we will explore the Brillouin zone of SWCNTs and describe the band structure for the three basic SWCNT types in the frame of the tight-binding model introduced in Section 4.2.

5.2 Electronic Structure of Single-Walled Carbon Nanotubes

The observations made in the previous section on structural characteristics of zigzag, armchair and chiral SWCNTs allow us to construct the Brillouin zones for these species. In each case we will assume periodic, i.e. one-dimensional, tubes. This implies that the Brillouin zones are defined by a **k**-space basis vector $\mathbf{K}_a$ where the subscript a stands for axial. This vector is reciprocal with respect to the translation vector of the considered SWCNT according to the relation $\mathbf{T} \cdot \mathbf{K}_a = 2\pi$. The wave vector in the circumferential direction is, in contrast to the case of graphene, naturally quantized for SWCNTs, and described by a basis vector $\mathbf{K}_c$ that is perpendicular to $\mathbf{K}_a$. The reciprocity between the SWCNT unit cell and Brillouin zone is summarized by the four relations

$$\mathbf{T} \cdot \mathbf{K}_a = 2\pi, \;\; \mathbf{T} \cdot \mathbf{K}_c = 0, \mathbf{C}_h \cdot \mathbf{K}_c = 2\pi, \;\; \mathbf{C}_h \cdot \mathbf{K}_a = 0, \tag{5.9}$$

using the definition of $\mathbf{C}_h$ in terms of the basis vectors of the graphene lattice, $\mathbf{a}_1$ and $\mathbf{a}_2$ as well as Eqs. (4.3) (5.1) and (5.8), we obtain from these equations the following expressions for $\mathbf{K}_a$ and $\mathbf{K}_c$:

$$\mathbf{K}_a = \frac{1}{N_{cell}}(m\mathbf{b}_1 - n\mathbf{b}_2), \mathbf{K}_c = \frac{1}{N_{cell}}(-t_2\mathbf{b}_1 + t_1\mathbf{b}_2). \tag{5.10}$$

A one-dimensional tube, i.e. a tube of infinite length, contains an infinite number of unit cells, N_{uc}. From the periodicity constraint imposed on the electronic wave function in position space, $\Psi(0) = \Psi(j|\mathbf{T}|)$, where j is an integer, in conjunction with Bloch's theorem, we conclude that

$$\Psi(0) = \Psi(N_{uc}|\mathbf{T}|) = e^{ik_a N_{uc}|\mathbf{T}|}\Psi(N_{uc}|\mathbf{T}|). \tag{5.11}$$

Consequently, $e^{ik_a N_{uc}|\mathbf{T}|} = 1$. This statement implies for the axial wave number k_a that

$$k_a = \pm \frac{2\pi j}{N_{uc}|\mathbf{T}|}, \quad j \leq N_{uc} - 1. \tag{5.12}$$

As N_{uc} is infinite, k_a is a continuous variable that varies between the limits $\frac{\pi}{|\mathbf{T}|}$ and $-\frac{\pi}{|\mathbf{T}|}$.

Analogous arguments lead to the quantization condition for k_c, the circumferential wave number. Since $2\pi/|\mathbf{C}_h|$ is the smallest wave number compatible with the condition of circumferential quantization, we obtain

$$k_c = j\frac{2\pi}{|\mathbf{C}_h|}, \tag{5.13}$$

and $j = 1, \ldots, j_{max}$. The maximum index, j_{max}, may be determined by realizing that the first Brillouin zone of an SWCNT is a sequence of line cuts through the first Brillouin zone of graphene. Since $N_{cell}\mathbf{K}_c$ is a reciprocal lattice vector of graphene (see Eq. (5.10) and Figure 5.5), connecting two equivalent points in $\mathbf{k}$-space, it holds that $|\mathbf{k}_c| < N_{cell}|\mathbf{K}_c|$. From this, we conclude that $j_{max} = N_{cell} - 1$.

Comparison with graphene yields a simple criterion for classifying any tube, defined by an index pair (n, m), as metallic or semiconducting. This follows from the observation that a metallic tube must have a Dirac point ($\mathbf{K}$, $\mathbf{K}'$) among its wave vectors. If this condition is satisfied, we conclude from applying Bloch's theorem to the $\mathbf{K}$ point that $e^{i\mathbf{K}\cdot\mathbf{C}_h} = 1$. From the coordinate representations of $\mathbf{K}$ and $\mathbf{C}_h$ (see Eqs. (4.2), (4.20), (5.2)), we obtain

$$\mathbf{K} \cdot \mathbf{C}_h = 2\pi\frac{n + 2m}{3}. \tag{5.14}$$

We distinguish two cases: (a) $n + 2m$ is a multiple of 3 ($n + 2m = 3l, l = 1, 2, 3\ldots$), (b) $n+2m$ is a not multiple of 3 ($n+2m = 3l\pm1$). If case (a) is realized, we conclude from Eq. (5.14) that $\mathbf{K} \cdot \mathbf{C}_h$ is an integer multiple of 2π. Thus, $\mathbf{K}$ is among the allowed wave vectors for the tube characterized by the index pair (n, m). This, in turn, implies that the considered tube is metallic. Note that this metallicity criterion may be rephrased in the following way: since $n + 2m = n - m + 3m$, *the tube is metallic if* $|n - m|$ *is a multiple of 3.*

An expression for the general SWCNT reciprocal lattice vector in the vicinity of the K point follows from the expansion $\mathbf{k} = k_a\frac{\mathbf{K}_a}{|\mathbf{K}_a|} + k_c\frac{\mathbf{K}_c}{|\mathbf{K}_c|}$ in combination with Eq. (5.10) and the quantization constraint on k_c,

$$\mathbf{k} = k_a\frac{\mathbf{K}_a}{2\pi/|\mathbf{T}|} + j\mathbf{K}_c, \tag{5.15}$$

where we have made use of $|\mathbf{K}_c| = \frac{2\pi}{|\mathbf{C}_h|}$ as well as Eq. (5.13). In Eq. (5.15), $|k_a| < \frac{\pi}{|\mathbf{T}|}$, and $j = 0, 1, \ldots, N_{cell} - 1$. This relation succinctly describes the structure of

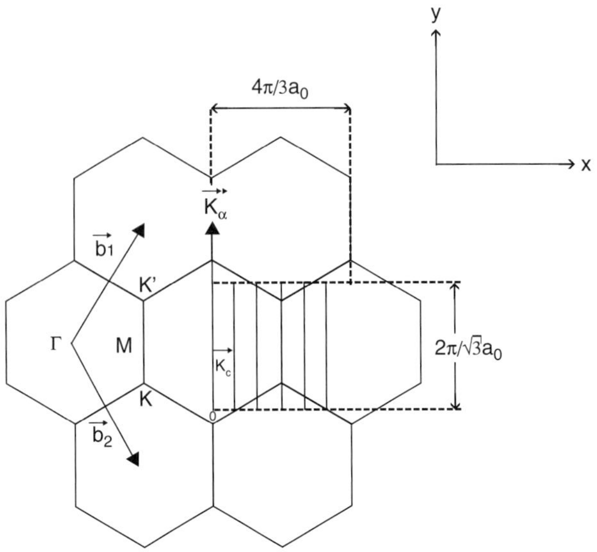

Figure 5.5 The first Brillouin zone of an aSWCNT, consisting of N lines, parallel to $\mathbf{K}_a$ and perpendicular to $\mathbf{K}_c$ within the first Brillouin zone of graphene.

the first Brillouin zone for a general SWCNT as a set of N parallel lines traversing the first Brillouin zone of graphene. In each one, the wave k_a number spans an interval of $\frac{2\pi}{|\mathbf{T}|}$, and the spacing between adjacent lines is $|\mathbf{K}_c| = \frac{2\pi}{|\mathbf{C}_h|}$.

5.2.1 Single-Electron Dispersion Relations

Having specified the first Brillouin zone for general SWCNTs as a subset of that of graphene, we proceed to derive the respective tight-binding single-electron dispersion relation for SWCNTs. In the *zone-folding approximation* [69], this is done by use of the graphene dispersion relation (4.14) for the selected wave vectors that persist in SWCNTs. Explicit expressions for the **k** vectors in the first Brillouin zone of SWCNTs are obtained from the graphene dispersion relation in conjunction with the formulae (5.10), (5.12) and (5.15), namely,

$$
\begin{aligned}
k_x &= \frac{2\pi l(n+m)|\mathbf{C}_h|3a_0 + \sqrt{3}k_a(n^3-m^3)3a_0^3}{2|\mathbf{C}_h|^3}, \\
k_y &= \frac{3a_0k_a(n+m)|\mathbf{C}_h| + 2\pi\sqrt{3}l(n-m)a_0}{2|\mathbf{C}_h|^2},
\end{aligned}
\tag{5.16}
$$

where $|k_a| < \frac{\pi}{|\mathbf{T}|}$ and $l = 0, 1, \ldots, N_{cell} - 1$. For the special cases $m = 0$ (zSWCNTs) and $n = m$ (aSWCNTs), compact closed formulae are derived from Eq. (5.16) that define the respective band structures in the tight-binding limit. Prior

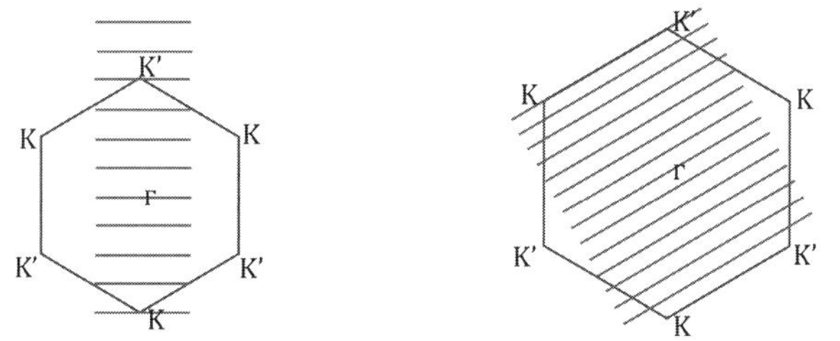

Figure 5.6 The first Brillouin zones of a zSWCNT (left) and an aSWCNT (right). For zSWCNTs, the line cuts through the first Brillouin zone of graphene are perpendicular to the line $\overline{\mathbf{\Gamma K}}$, for aSWCNTs, they are perpepependicular to the line $\overline{\mathbf{\Gamma}\boldsymbol{M}}$, with the $\mathbf{M}$ point bisecting $\overline{\mathbf{K}, \mathbf{K}'}$.

to specifying the one-electron dispersion relations for these two basic cases, we comment on the geometric difference between their first Brillouin zones, as readily inferred from (5.16), and indicated in Figure 5.6. Specifically, we observe that the line cuts defining the first Brillouin zone for zSWCNTs lie perpendicular to the vector $\overline{\mathbf{\Gamma K}}$ of the first Brillouin zone for graphene. For aSWCNTs, in contrast, the line cuts are perpendicular to the graphene vector $\overline{\mathbf{\Gamma}\boldsymbol{M}}$. The result is immediately obtained by inspecting Eq. (5.16), as the aSWCNT constraint $n = m$ implies that k_x depends only on the index l, but not on the variable k_a.

Inserting k_x and k_y according to Eq. (5.16) into Eq. (4.14) yields the general SWCNT single-electron dispersion relation in the form $E(\mathbf{k})_\pm = E_l(k_a)_\pm$, i.e. $2N_{cell}$ band energy functions that depend on the axial wave number k_a. Specifically, we obtain for zigzag tubes,

$$E_l(k_a)_\pm = \pm t\sqrt{1 + 4\cos(\frac{3}{2}k_a a_0)\cos(\frac{l\pi}{n}) + 4\cos^2(\frac{l\pi}{n})}, \tag{5.17}$$

with $l = 0, 1, \ldots, N_{cell} - 1$, and $|k_a| < \frac{\pi}{|3a_0|}$, where the latter condition is in keeping with the size of the translation vector (or the vector $\mathbf{K}_a$, see Table 5.1) for zSWCNTs. With reference to formula (4.14), we assume here that the overlap integral s is zero, and, as before, we set $E_{2p} = 0$. The sign of $E_l(k_a)_\pm$ then distinguishes the valence band of the considered SWCNT from its conduction band, consisting of π and π^* states, respectively. Thus, the sub-bands with negative energies are fully, i.e. each twofold, occupied, while the sub-bands with positive energies are empty. The dispersion relation (5.17) implies that the highest occupied and the lowest unoccupied sub-band coincide at $k_a = 0$ for some choices of n, but not for all. Specifically, the constraint that the distance $|\overline{\mathbf{\Gamma K}}|$ must be an integer multiple of $|\mathbf{K}_c|$ for metallic SWCNTs implies that $l = \frac{2}{3}n$ for these systems. Thus,

$$E_l(k_a = 0, n = \frac{3}{2}l)_- = E_l(k_a = 0, n = \frac{3}{2}l)_+ = 0. \tag{5.18}$$

The relation between the index l and the tube index n is obviously in keeping with the metallicity criterion derived earlier.

Similar arguments establish the nature of aSWCNTs. Once more, the relevant one-electron dispersion relation is obtained from Eqs. (4.14) and (5.16), resulting in

$$E_l(k_a)_{\pm} = \pm t\sqrt{1 + 4\cos(\frac{\sqrt{3}}{2}k_a a_0)\cos(\frac{l\pi}{n}) + 4\cos^2(\frac{\sqrt{3}k_a a_0}{2})}, \tag{5.19}$$

with $l = 0, 1, \ldots, N_{cell} - 1$, and $|k_a| < \frac{\pi}{|\sqrt{3}a_0|}$. In this case, points of coincidence between the valence and the conduction band are found for every tube index n, such that all aSWCNTs are predicted to be metallic within the tight-binding model.[1] Again, an intuitive explanation of this quantitative finding comes from comparing the first Brillouin zone of graphene with that of an aSWCNT, as shown in Figure 5.6. A graphene **K**-point falls on an aSWCNT line cut if

$$l|\mathbf{K}_c| = |\mathbf{\Gamma} \boldsymbol{M}|. \tag{5.20}$$

With $\mathbf{\Gamma M} = (\frac{2\pi}{3a_0}, 0)$, and $|\mathbf{K}_c| = \frac{2\pi}{3a_0 n}$ for aSWCNTs, this condition leads to $l = n$ as metallicity requirement. By this criterion, all aSWCNTs are metallic.

Figures 5.7–5.9 show examples for band structures of typical zSWCNTs and aSWCNTs.

It is to be noted that the zigzag and armchair tube display complementary behaviors at the Gamma point ($k_a = 0$). While the energy difference between the valence

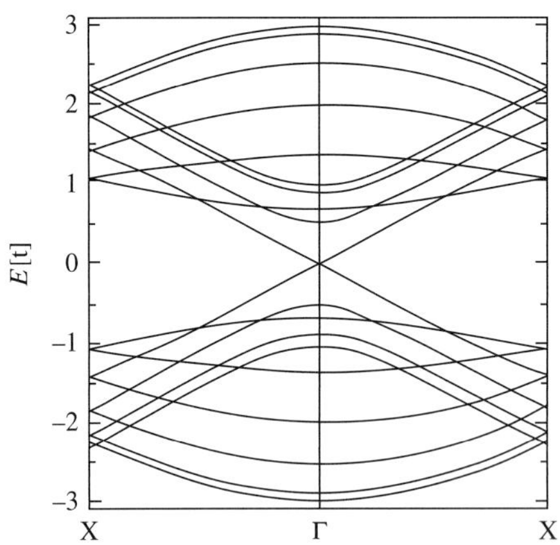

Figure 5.7 The band structure of zSWCNTs with n=9 (see (5.17)) from tight-binding calculation. Reprinted with permission from [118]. Copyright (2007) by the American Physical Society.

[1] At band indices with $l = n$, coincidences between the valence and the conduction band are found at $k_a = \pm\frac{2\pi}{3\sqrt{3}a_0}$.

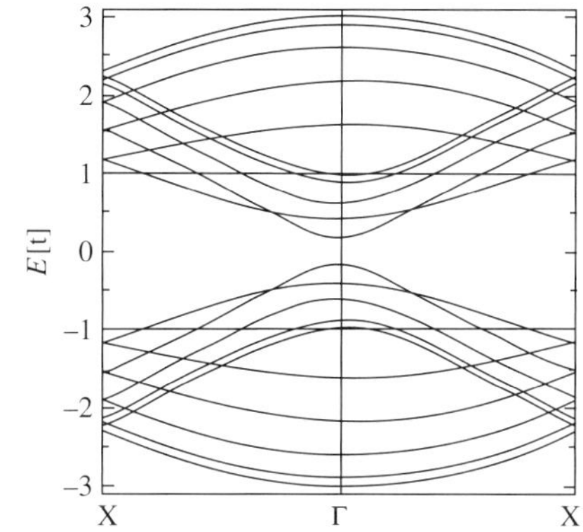

Figure 5.8 The band structure of zSWCNTs with n=10 from tight-binding calculation. Reprinted with permission from [118]. Copyright (2007) by the American Physical Society.

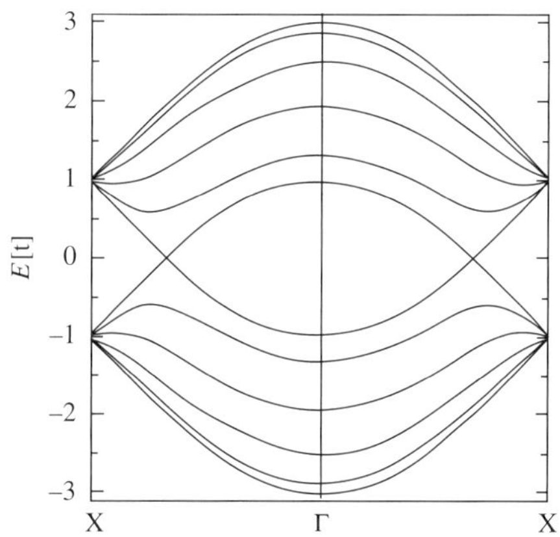

Figure 5.9 The band structure of aSWCNTs (see 5.19) with n=5 from tight-binding calculation. Reprinted with permission from [118]. Copyright (2007) by the American Physical Society.

and the conduction band is a minimum at this place for the former, it is a maximum for the latter species. This contrast is easily rationalized by the structure of the first Brillouin zones of the two SWCNT types, as compared with that of graphene (see Figure 5.6). For zSWCNTs, the Γ points of line cuts (i.e. the places where $k_a = 0$) lie along the direction $\overline{\mathbf{\Gamma K}}$ of the graphene reciprocal lattice, such that their energy gap vanishes at the **Γ** point whenever n is a multiple of three and assumes a minimal value at the same place in all other cases. For aSWCNTs, however, the **Γ** points lie along the $\overline{\mathbf{\Gamma}\boldsymbol{M}}$ direction of graphene, at locations of maximum distance from the adjacent **K**, **K**′ points and thus at places of energy gap maxima.

The most essential quantities that describe the direct and reciprocal lattice of SWCNTs are summerized in Table 5.1.

Table 5.1 *Important parameters for the geometric and electronic structure of single-walled carbon nanotubes*

Quantity	*Symbol*	*cSWCNT*	*zSWCNT*	*aSWCNT*
Direct lattice vectors	$\mathbf{a}_1$	$\frac{\sqrt{3}a_0}{2}(\sqrt{3}, 1)$		
	$\mathbf{a}_2$	$\frac{\sqrt{3}a_0}{2}(\sqrt{3}, -1)$		
Reciprocal lattice vectors	$\mathbf{b}_1$	$\frac{2\pi}{\sqrt{3}a_0}(\frac{1}{\sqrt{3}}, 1)$		
	$\mathbf{b}_2$	$\frac{2\pi}{\sqrt{3}a_0}(\frac{1}{\sqrt{3}}, -1)$		
Chiral vector	$\mathbf{C}_h$	$n\mathbf{a}_1 + m\mathbf{a}_2$	$(n, 0)$	(n, n)
	$\lvert\mathbf{C}_h\rvert$	$\sqrt{3}a_0\sqrt{n^2 + nm + m^2}$	$\sqrt{3}a_0 n$	$3a_0 n$
Translation vector	$\mathbf{T}$	$\frac{2m+n}{gcd}\mathbf{a}_1 - \frac{2n+m}{gcd}\mathbf{a}_2$	$\mathbf{a}_1 - 2\mathbf{a}_2$	$\mathbf{a}_1 - \mathbf{a}_2$
	$\lvert\mathbf{T}\rvert$	$\frac{\sqrt{3}\lvert\mathbf{C}_h\rvert}{gcd}$	$3a_0$	$\sqrt{3}a_0$
Greatest common divisor	gcd	$gcd(2m + n, 2n + m)$	n	$3n$
Number of hexagons per cell	N_{cell}	$\frac{2\lvert\mathbf{C}_h\rvert^2}{3a_0^2 gcd}$	$2n$	$2n$
Chiral angle	θ	$\theta = \cos^{-1}(\frac{2n+m}{2\sqrt{n^2+nm+m^2}})$	0°	30°
Circumferential lattice vector	$\mathbf{K}_c$	$\frac{(m+2n)\mathbf{b}_1+(n+2m)\mathbf{b}_2}{2(n^2+nm+m^2)}$	$\frac{2\mathbf{b}_1+\mathbf{b}_2}{2n}$	$\frac{\mathbf{b}_1+\mathbf{b}_2}{2n}$
	$\lvert\mathbf{K}_c\rvert$	$\frac{2\pi}{\lvert\mathbf{C}_h\rvert}$	$\frac{2\pi}{\sqrt{3}a_0 n}$	$\frac{2\pi}{3a_0 n}$
Axial lattice vector	$\mathbf{K}_a$	$\frac{m\mathbf{b}_1-\mathbf{b}_2}{N}$	$-\frac{\mathbf{b}_2}{2}$	$\frac{\mathbf{b}_1-\mathbf{b}_2}{2}$
	$\lvert\mathbf{K}_a\rvert$	$\frac{2\pi}{\lvert\mathbf{T}\rvert}$	$\frac{2\pi}{3a_0}$	$\frac{2\pi}{\sqrt{3}a_0}$

5.3 Densities of States

Dispersion relations, as given by Eqs. (5.17) and (5.19) for zSWCNTs and aSWCNTs, respectively, allow us to calculate the *densities of state (DOS)* for the various types of nanotubes, as these relations combine a factor that reflects the dimensionality of the considered system and a dispersion-dependent factor. As nanotubes are one-dimensional systems, their density of states is defined as the number of electron states per length (L) and energy interval (see Section 2.3.2). Specifically, we ask for the quantity $\frac{d\tilde{N}(E)}{dE}$, where $\tilde{N}(E)$ stands for the number of quantum states with an energy smaller than E. If the dispersion relation is known, however, the number of states may be expressed as a function of k rather than E (see Eq. (2.93)), such that the DOS ($g(E)$) is evaluated as

$$g(E) = \frac{1}{L}\frac{dN(k)}{dE} = g(k)\frac{dk}{dE}. \tag{5.21}$$

Extending Eq. (2.93) to the one-dimensional case, one finds

$$g(k) = \frac{1}{L}2\frac{L}{2\pi} = \frac{1}{\pi}, \tag{5.22}$$

where the factor of 2 reflects the spin multiplicity per k-state. If the dispersion relation for the considered system is known, this equation may be combined with formula (5.21) to derive $g(E)$ for one-dimensional structures. If, for instance, the energy of electronic states close to the minimum of the conduction band, E_{cond}^{min}, can be approximated as

$$E = \frac{\hbar^2}{2m}k^2 + E_{cond}^{min}, \quad \text{and} \quad E > E_{cond}^{min}, \tag{5.23}$$

the density of states turns out to be

$$g(E) = \frac{1}{h}\sqrt{\frac{2m}{E - E_{cond}^{min}}}, \quad \text{if} \quad E > E_{cond}^{min}. \tag{5.24}$$

It should be noted that no states exist at the point $E = E_{cond}^{min}$, at which $g(E)$ becomes singular, related to the fact that E_{cond}^{min} limits the energy domain. This gives an example for *Van–Hove singularities*, characteristic features of the density of states for one-dimensional cases.

Using the dispersion relation (5.17), we obtain the following expression as the density of states of zSWCNTs for any given sub-band with index l [69, 104],

$$g_l(E) = \frac{4d}{3a_0\pi}\frac{|E|}{\sqrt{(E^2 - E_{VHS_{1,l}}^2)(E_{VHS_{2,l}}^2 - E^2)}}, \tag{5.25}$$

where the factor d keeps track of the degeneracy associated with the Brillouin-zone mirror symmetry. Thus, d is 1 at the Γ point ($k_a = 0$) but, due to the coexistence of a positive and a negative branch in the dispersion relation (5.17), 2 if ($k \neq 0$). The symbols VHS_i, $i = 1, 2$ stand for Van–Hove singularities. These are given by the following expressions:

$$\begin{aligned} E_{VHS_{1,l}} &= \pm t|1 + 2\cos(\frac{l\pi}{n})|, \\ E_{VHS_{2,l}} &= \pm t|1 - 2\cos(\frac{l\pi}{n})|. \end{aligned} \tag{5.26}$$

As readily verified, $E_{VHS_{1,l}}$ is located at the Γ point, while $E_{VHS_{2,l}}$ lies in the second Brillouin zone. The positions of both critical energies vary with the tube diameter. The energy variable in (5.25) is confined to the zSWCNT valence and conduction band, ranging between the bottom and the top of either band. Formulae (5.17) and (5.25) allow us to derive the density of states at the Fermi energy for metallic zSWCNTs. From Eq. (5.17), the index of the sub-band that touches the Fermi energy ($E = 0$) is $l_1 \equiv 2n/3$. As pointed out earlier, the tube index of metallic zSWCNTs is an integer multiple of three, implying that l_1 is an integer. Inserting $l = l_1$ into Eq. (5.25) yields

$$g_{l_1}(E_F) = \frac{8}{3a_0\pi t} \approx 2nm^{-1}eV^{-1}, \tag{5.27}$$

where a factor of two has been applied, accommodating an additional degeneracy due to the coincidence of the valence and the conduction band at $E = E_F$.

An analogous development for aSWCNTs leads once more to a closed formula describing $g_l(E)$ [69].

For chiral SWCNTs, the DOS can be determined numerically [105], but no closed formulae are available. However, for systems with $(n, m) > (7, 0)$ that are not too strongly curved, and for bands not too far from the Fermi energy, the DOS has turned out to be determined predominantly by the tube diameter [106]. Thus, an approximation to the DOS of SWCNTs with arbitrary chirality in this regime may be obtained by recasting relations (5.25) and (5.26) as functions of the tube diameter. Within a small margin of error, the resulting DOS distributions may be regarded as universal within an energy domain that spans from the top of the third valence band to the bottom of the third conduction band. As shown by calculations for a wide variety of SWCNTs, the chirality dependence of the DOS is rather weak in this interval [69].

5.4 Nanotubes in Magnetic Fields

In Section 4.4, we described the impact of an external magnetic field on the band structure of graphene. While there is one naturally distinguished direction of an external magnetic field for graphene, namely perpendicular to the graphene sheet, two preferential directions exist for carbon nanotubes: along the tube axis and perpendicular with respect to the tube. While the axis-parallel case gives rise to a *Bohm–Aharonov effect*, perpendicular orientation of the field results, for sufficiently high field strengths, in the formation of Landau levels, in analogy to the behavior of graphene exposed to a magnetic field. In what follows, we discuss these two alternatives in separation from each other. The marked difference between the effects observed in the two cases is reflected by a large anisotropy of the *orbital susceptibility* of carbon nanotubes. We will address this finding in Section 5.4.3.

5.4.1 Axial External Fields: the Aharonov–Bohm Effect in Carbon Nanotubes

If a magnetic field $\mathbf{B}_a$ is applied along the axis of an SWCNT, the tube is threaded by a magnetic flux Φ. The vector potential $\mathbf{A}$ that generates the field may be expressed in terms of this flux and of the chiral vector $\mathbf{C}_h$ by setting

$$\mathbf{A} = \mathbf{A}_\varphi = \frac{\Phi}{|\mathbf{C}_h|}\frac{\mathbf{C}_h}{|\mathbf{C}_h|}, \tag{5.28}$$

as the curl of the azimuthal component of a vector in cylindrical coordinates yields a vector pointing in axial direction, and $\Phi = B_a \pi r^2$, with r as tube radius. For convenience, we adopt the basis $\{\frac{\mathbf{T}}{|\mathbf{T}|}, \frac{\mathbf{C}_h}{|\mathbf{C}_h|}\}$ for further evaluation of the effect of the axial magnetic field on the SWCNT electronic structure. In this basis,

$$\mathbf{A} = (0, \frac{\Phi}{|\mathbf{C}_h|}). \tag{5.29}$$

As the presence of $\mathbf{A}$ changes the kinetic energy Hamiltonian according to Eq.(2.3),

$$\frac{1}{2m^*}\hat{\mathbf{p}}^2 \to \frac{1}{2m^*}(\hat{\mathbf{p}} - e\mathbf{A})^2, \tag{5.30}$$

with m^* as effective electron mass, it modifies, by gauge invariance, the tight-binding approach as applied to graphene in Section 4.2 by introducing a phase factor. The eigenfunctions (4.7) of the tight-binding Hamiltonian (4.5) are modified by a phase factor, according to

$$\phi(\mathbf{k}, \mathbf{x}) = \frac{1}{\sqrt{N}} \sum_i^N e^{i\mathbf{k}\mathbf{R}_i + i\frac{e}{\hbar}G_{\mathbf{R}_i}} \varphi(\mathbf{x} - \mathbf{R}_i), \tag{5.31}$$

with $\mathbf{R}_i$ as a lattice vector. The phase factor is found from

$$G_{\mathbf{R}_i} = \int_{\mathbf{R}_i}^{\mathbf{x}} \mathbf{A}(\mathbf{x}') \cdot d\mathbf{x}'. \tag{5.32}$$

A plausible choice of the path that connects the integration limits in Eq. (5.32) is given by the straight line

$$\mathbf{x}' = \mathbf{R}_i + \lambda(\mathbf{x} - \mathbf{R}_i). \tag{5.33}$$

Here, λ is a real-valued parameter with $0 \leq \lambda \leq 1$. Thus,

$$G_{\mathbf{R}_i} = \int_0^1 (\mathbf{x} - \mathbf{R}_i) \cdot (\mathbf{A}(\mathbf{R}_i + \lambda(\mathbf{x} - \mathbf{R}_i)))d\lambda = (y - Y_i)\frac{\Phi}{|\mathbf{C}_h|}. \tag{5.34}$$

We can now ask for the impact of an axial magnetic field on the Hamiltonian matrix elements, as defined in Eq. (4.10), within the tight-binding formalism. Using $\mathbf{k}, \mathbf{k}'$ as labels for Bloch functions $\phi_\mathbf{k}, \phi_{\mathbf{k}'}$, we write [68]

$$H_{\mathbf{k},\mathbf{k}'} = \langle \phi_{\mathbf{k}} | \hat{H} | \phi_{\mathbf{k}'} \rangle = \frac{1}{N} \sum_{i,j} e^{i(\mathbf{k}'\cdot\mathbf{R}_j - \mathbf{k}\cdot\mathbf{R}_i - i\frac{e}{\hbar}(G_{\mathbf{R}_i} - G_{\mathbf{R}_j}))} H_{ij}, \tag{5.35}$$

and

$$H_{ij} = \langle \varphi(\mathbf{x} - \mathbf{R}_i | \hat{H} | \varphi(\mathbf{x} - \mathbf{R}_j) \rangle. \tag{5.36}$$

The sites i and j are here understood to be adjacent to each other. Importantly, the Hamiltonian in this relation does not contain the vector potential, which is absorbed into the phase factor of Eq. (5.35). In view of the result (5.34), the effect of the phase factor induced by the vector potential is a *shift of the circumferential wave number* in the expression for the Hamiltonian matrix element. Explicitly,

$$k_y \rightarrow k_y + \frac{e}{\hbar} \frac{\Phi}{|\mathbf{C}_h|} = k_y + 2\pi \frac{\Phi}{\Phi_0 |\mathbf{C}_h|}, \tag{5.37}$$

keeping in mind that we align here the y-coordinate with the chiral vector, $\mathbf{C}_h$.

Exercise 5.2 Use Eq. (5.14) to show

$$k_y = \frac{2\pi}{|\mathbf{C}_h|} (j \pm \frac{\nu}{3} + \frac{\Phi}{\Phi_0}),$$

with $\nu = 0$ for metallic and $\nu = \pm 1$ for semiconducting tubes, and $j = 0, 1, 2, \ldots$ In terms of Φ_0, indicate the minimal amount of magnetic flux that makes an SWCNT with tube indices (9,8) metallic. Show: within the tight-binding approach, an axial magnetic field makes a metallic tube semiconducting unless Φ is an integer multiple of Φ_0.

Overall, the phase shift in the Hamiltonian matrix due to the Bohm–Aharonov effect becomes

$$(k_y + 2\pi \frac{\Phi}{\Phi_0 |\mathbf{C}_h|}) Y_i - (k'_y + 2\pi \frac{\Phi}{\Phi_0 |\mathbf{C}_h|}) Y_j = (k_y Y_i - k'_y Y_j) + (Y_i - Y_j) 2\pi \frac{\Phi}{\Phi_0 |\mathbf{C}_h|} . \tag{5.38}$$

The magnetic contribution to this phase shift implies that, as a function of the magnetic flux, metallic SWCNTs may become semiconducting, and vice versa. Theoretical considerations and recent experimental data confirming this prediction will be discussed in Chapter 8.

5.4.2 Perpendicular External Fields: Landau Sub-bands in Carbon Nanotubes

If the magnetic field is oriented in a direction perpendicular to the tube axis, the net magnetic flux through the tube vanishes and the expression for the phase shift $\Delta G = G_{\mathbf{R}_i} - G_{\mathbf{R}_j}$ that accomodates the magnetic effect has to be modified accordingly. In analogy to the axis-parallel case, we define a magnetic vector potential in the adapted two-dimensional coordinate system $\{\frac{\mathbf{T}}{|\mathbf{T}|}, \frac{\mathbf{C}_h}{|\mathbf{C}_h|}\}$, making the following choice:

$$\mathbf{A} = (\frac{|\mathbf{C}_h|B}{2\pi} \sin \frac{2\pi}{|\mathbf{C}_h|} y, 0). \tag{5.39}$$

Once more, the variable y runs along the chiral direction and is thus an arc length, represented in cylindrical coordinates as the product of the radial and the azimuthal coordinate, $r\varphi$. Substituting Eq. (5.39) into Eq. (5.32), we find

$$\Delta G = (\frac{|\mathbf{C}_h|}{2\pi})^2 B \frac{\Delta X}{\Delta Y} (\cos \frac{2\pi}{|\mathbf{C}_h|} Y - \cos \frac{2\pi}{|\mathbf{C}_h|} (Y + \Delta Y)), \quad \text{if } \Delta Y \neq 0, \tag{5.40a}$$

$$\Delta G = (\frac{|\mathbf{C}_h|}{2\pi}) B \Delta X \sin \frac{2\pi}{|\mathbf{C}_h|} Y, \quad \text{if } \Delta Y = 0. \tag{5.40b}$$

Here, $\Delta \mathbf{R} = \mathbf{R}_i - \mathbf{R}_j = (\Delta Y, \Delta X)$. These differences between the coordinates of nearest-neighbor atoms are readily obtained from the atomic locations in the hexagonal lattice (see, e.g. Table 5.1). While the vector $\Delta \mathbf{R}$ adopts only three different values, which are identical for A and for B sites, ΔG depends, for any given SWCNT in a magnetic field of constant magnitude, only on the coordinate Y. The k_x components in the Hamiltonian (5.35) remain thus unaffected by the presence of the perpendicular magnetic field, while the k_y components are mixed by it. Since there are N_{cell} discrete values for k_y where N_{cell} is the number of hexagons per nanotube cell (see Section 5.2), and two inequivalent sites, A and B, a $2N_{cell} \times 2N_{cell}$ Hamiltonian matrix must be diagonalized for every value of k_x to evaluate the band structure of a nanotube exposed to a perpendicular magnetic field.

For sufficiently small **B**, the impact of the field may be regarded as a minor perturbation. As **B** increases in magnitude, however, the onset of a novel phenomenon is observed, namely the formation of Landau levels (see Section 4.4). For metallic SWCNTs, and close to the Dirac points, electronic wave functions can be derived by analytical means, making it possible to describe the evolution of Landau levels quantitatively. Extending the formalism (4.44) – (4.47) to nanotubes, one obtains two decoupled equations at the K point [107]:

$$\begin{aligned} [-\frac{\partial}{\partial y} + \frac{|\mathbf{C}_h|}{2\pi l_B^2} \sin \frac{2\pi y}{|\mathbf{C}_h|}] \psi_{AK}(y) = 0, \\ [+\frac{\partial}{\partial y} + \frac{|\mathbf{C}_h|}{2\pi l_B^2} \sin \frac{2\pi y}{|\mathbf{C}_h|}] \psi_{BK}(y) = 0. \end{aligned} \tag{5.41}$$

Conveniently, the second expression on the left-hand side ($=\frac{e}{\hbar}\mathbf{A}(y)$) is written in terms of the magnetic length, as introduced in Section 4.4. The solutions for ψ_{Ak} and ψ_{Bk} are

$$\psi_{AK}(y) = \frac{1}{\sqrt{|\mathbf{C}_h| I_0(2\nu^2)}} \exp\{-\nu^2 \cos(\frac{2\pi y}{|\mathbf{C}_h|})\},$$
$$\psi_{BK}(y) = \frac{1}{\sqrt{|\mathbf{C}_h| I_0(2\nu^2)}} \exp\{+\nu^2 \cos(\frac{2\pi y}{|\mathbf{C}_h|})\}. \tag{5.42}$$

This solution for $\mathbf{k} = \mathbf{K}$ is readily continued into the vicinity of the Dirac point [107]. The symbol I_0 stands here for the *modified Bessel function of the first kind*, defined as

$$I_0(z) \equiv \int_0^{\pi} \frac{d\theta}{\pi} \exp(z\, cos\theta), \tag{5.43}$$

and the constant ν is given by $\frac{|\mathbf{C}_h|}{2\pi l_B}$. Since $l_B = \sqrt{\frac{\hbar}{eB}}$, ν^2 is proportional to the magnitude of the magnetic field. An interesting feature of the wave functions (5.42) is that they are increasingly localized as the parameter ν increases. This tendency can be directly monitored if the condition $\nu \gg 1$ is satisfied. In this case,

$$I_0(2\nu^2) \approx \frac{e^{2\nu^2}}{\sqrt{4\pi\nu^2}}. \tag{5.44}$$

In the high-field regime, $\psi_{AK}(y)$ ($\psi_{BK}(y)$) localizes around $y = \pm\frac{|\mathbf{C}_h|}{2}$ ($y = 0$), where the function attains its maximum. As the magnetic field increases in strength, the electronic states at $\mathbf{k} = \mathbf{K}$ are more and more sharply defined in space. The signature of this trend in the bandstructure is the gradual formation of dispersion-free bands (or *flat bands*). This behavior is clearly reflected in Figure 5.10 by bands calculated for three different strengths of a magnetic field applied at a right angle with respect to the tube axis.

From the definition of the crucial quantity ν, both a high magnetic field strength and a large tube diameter are factors that favor the emergence of Landau levels, and their eventual dominance over the nanotube band structure. To close this section with a quantitative example, if $\nu = 1$, a tube diameter of 10 nm corresponds to a field strength of 26 T.

5.4.3 The Orbital Magnetic Susceptibility of Carbon Nanotubes

As outlined in the two previous sections, the band structure of carbon nanotubes depends sensitively on the direction of the magnetic field with respect to the tube axis. This orientational effect has been investigated in terms of the *orbital magnetic susceptibility* (χ_{orb}), as a quantity that measures the influence of the field on the band energies. We distinguish the orbital from the Pauli susceptibility, χ_P, where the latter refers to the response of the spin system to the field. For regular SWCNTs, the orbital component has been found to dominate the spin component by several orders of magnitude [109] so that the approximation $\chi \approx \chi_{orb}$ is justified.

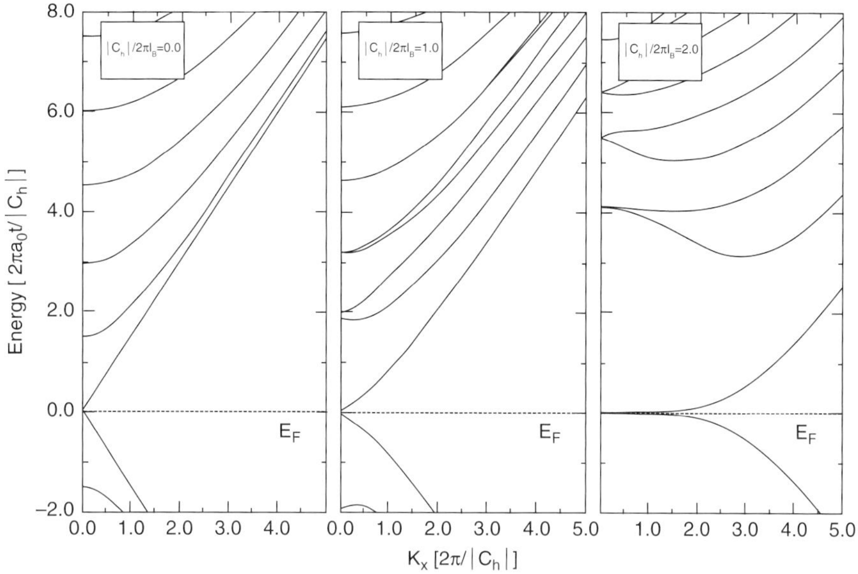

Figure 5.10 The formation of Landau sub-bands as a metallic SWCNT is exposed to a magnetic field oriented perpendicular to the tube axis. The left panel corresponds to vanishing, the right panel to maximum magnetic field strength [108]. The (axial) wave number is indicated in units of $\frac{2\pi}{|\mathbf{C}_h|}$, the energy in units of $\frac{2\pi}{|\mathbf{C}_h|} t a_0$. Reprinted with permission from [108]. Copyright (1995) by the Physical Society of Japan.

The quantity χ_{orb} has been derived from the relation [80]

$$\chi_{orb} \sim -\frac{\partial^2 FE(B,T)}{\partial B^2}|_{B=0}, \tag{5.45}$$

with FE as the thermodynamic potential of the grand canconical ensemble for noninteracting Fermions with energies E_n, defined as

$$FE(B,T) = -k_B T \sum_{n,k_a} \ln\{1 + e^{-\frac{E_n(k_a,B)-\mu}{k_B T}}\}. \tag{5.46}$$

The band energy E is here described as a function of the magnetic field. As implied by relations (5.45) and (5.46), the orbital susceptibility is due to band energy changes caused by the presence of a magnetic field. Analytical expressions for $\chi_{orb}^{\parallel}$ ($\chi_{orb}^{\perp}$), i.e. the susceptibility in case of a field oriented parallel (perpendicular) to the tube axis have been derived by Ajiki and Ando [108, 110] from formula (5.45). Lu [109] found a linear increase of χ_{orb} with the tube radius R and inspected the overall trend of the scaled orbital susceptibility χ_{orb}/R with the angle θ between the field direction and the tube axis. He arrived at a universal relation, namely

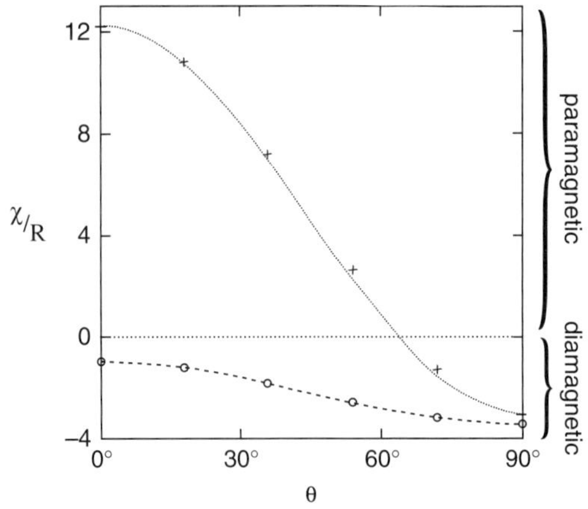

Figure 5.11 Calculated magnetic susceptibilities [109] for metallic (dotted line) and semiconducting (dashed line) SWCNTs. Shown is χ_{orb}/R, the scaled orbital susceptibility as a function of the angle between the direction of the magnetic field and the tube axis. While metallic tubes change from paramagnetic to diamagnetic behavior as θ increases from 0° to 90°, semiconducting tube are diamagnetic throughout. A marked magnetic anisotropy is found in both cases. The quantities χ_{orb} and R are in emu/g and Å, respectively. Reprinted with permission from [109]. Copyright (1995) by the American Physical Society.

$$\chi_{orb}(\theta) = a + b\cos(2\theta). \tag{5.47}$$

The two universal curves in Figure 5.11 refer to the angular susceptibility profiles of metallic ($n - m = 3l$, $l = 1, 2 \ldots$, as discussed earlier) and semiconducting ($n - m = 3l \pm 1$) tubes. As a general observation, all SWCNTs are diamagnetic for $\theta = 90°$, in accordance with the behavior of pristine graphene, as described in Section 4.4. In case of axis-parallel orientation, however, the two SWCNT types react to the field in qualitatively different ways. While semiconducting tubes remain diamagnetic throughout, metallic tubes are paramagnetic below a critical angle θ_{crit}.

The signature feature of sizably greater susceptibility anisotropy for metallic SWCNTs than for their semiconducting counterparts has been experimentally confirmed by means of magnetic linear dichroism spectroscopy [111].

5.5 Quantum Transport in Carbon Nanotubes

Nanotubes can be employed as transmission elements in nanocircuits [112, 113]. At sufficiently low temperatures, a conducting nanotube may be modeled as an ideal, one-dimensional *ballistic wire*. The latter is to be understood as a device with quantized transversal (or circumferential) and continuous longitudinal (or axial) motion. The ballistic condition involves free axial propagation, unhindered by scattering events. Intuitively, one might expect that this premise implies vanishing resistance, or infinite conductance. This, however, is not the case. The reason for the appearance of a finite conductance in ballistic transport systems lies in the interaction of the charge carriers with the electrodes, which may be described, in the most elementary approach, as electron reservoirs at local equilibrium, and thus characterized by constant chemical potentials μ_S and μ_D. The subscripts S and D denote here the *source* and the *drain* of the electronic connection, respectively. Thus, if the transmission channel is at the electric potential of the source, scattering will occur at the site of the drain that is at lower potential. Indeed, it has been confirmed by experimental inspection that there is no voltage drop inside a ballistic quantum channel [114]. Working within the *Landauer–Büttiker formalism* [115, 116], one represents the source and the drain by their associated Fermi–Dirac distributions, i.e. by

$$F(E, \mu_i) = \frac{1}{1 + e^{\frac{E-\mu_i}{k_B T}}} \tag{5.48}$$

and $i = S, D$. To obtain the current through the transmission channel, we sum over the current contributions of all nanotube sub-bands with energies E_i, such that $\mu_S > E_i > \mu_D$. Considering that a single electron in a sub-band i is associated with a current $I_i = e/t_i$, with t_i as the transit time of this electron, and that $v_i = \hbar^{-1}[\partial E_i(k_a)/\partial k_a]$, we approximate the net current as

$$I = \frac{e}{L} \sum_i \int \hbar^{-1} \frac{\partial E_i(k_a)}{\partial k_a} (F(E, \mu_S) - F(E, \mu_D)) dN. \tag{5.49}$$

The integration is here over the electronic states populating the sub-bands. Including the density of states (5.22), we turn this expression into

$$\begin{aligned} I &= \frac{e}{L} \frac{2L}{2\pi} \sum_i \int \hbar^{-1} \frac{\partial E_i(k_a)}{\partial k_a} (F(E, \mu_S) - F(E, \mu_D)) dk_a \\ &= \frac{2e}{h} \int (F(E, \mu_S) - F(E, \mu_D)) N_{ch}(E) dE. \end{aligned} \tag{5.50}$$

Assuming that the number of conduction channels, N_{ch}, is constant over the interval $[\mu_D, \mu_S]$ finally yields

$$I = N_{ch} \frac{2e^2}{h} \frac{\mu_S - \mu_D}{e}. \tag{5.51}$$

The chemical potentials may be defined with respect to the Fermi energy, E_F, as

$$\begin{aligned} \mu_S &= E_F + \frac{eV}{2} \\ \mu_D &= E_F - \frac{eV}{2}, \end{aligned} \tag{5.52}$$

and $\Delta\mu \equiv \mu_S - \mu_D = eV$. With these substitutions we arrive at the compact formula

$$I = \frac{2e}{h} N_{ch} \Delta\mu = \frac{2e^2}{h} N_{ch} V. \tag{5.53}$$

This relation defines the *quantum conductance*, $G_q \equiv \frac{dI}{dV}$:

$$G_q = \frac{2e^2}{h} N_{ch} = G_0 N_{ch}, \tag{5.54}$$

with $G_0 \equiv \frac{2e^2}{h}$ as basic unit. Accordingly, $R_0 \equiv 1/G_0 = 12.9 k\Omega$ is the basic unit of quantum resistance. Relation (5.54) is only valid under the assumption of *perfectly transparent contacts*. This is equivalent to setting all transmission coefficients T_i that determine the rate of transmitted versus reflected charge carriers in the channel i equal to unity. Generalizing Eq. (5.54), we write

$$G_q = \frac{2e^2}{h} \sum_i^{N_{ch}} T_i(E_F). \tag{5.55}$$

This statement is known as the *Landauer formula for a ballistic conductor* [117]. The number of channels available for transport in the low-energy regime of metallic SWCNTs, $E \approx E_F$, is two. This is obvious from, e.g. Figures 5.7 and 5.9, which show the degeneracy of two bands at $E = E_F$. As the energy rises, more channels open up for conduction, where the allowed values of the quantized circumferential wave number k_c define the threshold energies for the population of new transport channels, and thus, conductance increase. The emerging step profile of the conductance in metallic SWCNTs is shown in Figure 5.12.

5.6 Carbon Nanotubes as Quantum Dots

While ideal, periodic nanotubes are naturally one-dimensional, nanotubes used as transmission elements are axially confined and thus reduced to zero dimensions. If electrons can be trapped in these structures, they may act as *quantum dots*, with interesting perspectives for fabricating nanoelectronic devices, and also with features of fundamental relevance for many-particle physics. To construct a carbon nanotube quantum dot, one may proceed by placing the tube between metallic source and drain contacts, as shown in Figure 5.13. A gate that is capacitively coupled to the nanotube provides a third elecrode. Through the interplay of the Schottky barriers in the contact zone between the nanotube and the metal leads and the gate voltage that tunes the electrostatic potential in the center of the tube, this

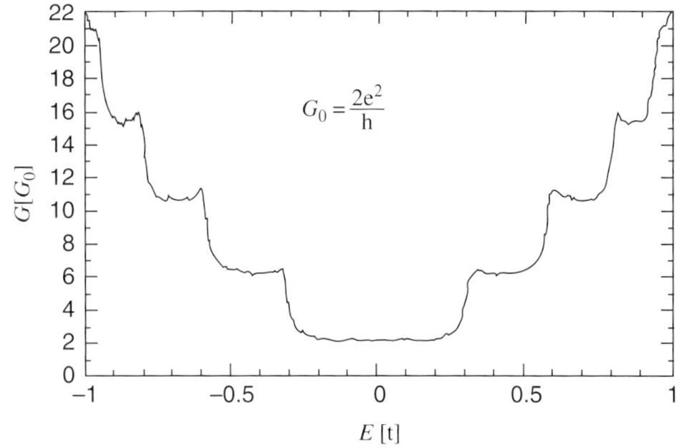

Figure 5.12 The maximum conductance of a metallic SWCNT in the ballistic limit versus energy, in accordance with Eq. (5.54). The step profile reflects the discrete nature of the sub-bands due to the quantization of the circumferential degree of freedom in nanotubes. Reprinted with permission from [118]. Copyright (2007) by the American Physical Society.

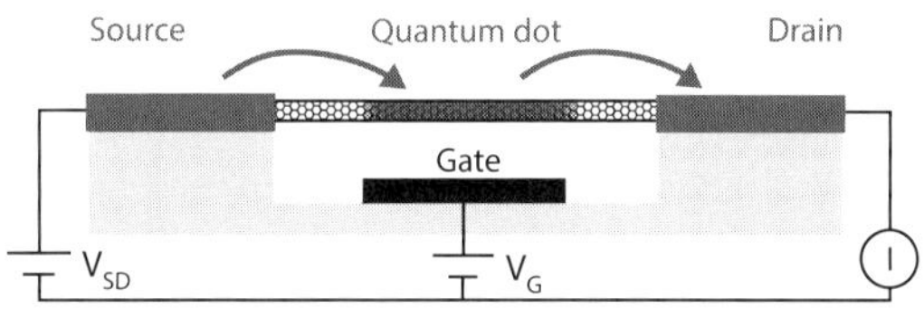

Figure 5.13 Sketch of a quantum dot device based on a nanotube transmission element. Schottky barriers between the tube and the two metallic leads, in conjunction with a capacitively coupled gate, confine charge carriers in a segment of the nanotube, which thus acts as a quantum dot. Reprinted with permission from [121]. Copyright (2015) by the American Physical Society.

configuration may realize a quantum dot, confining a discrete number of charge carriers to a segment of the tube [119, 120].

Charge carriers that flow from the source to the drain have to tunnel through the contact zones between the tube and the electrodes, at tunneling rates Γ_S and Γ_D for the interfaces at the source and the drain side, respectively. If the *charging energy* E_C dominates the tube-electrode coupling energy $\hbar\Gamma_{S,D}$ as well as the thermal energy $k_B T$, the electron population of the dot adopts an integer value [121]. The charging energy is obtained from the overall capacitance of the dot, C, by $E_C \equiv e^2/C$. The capacitance C, in turn, is understood as the sum of the source, drain, and gate capacitances: $C = C_S + C_D + C_G$. The gate-induced deviation of the electron number from its equilibrium value leads to an increase of the electrostatic energy of the device and blocks the current. This so-called *Coulomb blockade* effect interrupts the charge flow from source to drain, except in cases where the electrochemical potential associated with the Nth occupied discrete level of the dot, $\mu(N)$, lies between the electrochemical potential of the source and the drain: $\mu_S > \mu(N) > \mu_D$.

The electrochemical potential $\mu(N)$ is given here by the difference between the ground state energies U of the dot with an occupation of N and $N - 1$ electrons,

$$\mu(N) = U(N) - U(N-1). \tag{5.56}$$

Evaluating the quantum dot ground state energy from first principles is, in general, a challenging many-particle problem. The *constant-interaction model* however, leads to approximate solutions that have been found to capture the essential features of the quantum dot device sketched in Figure 5.14 [122]. Its premises are (1) the overall capacitance C accounts for all Coulomb interactions between the electrons in the dot as well as between these electrons and the environment and (2) the single-particle spectrum of the electrons is not altered by these interactions. In the framework of this model, the ground state energy for a dot with N electrons becomes [123]

$$U(N) = \frac{(-e(N-N_0) + C_S V_S + C_D V_D + C_G V_G)^2}{2C} + \sum_i^N E_i, \tag{5.57}$$

with N_0 as the electron occupation of the dot in the limit of vanishing voltages ($V_S = V_D = V_G = 0$) and $\{E_1, \ldots, E_i, \ldots E_N\}$ as the succession of single-electron levels. Note that $C_j V_j$ ($j = S, D, G$) signifies the charge induced on the dot by the agent j. Combining the definition of the charging energy E_C with formulae (5.56) and (5.57), we obtain the electrochemical potential for a dot filled with N electrons:

$$\mu(N) = (N - N_0 - \frac{1}{2})E_C + \frac{E_C}{e}(C_S V_S + C_D V_D + C_G V_G) + E_N. \tag{5.58}$$

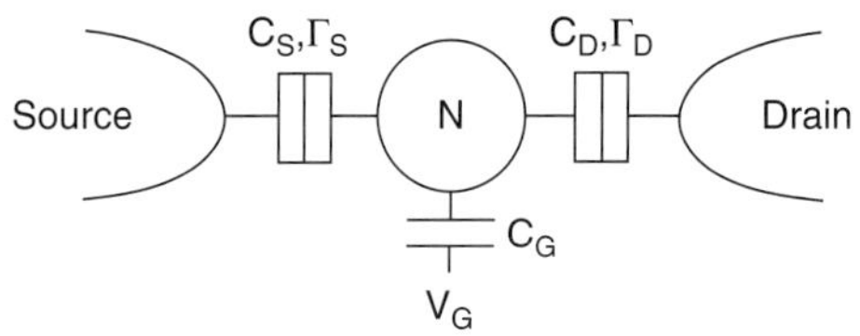

Figure 5.14 Electrical scheme of the nanotube quantum dot (N). The interface between the dot and the source/drain is defined by the tunneling rates Γ_S/Γ_D, respectively. Each of the confining agents, the source, the drain and the gate are specified by a characteristic capacitance C_j ($j = S, D, G$). Reprinted with permission from [121]. Copyright (2015) by the American Physical Society.

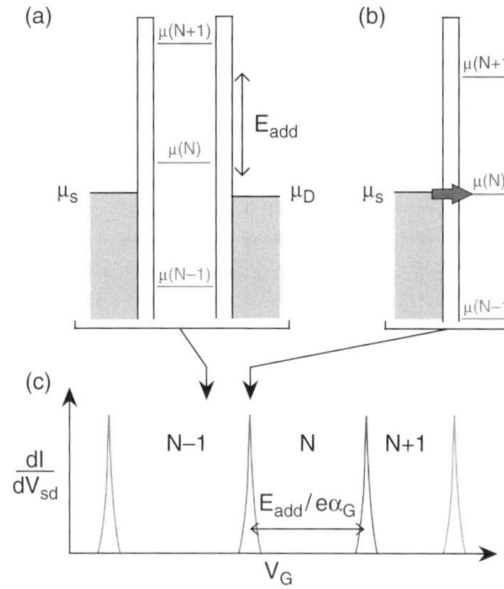

Figure 5.15 A nanotube as a quantum dot device under conditions of low-bias spectroscopy. (a) Sequence of electrochemical potential levels for electrons $N-1$, N, and $N+1$: the case of Coulomb blockade. (b) The level $\mu(N)$ is located within the bias window: the case of dot conductance. (c) The succession of transport peaks as a function of the gate voltage. The *addition energy*, E_{add}, refers to the difference between subsequent electrochemical potentials. Reprinted with permission from [121]. Copyright (2015) by the American Physical Society.

All levels lower than that of the drain, μ_D, are filled. For the case of low-bias spectroscopy ($\mu_s \approx \mu_D$), the hierarchy of electrochemical potential levels, as described by Eq. (5.58), is illustrated in Figure 5.15. In this case, $\Delta\mu(i) \equiv \mu(i) - \mu(i-1) >> \mu_s - \mu_D$, which guarantees that maximally one electrochemical potential level of the dot lies within the bias window. From Eq. (5.58) we infer the difference between subsequent electrochemical potentials, E_{add}:

$$\Delta\mu(N) = E_C + E_N - E_{N-1} \equiv E_{add}. \tag{5.59}$$

The constant-interaction model predicts that the sequence of the levels $\mu(i)$, as sketched in panel (a) of Figure 5.15, may be rigidly shifted up or down by manipulating the gate voltage V_G. This makes it possible to satisfy the "cascading condition" $\mu_S > \mu(N) > \mu_D$, making the dot conductive. Figure 5.15 displays this in terms of electrochemical potential matching (5.15 (b)), and by a representation of the dot conductance versus the gate voltage (5.15 (c)). From the latter plot, the spectrum $\mu(i)$ can be traced by sweeping the gate voltage through the range of interest and recording the dot conductance. By relations (5.58) and (5.59), the gate voltage difference $\Delta V_G(i)$ between adjacent conductance peaks is, for $i = N$,

$$\Delta V_G(N) = \frac{E_{add}}{e\alpha_G}, \tag{5.60}$$

and $\alpha_G = \frac{C_G}{C}$.

More complex transport phenomena than mentioned so far may be investigated in the *high-bias spectroscopy* mode. This applies whenever the bias, i.e. the quantity eV_{SD}, exceeds the single-particle level spacing of the dot. If this condition is realized, an electron injected into the dot from the source may occupy an excited state of the dot. In this case, an additional electrochemical potential level, $\mu'(N)$, may open a new pathway for conduction [121]:

$$\mu'(N) = \mu(N) + E_{N+1} - E_N. \tag{5.61}$$

In the high-bias setting, monitoring the transport peaks, as indicated for the ground state configuration of the dot in Figure 5.15 (c), makes it possible to determine the excited-state energies of the dot as well as to explore electron transport in an excited state of the quantum dot.

6

Fullerenes

Any hollow, cage-like molecule consisting entirely of carbon atoms in a state of sp^2 hybridization and composed of five- and six-membered rings is referred to as a *fullerene* [124]. This class of carbon nanostructures thus comprises the quasi-spherical species that are popularly known as *buckyballs* as well as axially confined carbon nanotubes, among other systems. As every carbon atom on the fullerene surface is threefold coordinated, a fullerene may be understood as a rolled-up graphene sheet, as illustrated in Figure 6.1 by the example of a buckyball. *Euler's theorem*, applied to polyhedra that consist exclusively of hexagons and pentagons, confines the number of pentagons to twelve [125], while there may be an arbitary number of hexagons, excepting a single one [126]. This implies that the smallest fullerene of this type is a combination of twelve pentagons, C_{20}. *Quasi-fullerenes* contain rings beyond pentagons and hexagons, such as heptagons [127] or octagons [128]. The effect of these structural irregularities is a major enhancement of fullerene reactivity.

6.1 C_{60}

The most readily available fullerene species, and the prototypical molecular allotrope of carbon, is C_{60}, also known as *Buckminster fullerene*, making fully explicit the name patron of the fullerenes, the American architect, author and visionary Richard Buckminster Fuller (1895–1983), whose name is famously associated with designs based on geodesic dome structures. The C atoms of C_{60} are positioned at the vertices of a regular truncated icosahedron. The point group of C_{60} is I_h, the number of symmetry operations of the molecule, i.e. operations that map C_{60} into itself, is 120.

Fullerenes satisfy the *Isolated Pentagon* rule, stating that structures with pentagons entirely surrounded by hexagons are more stable than those that admit pentagon adjacency. The geometry of C_{60}, where all pentagons are isolated is

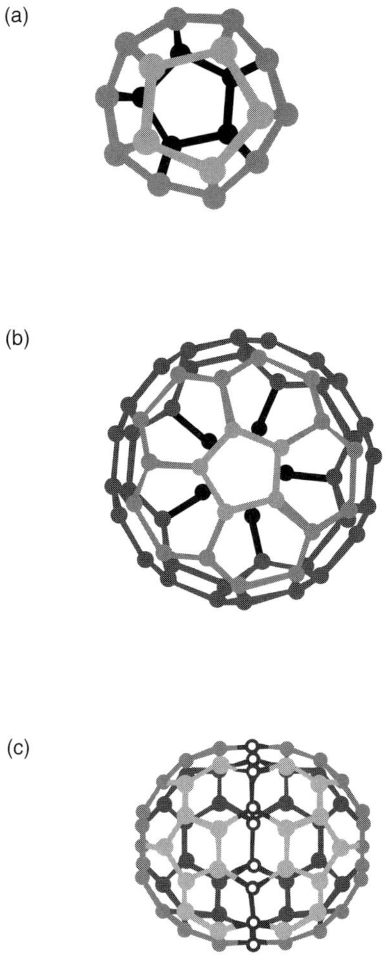

Figure 6.1 Some representative fullerene molecules: (a) C_{20}, the smallest fullerene, (b) C_{60}, or Buckminster fullerene, (c) C_{70}.

unique among the fullerenes C_n with $n \leq 70$. For topological reasons, all other species of this class involve adjacent pentagons. While pure cages containing these substructures are deemed too unstable to be fabricated in the laboratory [129], they can be stabilized by external or internal impurities (e.g. [130, 131]).

The average bond length between the atoms of C_{60} is 1.44 $\AA$ [132]. Each atom forms two single bonds and one double bond with adjacent atoms. As the double bonds are shared by two hexagons, and the single bonds by a hexagon and a pentagon, there is an alternation between both bond types around each of the hexagonal faces, associated with a periodic change of the bond length between 1.40 $\AA$ and 1.46 $\AA$ [133]. The absence of dangling, or unsaturated, bonds on the C_{60} surface implies that two C_{60} molecules interact by van der Waals forces.

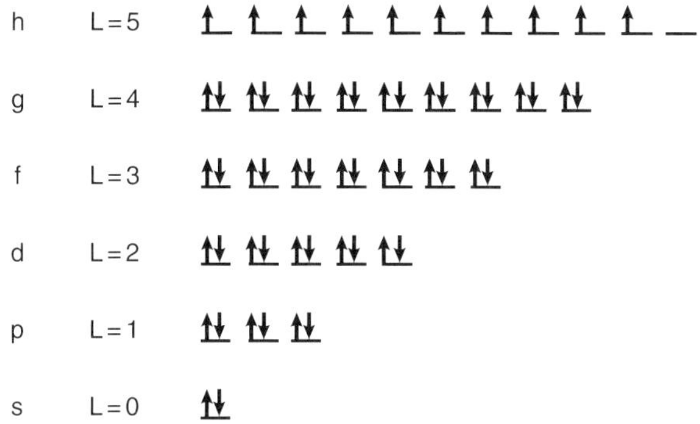

Figure 6.2 Scheme of energy levels for the π electron subsystem of C_{60}, based on the premise of a spherical potential [135]. The energy levels are labeled by their angular momentum, which is assumed to be a good quantum number.

Each double bond involves a bond of π character, and each carbon atom contributes one electron to the π subsystem of C_{60}. The presence of sixty π electrons and ninety edges implies an abundance of resonance structures (12,500). The electronic ground state solution of C_{60} might be understood as a superposition of numerous alternative electronic structures that differ from each other with respect to the spatial arrangement of the π subsystem, which therefore is largely delocalized. It is instructive to describe this subsystem in the framework of a *superatom approach*, assuming that the π electrons of C_{60} are exposed to a potential of perfect sphericality. As spin-orbit coupling is neglected, this approximation leads to the energy level scheme depicted in Figure 6.2. Hund's rule predicts a strongly paramagnetic ground state, in conflict with experimental evidence [134].

As the spherical symmetry is lowered to that of the truncated icosahedron, the degeneracy of the terms with $L = 3 - 5$ is lifted, and the observed non-paramagnetic, closed-shell arrangement is predicted.[1] The term with $L = 5$ is split into one sublevel of H and two sublevels of T symmetry, as indicated in Figure 6.3. The latter correspond to virtual orbitals that may be occupied by twelve π electrons shared among the twelve pentagons of C_{60}, converting each pentagon into a 6π aromatic ring [136]. These low-energy virtual states account for the high electron affinity of the molecule.

On the basis of both measurement and computation, C_{60} has been characterized as an aromatic molecule with a diamagnetic susceptibility of extremely small size, far below the susceptibilities of graphite or benzene. The almost vanishing diamagnetic effect is due to a near-cancellation of a diamagnetic susceptibility, originating

[1] Strictly speaking, the structure of C_{60} differs from that of a regular truncated icosahedron in that the two types of bonds present in the molecule have slighty different lengths.

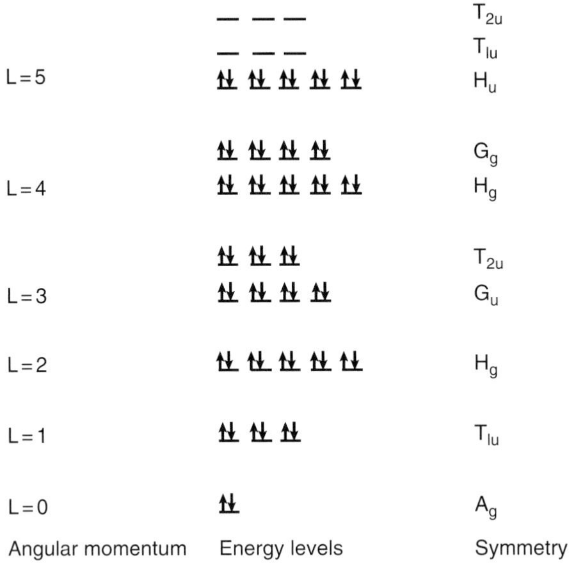

Figure 6.3 Scheme of energy levels for the π electron subsystem of C_{60}, assuming a potential of icosahedral symmetry [135].

from π electron ring currents, and a paramagnetic contribution due a vibronically coupled excited state of T_{1g} symmetry. This leaves an overall susceptibility of $\chi = -0.35 \times 10^{-6}$ emu/g, as obtained by SQUID magnetometry of a solid sample of pure C_{60} [137, 134]. This effect is specific to C_{60}. For C_{70}, for instance, a susceptibility on the order of twice that of C_{60} is reported [134, 137].

Representing the C atoms as points and adopting the two bond lengths indicated earlier, one arrives at a C_{60} diameter of 7.09 Å. As one includes the spatial extension of the π electron cloud, the resulting *Van der Waals diameter* may be estimated to be 10.34 Å. Here, the value for the size increment due to the presence of π electrons has been obtained from the vertical distance between adjacent graphite layers. This is in good agreement with experimental observation. Thus, an assessment of the C_{60} diameter based on measuring the electrical mobility of C_{60} in the gas phase yielded a value of 10.1 Å [138].

6.1.1 C_{70} and Larger Fullerenes

When fullerenes are fabricated in the laboratory from carbon-rich vapors or plasmas, a variety of C_n species is formed. While C_{60} tends to be the most abundant among them, it is not the most stable in terms of thermodynamic stability. Thus, the rugby-ball shaped molecule C_{70} (see Figure 6.1(c)), usually detected as the second most abundant fullerene, exceeds the C_{60} binding energy per carbon atom (7.40

eV/C atom) by about 0.02 eV. In spite of this energetic advantage over C_{60}, the formation of the latter is kinetically favored. While all atomic sites of C_{60} are symmetry equivalent, C_{70} contains five inequivalent sites [139]. Further, C_{70} exhibits eight distinct bonds, distinguished by different bond lengths, while only two types of bonds are present in C_{60}. One might imagine the emergence of C_{70} from C_{60} as the addition of five hexagons around an equatorial plane of C_{60} that is perpendicular to one of the fivefold axes of the molecule. This deformation is associated with a reduction of the symmetry group from I_h to D_{5h}.

The next fullerene that conforms with the isolated pentagon rule is C_{76}. This unit, as well as cages of greater size, is represented in carbon-rich vapors or plasmas with significantly lower abundance than C_{60} or C_{70}. On the other hand, the number of isomers for a given C_n species increases with increasing fullerene size. Thus, selecting structures that contain exclusively hexagons and pentagons and further obey the isolated pentagon rule, we obtain one isomer each for C_{60}, C_{70} and C_{76}, with symmetry groups I_h, D_{5h} and D_2, respectively, but seven isomers for C_{80}. Their symmetry groups are (listed in the order of decreasing stability): D_{5d}, D_2, C_{2v}, D_3, C_{2v}, D_{5h} and I_h [140].

6.2 Endohedral Fullerenes

Fullerene molecules encapsulating other species have been studied extensively since techniques of fabricating fullerenes in macroscopic amounts [141, 142] became available. Numerous computational and experimental studies have since dealt with these *endohedral fullerenes*, and a wide variety of potential technological applications have been identified for them, ranging from biomedical use as novel contrast agents in magnetic resonance imaging (MRI) [143] to the realization of qubits in quantum computing [144]. While most of the work on endohedral fullerene has dealt with metallofullerenes, i.e. carbon cages enclosing one or several metal atoms, significant efforts have also been devoted to other than metallic impurities. In the following, we summarize basic findings on metallofullerenes and fullerene-encaging nonmetal species.

6.2.1 Metallofullerenes

Composites of the form M@C_n, where C_n stands for a fullerene, were first synthesized by vaporization methods, such as laser ablation, applied to a metal-oxide/graphite mixture [145]. Since the first discovery of La@C_{82} in 1991 [146], it has been recognized that many M@C_{82} systems can be generated in macroscopic quantities and exhibit solubility in organic solvents as well as stability at ambient temperature. Further examples for monometallic guest species in C_{82} include

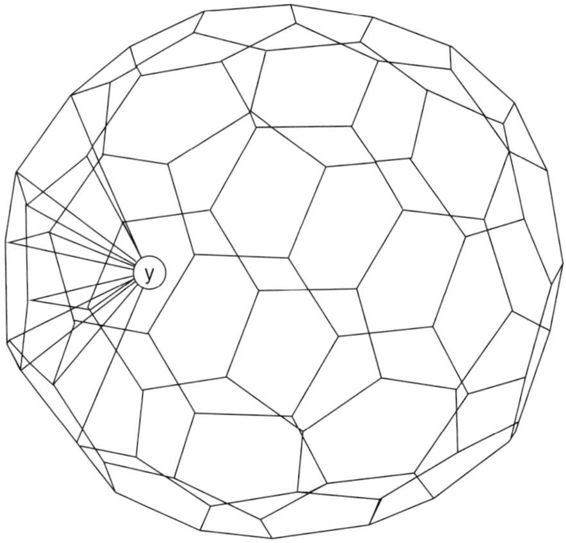

Figure 6.4 Sketch of the equilibrium structure of Y@C_{82}, as obtained by DFT geometry optimization. Reprinted from [154] with permission from Elsevier.

M = Ca, Tm, Sm, Yb, Er and Eu, and among the fullerene cages that were found to host single metal atoms are C_{60}, C_{72}, C_{74}, C_{80}, C_{84}, C_{92} and C_{94} [142].

As substantiated by various experimental and computational studies, the enclosed metal atom does not necessarily reside in the fullerene center. For instance, Rietveld analysis in conjunction with the maximum entropy method (MEM) gave evidence of a displacement of Y in Y@C_{82} from the center position toward the cage ([147], see Figure 6.4), as confirmed by computation [148], and ascertained for several other species, among them Sc@C_{82} [149]. Due to the interaction between the C_{82} cage and the enclosed metal species, the C_{82} geometry of M@C_{82} in the ground state deviates from the most stable structure of the pure C_{82} isomer. More specifically, the bonding between the metal core and the fullerene cage is strongly influenced by electron transfer between the metal atom and the C_{82} enclosure. For instance, in case of M@C_{82} with M = Y or La, this redistribution of charge results in the effective formula M^{3+}@C_{82}^{3-}, which has been verified using a wide array of experimental techniques [150, 151, 152].

With reference to La, this charge redistribution seems to result from the almost complete transfer of the three valence electrons ($5d^1$, $6s^2$) of the endohedrally connected atom to the fullerene cage, consequently causing the cage to adopt the structure of highest stability for the C_{82}^{3-} unit, which is characterized by C_{2v} symmetry, while the neutral C_{82} unit prefers C_2 symmetry [148]. However, experimental [153] as well as computational [154] results lead to the conclusion that the interaction between the fullerene cage and the metal core is not adequately described by a pure-charge transfer model. As demonstrated by DFT analysis, the two 6s electrons of the La atom are donated to the lowest unoccupied molecular orbital (LUMO) of C_{82}. The d electron, however, is primarily shared with the π orbitals of the cage, reflecting marked C_{82}-La(d) hybridization. The state occupied by the single unpaired electron of La@C_{82} consists predominantly of contributions from C_{82} and is delocalized, spreading over the fullerene cage. This finding agrees with resonant photoelectron spectroscopy (*resonant PES*) experiments [153] that yielded substantial La character in the occupied part of the La@C_{82} valence band. For lanthanide atoms (Ln) encapsulated by C_{82}, a formal charge state of +3e has been proposed for Ln = La, Ce, Pr, Nd, Gd, Tb, Dy, Ho, Er and Lu [152, 156, 157, 158, 159, 160]. Ln = Sm, Eu, Tm and Yb, in contrast, are seen to adopt a charge of +2e in the C_{82} cage [152, 157, 161]. The electronic configuration of Ln = La, Ce, Gd, and Lu may be described by [Xe]$6s^2 4f^n 5d^1$, with $n = 0$ (La), 1(Ce), 7 (Gd), or 14 (Lu). In these cases, the valence system beyond the xenon-like core comprises a lanthanum-like configuration of 6s and 5d electrons. Figure 6.5 illustrates the charge transfer scheme for these cases. In this simplified picture, the 5d electron of the metal impurity migrates into the one-dimensional LUMO+1 level of the pure cage in C_{2v} symmetry which becomes the highest occupied molecular orbital of the composite, while the two 6s electrons populate the LUMO level of the cage. Thus, as a consequence of electron transfer, the cage acquires a non-zero spin $S = \frac{1}{2}$.

Analogous observations have been made on M@C_{82} with M = Sc,Y where Sc(Y) has been assigned an effective charge of +2e(+3e) [146]. Further, DFT studies on other fullerene cages have resulted in findings very similar to those obtained for M@C_{82} (M = Ln, Y, Sc) [162, 155].

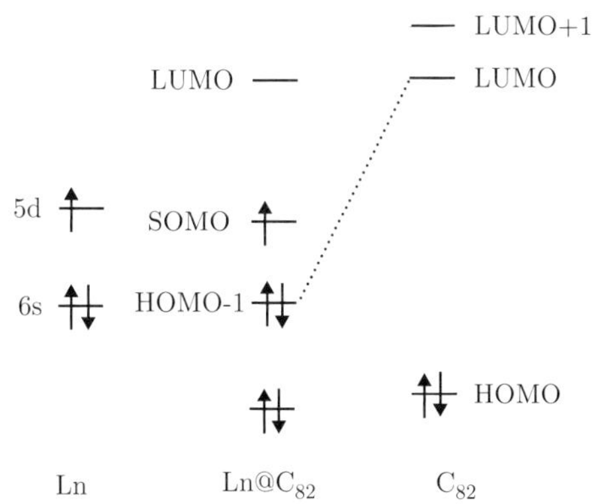

Figure 6.5 Charge transfer scheme for Ln@C_{82}, with Ln = La, Ce, Gd or Lu. The middle panel shows the three highest occupied molecular orbital levels of Ln@C_{82}, as result of the transfer of three electrons from the $6s^2$ and $5d^1$ orbitals of the lanthanide guest atom to the cage.

During the last two decades the research on fullerenes with endohedral metal systems has shifted its focus from monometallofullerenes to fullerenes enclosing metal atom clusters, acknowledging that fullerene cages are capable of accommodating more than one metal atom. Among the well-characterized group of dimetallofullerenes, the abundant species $La_2@C_{80}$ [163], $Sc_2@C_{74}$ or $Sc_2@C_{84}$ [164] can be regarded as prototypical. Experimental research [165] allows us to draw conclusions with respect to the intra-fullerene dynamics of the encaged metal atom pairs. More specifically, the two La ions of $La_2@C_{80}$ are free to circulate inside the C_{80} (I_h) cage with a very small activation energy of 5 kcal/mol. The magnetic field associated with two ^{139}La cations rotating inside the cage was shown to cause line broadening of the respective NMR signal with increasing temperature [166].

The trimetallofullerene $Sc_3@C_{82}$, detected as early as 1992 [167], was the first example of a polyatomic metal cluster enclosed in a fullerene cage. From synchrotron x-ray measurements [168], it has been established that the three endohedral Sc atoms form an equilateral triangle and are encapsulated in a C_{82} (C_{3v}) isomer. As demonstrated by electron spin resonance (ESR), the molecular charge state of $Sc_3@C_{82}$ can be approximated as $Sc_3^{3+}@C_{82}^{3-}$. However, computational analysis of the electronic structure of this system [169] yielded a considerable degree of cage-core hybridization.

The capacity of the fullerene enclosure to isolate and stabilize highly reactive chemical species has been manifested by a group of fullerenes that enclose trimetallic nitrides [170, 171]. Units of the form $(M_1)_x(M_2)_{3-x}N@C_{80}$ ($x = 0$-3), where M_1 and M_2 stand for two metallic elements, have been produced efficiently by introducing N_2 gas into a Krätschmer–Huffman generator utilized to manufacture the metallofullerenes. In this way, the species $Er_xSc_{3-x}N@C_{80}$ ($x = 0$-3) [172, 173, 174] have been created, with yields that surpassed those of $Sc_3@C_{82}$ by up to two magnitudes. The spectroscopic signatures of $Sc_3N@C_{80}$ (see Figure 6.6), as determined from ^{13}C NMR experiment, showed the whole unit to preserve the icosahedral symmetry of the C_{80} enclosure. This finding was ascribed to internal motion of the Sc_3N cluster in the carbon cage, leading to a time-averaged electronic environment that leaves the I_h symmetry of C_{80} invariant [170, 171].*Trimetallic nitride endohedral fullerenes* of composition $(M_1)_x(M_2)_{3-x}N@C_{80}$ (x = 0-3) with M_1, M_2 = Sc, Y, La, Gd, Ho, Er, Tm, Lu in various combinations are commercially fabricated [170, 175].

Fullerenes Encaging Nonmetal Atoms

Three major subgroups of nonmetal guest species in endohedral fullerenes may be distinguished: noble gas atoms, atomic radicals such as N or P, and nonmetal molecules. The methods of synthesizing these complexes differ from those used to

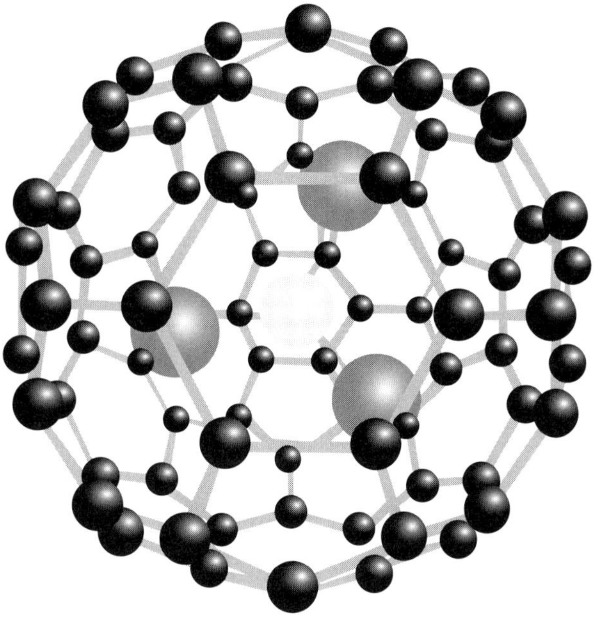

Figure 6.6 Equilibrium structure of $Sc_3N@C_{80}$.

generate metallofullerenes. In particular, ion bombardment has been employed for this purpose [176], besides high-pressure treatments, where fullerenes are heated in gas atmospheres at elevated pressure [177], and chemical protocols to open the carbon cage for insertion of the guest species [178].

Since 1993, several fullerenes with noble gas atoms in their interiors have been designed, among them the series $Ng@C_{60}$, with Ng = He, Ne, Ar, Kr and Xe. From ^{13}C NMR analysis of these species, the presence of noble gas atom impurities is manifested by a single downshifted line, indicating that the noble gas atom occupies the fullerene center. In all cases examined, the cage-core interaction has been found to be extremely small. This also holds for the radical-containing species $N@C_{60}$ and $P@C_{60}$, which, from ESR spectroscopy, leave the electronic structure of the encapsulated atoms essentially intact [176, 179].

Molecular endofullerenes are represented by $H_2@C_{60}$ [178] or $H_2O@C_{60}$ [180]. These systems offer the opportunity to study small molecules confined by an approximately spherical potential, as has been demonstrated by infrared (IR) spectroscopy applied to H_2 and D_2 in C_{60} and C_{70} [181].

6.2.2 Exohedral Fullerene Compounds

The C_{60} molecule is an electron acceptor, a property that is widely used in organic solar cell design [182]. This trend is reflected by a relatively high first vertical electron affinity of 2.65 (±0.05) eV, as experimentally determined [183], and also documented by the energy term scheme presented in Figure 6.3. The closed HOMO level subshell of H_u symmetry is followed by low-lying T_{1u} and T_{2u} level that have a combined capacity for twelve electrons, corresponding to the twelve pentagons in the fullerene surface that are electron-deficient, as compared with the hexagons (see the earlier discussion). This suggests the possibility of coating C_{60} with several electron donor atoms, where alkali metal atoms present themselves as the most obvious choice. Indeed, enhanced stability has been found for the species $Li_{12}C_{60}$, a fullerene with twelve Li atoms, such that each one is attached to a pentagon, by mass-spectrometric observation [184].

A different architecture is realized for alkaline earth metal atoms coating C_{60} [185]. One Ca or Ba atom, for example, has been found to attach to each polygon of the C_{60} surface. Accordingly, pronounced stability has been reported for $AE_{32}C_{60}$ [185], with AE = Ca, Ba, in photoionization time-of-flight mass-spectrometry experiments. This structural model is confirmed by the species $Ca_{37}C_{70}$. This stoichiometry is plausible, considering that C_{70} emerges from C_{60} by addition of five hexagons (see earlier, p. 135).

Higher exohedral fullerenes exhibit a multiply nested, onion-like structure. For C_{60} in combination with Ca atoms, the second metal layer is generated by adding another Ca atom to each pentagon site and three Ca atoms to each hexagon site, forming closely packed four-atoms substructures. Units as large as $Ca_{448}C_{60}$ have been isolated [185]. The latter species displays a hierachy of metal atom shells that may be described by the formula $C_{60}@Ca_{32}@Ca_{72}@Ca_{132}@Ca_{212}$ [186].

6.3 Fullerene Crystals

C_{60} molecules are the building blocks of face-centered cubic (FCC) fullerene crystals with a lattice constant of 14.17 Å and a center-to-center distance between adjacent units of 10.02 Å [187]. This solid emerges from vapor sublimation of C_{60} molecules. Other solid phases of Buckminster fullerene can be generated by use of other fabrication methods. Thus, single crystals consisting of C_{60} may also be manufactured from solution by use of solvents such as CS_2 and toluene. This leads to a crystal with orthorhombic structure and lattice constants a = 25.01 Å, b = 25.58 Å and c = 10.00 Å [188]. As these crystals are heated to higher temperatures than 373 K in vacuum, they transform into the FCC solid.

From NMR studies [189], the molecules in the FCC fullerene lattice at room temperature rotate rapidly about their midpoints, unhindered by the weak Van der Waals forces that act between them. However, as the temperature is reduced below 261 K, orientational ordering of the fullerene molecules is observed [190] corresponding to lattice symmetry lowering, as FCC order changes into the simple cubic structure [191]. More specifically, the molecules still locate at FCC sites, but the unit cell combines four of them, all differing with respect to their spatial orientations.

The weak forces between the C_{60} molecules imply that the electronic overlap between next neighbors is small. Therefore, the widths of the electronic bands are quite narrow, not exceeding 0.5 eV [192], and the molecular orbitals give rise to well-separated bands, excepting cases of near-degeneracy.

We add that the crystal structures of higher fullerenes tend to be of greater complexity than that of C_{60}. Solid C_{70} is the most extensively studied material among them [193]. In this case, the FCC phase has been found to be stable above 350 K. Below this threshold, C_{70} exists as mixture between FCC, rhombohedral, and monoclinic phases.

6.3.1 Alkali Metal Fullerides

If the shape of C_{60} is approximated by a sphere of well-defined radius $R_{C_{60}}$, the regions outside the spheres in the fullerene crystal amount to about 25 percent of the overall volume. The empty spaces surrounding octahedral and tetrahedral sites between the spheres have approximate radii of about 2.1 $\AA$ and 1.1 $\AA$, which is in the range of the alkali cation radii, which, from Na to Cs, vary between 1.0 and 1.7 $\AA$. As alkali atoms coating C_{60} donate electrons to the T_{1u} orbitals of the molecule, interstitial alkali impurities in the C_{60} crystal transfer electrons to its T_{1u} band. These atomic impurities may be located at tetrahedral or octahedral sites, as illustrated in Figure 6.7. At these places, the guest atom is coordinated to four or six next-neighbor molecules, respectively. As the sum over the tetrahedral and octahedral sites in the unit cell of the FCC lattice is twelve and the number of C_{60} molecules in the unit cell of the C_{60} crystal is four, the stoichiometry for an alkali-atom-doped C_{60} crystal is A_3C_{60} if each interstitial site is occupied by exactly

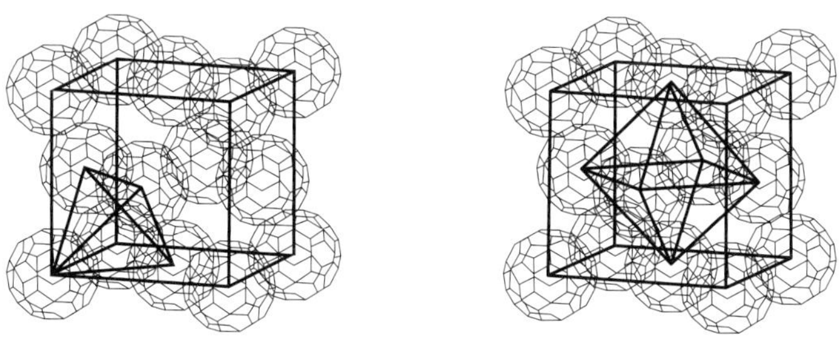

Figure 6.7 Representation of tetrahedral (a) and octahedral (b) sites in an FCC C_{60} crystal. The coordinates of these sites are $(\frac{1}{4}, \frac{1}{4}, \frac{1}{4})a$ and $(\frac{1}{2}, \frac{1}{2}, \frac{1}{2})a$, with a as the lattice constant. There are eight tetrahedral and four octahedral sites in the FCC unit cell.

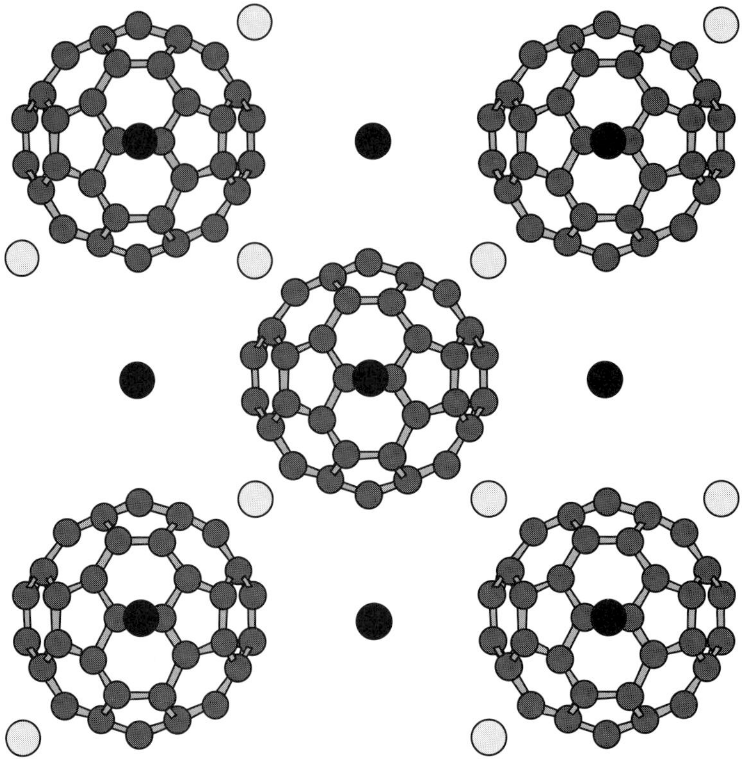

Figure 6.8 Sketch of K_3C_{60}. The potassium-atom-doped crystal maintains the FCC structure of the pure C_{60} lattice at room temperature. Interstitial atoms at octahedral (tetrahedral) sites are represented as black (grey) spheres.

one atom (see Figure 6.8). This formula is realized for *alkali metal fullerides* with A = K, Rb [194], where the matching between the ionic radius of the impurity and the size of the interstitial vacuum in the fullerene lattice is at its optimum.

The conductivity of fullerides is sensitively determined by the electron transfer within a compound. Thus, the A_3C_{60} structures with A = K, Rb, as well as Cs, are metals, as the guest atoms donate three electrons to each fullerene molecule, which results in a half-filled T_{1u} orbital. The metallic character of these composites emerges from electron hopping between neighboring C_{60} molecules, where conduction takes place within the narrow T_{1u}-derived bands.

Other stoichiometries than A_3C_{60} prevail for smaller or larger alkali atom impurities. Also, for potassium and rubidium dopants, alternative stoichiometries have been observed, as more than one atom may occupy a given interstitial site. Examples are provided by K_4C_{60} or K_6C_{60}. These compounds stabilize in other than the FCC structure, preferring a body-centered tetragonal (BCT) [195] or a body-centered cubic (BCC) [196] lattice, respectively.

Alkali metal fullerides may retain the FCC structure of the pure C_{60} crystal even when loaded with alkali atom moieties exceeding that of A_3C_{60}, as exemplified by the case A = Na. Thus, an FCC phase has been identified for Na_6C_{60} [197]. While each tetrahedral site in these compounds has been found to be occupied by a single Na atom, each octahedral site is shared among four Na atoms. In terms of electron donation to the fullerene host, A_6C_{60} compounds with A = K, Rb and Cs, where six electronic s levels turn into bands derived from the T_{1u} levels of C_{60}, are more efficient than Na_6C_{60}. Here, complete charge transfer is not favorable, since it would lead to high electrostatic repulsion within the octahedral sites. Only about two electrons are donated to the fullerene host molecules, as in Na_2C_{60}, a stable compound where the impurity atoms occupy the tetrahedral sites exclusively. The four Na atoms in the octahedral void of Na_6C_{60} form a metal cluster substructure within the fulleride crystal. The maximally stable arrangement is found for the geometry of a regular tetrahedron whose center coincides with that of the interstitial site. A fulleride with stoichiometry $Na_{11}C_{60}$ [198, 199] presents another example for an *interstitial metal cluster* composed of Na constituents. From x-ray analysis, this Na_9 cluster stabilizes in the geometry of a body-centered cube. Intercalated Li clusters have been identified inside a fulleride of composition $Li_{12}C_{60}$ [200]. As demonstrated by x-ray diffraction, all Li atoms assemble in the octahedral void of this composite that involves, at room temperature, a BCT (= body-centered tetragonal) arrangement of the C_{60} subsystem.

Of particular significance for this monograph are the magnetic alkali-metal fullerides AC_{60}, with A = Na, K, Rb and Cs [201, 202, 203, 204]. At high temperatures extending from about 410 to 460 K, these compounds form a rock-salt structure, where a C_{60} FCC lattice is intercalated by alkali atoms that populate the octahedral sites. At temperatures lower than 373 K, however, an orthorhombic phase with a low lattice constant a has been reported for A = K, Rb and Cs [205]. This structure is understood as a *polymerized phase*, involving chains of bonded fullerenes, as shown in Figure 6.9. In the polymer chains, the center-to-center distance between adjacent fullerenes shrinks to 9.14 $\AA$, to be compared with its value in the undoped FCC C_{60} crystal, 10.04 $\AA$, as indicated earlier. Information about the magnetic properties of AC_{60} is found in Section 12.3.2. Polymerized fullerene phases are discussed in the following section.

Various compounds of the form EA_xC_{60}, where AE stands for alkaline earth atom impurity, have been treated by experiment and computation [197, 206, 207, 208]. The specific interest in these systems lies in their divalence, leading to stronger electron transfer than observed for systems with alkali metal atom dopants. We mention the example of AE = Ca. In this case, the simple cubic structure that was identified as the C_{60} lattice geometry at temperatures below 261 K prevails up to 673 K. In analogy to the behavior of Na_xC_{60} with x=6, 11, the octahedral

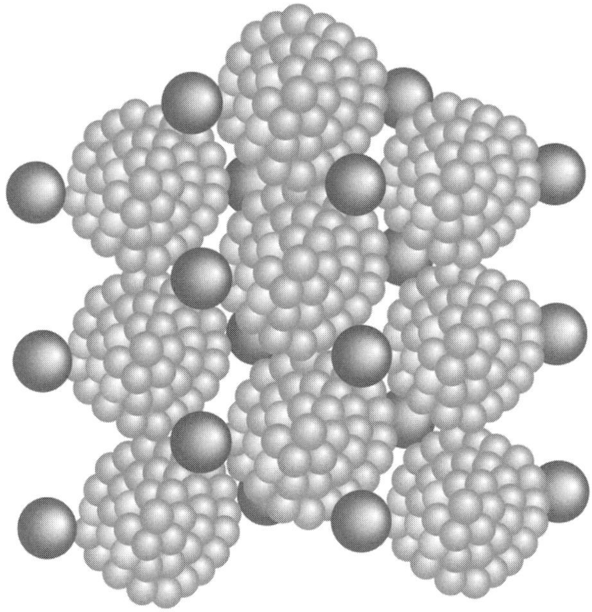

Figure 6.9 Structure of RbC_{60} [205]. Shown is a projection into the $a - c$ plane of the orthorhombic phase. The fullerene molecules exhibit polymerization.

sites of Ca_5C_{60} in the simple cubic phase can accommodate clusters of calcium atoms [187].

6.4 Crystalline Polymers of C_{60}

The FCC C_{60} crystal may be turned into various types of polymers involving chemical bonds between the C_{60} constituents. The presence of thirty unsaturated double bonds between adjacent carbon atoms in the C_{60} cage suggests that a wide range of different polymeric forms may be generated from the pristine solid. Several diverse procedures are known to initiate this transition, among them optical techniques involving visible light [209] and high-pressure or high-temperature treatment [210, 211]. The possibility of creating C_{60} polymers by doping with alkali metal atoms A at the stoichiometry AC_{60} was mentioned earlier, in the context of fullerene compounds with exohedral impurities. Application of pressure in conjunction with heating has given rise to a variety of C_{60} polymer structures, such as a one-dimensional orthohombic, a two-dimensional tetragonal and a two-dimensional rhombohedral phase. Sketches of the latter two structures are provided in Figure 6.10. All of these systems contain C atoms in both the sp^3- and the

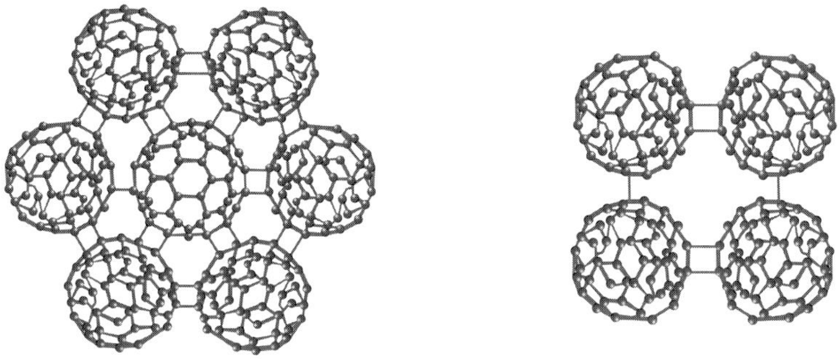

Figure 6.10 Geometries of the rhombohedral (left) and the tetragonal (right) phases of polymeric C_{60} after heat and pressure treatment.

sp^2-hybridized states, in contrast to the FCC structure, which contains C atoms solely in the latter state.

In the rhombohedral (tetragonal) phase, each C_{60} molecule is bonded to six (four) neighboring molecules. The bonding between two adjacent fullerenes is the result of *[2+2] cycloaddition*. This term refers to a bonding type that involves the reorganization of two neighboring double bonds into intermolecular single bonds. In the case of two bonding C_{60} molecules, these single bonds form a four-membered ring that joins the two cages. In both phases of polymeric C_{60}, the resulting connection between neighboring fullerenes involves two hexagon-hexagon bonds. As they are reduced from double to single bonds, two new single bonds emerge that link the two neighboring fullerene molecules, as illustrated in Figure 6.10. The response of C_{60} polymers to pressure, temperature and laser irradiation intensity has been thoroughly investigated. A survey is provided in [212].

Part III

Spin Effects in Graphene and Carbon Nanotubes

7

Magnetic Textures at Edges and Defect Sites

The previous three chapters presented a survey of typical carbon nanosystems in two, one and zero dimensions, namely graphene, carbon nanotubes and fullerenes, respectively. None of these systems is intrinsically magnetic. Magnetism, however, can arise through *dimensional reduction* of structures with one or more dimensions. Thus, as one truncates the infinite graphene sheet, one may generate graphene nanoribbons of the zigzag type (zGNRs), as introduced in Section 4.5.1. In the first section of this chapter, we will show that localized states at the zGNR edges give rise to antiferromagnetic ground states in these systems, and review experimental confirmation for this assertion. The analysis of zigzag edges will be extended to vacancies in graphene-based structures that lack magnetic edge states, such as aGNRs. In this context, we will emphasize the importance of *Lieb's theorem* as a principle underlying many magnetic phenomena in carbon nanostructures, albeit not all. Further, we will describe specific one-dimensional substructures in graphene-based systems, namely zGNR edge excitations (7.2.3), and topological line defects (7.3).

Intrinsic magnetism in truncated carbon nanotubes of the zigzag type (zSWCNTs) will be discussed in analogy to zGNR magnetism. Section 7.4.1 deals with aggregates of several zSWCNTs. The magnetic effects in these complexes turn out to be rooted in the electronic features of carbon surfaces with negative Gaussian curvature. We conclude with some remarks on the controversial topic of intrinsic magnetism in pure fullerene systems.

7.1 Systems Derived from Graphene

In what follows, we review magnetic effects associated with the rupture of the pristine graphene network, as it arises from the formation of edges, or vacancies, or voids as aggregations of vacancies. The focus is here on zigzag edge structures as the preferred loci of unpaired electrons, and thus the substructures of the graphene lattice from which spin interactions originate.

7.1.1 Magnetism in Zigzag Graphene Nanoribbons

Zigzag graphene nanoribbons (see Section 4.5.1) exhibit magnetism in their electronic ground states. In the following we will consider the physical origin of this phenomenon. Once more, we associate the x (y) coordinate with the periodic (finite) direction. Inspecting the zGNR ground state solutions (see Eqs. (4.71)–(4.77)), we found that the graphene wave function must be modified to allow for edge-localized states. Since the zigzag edges are formed by carbon atoms at equivalent sites, consisting of A or B locations exclusively, the tight-binding model predicts that the energy of a strictly edge-localized state must coincide with the Fermi energy. As the energy of this state is not determined by any bonding between adjacent sites, it holds that

$$E(\mathbf{k})_{\pm} = \frac{E_{2p} \pm t|S(k)|}{1 \pm s|S(k)|} = E_{2p} = E_F. \tag{7.1}$$

Figure 7.1 presents the zGNR band structure, displaying, along the periodic coordinate k_x, the sub-bands that result from k_y quantization. Comparing with the band structure of the infinite graphene sheet, as shown Figure 4.3, one finds that the point $k_x = \pi/a$ is the zGNR counterpart of the $\mathbf{M}$ point, where a is the unit cell length of the zigzag ribbon, $a = \sqrt{3}a_0$. In case of graphene, the highest valence and the lowest conduction band coincide at the $\mathbf{K}$ point and differ from each other

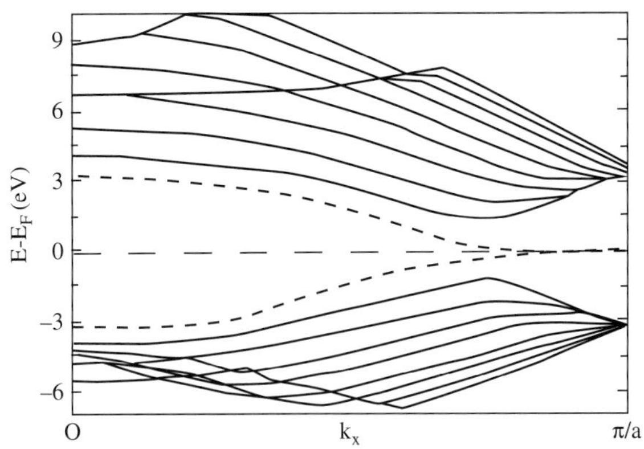

Figure 7.1 Band structure of the monohydrogenated zGNR (each C atom at the ribbon edge is terminated by one H atom, as opposed to two) with width M = 10, where a non-magnetic state has been assumed [213]. The flattening bands at $k_x = \pi/a$ (dashed lines) describe a state that is sharply localized at the zigzag edges of the ribbon. Reprinted with permission from [213]. Copyright (2007) by the American Physical Society.

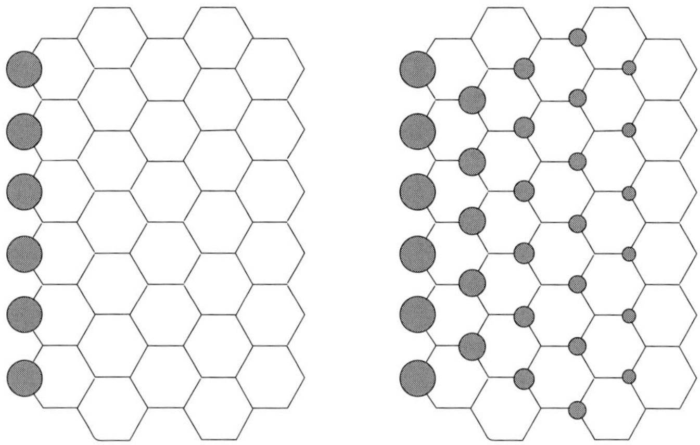

Figure 7.2 Electronic charge density distribution of a zGNR at $k_x = \frac{\pi}{a}$ (left) and for $\frac{2\pi}{3a} < k_x < \frac{\pi}{a}$ (right). In the latter case, the electronic density penetrates into the interior of the zGNR, while it is still mostly concentrated at the ribbon edge.

by a maximum margin at the point **M** (see Figure 4.4) [213]. The zGNR displays a markedly different behavior. Here the two bands are seen to converge gradually as one goes from $k_x = 2\pi/3a$ to $k_x = \pi/a$, finally coinciding at $E = E_F$. A *flat band* emerges that extends through the wave number interval $2\pi/3a \leq k_x \leq \pi/a$.[1] As the ribbon widens, the two bands at $k_x = 2\pi/3a$ approach E_F more and more closely. It is attained, however, only as the ribbon width tends to infinity, i.e. in the graphene limit. As this case is realized, the edge-localized states have vanished, and the degeneracy of the two bands for $k_x > 2\pi/3a$ is lifted [213].

The extreme $k_x = \frac{\pi}{a}$ corresponds to complete edge localization, as can be inferred from the analytical zGNR wave function (4.71) by deriving the electronic charge density for this case [214]. The result is shown in Figure 7.2 and contrasted with the electronic charge distribution for a choice of k_x within $[\frac{2\pi}{3a}, \frac{\pi}{a}]$. Evidently, the latter condition leads to a more diffuse distribution as the electronic charge density spreads into the zGNR lattice.

For further analysis of the flat band states in the limit of energy degeneracy between the valence band of highest and the conduction band of lowest energy, we turn once more to the basic tight-binding approximation (4.11) that led to the dispersion relation (4.14) for graphene. Within the tight-binding approach, the assumption of a dispersionless state located at $E = E_F$ implies that the sum over next-neighbor wave function components vanishes. Specifically, the π states localized at consecutive zGNR edge sites with labels $n - 1$, n and $n + 1$ differ by prefactors $e^{ik_x a(n-1)}$, $e^{ik_x an}$ and $e^{ik_x a(n+1)}$ [215]. This finding may be used to obtain relations between the wave function components at the edge sites ($m = 0$ in

[1] *Flattening band* may a more appropriate term, since the dispersionless portion of the interval increases with increasing ribbon width.

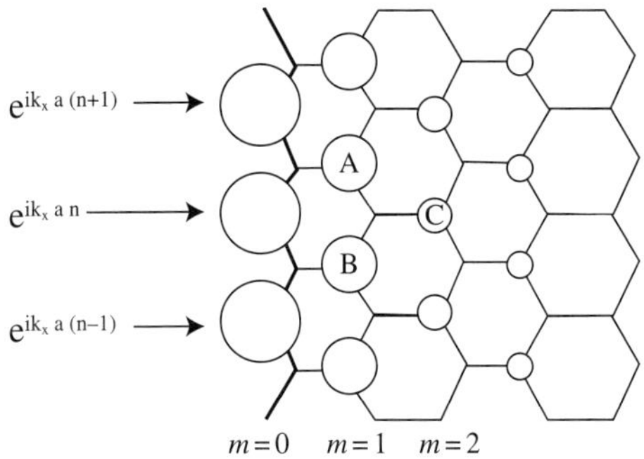

Figure 7.3 Sketch for aiding the application of the dispersion formula (4.14) to the zGNR edge. Phase factors for selected edge sites are indicated.

Figure 7.3) and corresponding sites in the adjacent zigzag chains with $m = 1$ and $m = 2$ [215].

With the assignments chosen as indicated in Figure 7.3, the tight-binding approach in conjunction with the assumed dispersionless behavior implies the following statements:

$$\begin{aligned} e^{ik_x a(n+1)} + e^{ik_x an} + A &= 0, \\ e^{ik_x a(n-1)} + e^{ik_x an} + B &= 0, \\ A + B + C &= 0. \end{aligned} \tag{7.2}$$

This system of equations is readily solved (see Exercise 7.1).

Exercise 7.1 Demonstrate that the system of equations (7.2) is solved by:

$$\begin{aligned} A &= D_k e^{ik_x a(n+\frac{1}{2})}, \\ B &= D_k e^{ik_x a(n-\frac{1}{2})}, \\ C = D_k^2 e^{ik_x an}, \quad \text{where } D_k &= -2\cos(\frac{k_x a}{2}). \end{aligned}$$

Further, show that the condition $|D_k| \leq 1$ implies $2\pi/3a \leq |k_x| \leq \pi/a$.

At each occupied site in the zigzag chain with index m, the electronic charge density is therefore proportional to $D_k^{2(m-1)}$. As all arguments supplied to derive this expression are also valid for the case of a semi-infinite graphene sheet, the condition

$$|D_k| \leq 1 \tag{7.3}$$

must hold, securing the convergence of the charge density as a function of the parameter m. This criterion indeed singles out the interval $2\pi/3a \leq |k_x| \leq \pi/a$ as the regime where the dispersionless band emerges. This structure corresponds to a high density of states at the Fermi energy. It is plausible that lifting the degeneracy at $k_x > 2\pi/3a$ or lowering the energy of the degenerate band would result in a stability gain for the system as a whole. As discussed by Pisani et al. [213], this stabilization is not achieved by geometric distortion, although edge reconstruction is observed, giving rise to a uniform contraction of all bonds within the edge chain, and a concomitant expansion of the bonds between this chain and the one adjacent to it. Removal of the electron density at $E = E_F$ proceeds through another mechanism, namely the *transition into a magnetic ground state*. This is demonstrated in Figure 7.4 once more by the example of a monohydrogenated zGNR with width

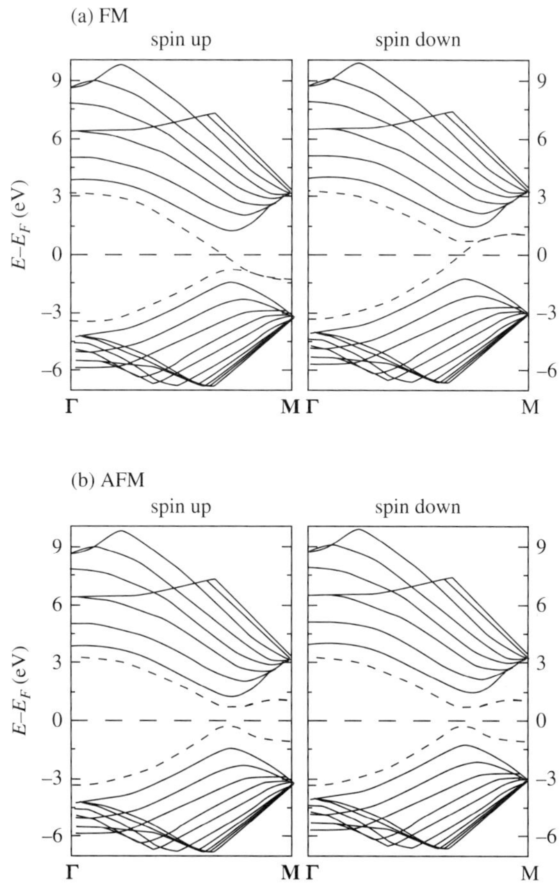

Figure 7.4 Band structure schemes for magnetic phases of the monohydrogenated zGNR with ten parallel zigzag chains (see Figure 7.5). The ferromagnetic (FM, (a)) solution is contrasted with the antiferromagnetic (AFM, (b)) solution. Both patterns are subdivided according to spin orientation. Reprinted with permission from [213]. Copyright (2007) by the American Physical Society.

M = 10. As inferred from the figure, the spin-polarized treatment yields significantly different profiles for the majority (spin up) and the minority (spin down) orientations. It is instructive to compare the band structure schemes for spin-up and spin-down electrons in the ferromagnetic (FM) case, as shown in Figure 7.4(a). In both situations, the degeneracy of the flat band is preserved. However, the band now deviates markedly from $E = E_F$. For the former spin orientation, the band is completely occupied, with $E < E_F$, for the other orientation, it is completely unoccupied ($E > E_F$). This discrepancy establishes an electronic ground state with non-zero spin.

Higher stability, however, is found for the antiferromagnetic (AFM) alternative displayed in Figure 7.4 (b). This solution involves two distinct magnetic sublattices, as opposite spins are present at the two sites A and B. In this case, the degeneracy between the conduction and the valence band is lifted. Once more, the formerly flat band acquires dispersion, and the valence band energy is reduced to values below E_F. The gap at $k_x = 2\pi/3a$ diminishes with increasing ribbon width and closes as the graphene limit is attained. We note that for both the FM and the AFM solution, the coordination within the ribbon edges is ferromagnetic.

The two magnetic states are associated with markedly different spin density profiles, as represented in Figure 7.5. We find a depletion of the spin density in the mid-region of the ribbon as a characteristic feature of the FM phase (Figure 7.5 (a)), while continuous alternation between spin-up and spin-down sites is observed in its AFM counterpart (Figure 7.5 (b)).

The size of the local magnetic moments of hydrogenated zGNRs is readily estimated by use of a heuristic argument. Let N_r be the number of edge sites included in the unit cell. Each C atom contributes $\frac{4}{3}$ electrons to each C-C bond. If the C atom is located at the zGNR edge it bonds with two C atoms of the opposite sublattice and with an H atom. This leaves one-third of an electron per edge carbon atom unsaturated. Each edge atom of a periodic zGNR is thus associated with a magnetic moment of one-third of a Bohr magneton. The total magnetic moment for a zGNR of finite length, comprising N_r carbon atoms along each edge, and assuming parallel coordination of the two edges, is therefore expected to approach $\frac{2N_r}{3}\mu_B$.

No edge-localized states exist in aGNRs. Correspondingly, non-magnetic ground state solutions have been found for these systems [216]. As reflected by relation (4.70), the energy gap for both nanomaterials, zGNRs and aGNRs, decreases with increasing ribbon width. While the gaps of zGNRs can be approximated with one set of parameters (α, W^*), three distinct sets have been employed to model the behavior of aGNRs as a function of the width. Specifically, the fitting parameters turned out to depend on the number of chains in the aGNR (see Figure 4.8), N_a. Different curves emerged for the three series $N_a = 3N, 3N + 1$

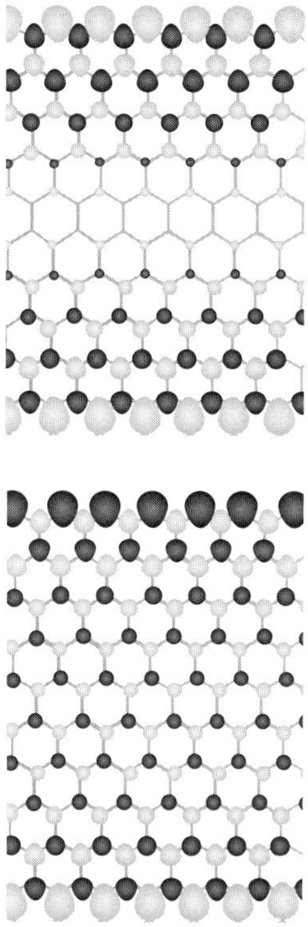

Figure 7.5 The spin densities of the monohydrogenated zGNR with width $M = 10$ for the two magnetic phases compared, FM (a) and AFM (b). The spin up (spin down) contribution is shown in black (grey). The isovalues used in (a) are -0.09 versus $0.28\ \mu_B/\mathring{A}^3$, those used for (b) -0.28 versus $0.28\ \mu_B/\mathring{A}^3$. Reprinted with permission from [213]. Copyright (2007) by the American Physical Society.

and $3N + 2$, with $N = 1, 2, 3...$ From tight-binding treatment, the last of these three groups is metallic. A density functional theory (DFT) analysis at the local density approximation (LDA) level, however, yielded semiconducting behavior for the $3N + 2$ group [216]. The band structures of zGNRs and aGNRs are, in some respect, complementary to each other. Consistently, the top of the zGNR conduction band as well as the bottom of the valence band are located at the $\mathbf{\Gamma}$ point. For aGNRs, in contrast, the $\mathbf{\Gamma}$ point is the place of the band gap minimum.

7.1.2 Zigzag Edge Magnetism: Experimental Results

Experimental evidence for vacancy-induced magnetism in graphene-based systems has been obtained by Ugeda et al. through scanning tunneling spectroscopy (STS) measurements on graphite [217]. In particular, highly ordered pyrolytic graphite (HOPG) samples were exposed to Sr^+ ion irradiation, followed by low-temperature

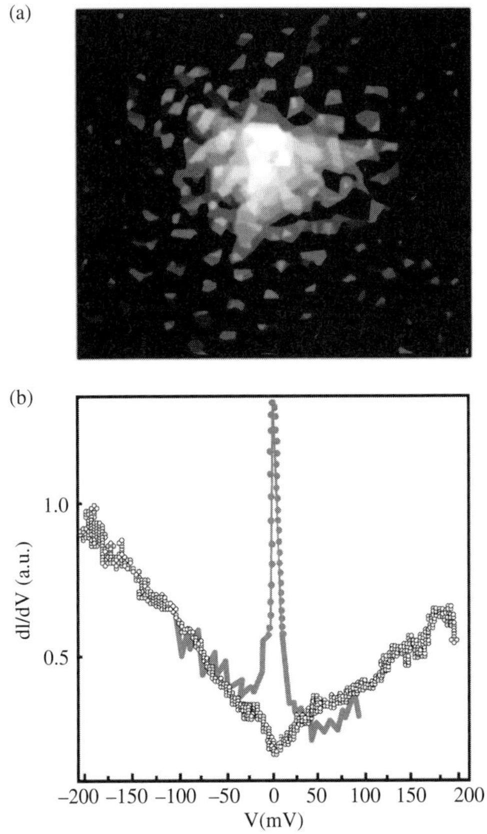

Figure 7.6 Experimental investigation of single vacancies in graphite, generated by ion irradiation. (a) Scanning tunneling microscopy (STM) image of an isolated single vacancy in the surface layer of an HOPG sample, recorded at a sample bias of +270 mV, and a tunneling current of 1 nA. (b) Differential conductance spectrum, recorded on top of the vacancy site (solid circles, forming a sharply peaked line) and on pristine graphite (open circles). Reprinted with permission from [217]. Copyright (2010) by the American Physical Society.

scanning tunneling microscopy (LT-STM). By recording the *differential conductance* $\frac{dI}{dV}$, a sharp resonance in the Fermi energy was found at the location of the vacancy. This is shown in Figure 7.6(b) as a function of the sample bias. No such feature is identified by measurements made on pristine HOPG, or in a sufficient distance from the defect, as demonstrated by the reference curve in the figure. An LT-STM image of a single isolated vacancy is shown in Figure 7.6(a). Density-of-states modeling based on the tight-binding approach confirms that the experimental signature is due to unpaired electrons at the vacancy site, prominently a single dangling bond of sp^2 character, in agreement with previous DFT computations [218] on irradiation-induced magnetism in graphite. While equivalence exists in graphene between vacancies of types A and B, the Bernal stacking order in graphite destroys this symmetry, as reflected by STS resonances at type-A vacancies that substantially exceed those associated with vacancies of type B. This implies magnetic moments of different magnitude at the two alternative sites, suggesting the possibility of tailoring graphite samples with ferrimagnetic order.

Prior to the experiments of Ugeda et al., magnetic force microscopy observations of the graphite surface yielded ferromagnetic ordering that was ascribed to electronic states localized at the grain boundaries of the examined sample [219].

Independent evidence for intrinsic graphene magnetism was provided by the experimental detection of the Kondo effect (see Section 2.2.2) in an ion-irradiated graphene sheet [220], manifesting the presence of vacancy-induced local magnetic moments that interact with the itinerant π electrons of graphene.

Likewise, the existence of spin-polarized edge states in GNRs has been subjected to experimental scrutiny [221, 222, 223]. Direct experimental observation of magnetic edge states was reported for multilayer GNRs [221] prepared by chemical vapor deposition (CVD). Specifically, near-edge x-ray absorption fine structure (NEXAFS) spectroscopy was combined with electron-spin resonance (ESR, see Section 3.3) measurement. Using the former procedure, the authors identified a distinct low-energy peak within a broader structure due to the $C(1s) \rightarrow \pi^*$ transition characteristic for the studied graphitic objects. This spectroscopic signature could be attributed to a GNR edge state. Correlating it with a distinct response from ESR observation, the authors demonstrated that both signals displayed the same trends when the sample was annealed. Both were seen to reduce as a function of the annealing temperature and to disappear as a temperature threshold was crossed. This parallel behavior was rationalized by a decrease of the zGNR edge state population in the annealing process, which induced connections between adjacent edges, thus quenching their spin polarization.

Tada et al. [224] inspected the edge magnetism of graphene-derived structures, namely *graphene nanopore arrays* (NPAs). The authors placed mechanically exfoliated graphene sheets on a nanoporous alumina template that was employed as an etching mask. In this way, a modified graphene sheet with regularly spaced hexagonal holes was fabricated. STM inspection suggested the presence of localized electronic states at the pore boundaries, consistent with the assumption of zigzag-structured edges. Annealing in conjunction with monitoring by use of Raman spectroscopy ensured a low volume of defects. The annealing proceeded in an H_2 atmosphere to allow for edge hydrogenation. The magnetization of the samples thus prepared was recorded by a superconducting quantum interference device (SQUID, see Section 3.5), preceding magnetic force microscopy (MFM) observations. The former measurements yielded a distinct hysteresis profile, in accordance with ferromagnetism, while the latter specified the nanopore edges as the sites of localized magnetic moments. This finding may be interpreted as direct experimental evidence for ferromagnetically coupled electronic states at zigzag edges of graphene. Simulations based on Lieb's theorem (see Section 7.2.1) led to an overall ferromagnetic rather than antiferromagnetic arrangement of the porous array as a whole.

Observations consistent with these findings have been made on another graphene-based species, one of considerably higher complexity than multi-layer zGNRs, namely activated carbon fibers (ACFs) [222], disordered networks of nanographite units. The distinct magnetic effects in these aggregates were attributed to unpaired π electrons at the edges of the nanographite constituents. A more detailed characterization of these units is found in Section 13.3.1, which focuses on composites of magnetic carbon nanostructures.

7.2 Voids and Edges

Like any other non-magnetic carbon nanostructure, an aGNR may acquire ground state magnetism due to the formation of internal voids. In the most general mode of description, a void in a graphene-derived structure may be classified in terms of the type (A versus B) and the number of atoms removed from this structure. The most crucially important parameter for the characterization of these voids is the *sublattice imbalance* N_I, defined as the difference between the number of type-A and type-B atoms removed from the lattice, $N_A - N_B$ [225]. In most cases, voids are understood as islands of connected vacancies in the bulk of the structure (see Figure 7.7), but they can also emerge at its edge, giving rise to notches. In a first approach, we associate the term void with the removal of a carbon atom or a group of carbon atoms from their sites, without taking into account any reconstruction effects. With reference to the configurations shown in Figure 7.7, readers may convince themselves that the values of the sublattice imbalance are indeed 0, -1, and 2, as indicated in the figure caption.

With basic arguments it may be established that the sublattice imbalance determines the number of solutions at $E = E_F$, or *zero modes* [226]. For clarification,

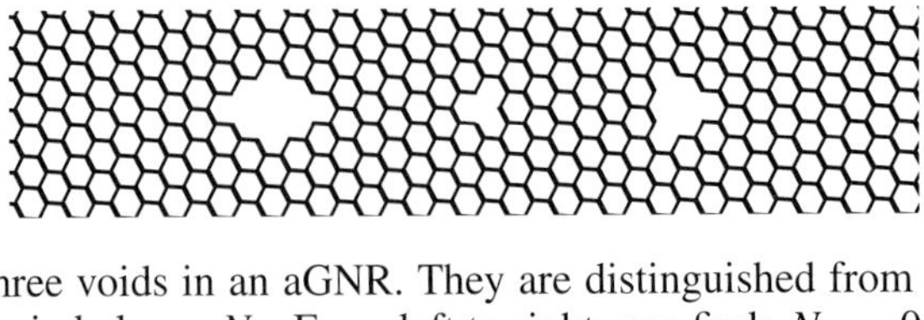

Figure 7.7 Three voids in an aGNR. They are distinguished from each other by their sublattice imbalance N_I. From left to right, one finds $N_I = 0, -1$ and 2. In the second case, the void is a single vacancy, arising from the removal of a type-B atom. Reprinted with permission from [225]. Copyright (2008) by the American Physical Society.

we inspect once more the tight-binding Hamiltonian (4.5). Setting again $t' = 0$, we cast the solution $|\Psi\rangle$ into the following form,

$$|\Psi\rangle = (\boldsymbol{\psi}_A, \boldsymbol{\psi}_B) = (\varphi_A^1, \varphi_A^2, \ldots, \varphi_A^{N_A}, \varphi_B^1, \varphi_B^2, \ldots, \varphi_A^{N_B}), \tag{7.4}$$

letting the amplitudes φ_A on sublattice A precede sublattice B amplitudes φ_B. Making allowance for vacancies, we might now assume that $N_A > N_B$. With these premises, we write the stationary Schrödinger equation in matrix form:

$$\begin{pmatrix} \varepsilon_A \mathbf{1}_{N_A} & \mathbf{h}_{AB} \\ \tilde{\mathbf{h}}_{AB} & \varepsilon_B \mathbf{1}_{N_B} \end{pmatrix} \begin{pmatrix} \boldsymbol{\psi}_A \\ \boldsymbol{\psi}_B \end{pmatrix} = E \begin{pmatrix} \boldsymbol{\psi}_A \\ \boldsymbol{\psi}_B \end{pmatrix}. \tag{7.5}$$

The dimension of the matrices $\mathbf{h}_{AB}$ and $\tilde{\mathbf{h}}_{AB}$ is $N_A \times N_B$ and $N_B \times N_A$, respectively. Splitting up Eq. (7.5) into two subequations, we write:

$$\begin{aligned} \mathbf{h}_{AB}\boldsymbol{\psi}_B &= (E - \varepsilon_A)\boldsymbol{\psi}_A \\ \tilde{\mathbf{h}}_{AB}\boldsymbol{\psi}_A &= (E - \varepsilon_B)\boldsymbol{\psi}_B. \end{aligned} \tag{7.6}$$

Therefore,

$$\tilde{\mathbf{h}}_{AB}\mathbf{h}_{AB}\boldsymbol{\psi}_B = (E - \varepsilon_A)(E - \varepsilon_B)\boldsymbol{\psi}_B. \tag{7.7}$$

The eigenvalues of $\tilde{\mathbf{h}}_{AB}\mathbf{h}_{AB}$ are non-negative. Calling them λ^2, we find

$$E_\pm = \frac{\varepsilon_A + \varepsilon_B}{2} \pm \sqrt{(\frac{\varepsilon_A - \varepsilon_B}{2})^2 + \lambda^2}. \tag{7.8}$$

Recalling the assumption $N_A > N_B$, we conclude that the rank of both $\mathbf{h}_{AB}$ and $\tilde{\mathbf{h}}_{AB}$ is N_B. As a further observation, the equation $\mathbf{h}_{AB}\boldsymbol{\psi}_B = \mathbf{0}$ implies $\boldsymbol{\psi}_B = \mathbf{0}$. Lastly, analyzing the matrix $\tilde{\mathbf{h}}_{AB}$, and using the rank-nullity theorem, we conclude:

$$\text{rank}\{\tilde{\mathbf{h}}_{AB}\} + \text{nullity}\{\tilde{\mathbf{h}}_{AB}\} = N_A. \tag{7.9}$$

The dimension of the solution space of

$$\tilde{\mathbf{h}}_{AB}\boldsymbol{\psi}_A = \mathbf{0}, \tag{7.10}$$

or the nullity of $\tilde{\mathbf{h}}_{AB}$, is thus equal to $N_A - N_B$. The space of solutions $\boldsymbol{\psi}_A^{\mathbf{0}}$ of Eq. (7.10) thus has the dimension $N_A - N_B$. By Eq. (7.6), the eigenenergy E of the eigenstate

$$|\Psi\rangle = |\Psi^0\rangle = \begin{pmatrix} {\boldsymbol{\psi}_A}^0 \\ \mathbf{0} \end{pmatrix} \tag{7.11}$$

is $E = \varepsilon_A$. The wave function $|\Psi^0\rangle$ encodes the peculiar feature that a vacancy induces a state that is localized on the majority sublattice exclusively. Enhancing the number of vacancies at a site of type B from one to several will change the number of solutions with $E = \varepsilon_A$, but not the fact that they populate only sites of type A. In the framework of the tight-binding model considered here, the state $|\Psi\rangle$ is seen to involve no interactions, as any A site only interacts with B sites, and the latter are unoccupied in their entirety. Therefore, $|\Psi\rangle$ is to be understood as a *zero-energy mode*, corresponding to the condition that $E = \varepsilon_A = 0$. As motivated in Section 4.2, the energy $E = 0$ can, within the tight-binding model, be identified with the Fermi energy. Voids in the B sublattice thus give rise to the emergence of zero modes located at the Fermi level.

Exercise 7.2 Reviewing Eqs. (7.4) to (7.11), prove the theorem: Whenever the sublattice imbalance $N_I = N_A - N_B$ is positive, there will be $N_A - N_B$ degenerate states with energy eigenvalue ε_A. All of these states are linearly independent and localized on the A sublattice exclusively.

As shown in [226], an approximate coordinate representation of the wave functions defining the zero modes can be obtained by analytical calculation, using arguments similar to those employed earlier in the context of deriving the electron charge distribution in zGNRs. If a void of type B is located at ($x = 0$, $y = 0$), the zero mode can be expressed in the following way:

$$\Psi(x, y) \approx \frac{e^{i\mathbf{K}'\cdot\mathbf{x}}}{x + iy} + \frac{e^{i\mathbf{K}\cdot\mathbf{x}}}{x - iy}. \tag{7.12}$$

The coordinate pair (x, y) denotes here the position of a type-A atom. From the form of the wave function (7.12), it can be inferred that the amplitude of the zero mode decreases as $1/r$, with r as the distance from the place of the void. Two such states, corresponding to two voids of the same type, will undergo hybridization. On account of the *quasi-localized character* of the solutions (7.12), however, this effect will be small if the distance between the voids is large. We emphasize that the existence of these void-induced long-range quasi-localized states is an experimentally well-established phenomenon. Specifically, scanning tunneling microscopy observations of electronic states on graphite surfaces with voids and other defects yielded empirical evidence for a $\sqrt{3} \times \sqrt{3}R30^\circ$ superlattice formed by these states, reflecting the symmetry of the point defects modulating the surface [227, 228, 229].

To summarize, as admission is made for voids in the graphene sheet, causing an imbalance N_I between the sublattices A and B, zero modes emerge whose multiplicity equals N_I. We point out that this result has also been derived by applying bezinoid graph theory to the tight-binding description of the honeycomb lattice [6, 230]. This approach yields

$$\eta = 2\alpha - N, \tag{7.13}$$

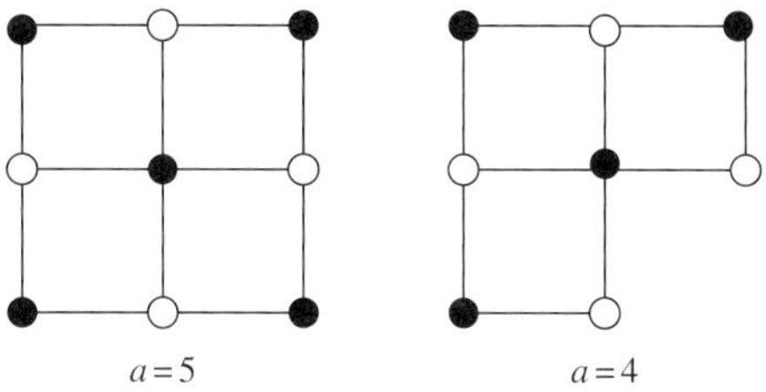

Figure 7.8 Two cases illustrating the counting rule (7.13). The maximum number of non-adjacent sites α is given as 5 in the model shown on the left, and as 4 in the model on the right side. After [230].

where α stands for the maximum number of non-adjacent sites, while the overall number of sites is denoted by N. The quantity η is the *nullity of the graph* representing the lattice, a concept defined by the adjacency matrix characterizing the lattice. More specifically, η refers to the degree of degeneracy associated with the zero eigenvalue of this matrix [231]. This number has been shown to be identical with that of the zero-energy states present in the lattice [230]. For two-dimensional lattices with two alternative sites, A and B, the right-hand side of Eq. (7.13) reduces to the sublattice imbalance, as is clarified by the two examples shown in Figure 7.8.

7.2.1 Lieb's Theorem

The arguments presented in the previous section lead to the question for a general rule that allows us to predict the spin ground state of a graphene-based lattice with edges and voids. *Lieb's theorem* [232], the most general analytical result related to this problem, is based on the Hubbard model and holds for bipartite systems. The latter are defined in terms of the hopping integrals of the Hubbard Hamiltonian (see Section 3.1), and, in more detail, by the condition that the sites of the lattice can be subdivided into two disjoint sets, A and B, such that t_{ij} vanishes if the indices i and j label both A sites or both B sites [232]. Further, the lattice is assumed to be connected, in the sense that between any two sites there is a connected path of site pairs (k, l), such that $t_{kl} \neq 0$. Adding to these premises the further stipulation that the Coulomb integral is positive ($U > 0$), Lieb showed the following: in the case of a *half-filled band* where the number of electrons, N, equals the number of sites ($N = N_A + N_B$), with $N_A \geq N_B$, the electronic ground state is, apart from the ($2S_{tot}$+1)-fold spin degeneracy, unique, with a total spin quantum number S_{tot} = $\frac{1}{2}(N_A - N_B)$. We stress that this finding is valid for any network of sites A and B, i.e. no periodicity constraint is applied.

Lieb's theorem does not only establish the existence of magnetic ground states in systems with non-zero sublattice imbalance, it also makes a profound assertion for situations of vanishing N_I. For those cases, it predicts S_{tot} = 0. Physically, this condition is trivially realized for the intact graphene sheet, where the ground state is non-magnetic. It is, however, also realized for systems with non-interacting voids if $N_A = N_B$, and for zGNRs, as discussed earlier in this section. The vanishing total spin, as derived from Lieb's theorem, in conjunction with the presence of localized unpaired electrons, results in an *antiferromagnetic ground state solution* for these species. As described above, this expectation has been confirmed for zGNRs by DFT computation. Although the predictions of Lieb's theorem rest on Hubbard theory, they display a certain robustness with respect to modifications of the underlying theoretical formalism.

An ambiguity arises for a graphene sheet with several voids and overall vanishing sublattice imbalance, as Lieb's theorem does not distinguish between non-magnetic and antiferromagnetic order. If the separation between two voids, one of type A, and the other of type B, is large, one will expect to find the latter arrangement. Can, however, a critical distance be identified at which the non-magnetic solution becomes preferred over the antiferromagnetic one? Addressing this question, Palacios et al. [225] considered several scenarios involving two voids located in aGNRs at varying distances from each other. Specifically, they evaluated a quantity analogous to a magnetic-moment standard deviation, although not normalized to the number of lattice sites i,

$$\sigma_n = \sqrt{\sum_i \langle \mu_i \rangle^2}, \tag{7.14}$$

with $\langle \mu_i \rangle$ as the expectation value of the magnetic moment at site i. The label n indicates the number of voids included in the model. We emphasize that σ_2 allows to discriminate between antiferromagnetic and non-magnetic configurations. Numerical solutions were obtained by use of a mean-field Hubbard model. Figure 7.9 displays the calculated results for σ_2 normalized to $\sqrt{2}\sigma_1$, a variable that yields one in the limiting case of a two-void ensemble behaving like two non-interacting single voids.

From Figure 7.9, the two-void standard deviation merges for large distances into that for two fully separated voids. In the first case, involving two voids of the same type, σ_2 tends towards higher values below a certain critical distance, corresponding to enhanced localization of the magnetic moment at reduced separation. In the three remaining situations, combining voids of types A and B, a phase transition is observed, involving a change from antiferromagnetic to non-magnetic order at a sufficiently small distance. This phenomenon is ascribed to the hybridization of zero-energy states at a well-defined threshold concentration of voids. The decoupling distance diminishes with increasing on-site Coulomb repulsion, and with decreasing ribbon width. The latter, less-intuitive effect is rationalized by the

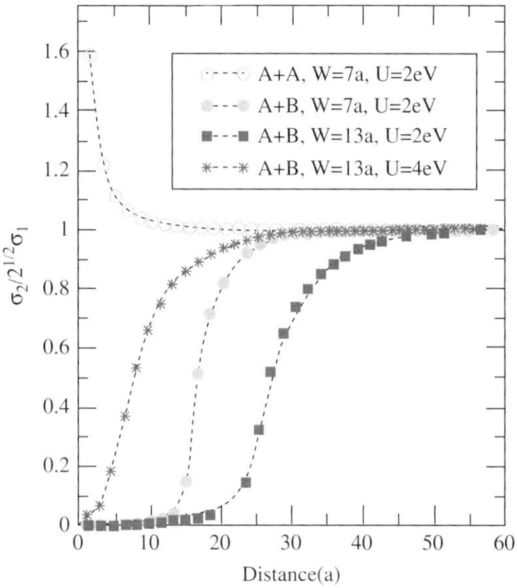

Figure 7.9 The normalized standard deviation of the magnetic moment according to (7.14) for two voids in aGNRs versus the distance between the vacated sites in units of $a = a_0\sqrt{3}$. The on-site Coulomb repulsion U as well as the ribbon width are treated as parameters. Four cases are summarized. Open circles: Both voids are of type A, the ribbon width is 7 a, and U is 2 eV. The three remaining cases are defined likewise. Reprinted with permission from [225]. Copyright (2008) by the American Physical Society.

observation that zero-energy states are more delocalized in wider ribbons, such that their hybridization survives at greater distances between the generating voids.

The addition rules that may be extracted from these observations, determining the overall magnetic moment that results from a two-void ensemble, draw a parallel to the behavior of vortices, as they are found, for instance, in superconductors [233]. In sufficient distance from each other, vortices, as much as magnetic moments induced by local sublattice imbalances, are independent of each other, but both undergo mutual annihilation when they come into close proximity of each other, as defined by a critical distance.

7.2.2 Void Magnetism from First-Principles Computation

A quantitative assessment of local magnetic moments associated with voids in graphene is possible by means of atomistic simulation. Yazyev and Helm [234] performed DFT calculations, using a numerical atomic orbital basis set, to study the magnetic effects caused by two types of defects in graphene: (1) a hydrogen atom chemisorbed to an individual carbon atom of the graphene sheet and (2) a single

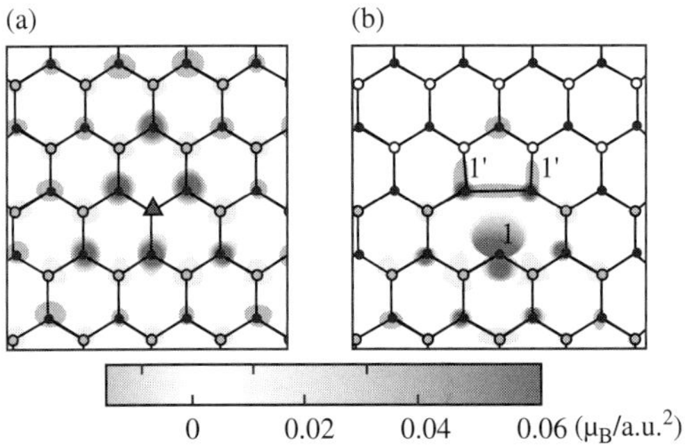

Figure 7.10 Schematic representation of the graphene sheet in the vicinity of a defect. Open (closed) circles indicate lattice sites of type A(B). (a) The case of hydrogen atom adsorption, involving hydrogenation of the central carbon atom. (b) The case of a single-atom vacancy, giving rise to major geometric rearrangement, specifically the emergence of a nine-membered ring structure. For both situations, the spin density, projected on the graphene plane, is indicated in units of $\mu_B/a.u.^2$ The darkly (lightly) shaded regions correspond to positive (negative) values of the spin density. Reprinted with permission from [234]. Copyright (2007) by the American Physical Society.

vacancy. The former type is produced chiefly by hydrogen plasma treatment [229]; the latter by irradiation with high-energy particles [218]. The geometric structures in the vicinity of the two defects are schematically shown in Figure 7.10. Little rearrangement of the graphene network is observed for hydrogen atom absorption (see Figure 7.10(a)), excepting a small vertical displacement of the hydrogenated carbon atom toward the adsorbate. In the single-vacancy case, on the other hand, the graphene sheet reconstructs, resulting in the formation of a nine-membered ring (see Figure 7.10(b)). This pattern arises from Jahn–Teller distortion[2] of the carbon mesh following the removal of one atom. Specifically, the two atoms closest to the vacancy develop a weak covalent bond [218].

The magnetic effects due to a single vacancy, represented in Figure 7.10 by the planar spin density, are strongly determined by the one remaining unsaturated sp^2 dangling bond, which contributes a magnetic moment of size $1\mu_B$. The overall magnetic moment ranges between 1.45 and 1.53 μ_B per defect, provided the defect density is below a certain threshold, corresponding to a next-neighbor distance of approximately 4.0 $\AA$. The deviation of this finding from unit value is ascribed to an admixture that stems from the p_z orbitals of the three atoms closest to the vacated site and involves a certain margin of next-nearest neighbor interaction. In the chemisorption case, however, an overall magnetic moment of $1\mu_B$ is observed throughout. Here the p_z orbital of the hydrogenated carbon atom forms a bond with

[2] Section 12.3.1 contains a condensed introduction to this effect.

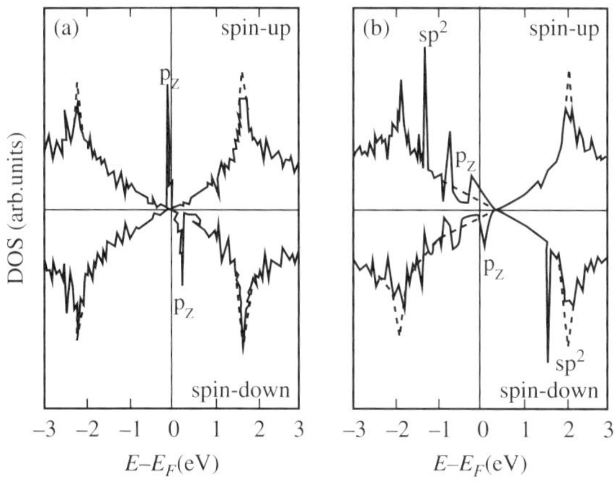

Figure 7.11 The density-of-states distributions for the two defects. As in Figure 7.10, the left panel represents the case of H atom adsorption on top of a graphene C atom, and the right a single vacancy. The dashed line refers to the intact graphene sheet. The labeled substructures are attributed to the defects. Reprinted with permission from [234]. Copyright (2007) by the American Physical Society.

the adsorbed hydrogen atom. Thus the magnetism is carried by p_z contributions alone that are due to ruptured π-bonding between the hydrogenated atom and those adjacent to it. The different nature of the two defect types is manifested by the two density-of-states distributions shown in Figure 7.11. For the hydrogenated system, the signatures of the defect are much more sharply localized than for the vacancy-containing system, in accordance with more regular defect-generated features in the former, and more extended features in the latter case, as is apparent from a comparison between the spin densities displayed in Figure 7.10 for both scenarios.

More information about magnetic effects associated with atomic impurities will be supplied in Chapter 12. In the present context, hydrogen chemisorption is included for the sake of clarifying the distinguishing traits of vacancy-induced magnetism in graphene sheets.

From transmission electron microscopy (TEM) studies of graphene [235], it has been concluded that electron-beam irradiation of graphene gives rise to the formation of large holes in the carbon network rather than widely distributed multitudes of smaller voids. This may be rationalized by comparing voids of different extensions in terms of their stability. To this end, Cui et al. [236] investigated a series of extended voids, termed *nanoholes*, by DFT computation of their *formation energies*, E_{form}. This quantity is defined as

$$E_{form} = E_{def} - E_{int} + n\mu_C, \tag{7.15}$$

where E_{def} and E_{int} stand for total supercell energies of the graphene sheet with defects and the intact sheet, respectively, μ_C for the chemical potential of a carbon atom, and n for the number of carbon atoms removed. The overall tendency of E_{form} as a function of n in the studied interval, ranging from n=2 to n=60, is a marked decrease from about 8 eV to about 1 eV, indicating that the energy needed for consecutive removal of edge atoms diminishes with the size of the nanohole. This trend makes it likely that extended voids, as experimentally observed, originate from progressive elimination of edge atoms in a given void rather than from the fusion between smaller vacancies. The decrease of the function E_{form} with increasing n is modulated by a series of local maxima, corresponding to *magic void sizes*, and thus structures of exceptionally high stability, and characterized by magic numbers of removed atoms. The magic status of these systems is also reflected by high values of the *nanohole dissociation energy*, which indicates the energy needed to eliminate one more atom from a hole of size n.

From TEM spectroscopy, a preference for zigzag lines at nanohole boundaries has been noticed [235]. Consequently, many of these voids turn out to be magnetic. A finite energy band gap is reported for these magnetic structures [236], which might account for the observation of room-temperature ferromagnetism in graphene [237]. This phenomenon, secured by SQUID magnetometer measurement, might be a manifestation of nanohole magnetism. From the study of Cui et al., confirming earlier evidence [238], Lieb's theorem is not a reliable guide for exploring the magnetic moments associated with these large holes in the graphene mesh. The discrepancies found between the prediction from Lieb's theorem ($\mu = N_I \mu_B$) and the magnetic moments obtained from spin-polarized DFT calculation underscore the need to go beyond the confines of the Hubbard model when treating graphene-based systems that involve major structural modifications and significant reconstruction.

7.2.3 Edge Excitations in Zigzag Graphene Nanoribbons

Many conclusions drawn for the bounding structures of voids in the graphene network carry over to zGNR edges. The unpaired electrons at these sites may be viewed as occupying connected and ferromagnetically ordered single-vacancy states. Their continuity likens them to one-dimensional spin chains. As a conclusion from the *Mermin–Wagner theorem* [239], however, perfect ferromagnetic order in one dimension is not possible at any finite temperature. Thus, for $T > 0$, fluctuations modify the rigid alignment of spins at zGNR edges. More specifically, the spin chain is capable of sustaining collective magnetic excitations, which may assume the form of transversal or longitudinal waves [240], as illustrated in Figure 7.12.

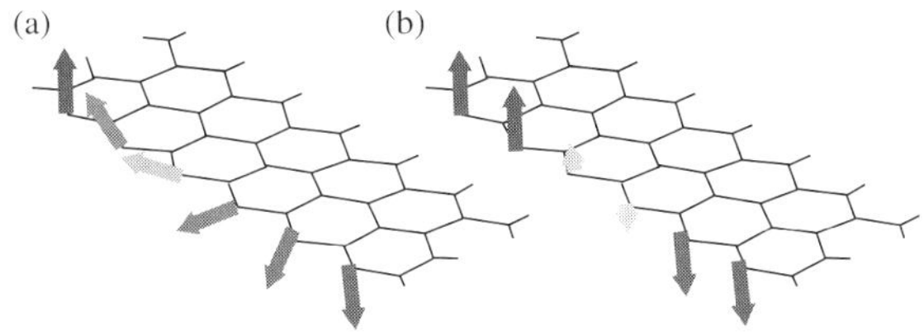

Figure 7.12 Collective magnetic excitations of zGNR edges. (a) Transversal fluctuations: a series of coordinated, phase-shifted rotations of the spins localized at the zGNR edges (spin waves). (b) Longitudinal fluctuations: variation of zGNR edge spins oriented in a direction perpendicular to the ribbon plane. Reprinted with permission from [240]. Copyright (2008) by the American Physical Society.

Yazyev and Katsnelson performed constrained self-consistent computations within the spin-polarized DFT formalism to obtain the spectrum of magnetic excitations at zGNR edges, employing a non-collinear approach [241, 242]. Combining these results with the quadratic spin-wave dispersion relation (see the supplementary section indicated on p.52), they derived the spin-wave stiffness for zGNR edge states. This quantity turned out to be higher than the spin-wave stiffness of bcc iron by one order of magnitude. Further, the authors investigated the zGNR spin wave coherence, adopting a classical model Hamiltonian of the anisotropic Heisenberg form (compare with Eq. (3.33)):

$$\hat{H} = -a \sum_{\langle i,j \rangle; i \neq j} \mathbf{S}_i \cdot \mathbf{S}_{i+1} - d \sum_i S_{z,i} \cdot S_{z,i+1}. \tag{7.16}$$

Admission is made for axial anisotropy, as measured by the ratio of the coefficients d and a. Since the spins are here conceived as semi-classical entities and the second expression on the right-hand side of (7.16) involves only spin z-components, this correction may be understood as an *Ising anisotropy* (see Section 3.1.2). The small spin-orbit interaction strength in graphene, which is on the order of 0.01 meV ([243, 244, 245], see Chapter 8), as well as experimental information about two-dimensional magnetic correlations in graphite [246] yield an estimate of 10^{-4} for d/a. From modeling the magnetic order of the system as a function of temperature, the anisotropy term was found to be of negligible influence above a critical (or *crossover*) temperature [247] around 10 K. All three components of the correlation length vector $\boldsymbol{\xi}$, implicitly defined through the relation $\langle S_{a,i} S_{a,i+l} \rangle = \langle S_{a,i} S_{a,i} \rangle \exp(-\frac{l}{\xi_a})(a = x, y, z)$, match the functional form $\xi_a \approx \frac{300}{T}$, which is the expected outcome for the isotropic Heisenberg model [250]. Below the crossover temperature, however, ξ_z is seen to diverge exponentially as $T \to 0$, which is a fingerprint of the Ising model. At room temperature, the spin correlation length was found to be on the order of one nanometer, confining long-range magnetic order in zGNRs, and thus the dimensions of spintronics devices, to the nanometer regime. The constraint of operating these devices at low temperatures can be eased by exploring ways to increase the crossover temperature. This may be achieved by strengthening the spin-orbit coupling within the zGNR, which in turn will enhance the axial anisotropy.

7.3 Topological Defects in Graphene

The *topological defects* to be discussed in this subsection involve structural alterations of the graphene network that preserve the connectivity of the graphene lattice, leaving every C atom with three nearest neighbors [248]. The best-known topological defect in carbon networks arises from the *Thrower–Stone–Wales* transformation [249, 252], which involves a rotation of a C-C bond by 90° in the graphene network, generating a heptagon-pentagon pair as shown in Figure 7.13.

Much recent work has focused on extended topological defects in graphene [248, 253, 254, 255, 256, 257]. A natural starting point of this search is to ask for one-dimensional counterparts of grain boundaries in graphite [258].

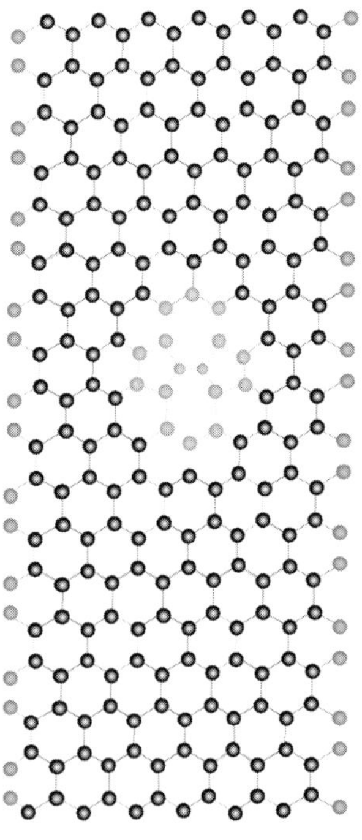

Figure 7.13 Thrower–Stone–Wales defect in an aGNR. The substructure originates from turning a C-C bond by 90°, which results in two connected pentagon-hexagon defects. Reprinted from [248], with permission from IOP Publishing.

One-dimensional topological defects (or *topological line defects*) have been predicted by computer modeling [253, 254] and detected in the laboratory [255]. STM investigation yielded a defect line formed by the periodically repeating motif of an octagon adjacent to two fused pentagons, termed *558 defect*. This geometric motif can be designed in a controlled way by anchoring two graphene half-lattices on an Ni(111) surface, such that every other carbon atom of the first half-lattice is located above an Ni fcc-hollow site, and every other carbon atom of the second above an hcp-hollow site. The boundary separating the two graphene domains then reconstructs, and the 558-pattern emerges [255]. DOS analysis of the defect ascertained the coincidence of $\frac{1}{\sqrt{E}}$ singularities with the maxima of the defect-induced energy bands, which is the signature of a one-dimensional structure (see Eq. (5.24)). This line defect is oriented along the zigzag direction of the graphene sheet, as shown in Figure 7.14 (a). Analogous substructures, extending along the armchair or various chiral directions, have been discussed [2].

The presence of the line defect transforms graphene from a semi-metal into a metal [255]. In terms of the graphene band structure, this effect is rooted in the appearance of a band with an almost flat regime in the center of the first Brillouin zone, giving rise to a spike in the DOS at the Fermi energy. The corresponding electronic states extend along the defect in a narrow interval. In summary, a line defect of the 558 type acts as a *one-dimensional wire*.

The 558 defect may be viewed as a junction of two zigzag lines connected by a line of periodic C_2 units, as shown in Figure 7.14(b,c). Computational investigations of the magnetism imported into the graphene sheet by this structural modification yielded ferromagnetic coordination across the connecting C_2 line ([259], see Figure 7.14(a)). The reason for the emergence of a magnetic ground state lies, as in the case of zGNRs, in the splitting of a flat band into two separate bands. In contrast to zGNRs, which display antiferromagnetic order, defect-modified graphene is not a bipartite system, and, therefore, prevalance of an antiferromagnetic configuration at the site of the defect cannot be expected.

As the 558 defect is incorporated into a zGNR, however, a rich variety of electronic and magnetic phases can be generated, as suggested by DFT modeling [256]. Specifically, the physical properties of the zGNR may be manipulated by selecting the position of the line defect within the ribbon. This position is defined by the notation N_1-N_2, where the labels N_1 (N_2) stand for the numbers of zigzag lines parallel to the defect and on the left (right) of it, as illustrated in Figure 7.14(c) for the example of $N_1 = 7$ and $N_2 = 3$. Keeping the sum of N_1 and N_2 fixed (Lin and Ni consider the case of $N_1 + N_2 = 10$ [256]), one finds that the band structure of the ribbon depends sensitively on the choice of N_1. For $N_1 = N_2 = 5$, for instance, the system turns out to be semiconducting. Half-metallicity, however, emerges for ($N_1 = 9$, $N_2 = 1$), corresponding to a finite gap between the valence

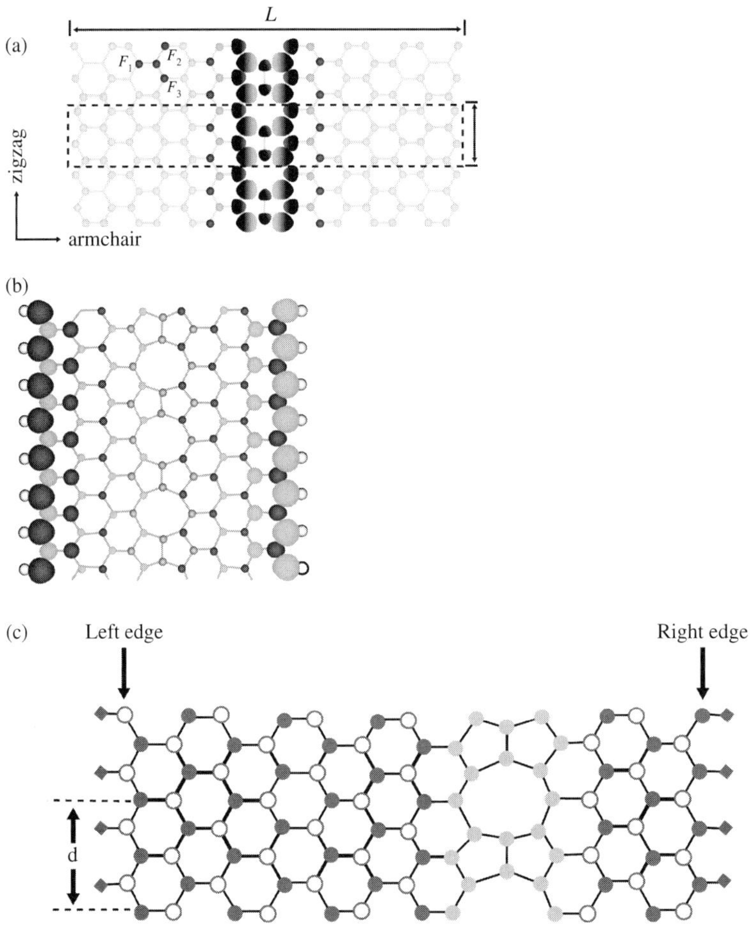

Figure 7.14 Topological line defect of the 558 type. (a) Geometric structure of the defect, extending along the zigzag direction of the hexagonal network [259]. The different shades refer to the two spin orientations. (b) A 558 line defect at the center of a nine-layer zGNR [257]. Again, the spin density is represented. (c) A 558 line defect embedded into a zGNR, asymmetrically placed. The defect is of the 7-3 type, having 7(3) parallel zigzag lines on its left (right) side [256]. Reprinted with permission from [259] (a). Copyright (2011) by the American Chemical Society. Also, reprinted with permission from [257] (b) and [256] (c). Copyright (2012) and (2011) by the American Physical Society.

and the conduction band for one spin channel and a vanishing gap[3] for the other. In terms of their magnetism, both configurations turn out to be antiferromagnetic. This assignment changes, however, as one proceeds to zGNRs with greater width.

[3] *Vanishing gap* in the sense of $\Delta E_{cb-vb} \leq 0$, with ΔE_{cb-vb} as the eigenvalue difference between the bottom of the conduction band and the top of the valence band.

For $N_1 + N_2 = 12$, a crossover from the antiferromagnetic to the ferromagnetic state was found for the index pair ($N_1 = 11$, $N_2 = 1$). The magnetic phase is thus governed by an elementary geometric parameter, the absolute width of the ribbon.

Other strategies of employing topological line defects to tune the magnetic phase of a graphene sheet have been described [257, 259]. As suggested by DFT results [259], defect-induced magnetism in graphene can be substantially increased by applying tension along the zigzag direction, i.e. along the defect line. A critical tension that would cause a sharp sudden drop of the magnetic moments could be identified. Straining the structure along the armchair direction, in contrast, was shown to reduce the magnetism. In zGNRs, applying tensile strain was seen to cause transitions of the magnetic phase [257]. Kan et al. [257] investigated a 558 line defect at the center of a nine-layer zGNR, displayed in Figure 7.14(b), by DFT computation. In keeping with regular zGNRs, this system is antiferromagnetic in its equilibrium geometry. Strain applied along the zigzag direction, however, stabilizes the ferromagnetic and destabilizes the antiferromagnetic state. The crossover point between the competing phases is reached at a lattice elongation of 2.7 percent. This change is related to the magnetization of the connecting C_2 line, which at equilibrium is entirely non-magnetic, as demonstrated by the spin density distribution shown in Figure 7.14 (b). In the process of lattice expansion along the direction of the defect, the interatomic distance of the C_2 units increases, endowing them with a magnetic moment and changing the magnetic balance of the system as a whole.

Further DFT simulations imply that a similar feat of magnetic tuning can be accomplished by substituting one of the carbon atoms in a C_2 unit with a foreign atom having a valence electron number that differs from that of carbon by one [257]. This manipulation exchanges carbon for another atom with an intrinsic magnetic moment and, once more, increases the interatomic next-nearest-neighbor distance in the connecting line of the 558 defect. The effect of substituting with B, N, P and Al atoms was examined. In all cases, atomic doping led to ferromagnetic ground state configurations, where Al substitution resulted in the strongest preference of the ferromagnetic over the antiferromagnetic phase. Applying tensile strain and atomic substitution at selected defect sites turn out to be alternative strategies to tune the magnetism of line-defect-modified zGNRs, turning antiferromagnetic into ferromagnetic coordination.

7.4 Truncated Nanotubes

Extending the foregoing arguments from GNRs to nanotubes, one may conjecture that tubes of the zigzag type display ground state magnetism, in analogy to the zGNRs from which they originate. However, complicating factors obscure this inference. First, the impact of nanotube curvature on the ground state magnetic moment is not clear, *a priori*. A second direction of inquiry pertinent to zigzag tubes, but not to the underlying zGNRs, is the role of the tube termination mode in defining the magnetic properties of the tube. While zGNRs are assumed to be hydrogenated, nanotubes are frequently capped by fullerene hemispheres. Alternatively, their ends may be left without terminating structure and instead be subjected to geometric reconstruction.

DFT computations on axially confined zSWCNTs of the (10,0) type, with lengths ranging from three to twenty C atom layers, verified the expectation of a magnetic ground state unanimously [260]. This preference was confirmed for (10,0) tubes with the three termination structures mentioned earlier, namely hydrogenation, truncation with edge reconstruction, and capping by fullerene segments, as shown in Figure 7.15. The magnetic properties of the two former systems were seen to be quite similar. For the capped tube, in contrast, local magnetic moments of diminished magnitude were obtained, associated with a more delocalized spin density distribution along the tube axis.

The *linear spin density* is an important tool for analyzing the magnetic effects associated with both zSWCNTs and zGNRs [260] as a function of their length or

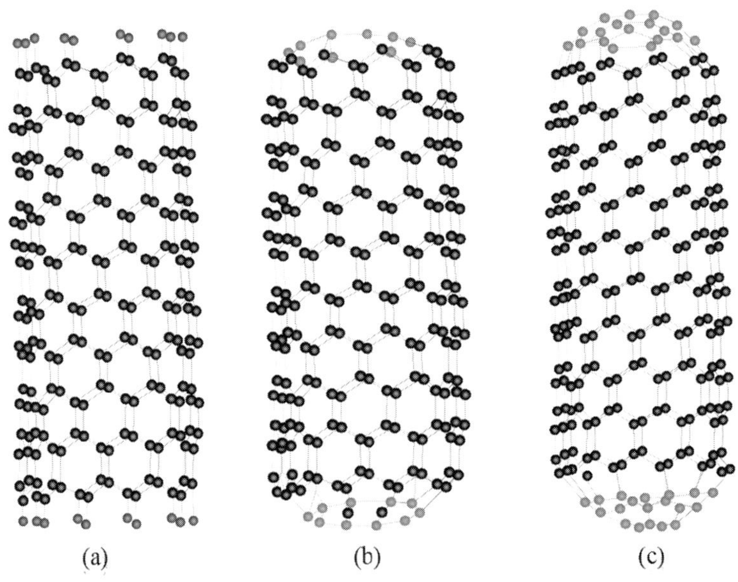

Figure 7.15 Equilibrium structures of a zSWCNT of the (10,0) type with (a) H-termination ($C_{200}H_{20}$), (b) truncation and edge reconstruction (C_{200+20}) and (c) capping with fullerene hemispheres (C_{200+40}). The light grey symbols represent H atoms (a), C atoms in the terminating decagons (b) or fullerene hemispheres (c). Reprinted with permission from [260]. Copyright (2009) by the American Physical Society.

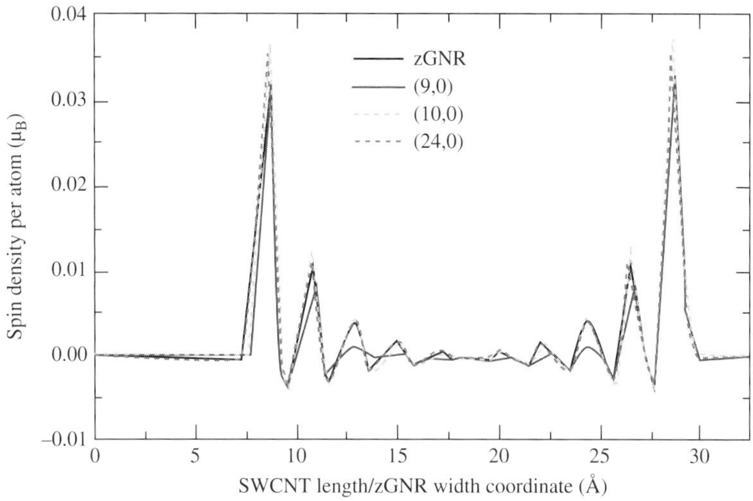

Figure 7.16 Linear spin density distribution as a function of the zSWCNT length (zGNR width) coordinate for the case of FM coordination. SWCNTs of the (n,0) type with n = 9, 10 and 24 are included. The full length of all systems is ten atomic layers. Reprinted from [261] with permission from Elsevier.

width coordinates, respectively. This quantity describes a spin polarization average, taken over the C atoms of an atomic layer perpendicular to the tube axis for zSWCNTs, and over the ribbon length for zGNRs. As demonstrated in Figure 7.16 by the results of DFT computations for zSWCNTs with three different tube indices and a zGNR, all systems comprising ten layers of C atoms, both species display very similar linear spin polarization profiles [261]. The case of ferromagnetic coordination is displayed here. Analogous conclusions have been drawn for the antiferromagnetic phase. In accordance with the zGNR spin density profiles shown in Figure 7.5, the spin polarization is not confined to the edges of the respective nanostructure, but shows an oscillatory pattern, fading out toward the middle of the structure. This behavior is analogous to that of the spin polarization in a non-magnetic material sandwiched by layers with itinerant magnetism [260, 262].

In recent DFT computations [261], both ferromagnetic and antiferromagnetic solutions, related to parallel and antiparallel orientation of the spins at the two nanotube edges, were identified for a large set of zSWCNTs, with tube indices ranging from seven to twenty-four. In accordance with the zGNR paradigm, the latter phase was seen to prevail consistently over the former. This observation has been made by several authors, e.g. for zSWCNTs of the (14,0) type [263] and for short zSWCNTs with tube indices varying from four to fourteen [264]. A DFT study on the variation of the magnetic zSWCNT properties with the length of the tube, however, identified a threshold tube index (n=7) below which the preferred magnetic phase changes sensitively as a function of the tube length [261]. This behavior establishes another similarity between zSWCNTs of sufficiently high curvature and layered transition-metal compounds that consist of two ferromagnetic segments enclosing a spacer unit, such as Fe/Cr/Fe or Co/Ru/Co [265]. By experimental observation, changing the spacer length was shown to induce oscillations of both the sign and the magnitude of the magnetic exchange coupling between the two boundary layers, with periods between about 12 to about 21 $\AA$. Aspects of this phenomenon were explained by use of quantum-well theory [266] and RKKY theory [267]. Quantitative modeling of the oscillation periods succeeded in the framework of a more generalized approach [268]. While the qualitative difference between the complexes compared here, i.e. layered transition-metal systems involving a spacer unit that differs from the ferromagnetic boundary segments, and homogeneous zigzag SWCNTs with sufficiently high curvature, must be kept in mind, both materials allow for controlled variation of the magnetic phase by manipulating the length of the system.

7.4.1 Structures with Negative Gaussian Curvature

The building block of the *carbon Schwarzites*, materials assembled from units with hyperbolic surfaces, is a *carbon tetrapod*, as shown in Figure 7.17. Due to the geometric constraint imposed by the saddle-like shape of the regions between the three tubular extensions, three heptagons appear in these regions. As clarified by the bonding scheme displayed in the image, this arrangement requires the presence of three sp^2 carbon radicals in the tetravalent carbon network. Computations at the local spin density approximation (LSDA) level [269] yield a ground state spin multiplicity of five, corresponding to four unpaired spins, for a carbon tetrapod with saturated edges. This finding is plausible if one assumes that each trivalent center in each region of negative curvature contributes one electron, and that two of these electrons in each negative-curvature region bond with each other, leaving a balance of four unsaturated electrons.

Combining the geometric paradigm of negative Gaussian curvature with the zSWCNT motif to form novel carbon nanostructures, one arrives at composites of fused zSWCNTs, such as T- [270], Y- [271] or H-shaped [272] units. Commenting in more detail on the latter, we focus on *cross-linking carbon architectures* or *cross-linking carbon nanotubes* (CLCNTs), as exemplified by the species shown in Figure 7.18, which involves a regular zigzag nanotube as basic structural element. These structures arise from merging three SWCNTs. Adjacent pentagons, heptagons or octagons [273] have to be admitted into the contact regions between the three segments in order to reduce the strain associated with fusing them. Figure 7.18 shows a specific structural CLCNT type, composed of (10,0) nanotubes, and characterized by an octagon-pentagon-heptagon-pentagon-octagon (8-5-7-5-8) motif in the contact regions.

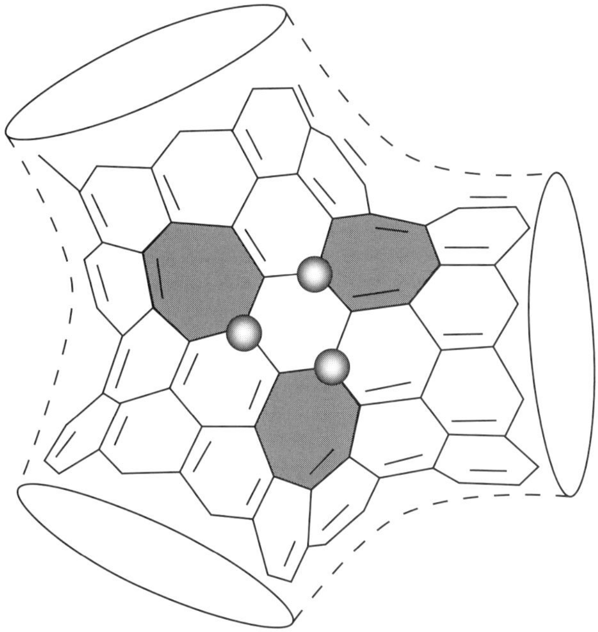

Figure 7.17 Sketch of a carbon tetrapod, the building block of Schwarzites. In the presented view, the fourth tubular extension is not visible. Emphasized is a region of negative Gaussian curvature, containing three heptagons and, correspondingly, three carbon radicals. Reprinted with permission from [269]. Copyright (2003) by the American Physical Society.

In terms of CLCNT magnetism, three basic configurations have to be distinguished: the two vertical segments may both display ferromagnetic (FM-FM) or antiferromagnetic (AFM-AFM) order, or a mixed solution may be realized, in the sense that one is ferromagnetically, and the other one antiferromagnetically (FM-AFM) coordinated. In most investigated cases, the AFM-AFM configuration prevails over its alternatives, which is in accordance with the behavior of an elementary zSWCNT of the (10,0) type (see above). For the model shown in Figure 7.18, DFT computations proved the AFM-AFM solution to be most stable, followed, with an energy difference of 6 meV, by the FM-AFM, and lastly, with a difference of 10 meV, the FM-FM arrangement. Not all CLCNTs analyzed, however, conform with this hierarchy of magnetic states. Other departures from expectations based on zSWCNT magnetism include the FM-FM configuration of a 5-7-5 CLCNT. The (10,0) constituent tube carries a magnetic moment of 3 μ_B at each end, from which an overall magnetic moment of 12 μ_B might be inferred for the ferromagnetic state of the CLCNT. The computational result for this, in contrast, is 10 μ_B. Both irregularities are rooted in the same effect, namely the appearance of additional, local magnetic moments in the geometrically complex contact regions between the three system parts, in accordance with the

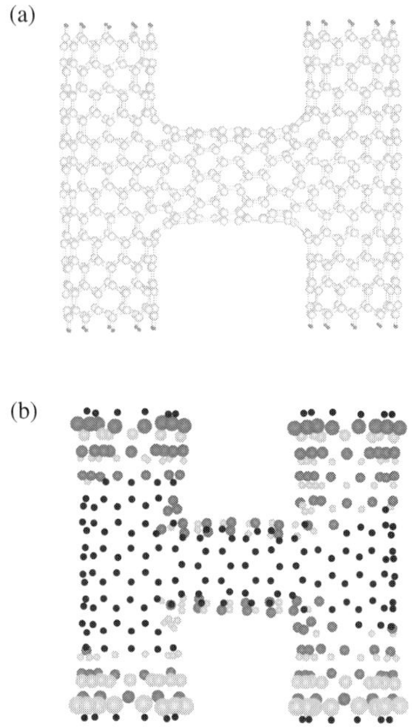

Figure 7.18 (a) Structure of a hydrogenated cross-linking carbon nanotube (CLCNT) with an 8-5-7-5-8 motif in the contact region joining the nanotube segments, (b) spin density of the species shown in panel (a) for the case of AFM-AFM coordination, which is, from DFT calculation, the maximally stable magnetic state. Spin up (down) contributions to the total spin density are indicated in dark grey (light grey). This species consists of three zSWCNTs of the (10,0) type. Reprinted with permission from [272]. Copyright (2013) by the American Chemical Society.

negative Gaussian curvature of these zones. These moments might form a ferro- or antiferromagnetic connection across the bridge, giving rise to a large diversity of different magnetic phases in CLCNTs. This is exemplified by the model shown in Figure 7.18, where both sidewalls are coordinated antiferromagnetically, and the local moments in the contact regions exhibit the same type of order.

CLCNTs, as much as T- and Y-shaped composites, may be of interest as junction elements on nanoelectronic or spintronic applications. They also might find use as highly durable fibers in nanomechanics, as the strongest connection between two nanotubes is provided by another nanotube.

7.5 Intrinsic Fullerene Magnetism

Research, both experimental and theoretical, on fullerene magnetism has mostly focused on polymers based on C_{60} [274]. As mentioned in Section 6.3, solid FCC C_{60} forms under the influence of the Van der Waals force. Crystalline C_{60}

may be turned into a polymer, involving chemical bonds between C_{60} molecules. Specifically, a one-dimensional orthorhombic C_{60} polymer phase has been generated by chemical doping with alkali atoms [275], and a diversity of C_{60} polymer structures, such as a one-dimensional orthohombic, a two-dimensional tetragonal and a two-dimensional rhombohedral phase (see Figure 6.10), has resulted from pressure and temperature treatment.

Some authors reported the emergence of ferromagnetism in compressed and heated C_{60} polymers, with Curie temperatures in the regime of 500–800 K [276, 277, 278, 279]. This has been observed for the two-dimensional tetragonal and rhombohedral phase after heat treatment and application of pressure. These results were greeted with great interest, since the practical use of purely organic molecular magnets is compromised by their low Curie temperatures.[4]

Models have been put forward to rationalize the appearance of ferromagnetism in two-dimensional polymeric phases of C_{60} [274, 281, 282]. These approaches rest on the nature of the fullerene-fullerene bond realized in these systems. We recall that the bonding between two adjacent C_{60} molecules is the result of a [2+2] cycloaddition process, linking them by two single bonds that emerge from the rupture of two hexagon-hexagon double bonds on each molecule. Performing DFT investigations on the C_{60} dimer by use of the B3LYP functional, Ribas-Ariño et al. [282] identified a critical distance between the fullerene molecules, below which a single bond between them led to higher stability than a double bond. The reason for this behavior is simple: if the molecules are sufficiently close together, compressing them even further takes less energy if the fragments are joined by one rather than by two C-C bonds. One of the two bonds that originate from [2+2] cycloaddition thus breaks, resulting in one unpaired spin on each fullerene. From DFT computation, the two spins configure to form a spin triplet state. These observations provide a quantum-chemical argument for the emergence of magnetism in polymerized C_{60} if a sufficiently high amount of pressure is applied.

While this and competing models make it plausible that exposing polymerized C_{60} to pressure might induce a magnetic phase, the experimental evidence for this phenomenon has been called into question. Closer experimental scrutiny revealed that the magnetic effects in polymerized C_{60} had to be, at least in part, attributed to the presence of magnetic metal traces in the analyzed materials. For instance, a laterally resolved particle induced x-ray emission (PIXE) and magnetic force microscopy (MFM) examination of a C_{60} polymer [283] found substantantial amounts of iron impurities in the sample. A combined experimental-computational investigation [284] arrived at the conclusion that the pure rhombohedral phase of C_{60} is not ferromagnetic. Specifically, a local spin density approximation (LSDA)

[4] Pressure increase has been shown to lead to an enhanced Curie temperature when applied to organic molecular magnets, see [280].

study yielded a non-magnetic ground state solution for this phase. Magnetization measurements carried out on a rhombohedral C_{60} sample revealed disordered magnetism that was ascribed to the presence of isolated ferromagnetic clusters within the fullerene-polymer phase. Also, C_{60} polymer samples that were mixed with iron prior to polymerization [285] turned out to have a Curie temperature close to that measured in the initial experiments on C_{60} polymer magnetism [276]. This was ascribed to the formation of the compound cementite (FeC_3) in the mixture [285].

Ferromagnetism was found to appear as a result of ion irradiation of thin polymeric C_{60} films [286, 287], where beams consisting of Si [286], O, Ti or Au [287] projectiles were employed. Raman spectra, recorded for the pristine as well as the irradiated films, gave evidence of a transformation of the C_{60} molecules into amorphous carbon when a certain threshold of the irradiation fluence was crossed. As the initial polymer is restructured into an amorphous network formed by carbon atoms in sp^2 and sp^3 states of hybridization, ferromagnetism emerges in the sample, as was demonstrated by SQUID as well as ESR analysis of the pristine and the irradiated fullerene film. The detected increase of the magnetic signal with the ion fluence suggests that the magnetic effects are due to the creation of defects by the ion beam.

A great variety of magnetic materials has been manufactured by joining fullerenes with endohedrally or exohedrally coordinated impurities. A survey of these system will be provided in Chapter 12.

8

Spin-Orbit Coupling in Carbon Nanostructures

Among the advantages of graphene as medium for spin currents is the weak spin-orbit coupling in this material, in accordance with the low atomic number of carbon. This feature is beneficial wherever long spin lifetimes are relevant, as in spintronics networks. Assessing and optimizing the efficiency of graphene and other carbon nanostructures as elements of spintronic circuits requires a detailed understanding of spin-orbit coupling mechanisms. Beyond practical considerations, examination of spin-orbit interactions also provides insight into the basic architecture of graphene and nanosystems derived from it. As outlined in Chapter 4, the two Dirac points are each characterized by an electronic state of fourfold degeneracy, corresponding to two spin and two sublattice degrees of freedom. Addition of the spin-orbit effect to the description of graphene lifts the spin degeneracy and thus opens a gap at the points K and K', separating two states with twofold degeneracy. Spin-orbit interaction thus modifies the fundamental electronic structure of graphene.

As specified in Section 8.1, the *intrinsic spin-orbit gap* in graphene is extremely small. Use of the *Rashba effect*, on the other hand, makes it possible to tune the gap by applying an external electric field to the graphene sheet. This device allows us to close the gap, or to widen it by several orders of magnitude. Further, a physical sheet of carbon atoms is never ideally flat, and curved segments of the graphene layer turn out to be associated with effective electric fields. This feature implies a critical dependence of the spin-orbit effect on the nanostructure geometry. It acquires major importance as one goes from graphene to carbon nanotubes. Section 8.3 is dedicated to spin-orbit coupling in SWCNTs and emphasizes the geometric origin of this effect. In contrast to graphene, this interaction is very pronounced in SWCNTs, inducing a marked energy splitting between two Kramers doublets at the Dirac point. As it is masked by competing effects, its definite experimental characterization was a relatively recent event [288].

Section 8.2 deals with orbital magnetic moments in SWCNTs and an *orbital shift* undergone by the Dirac cones as a consequence of magnetic interactions involving these moments. This mechanism turns out to be essential for a proper appreciation of spin-orbit processes in nanotubes, as will be outlined in Section 8.3.

8.1 Spin-Orbit Coupling in Graphene

The operator of spin-orbit coupling is given by

$$\hat{H}_{SO}(r) = \frac{\hbar}{4m_e^2c^2}(\nabla V \times \hat{\mathbf{p}}) \cdot \boldsymbol{\sigma}, \tag{8.1}$$

where V stands for the relevant electrostatic potential energy function, $\hat{\mathbf{p}}$ for the momentum operator, m_e for the electron mass, and $\boldsymbol{\sigma}$ for the Pauli matrices. Applying Eq. (8.1) to the carbon atom, one may write

$$\hat{H}_{SO} = \xi \hat{\mathbf{L}} \cdot \boldsymbol{\sigma}, \tag{8.2}$$

as a representation that is valid in an intra-atomic basis, such as provided by the (2s, 2p) valence electrons of the carbon atom. Here $\hat{\mathbf{L}}$ denotes the angular momentum operator. The intra-atomic spin-orbit coupling constant ξ is obtained from

$$\xi = \frac{\hbar^2}{4m_e^2c^2}\langle\frac{1}{r}\frac{\partial V}{\partial r}\rangle. \tag{8.3}$$

To estimate the energy splitting on account of intra-atomic spin-orbit coupling in carbon, we substitute the Coulomb potential ($V(r) = \frac{Ze^2}{4\pi\varepsilon_0 r}$) into formula (8.3) and make use of the of the relation

$$\langle\frac{1}{r^3}\rangle = \frac{Z_{eff}^3}{L(L+\frac{1}{2})(L+1)n^3a_{Bohr}^3}, \tag{8.4}$$

with a_{Bohr} as the first Bohr radius, and Z_{eff} as effective nuclear charge. This yields, for an electron with the main quantum number n and angular momentum quantum number L, the following energy correction:

$$\Delta E_{SO} = Z_{eff}^4 m_e c^2 \frac{\alpha^4}{4n^3}\frac{\{J(J+1) - L(L+1) - S(S+1)\}}{L(L+\frac{1}{2})(L+1)}. \tag{8.5}$$

The expression provides a measure of the spin-orbit splitting in a C atom. For $n = 2$, $L = 1$, and $Z_{eff} \approx 4$, one finds a margin of $\Delta E_{SO} \approx 12$ meV between the states with total angular momentum quantum numbers $J = \frac{1}{2}$ and $J = \frac{3}{2}$, as confirmed by recent computations of higher accuracy [288, 289].

An alternative way of partitioning the spin-orbit Hamiltonian (8.2) is given by:

$$\hat{H}_{SO} = \xi[\frac{1}{2}(\hat{L}_{+}\boldsymbol{\sigma}_{-} + \hat{L}_{-}\boldsymbol{\sigma}_{+}) + \hat{L}_{z}\boldsymbol{\sigma}_{z}], \tag{8.6}$$

clarifying that $\hat{H}_{SO}$ may induce transitions between spin and orbital states. The subscript $+(-)$ refers here to the step-up (step-down) operator derived from the spin or the angular momentum operator.

What is the impact of spin-orbit interaction on the band structure of graphene? In the first place, does this effect lift the degeneracy at the Dirac points?

Inquiring about spin-orbit coupling in the low-energy regime, i.e. the graphene π band, we find that it is an effect of second order. Symmetry under reflection with respect to the graphene plane (*space inversion symmetry*) implies that the terms proportional to $\hat{L}_i \cdot \boldsymbol{\sigma}_i$, with $i = x, y$, in Eq. (8.2) do not contribute to the spin-orbit effect as reflection by this plane maps $\{\hat{z}, \hat{p}_z\}$ into $\{-\hat{z}, -\hat{p}_z\}$ and thus $\{\hat{L}_x, \hat{L}_y\}$ into $\{-\hat{L}_x, -\hat{L}_y\}$. The remaining term $\hat{L}_z \cdot \boldsymbol{\sigma}_z$ also vanishes due to a further reflection symmetry, with respect to a vertical plane that bisects bonds between nearest neighbors. This feature is referred to as *pseudospin inversion symmetry*, since the underlying operation interchanges the two sublattices, A and B. From an argument analogous to that used in the case of horizontal reflection, one infers that this interchange maps $\hat{L}_z$ into $-\hat{L}_z$, which leads to the conclusion that there is no spin-orbit effect of first order in the π band of graphene, understood as a perfectly planar system, free of impurities.

The leading order of spin-orbit interactions within the π system of graphene is thus the second. A first-order effect in $\hat{H}_{SO}$, on the other hand, is the spin-orbit coupling between the π and the σ subsystem of graphene [290]. To evaluate the impact of this interaction on the π system at $E = 0$, i.e. at the Dirac points K and K', we treat $\hat{H}_{SO}$ as perturbation operator. Writing this operator in matrix representation at second order [75, 291], we find

$$H^{(2)}_{SO,\pi_0,m,n} = \sum_{l \neq \pi_0} \frac{\langle m^{(0)} | \hat{H}_{SO} | l^{(0)} \rangle \langle l^{(0)} | \hat{H}_{SO} | n^{(0)} \rangle}{E_{\pi_0} - E^{(0)}_l}, \tag{8.7}$$

with $E^{(0)}_l$ as the eigenvalues of the unperturbed graphene Hamiltonian. The symbol π_0 stands here for the eigenstates at $E = 0$, which are labelled by the indices m, n. Their multiplicity is four, corresponding to two sublattices (A versus B) as well as two spin degrees of freedom. The second-order process described by this Hamiltonian may be understood as a virtual transition from the π to the σ system, followed by the reverse transition, both mediated by spin-orbit interaction.

A more general expression for the spin-orbit interaction, incuding both Dirac points, can be obtained in a compact fashion by use of Pauli matrices that refer to the spin ($\boldsymbol{\sigma}$), the sublattice ($\boldsymbol{\eta}$) and the valley ($\boldsymbol{\tau}$) degrees of freedom. Like $\boldsymbol{\eta}$ and $\boldsymbol{\sigma}$, the matrix array $\boldsymbol{\tau}$ is represented by the Pauli matrices. Forming products of these three operators, one may write an effective low-energy graphene Hamiltonian as an 8×8 matrix that includes the intrinsic spin-orbit and the Rashba interaction in a clearly partitioned form. Specifically, tight-binding calculations of the intrinsic spin-orbit effect in the basis given by the orbital states $|A\uparrow\rangle, |A\downarrow\rangle, |B\uparrow\rangle, |B\downarrow\rangle$ $(= |p_z\rangle^A|\uparrow\rangle, |p_z\rangle^A|\downarrow\rangle, |p_z\rangle^B|\uparrow\rangle$ and $|p_z\rangle^B|\downarrow\rangle)$ arrived at the following result [291, 292]:

$$\mathbf{H}^{(2)}_{SO,\pi_0} = \xi_I \boldsymbol{\eta}_z \boldsymbol{\tau}_z \boldsymbol{\sigma}_z. \tag{8.8}$$

The parameter ξ_I is given by

$$\xi_I \equiv 2\frac{E_{2p} - E_{2s}}{9V^2_{sp\sigma}}\xi^2, \tag{8.9}$$

where E_{2s}, E_{2p} refer to the the $2s$ and $2p$ energy levels in the carbon atom, respectively, and $V_{sp\sigma}$ is defined within the tight-binding formalism as a *Koster–Slater hopping matrix element* [293, 294], a two-center Hamiltonian matrix element that connects s and p states of neighboring C atoms, where σ bonding is assumed. The components $\tau_{z,1,1} = +1$ and $\tau_{z,2,2} = -1$ are associated with the points K and the K', respectively.

The approximation (8.8) for the low-energy regime has been derived in analogy to the Dirac Hamiltonians (4.22) and (4.27) [8], with the crucial difference that hopping between next-nearest-neighbor (NNN) sites, characterized by the parameter t' in the tight-binding Hamiltonian (4.5), was considered, in contrast to nearest-neighbor interaction. Transitions between lattice sites are thus confined to a single sublattice. The spin-orbit effect is implemented by a spin-dependent hopping amplitude that also includes a signature of the rotational sense of the hopping process. Expansion around the Dirac points, as decribed in Section 4.2.1, leads to expression (8.8).

We add a comment on the peculiar structure of the right-hand term of Eq. (8.8), consisting of a threefold product of spin or pseudo-spin operators. A term proportional to $\boldsymbol{\eta}_z$ alone would be odd with respect to the parity operation applied to the sublattice degree of freedom and thus break with the corresponding symmetry of the low-energy Hamiltonians (4.22) and (4.27), i.e. invariance with respect to interchange between A and B sites in the graphene sheet. The product $\boldsymbol{\eta}_z\boldsymbol{\tau}_z$ is, in contrast, parity-invariant, as both of its factors change their sign upon sublattice inversion, and so is $\boldsymbol{\sigma}_z$. Since $\boldsymbol{\sigma}_z$ and $\boldsymbol{\tau}_z$ are both odd under time reversal, while $\boldsymbol{\eta}_z$ is even, the expression $\boldsymbol{\sigma}_z\boldsymbol{\tau}_z\mathbf{s}_z$ is invariant with respect to both operations.

Adding the correction (8.8) to the Hamiltonians (4.22) and (4.27), and evaluating the eigenenergies, we find

$$E = \pm\sqrt{(\hbar v_F q)^2 + \xi_I^2}. \tag{8.10}$$

Therefore, the gap induced by the spin-orbit interaction in graphene is

$$\Delta_{SO} = 2|\xi_I|. \tag{8.11}$$

The numerical value of Δ_{SO}, as derived from this formula, has been found to be exceedingly small, on the order of 1 μeV, as rationalized by the appearance of the squared spin-orbit parameter ξ in expression (8.9). *Ab initio* calculations, on the other hand, arrive at values that range from 24 μeV [295] to 50 μeV [296]. This discrepancy is explained by carbon d orbitals that were included in the *ab initio*, but not in the earlier tight-binding calculations. Specifically, d orbitals of d_{xz} and d_{yz} symmetry hybridize with the p_z states, and the intrinsic gap turns out to depend linearly on the splitting between orbitals of the form $d_{xz} \pm i d_{yz}$ [295]. Inclusion of the d orbital contributions increases the tight-binding result for the intrinsic gap to 24 μeV, thus bridging the rift between the *ab initio* and the tight-binding approach [290]. Experiment yielded an upper bond of about 100 μeV on the value of Δ_{SO} [297].

The strength of the spin-orbit effect may be modified, as reflection symmetry with respect to the $x - y$ plane (the plane of the graphene sheet) is broken by an electric field $\mathbf{E}$ applied along the z-axis. In this situation, the intrinsic spin-orbit interaction is augmented by Stark coupling, and the overall perturbation Hamiltonian is given by

$$\Delta\hat{H} = \hat{H}_{SO} + \hat{H}_{SE}, \tag{8.12}$$

where $\hat{H}_{SE}$ is a Stark Hamiltonian of the form

$$\hat{H}_{SE} = eE\sum_i z_i, \tag{8.13}$$

with i as site index. Numerical results for the combined effect of intrinsic spin-orbit and Stark interaction are obtained by replacing $\hat{H}_{SO,\pi_0}$ in Eq. (8.7) with $\Delta\hat{H}$ (here we do not reiterate the reference to the π_0 system). The terms of the second-order Hamiltonian (8.7) that contain both components of $\Delta\hat{H}$, $\hat{H}_{SO}$ and $\hat{H}_{SE}$, generate the *Rashba effect*. From second-order perturbation theory, the Rashba effect may thus be understood as a hopping event followed by an on-site spin flip. We point out that substituting $\Delta\hat{H}$ into the Hamiltonian (8.7) also gives rise to an admixture, due to the quadratic Stark effect. The corresponding term is neglected here, as it does not contribute to spin relaxation.

From the symmetry of the Hamiltonian H_{SE}, it couples s and p_z states within the same atom. The strength of the Rashba interaction is thus determined by the coupling constant ξ_{SE} [291], as defined by

$$\xi_{SE} \equiv \frac{eE\langle p_z|z|s\rangle}{3V_{sp\sigma}}\xi. \tag{8.14}$$

We note that this term contains both the electric field parameter and the spin-orbit parameter ξ at linear order, while the constant ξ_I contains the spin-orbit parameter as a square. For Rashba interaction, in contrast to the pure spin-orbit coupling, $\sigma - \pi$ hybridization within the subsystem of p orbitals plays the determining role, while the effect of d orbitals is negligible [290]. Enhancing the intrinsic spin-orbit effect by a Rashba contribution, we obtain an effective spin-orbit Hamiltonian in the Dirac regime [8, 75, 291]. For the K valley,

$$\Delta\mathbf{H} = \begin{pmatrix} -\xi_I & 0 & 0 & 0 \\ 0 & \xi_I & 2i\xi_{SE} & 0 \\ 0 & -2i\xi_{SE} & \xi_I & 0 \\ 0 & 0 & 0 & -\xi_I \end{pmatrix}, \tag{8.15}$$

In analogy to Eq. (8.8), we may write a spin-orbit Hamiltonian that makes the spin, the sublattice and the valley degrees of freedom explicit, namely:

$$\mathbf{H}_{SO}^{eff} = -\xi_I \boldsymbol{\eta}_z \boldsymbol{\tau}_z \boldsymbol{\sigma}_z + \xi_{SE}(\boldsymbol{\eta}_x \boldsymbol{\tau}_z \boldsymbol{\sigma}_y - \boldsymbol{\eta}_y \boldsymbol{\sigma}_x). \tag{8.16}$$

As implied by the foregoing discussion, the second term vanishes unless the mirror symmetry about the plane of the graphene sheet is broken. Investigating the spectrum of $\mathbf{H}_{SO}^{eff}$, one finds a finite gap at these points if $|\xi_I| > |\xi_{SE}|$. In this case, the gap size turns out to be

$$\Delta_{SO} = 2(|\xi_I| - |\xi_{SE}|). \tag{8.17}$$

The electric field thus might be viewed as a handle that makes it possible to tune the gap at the K/K' points within the small scale set by the parameter ξ_I. The purely extrinsic gap, as obtained by setting ξ_I equal to zero, varies linearly with the electric field. It has been found to be about 10 μeV at a field of 1.0 V/nm [290].

The arguments that establish intrinsic spin-orbit coupling as an effect of second order are strictly valid only for graphene as a perfect two-dimensional lattice. Under realistic conditions, however, various physical factors are present that enhance the spin-orbit effect. Thus, curvature breaks the mirror symmetries of the graphene sheet, inducing a Rashba-analogous field with a coupling constant ξ_{curv} where

$$\xi_{curv}^{C} \propto \xi C, \tag{8.18}$$

with C as the mean curvature of the surface [243]. Avsar et al. explored the effect of *nanoripples* as they typically emerge when graphene monolayers are manufactured by chemical vapor deposition (CVD) [298]. Quasi-periodic ripples with radii of curvature around 200 nm were observed and associated with a rather low contribution to the spin-orbit effect of 3.3 μeV.

Spin-orbit coupling within the π_0 system has turned out to be an effect of first order in the presence of curvature, making ξ_R^C typically an order of magnitude larger than the intrinsic spin-orbit effect in flat graphene, as measured by ξ_I [243].

Further, adsorbates covalently bonded to the graphene sheet, such as hydrogen adatoms, change the hybridization of the carbon atoms locally from sp^2 to sp^3 [299]. In this case, the spin-orbit effect on π bands at the K and K' points is of first order in the coupling constant ξ. Chemisorption of external species to the graphene surface may thus greatly enhance its spin-orbit scattering efficiency. This is in agreement with recent experimental data [300].

The very small size of the intrinsic spin-orbit effect in graphene suggests long spin relaxation times in this medium, a prediction of great relevance for possible applications of graphene in spintronics.

8.2 Orbital Magnetic Moments of Single-Walled Carbon Nanotubes

The orbital magnetic moment of carbon nanotubes is a direct consequence of their circumferential momentum $\hbar k_c$ (see Chapter 5). The size and the orientation of this moment follows from the dispersion theory of single-walled nanotubes, as outlined in Section 5.2, in conjunction with the quantization constraint on k_c (Eq. (5.13)). In Figure 8.1, four pairwise-degenerate sub-bands are represented as hyperbolic sections of the double-cones around the Dirac points K, K'. Both the conduction as well as the valence electron dispersion curves are mirror images of each other. All four bands are associated with orbital velocities which may be determined by use of the dispersion relation,

$$v_c = \frac{1}{\hbar}\frac{dE}{dk_c}_{|k_c=\frac{2\pi j}{|\mathbf{C}_h|}}, \tag{8.19}$$

where $j = \pm|j|$ is to be chosen such that k_c $(-k_c)$ is closest to K (K'), as states with energies closest to E_F determine the transport properties of the system. With reference to Figure 8.1, the energy derivative in Eq. (8.19) turns out to be positive (negative) for valence (conduction) electrons around the point K, while these asignments switch for the point K'. This distinction corresponds to orbital motion that winds around the nanotube axis in clockwise or counterclockwise sense, as illustrated in Figure 8.1(a).

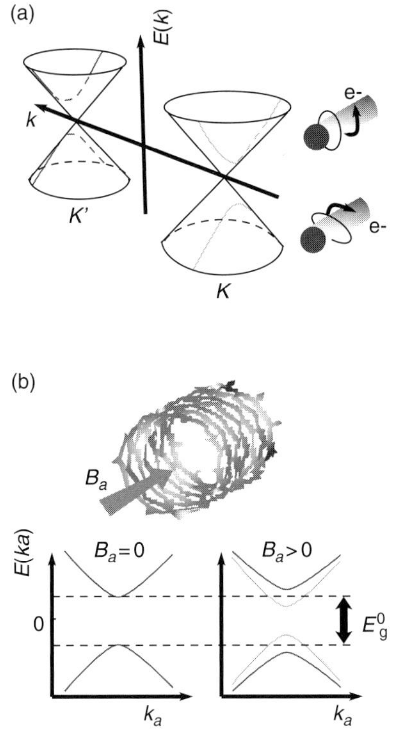

Figure 8.1 (a) Representation of the electron dispersion cones around the Dirac points K and K'. The hyperbolic curves refer to intersections at k_c values allowed by the quantization condition for the circumferential momentum in single-walled carbon nanotubes. The corresponding electron or hole states are characterized by clockwise ($\frac{dE}{dk}_{|k=k_c} > 0$) or counterclockwise ($\frac{dE}{dk}_{|k=k_c} < 0$) motion around the tube axis, where the energy derivatives are taken at one of the discrete k_c values that characterize the nanotube. (b) If the points K (K') do not coincide with any allowed choice of k_c, an energy gap opens between the valence and conduction states. A magnetic field applied parallel to the axis lifts the degeneracy between the K and the K' cone states. Reprinted with permission from [301]. Copyright (2004) by Nature Publishing Group.

This motion is associated with an orbital magnetic moment, $\boldsymbol{\mu}_{orb}$. With D as the diameter of the loop traversed by the rotating electron, μ_{orb} may be described as $\frac{Dev_c}{4}$. As admission is made for magnetic interaction by applying a magnetic field $\mathbf{B}_a$ along the nanotube axis, the degeneracy of conduction and valence states in metallic nanotubes is lifted and, likewise, the size of the energy gap characteristic of semiconducting tubes is changed. The sign of this change depends on the relative orientaton of $\boldsymbol{\mu}_{orb}$ and $\mathbf{B}_a$:

$$E_{mag} = -\boldsymbol{\mu}_{orb} \cdot \mathbf{B}_a = \pm \frac{Dev_c B_a}{4}. \tag{8.20}$$

A comparison between the cases $B_a = 0$ and $B_a > 0$ is shown in Figure 8.2(b) for the case of a semiconducting SWCNT. Here, a degeneracy between the conduction states in the vicinity of the two Dirac points persists if no external field is applied, and the same holds for the valence states. A finite magnetic field will lift this degeneracy and thus reduce the zero-field energy gap E_g^0, as is clear from the two schemes juxtaposed in Figure 8.2(b). With Δk_c as the difference between K and the allowed value of k_c that is closest to it, one may write E_g^0 as $2\hbar v_F \Delta k_c$. The presence of a magnetic field narrows (widens) this gap according to

$$E_g^B = 2\hbar v_F \Delta k_c \mp \frac{Dev_c}{2} B_a, \tag{8.21}$$

in the vicinity of the $K(K')$ point. We note that the energy dispersion close to the Dirac points remains unchanged as an external magnetic field is added. The position of these points, however, is shifted due to the field by a margin Δk_c^B. In order to determine this margin, we extend the discussion in Section 5.4.1 on the quantum phase in SWCNT wave functions by making admission for a vector potential **A** from which the axial magnetic field is derived [118]. This leads to the following modification of the Bloch state made explicit in formulae (5.31, 5.32):

$$\Psi_k(\mathbf{x} + \mathbf{C}_h) = e^{i\mathbf{k}\cdot\mathbf{C}_h} e^{i\frac{e}{\hbar}\int_{\mathbf{x}}^{\mathbf{x}+\mathbf{C}_h} \mathbf{A}(\mathbf{x}')d\mathbf{x}'} \Psi_k(\mathbf{x}) = \Psi_k(\mathbf{x}). \tag{8.22}$$

This periodic boundary condition accommodates the axial magnetic field by an *Aharonov–Bohm* phase expression, where the relation between the vector potential and the field is given by $\mathbf{B}_a = \nabla \times \mathbf{A}$. Using the Stokes theorem, the Bohm–Aharonov phase may be recast into the term $2\pi\Phi/\Phi_0$, where Φ denotes the magnetic flux threaded through the SWCNT cross section and Φ_0 is the quantum unit of magnetic flux, h/e. Using this compact substitution, we write the quantization condition for the circumferential wave number in the vicinity of the K point as (see Exercise 5.2)

$$\Delta k_c = \frac{2\pi}{|\mathbf{C}_h|}\left(j \pm \nu\frac{1}{3} + \frac{\Phi}{\Phi_0}\right), \tag{8.23}$$

with Φ_0 as the quantum unit of magnetic flux, h/e, and $\nu = 0, \pm 1$. Setting $\Phi = \pi(\frac{D}{2})^2 B_a$, the shift that is caused by the magnetic field is readily evaluated and found to be

$$\Delta k_c^B = \frac{2\pi}{|\mathbf{C}_h|}\frac{\Phi}{\Phi_0} = \frac{De}{4\hbar} B_a. \tag{8.24}$$

The direction of this magnetic shift is the same for both the K and the K' valley. As illustrated in Figure 8.2 (a), this shift opens a gap in metallic SWCNTs and leads to a valley-dependent increase or decrease of the gap in semiconducting SWCNTs. The latter effect is consistent with formula (8.21) for E_g^B, since the

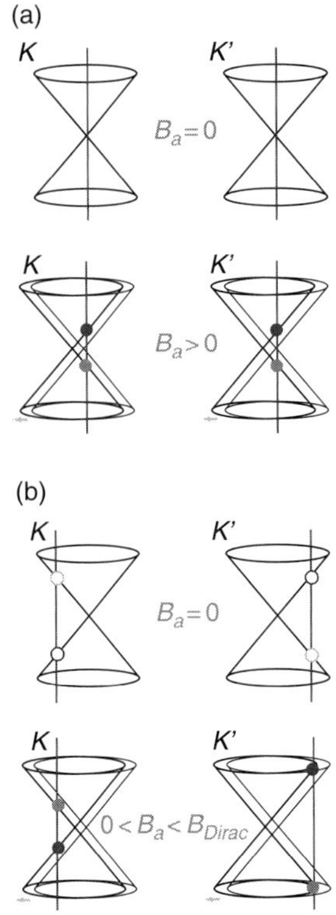

Figure 8.2 Effect of an axial magnetic field on SWCNT energy gaps. The Dirac point is shifted in **k** space by the same margin for both the K and the K' valley. (a) Metallic SWCNTs: the magnetic shift leads to an energy gap opening of the same size in both cases. (b) Semiconducting SWCNTs: the shift causes a narrowing (widening) of the energy gap in the K (K') valley [121]. Reprinted with permission from [121]. Copyright (2015) by the American Physical Society.

energy gap widening (narrowing) due to the shift of the K point (K' point) amounts to $2\Delta k_c^B \hbar v_F = \frac{Dev_F}{2} B_a$, which agrees with Eq. (8.21) for $v_c = v_F$.

From Figure 8.2(b), a critical strength B_a^{Dirac} of the axial magnetic field exists at which the energy gap in the K-valley is closed. This value is attained as the field is chosen as [301]

$$B_a^{Dirac} = \frac{2E_g^0}{eDv_F}. \tag{8.25}$$

This is easily verified by recourse to Eq. (8.21): the condition $B_a = B_a^{Dirac}$ implies a change of the single-electron energy by $E_g^0/2$ and thus closes the energy gap in the K valley while doubling it in the K' valley.

Defining the *orbital g factor* as $g^{orb} \equiv \frac{\mu_{orb}}{\mu_B}$, we include the spin degree of freedom into the description of the magnetic energy [121]:

$$E^{\pm}_{\tau,s_z,mag} = (\mp g_{orb}\tau + \frac{1}{2}g_s s)\mu_B B_a. \tag{8.26}$$

Here the parameter τ labels the valley degree of freedom, with $\tau = \{+1, -1\}$ for $\{K, K'\}$, and s the spin, where $s = \{+1, -1\}$ for $\{\uparrow, \downarrow\}$. The +(−) sign in Eq. (8.26) is valid for electrons (holes).

In the low-energy regime, the orbital g_{orb} factor can be derived from the relation [121]

$$\mu_{orb} \equiv \frac{d|E^{Dirac}|}{dB_a}, \tag{8.27}$$

with E^{Dirac} as the energy in the vicinity of the Dirac points:

$$E^{Dirac} \equiv \pm\hbar v_F\sqrt{k_a^2 + (\Delta k_c)^2}. \tag{8.28}$$

From the shift Δk_c^B (see Eq. (8.24)), induced by the magnetic field in the band structure close to the Dirac points, one can readily draw a conclusion with respect to the orbital magnetic moment μ^0_{orb} for small axial wave numbers ($k_a \approx 0$). For this quantity, one finds

$$\mu^0_{orb} = \frac{Dev_F}{4}. \tag{8.29}$$

For $|k_a| > 0$, we expand Eq. (8.28) up to first order and arrive at

$$\mu_{orb} = \frac{Dev_F}{4\sqrt{1 + (\frac{k_a}{\Delta k_c})^2}}, \tag{8.30}$$

and therefore

$$g_{orb} = \frac{g^0_{orb}}{\sqrt{1 + (\frac{k_a}{\Delta k_c})^2}} = \frac{g^0_{orb}}{\sqrt{1 + (\frac{2E_a}{E^0_g})^2}}, \tag{8.31}$$

where $g^0_{orb} \equiv \frac{Dev_F}{4\mu_B}$ and $E_a \equiv \hbar v_F k_a$.

Reducing SWCNTs to finite length or turning them into nanotube quantum dots imposes a quantization condition on the axial dimension as well. Discrete energy levels can now be discerned within the dispersion curves $E(k_a)$ (see Figure 8.1). The higher the energy of the levels that emerge from axial tube confinement, the lower the orbital magnetic moment associated with these levels, as predicted by formula (8.31). Since their spacing increases with decreasing tube length [113, 302], this effect will be more pronounced in strongly than in weakly confined systems.[1]

[1] In the simplest case of axial confinement, the electrons may be described as massive particles in an infinte-square-wall potential, and the corresponding energy levels are given by $E_j = j^2\frac{h^2}{8m_{eff}L^2}$, with L as tube length.

8.3 Spin-Orbit Coupling in Single-Walled Carbon Nanotubes

As pointed out earlier (see p. 181), the small magnitude of the the intrinsic spin-orbit effect in pristine, two-dimensional graphene is a consequence of the high degree of symmetry of this system. Specifically, the vanishing of the first-order perturbation theory in the spin-orbit interaction due to the space reflection properties of the graphene sheet makes this interaction a second-order effect. The reflection symmetry of graphene, however, lowers as the graphene sheet is rolled up into a nanotube, increasing the size of spin-orbit coupling. The curvature now induces the direct spin-orbit interaction between p_z orbitals of adjacent carbon atoms belonging to different sublattices.

Thus, the spin-orbit effect in SWCNTs can be analyzed by extending the treatment of this effect in graphene (see Section 8.1) [243, 304] to the case of curved hexagonal carbon networks. Due to the curvature of SWCNT surfaces, spin-orbit coupling between adjacent atomic centers can be efficient as a direct process, i.e. unassisted by concomitant Stark interaction. This is illustrated in Figure 8.3 for the case for two sets of p orbitals localized on neighboring carbon atoms along the circumference of the tube.

The spin-orbit effect in SWCNTs may be described in terms of an effective electric field **E**, associated with the curvature of the rolled-up graphene sheet and originating from the carbon ions. This field is oriented perpendicular to the two-dimensional carbon network, i.e. along the z direction indicated in Figure 8.3. The effect is thus of the Rashba type. As **E** points into the radial direction, it becomes the source of an effective magnetic field felt by an electron in circumferential motion with velocity **v**, as described by the relativistic formula

$$B_{SO} = -\frac{\mathbf{E} \times \mathbf{v}}{c^2}. \tag{8.32}$$

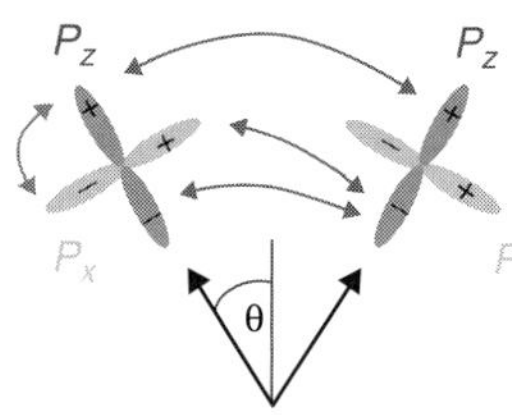

Figure 8.3 Hopping channels relevant for spin-orbit coupling in a curved section of graphene [243]. The arrows between the two orbital profiles refer to next-neighbor hopping interactions present in SWCNTs, but not in the flat graphene sheet. With reference to the orbital assignments in the figure, the curvature modifies the hopping between p_x and p_z orbitals, but leaves that between p_y orbitals unaffected. Reprinted with permission from [243]. Copyright (2006) by the American Physical Society.

As this field interacts with the spins of electrons in the low-energy regime, it modifies the respective region of the SWCNT band structure by inducing a spin-orbit term beyond the intrinsic spin-orbit coupling [305]. To evaluate its effect in a tight-binding framework, one needs to inspect the hopping channel that is opened due to the canted orientation of the p_x and p_z orbitals on adjacent atoms, as shown in Figure 8.3. In contrast to the case of the flat graphene sheet, hopping between these orbitals will now take place. This effect gives rise to an interaction described by

$$\mathbf{H}_{SO}^{curv} = \xi_{curv}(\boldsymbol{\eta}_x \boldsymbol{\tau}_z \boldsymbol{\sigma}_y - \boldsymbol{\eta}_y \boldsymbol{\sigma}_x). \tag{8.33}$$

The constant ξ_{curv} is here a function of the radius (or radii) of curvature, the energies $E(2s)$ and $E(2p)$, as well as the Koster–Slater hopping matrix elements. Comparison with Eq. (8.16) shows that the structure of the Hamiltonian (8.33) is indeed strictly analogous to the component of $\mathbf{H}_{SO}^{eff}$ that accounts for the impact of a surface-perpendicular electric field on the spin-orbit coupling. Both constants, ξ_{SE} and ξ_{curv}, can thus be summed into a *Rashba coefficient*, defined as $\xi_R = 2(\xi_{SE} + \xi_{curv})$, that determines the strength of spin-orbit interactions induced by external electric fields or geometric distortions of the graphene fabric.

For numerical evaluation of the generalized spin-orbit coupling, is it necessary to align the spins along the quantization axis of the considered structure, such as the axis of a nanotube. So far, a local coordinate system has been adopted, involving perpendicular orientation with respect to the carbon surface. In the appendix to this chapter, we describe the generalized spin-orbit effect as a function of the angle θ included with the quantization axis, as well as the azimuthal angle ϕ. The matrix representations of the spin-orbit Hamiltonians in the Dirac regimes of both graphene and SWCNTs are derived from this approach.

Carrying out this transformation, one arrives at the following Hamiltonian matrix for the low-energy $\mathbf{k}$ space regime in the vicinity of the Dirac points:

$$\hat{H} = \begin{pmatrix} \mathbf{0} & \hbar v_F(k_a - i\Delta k_c)\mathbf{1} - i\tau\frac{\xi_R}{2}\boldsymbol{\sigma}_z \\ \hbar v_F(k_a + i\Delta k_c)\mathbf{1} + i\tau\frac{\xi_R}{2}\boldsymbol{\sigma}_z & \mathbf{0} \end{pmatrix}. \tag{8.34}$$

The form of the spin-orbit admixture to the Dirac Hamiltonian is motivated in the second special-topic section of this chapter. The solution $\boldsymbol{\Psi}_\tau$ of the corresponding eigenvalue equation is a four-component spinor. As before, $\tau = 1$ ($\tau = -1$) designates the K (K') point. It is to be noted that integrating over the SWCNT circumference eliminates any dependence on the intrinsic spin-orbit effect that is explicit in the spin-orbit Hamiltonian for graphene, Eq. (8.16). The energy eigenvalues of the Hamiltonian (8.34) are [288]

$$E_{SO,s}^{\pm} = \pm\hbar v_F\sqrt{k_a^2 + (\Delta k_c - s\Delta k_{SO})^2}. \tag{8.35}$$

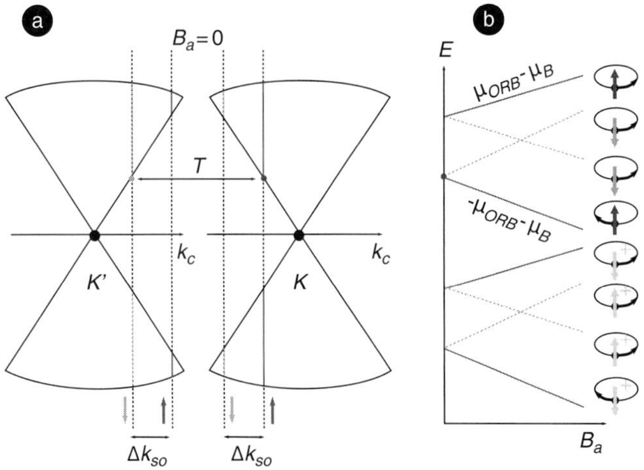

Figure 8.4 SWCNT energy quantization in the presence of spin-orbit interaction. (a) Four Kramers doublets emerge, involving degenerate K and K' states. (b) A magnetic field along the nanotube axis lifts the degeneracy. The four states of lowest energy are labeled as hole states, implying charge and spin reversal while the sense of the orbital motion remains unaffected. From [288], with permission from the author.

Here, $\Delta k_{SO} = \frac{\xi_R}{2\hbar v_F}$ and $s = \pm 1$. As admission is made for spin-orbit interaction, the nanotube band gap becomes spin dependent. In particular, for both Dirac points, two band gaps result, separated by a difference of Δk_{SO} in the circumferential wave number and distinguished by opposite electron spin orientations. This is illustrated in Figure 8.4. The electronic structure generated by spin-orbit coupling may be described as a set of four Kramers doublets, i.e. two electron and two hole states along with their time- and, consequently, spin-reversed counterparts. The Kramers degeneracy is lifted as a magnetic field is applied along the tube axis, in accordance with formulae (8.20) and (8.21).

Further analysis of the spin-orbit field defined by Eq. (8.32) reveals that, in some cases, the coupling types considered so far must be augmented by a term proportional to

$$B_{jj} = \langle \Phi_j | (\mathbf{E} \times \mathbf{p})_{\parallel} | \Phi_j \rangle, \tag{8.36}$$

with $j = A, B$. The index $\parallel$ refers here to the axis-parallel component of the vector $\mathbf{E} \times \mathbf{p}$. This expression vanishes unless the field $\mathbf{E}$ exhibits spatial variation. As for a constant electric field, the term disappears for any field that displays C_3 symmetry [121]. Investigating the symmetry properties of the scalar product (8.36), one finds that it depends on the chirality of the SWCNT. Specifically, the fact that armchair tubes have an axis-perpendicular mirror plane implies vanishing of the term in this case. For zigzag tubes, on the other hand, the term yields a finite value on account of their axis-parallel, rather than perpendicular mirror plane. Wherever expression (8.36) assumes a non-zero value, a *diagonal interaction* is present in the respective SWCNT, adding to the non-diagonal coupling described by the spin-orbit Hamiltonian (8.34). As a diagonal spin interaction with a magnetic field, this component is interpreted as a *Zeeman term*, changing the energies of states with opposite spin orientations in opposite directions. While the spin-orbit interactions introduced so far are non-diagonal and thus induce a purely horizontal shift of the band structure in the low-energy regime the Zeeman contribution gives rise to a purely vertical displacement of the band structure. Two basic cases are to be distinguished, involving the relative orientation of the spin magnetic moment and the field $\mathbf{B}_{SO}$ (see Eq. (8.32)) in identical and in opposite directions. As shown in Figure 8.4, the first (second) condition is realized for the states $K \downarrow$ and $K' \uparrow$ ($K \uparrow$ and $K' \downarrow$). The two states in each of these two pairs are degenerate if there is no external magnetic field. Adjusting the notation for this Zeeman-analogous interaction to match that for the intrinsic and the Rashba spin-orbit coupling, we write

$$E_{SO,Z} = \tau s \xi_Z. \tag{8.37}$$

As the numerical value of ξ_Z is negative, the Zeeman-type spin-orbit interaction increases (lowers) the energy of orbitals with down (up) spin orientation. Figure 8.5 suggests a geometric interpretation of the difference between the Rashba, or *orbital-like*, and the *Zeeman-like* component of the spin-orbit effect in SWCNTs.

Exercise 8.1 For a vanishing axial magnetic field ($B_a = 0$), each one of the levels due to confinement is fourfold degenerate, as a consequence of valley- and spin symmetry. Show: This degeneracy is lifted if a finite axial magnetic field is applied ($B_a \neq 0$). Beyond a critical field strength, states in the vicinity of Dirac points are filled in the following order: $K \downarrow, K \uparrow, K' \downarrow, K' \uparrow$, as indicated by Figure 8.5(d). Assume that $\xi_Z = 0$.

We emphasize that the magnitude of the parameter ξ_Z is chirality-dependent. Thus, it vanishes in armchair tubes and assumes a finite value in zigzag tubes [121].

Summarizing, we write the eigenenergies of the SWCNT Hamiltonian in the Dirac regime, including the Zeeman term as well as an axial magnetic field (see Figure 8.5(b,d)). The energies are here given in a basis of spin and valley eigenstates:

$$E^{\pm}_{SO,\tau,s} = \pm\sqrt{(-\tau\hbar v_F \Delta k_c + \mu^0_{orb} B_a + s\frac{\xi_R}{2})^2 + (\hbar v_F k_a)^2} + \tau s \xi_Z + \frac{1}{2} s g_s \mu_B B_a. \tag{8.38}$$

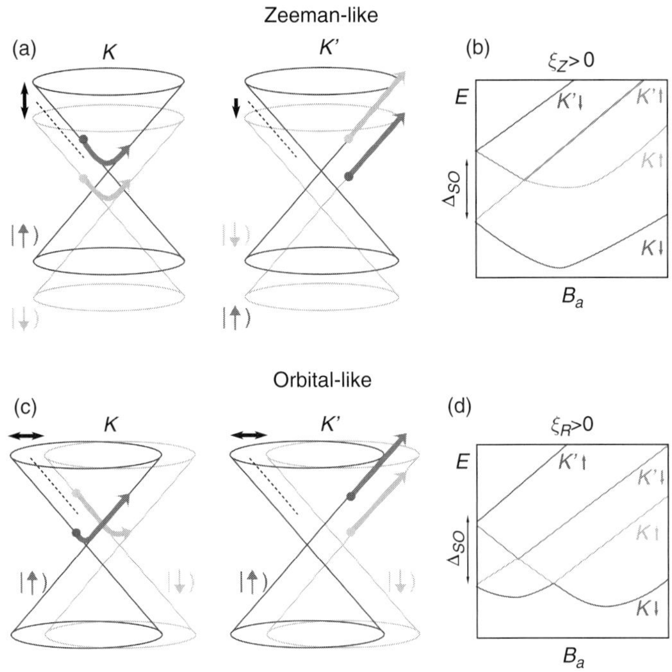

Figure 8.5 Characteristic band structure profiles of Zeeman-like and orbital-like spin-orbit interactions in SWCNTs. (a) and (c): The Zeeman-like effect (a) causes a vertical displacement of the band structure, where the shift proceeds in opposite directions in the two valleys. The orbital-like effect (c) leads to a horizontal displacement. The direction of the shift is the same for both valleys. The arrows in the upper halves of the energy double-cones refer to the evolution of the eigenenergies of the four states as an axial magnetic field is applied to the SWCNT, shown in greater detail in (b) and (d). As a distinguishing feature, one finds that Zeeman-like coupling yields an energy minimum in both the first and the second energy levels. Orbital-like coupling, however, generates two minima in the first level. Reprinted with permission from [121]. Copyright (2015) by the American Physical Society.

Exercise 8.2 Conclude from Eq.(8.38): The magnitude of the *zero field splitting* $\xi_{SO,B_a=0}$ between the spin-up and spin-down moieties is, to linear order in the coupling constant ξ_R:

$$\xi_{SO,B_a=0} = 2\xi_Z \mp \xi_R \frac{g_{orb}}{g^0_{orb}}.$$

So far, we have ascribed the mixing of spin indices ($s = \pm 1$) to spin-orbit interactions. This finding, however, hinges on the chosen orientation of the external magnetic field, along a direction parallel to the tube axis. Deviations from the axial alignment of the field give rise to an additional coupling of the spin states (this can be readily seen from the SWCNT spin Hamiltonian (8.40) in the first special section of this chapter). In the extreme case of axis-perpendicular orientation, the external field couples to a linear combination of $\boldsymbol{\sigma}_x$ and $\boldsymbol{\sigma}_y$, inducing naturally the interaction between $|\uparrow\rangle$ and $|\downarrow\rangle$ states, where we always assume that *up* and *down* refer to the tube axis.

The spin coupling generated by changing the field direction leaves states with different valley indices ($\tau = \pm 1$) uncoupled. A signature of $K - K'$ interactions, however, has been observed in experiment [5]. A phenomenological constant $\Delta_{K-K'}$ has been associated with the *intervalley scattering* effect, which mixes K and K' while preserving the spin orientation. Figure 8.6 displays the energies $E^{\pm}_{SO,\tau,s_z}$ (see Eq. (8.38)) as the external magnetic field is varied in axial and in radial directions [121]. In the absence of spin-orbit interaction as well as intervalley scattering, one obtains the profile shown in panel (a). The Kramers doublets($|K\uparrow\rangle, |K'\downarrow\rangle$) and ($|K\downarrow\rangle, |K'\uparrow\rangle$) remain degenerate in a radial field and the energy is entirely determined by the spin direction as well as the magnitude of the field, resulting in two lines. Four lines emerge in case of an axis-parallel field, as the latter couples to both the spin and the orbital magnetic moment. In panel (b), allowance is made for a non-zero spin-orbit effect, giving rise to an energy shift between the two Kramers doublets at $B_a = 0$. Panel (c) adds the impact of intervalley scattering: $\Delta_{K-K'} > 0$. Both Kramers doublets are now split not only for the axial, but also for the radial field direction. Further, the $K - K'$ interaction due to intervalley scattering causes an avoided crossing between the energy curves of the $|K\uparrow\rangle$ and the $|K'\downarrow\rangle$ states.

The first experimental evidence for spin-orbit coupling in SWCNTs [5] appeared about one decade after the first fabrication of axially confined carbon nanotubes, or carbon nanotube quantum dots [121]. This considerable delay has been ascribed to the dominance of intervalley scattering, induced by disorder in supported SWCNT samples, and masking the spin-orbit effect in these samples. Operating with ultra-clean suspended carbon nanotube quantum dots, however, provided the experimental basis for extracting the spin-orbit interaction and for clearly distinguishing between the two rivaling effects.

The pioneering experiment that established spin-orbit coupling for SWCNTs [5] involved measurements of the linear conductance G ($= \frac{dV_{sd}}{dI}$) in a carbon nanotube quantum dot used as a transmission element. This device was subjected to a source-drain voltage V_{sd} as well as two gate voltages applied from below, as shown in Figure 8.7 (a,b). In the example displayed in panel (a), this arrangement is used to create a quantum dot above the right gate electrode. Controlled variation of the gate voltage allows us to inject selectively a single charge carrier into the device and thus avoid effects from electron-electron interaction. A magnetic field was applied in the direction of the SWCNT axis. Panel (b) shows the conductance

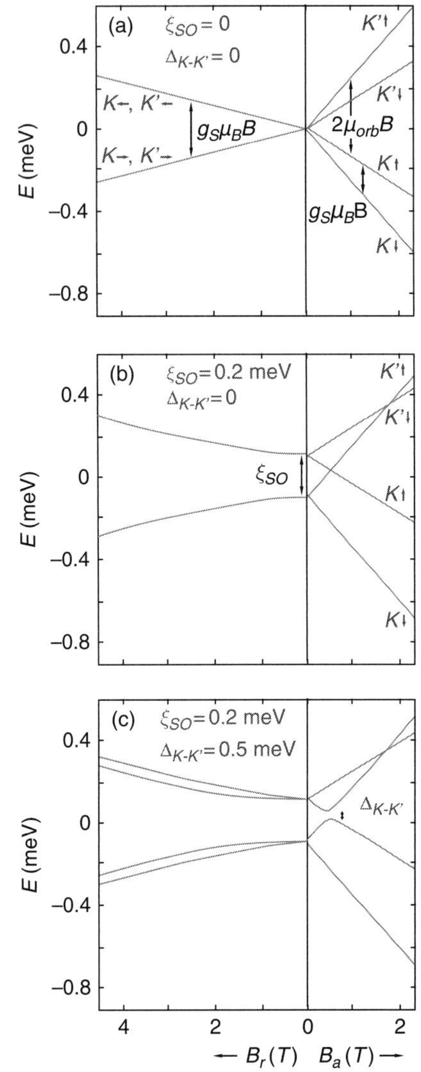

Figure 8.6 The energy eigenvalues $E^{\pm}_{SO,\tau,s_z}$ versus the external magnetic field. The panels contain the dependence of the energies on the axial (B_a see Eq. (8.38)) and the radial (B_r) magnetic field for vanishing spin-orbit coupling ($\xi_{SO} = 0$) and intervalley scattering ($\Delta_{K-K'} = 0$) (a), $\xi_{SO} > 0$, $\Delta_{K-K'} = 0$ (b) and $\xi_{SO} > 0$, $\Delta_{K-K'} > 0$ (c). The constant ξ_{SO} refers here to the combined spin-orbit term $\xi_{SO} \equiv 2\xi_Z \mp \xi_R \frac{g_{orb}}{g^0_{orb}}$, with $g_{orb} = \frac{g^0_{orb}}{\sqrt{1+(\frac{ka}{\Delta k_c})^2}}$ and $g^0_{orb} = \frac{Dev_F}{4\mu_B}$. The latter expression is obtained from Eq. (8.20) with $v_c = v_F$. The calculations are based on the assumption that the SWCNT band gap is substantially larger than the magnetic and the spin-orbit energy terms that characterize the problem ($2\hbar v_F \Delta k_c >> \mu^0_{orb}|\mathbf{B}|, g_s\mu^0_B|\mathbf{B}|, \xi_R, 2\xi_Z$). Reprinted with permission from [121]. Copyright (2015) by the American Physical Society.

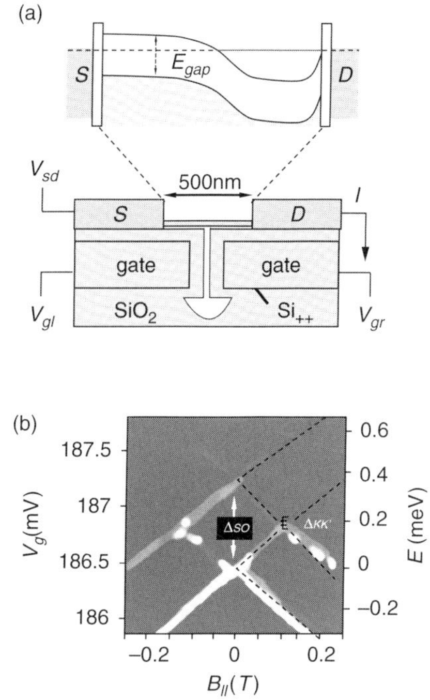

Figure 8.7 Experimental determination of the spin-orbit effect in SWCNTs by observing single-electron levels in a carbon nanotube quantum dot. The SWCNT is used as a device between source and drain electrodes and in contact with two gate electrodes. The configuration displayed in panel (a) involves a quantum dot above the right gate electrode. A cross-section of the setup is shown. (b) The measured linear conductance of the quantum dot as a function of an axial magnetic field, the gate voltage, and the energy of the single-electron level. Shown are the zero-field splitting, which indicates the spin-orbit interaction, and the avoided crossing due to $K - K'$ intervalley scattering. Reprinted with permission from [5]. Copyright (2008) by Nature Publishing Company.

resonances identified for the single-electron case as a function of the magnetic field magnitude (horizontal scale) and the gate voltage (left vertical scale) as well as the derived energy of the single-electron level (right vertical scale). From the zero-field energy difference, attributed to spin-orbit coupling, and from the size of the avoided crossing, associated with intervalley scattering, one may recognize this profile as a realization of the prototype shown in Figure 8.6(c). The authors reported a spin-orbit splitting $\Delta_{SO}(= \xi_{SO})$ of 0.37 ± 0.02 meV, and $K - K'$ mixing term $\Delta_{K-K'}$ of 65 μeV. More recent experiments have yielded results for the spin-orbit effect in SWCNTs that exceeed the original finding by an order of magnitude [307, 308]. Proper incorporation of electron interaction into the prevailing models of SWCNT magnetism may help to remove the uncertainty about the size of this effect. Further, the deviations between the measurements may be related to environmental factors, involving the impact of gate dielectrics or external adsorbates [121].

We add two special sections. The first is a brief introduction to the Bloch sphere, a topic of relevance also for Chapters 9 and 12. In the following section, a spin-orbit interaction Hamiltonian (8.16) is cast into a basis of spinors on the Bloch sphere. The result is used to derive the Dirac Hamiltonian (8.34) for nanotubes.

8.4 Special Topic I: The Bloch Sphere

The *Bloch sphere* provides a three-dimensional representation of a quantum state with spin $\frac{1}{2}$. Any two-state system in quantum physics may we written as

$$|\psi\rangle = \alpha|+\frac{1}{2}\rangle + \beta|-\frac{1}{2}\rangle, \tag{8.39}$$

where α and β are complex-valued coefficients with $|\alpha|^2 + |\beta|^2 = 1$. We consider the eigenstates of the "spin Hamiltonian"

$$\hat{H}_{spin} = A\mathbf{u}\cdot\boldsymbol{\sigma}, \tag{8.40}$$

where the array $\boldsymbol{\sigma}$ contains the Pauli matrices $\boldsymbol{\sigma}_x, \boldsymbol{\sigma}_y, \boldsymbol{\sigma}_z$, and $\mathbf{u}$ is a unit vector:

$$\mathbf{u} = \begin{pmatrix} \sin\theta\cos\phi \\ \sin\theta\sin\phi \\ \cos\theta \end{pmatrix}. \tag{8.41}$$

Note that the choice of $-g_s\mu_B B_0/2$ for the constant A generates the Hamiltonian for the interaction of a spin $\frac{1}{2}$ with an homogeneous magnetic field of magnitude B_0. Apart from a common phase factor, an eigenstate of $\hat{H}_{spin}$ is given by

$$|\psi\rangle_+ = \cos(\frac{\theta}{2})\begin{pmatrix}1\\0\end{pmatrix} + \sin(\frac{\theta}{2})e^{i\phi}\begin{pmatrix}0\\1\end{pmatrix}. \tag{8.42}$$

The angular range of this spinor is $0 \le \theta \le \pi$ and $0 \le \phi \le 2\pi$. Thus, the state (8.42) can be uniquely mapped into a unit vector $\mathbf{u}$ and thus be represented on the surface of a three-dimensional sphere of unit radius, as shown in Figure 8.8. The choice $\theta = \frac{\pi}{2}$ yields *equatorial states*, with the superpositions $\frac{1}{\sqrt{2}}[\begin{pmatrix}1\\0\end{pmatrix} \pm \begin{pmatrix}0\\1\end{pmatrix}]$, resulting for $\phi = 0, \pi$, respectively.

A more general prescription for assigning a spin $\frac{1}{2}$ system to a point on the Bloch sphere is obtained by use of the density matrix formalism. *Mixed states* are conveniently described in terms of the density operator $\hat{\rho}$, defined as

$$\hat{\rho} = \sum_i p_i|\psi_i\rangle\langle\psi_i|, \tag{8.43}$$

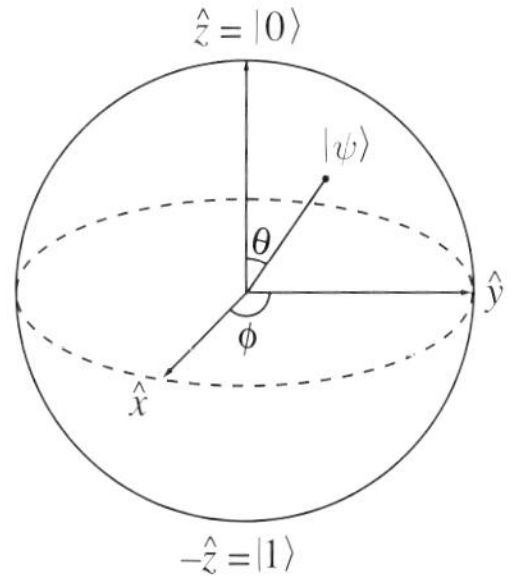

Figure 8.8 Bloch sphere. The spin $-\frac{1}{2}$ quantum state with $\theta = 0$ is mapped on the North pole of the sphere, that with $\theta = \pi$ on the South pole. The choice of $\theta = \frac{\pi}{2}$ yields superpositions that contain the alternatives $|+\frac{1}{2}\rangle$ and $|-\frac{1}{2}\rangle$ with equal weights. These are assigned equatorial locations on the Bloch sphere.

with p_i as probability for a spin $\frac{1}{2}$ system to be in the state ψ_i. Consequently,

$$Tr\{\boldsymbol{\rho}\} = 1. \tag{8.44}$$

The density matrix (8.43) reduces to a representation of the pure state ψ_k when p_i is chosen as δ_{ik}. In this case, $Tr\{\boldsymbol{\rho}^2\} = 1$. Any arbitrary density matrix for a spin $\frac{1}{2}$ system can be written as

$$\hat{\rho} = \frac{1}{2}(\mathbf{1} + \mathbf{b} \cdot \boldsymbol{\sigma}). \tag{8.45}$$

The reader readily verifies that the unit trace constraint is satisfied by this expression if $|\mathbf{b}| = 1$. Further,

$$Tr\{\boldsymbol{\rho}\boldsymbol{\sigma}\} = \mathbf{b}. \tag{8.46}$$

As implied by the parametrization (8.45), any spin $\frac{1}{2}$ density matrix is uniquely determined by the vector $\mathbf{b}$, also referred to as the *Bloch vector*. If mixed states are included in the manifold described by the matrix (8.45), the magnitude of $\mathbf{b}$ is not constrained to one; in fact, $0 \leq |\mathbf{b}| \leq 1$. In view of Eq. (8.46), the completely umpolarized state is obtained by choosing $\mathbf{b} = \mathbf{0}$. The corresponding density matrix $\boldsymbol{\rho}_0$ is

$$\boldsymbol{\rho}_0 = \frac{1}{2}(|+\frac{1}{2}\rangle\langle+\frac{1}{2}| + |-\frac{1}{2}\rangle\langle-\frac{1}{2}|), \tag{8.47}$$

thus describing the state of vanishing spin polarization. Its geometric locus is at the center of the Bloch sphere. Note that for pure states, we have

$$\boldsymbol{\rho}^2 = \boldsymbol{\rho}. \tag{8.48}$$

For mixed states, in contrast, we find

$$\rho^2 \neq \rho. \tag{8.49}$$

8.5 Special Topic II: The Spin-Orbit Hamiltonian in Spherical Coordinates

In a global coordinate system, defined by a fixed set of basis vectors $\hat{\mathbf{x}}$, $\hat{\mathbf{y}}$, $\hat{\mathbf{z}}$, the spin is invariant with respect to hopping between lattice sites. When a local system is used, however, the spin must be adjusted when the electron moves from one site to another. This needs to be taken into account when the transition from a flat graphene sheet to an SWCNT is made. In the case of graphene, a natural choice for the spin quantization axis is the direction perpendicular to the sheet, providing the z-axis of a global coordinate system. As the graphene plane is rolled into the cylinder of an SWCNT, this global system changes into a local one, as the z-axis now indicates the radial direction. If the x coordinate labels the longitudinal axis of the tube and y is adopted as circumferential coordinate, the locally defined spin functions $|\uparrow_\perp\rangle$ and $|\downarrow_\perp\rangle$ are represented in terms of the spin projections $|\theta, \phi, \uparrow\rangle$ and $|\theta, \phi, \downarrow\rangle$ by use of the following rotation on the Bloch sphere:

$$\begin{aligned} |\uparrow_\perp\rangle &= \cos(\frac{\theta}{2}) e^{-i\frac{\phi}{2}} |\theta, \phi, \uparrow\rangle - \sin(\frac{\theta}{2}) e^{i\frac{\phi}{2}} |\theta, \phi, \downarrow\rangle \\ |\downarrow_\perp\rangle &= -\sin(\frac{\theta}{2}) e^{-i\frac{\psi}{2}} |\theta, \phi, \uparrow\rangle + \cos(\frac{\theta}{2}) e^{i\frac{\phi}{2}} |\theta, \phi, \downarrow\rangle. \end{aligned} \tag{8.50}$$

Here, θ is the angle enclosed with the tube axis, and ϕ the azimuthal angle. We express the Hamiltonian (8.16) that contains both the intrinsic spin-orbit effect and the Rashba interaction in terms of the spin functions $|\theta, \phi, \uparrow\rangle$ and $|\theta, \phi, \downarrow\rangle$. As in Section 8.1, the Hamiltonian matrix acts on the vector $|A\uparrow\rangle, |A\downarrow\rangle, |B\uparrow\rangle, |B\downarrow\rangle$. At the K point, we find

$$\hat{H}_{SO} = \begin{pmatrix} -\xi_I \cos\theta & -\xi_I \sin\theta e^{-i\phi} & -i\xi_R \sin(\frac{\theta}{2})\cos(\frac{\theta}{2}) & -i\xi_R \sin^2(\frac{\theta}{2}) e^{-i\phi} \\ -\xi_I \sin\theta e^{i\phi} & \xi_I \cos\theta & i\xi_R \cos^2(\frac{\theta}{2}) e^{i\phi} & i\xi_R \sin(\frac{\theta}{2})\cos(\frac{\theta}{2}) \\ i\xi_R \sin(\frac{\theta}{2})\cos(\frac{\theta}{2}) & -i\xi_R \cos^2(\frac{\theta}{2}) e^{-i\phi} & \xi_I \cos\theta & \xi_I \sin\theta e^{-i\phi} \\ i\xi_R \sin^2(\frac{\theta}{2}) e^{i\phi} & -i\xi_R \sin(\frac{\theta}{2})\cos(\frac{\theta}{2}) & \xi_I \sin\theta e^{i\phi} & -\xi_I \cos\theta \end{pmatrix}, \tag{8.51}$$

with an analogous expression for the K' point. Setting $\theta = \phi = 0$, we recover the matrix representation of the spin-orbit Hamiltonian for the two-dimensional graphene sheet. The choice $\theta = \pi/2$ yields the equivalent expression for an SWCNT:

$$\hat{H}_{SO} = \begin{pmatrix} 0 & -\xi_I e^{-i\phi} & -i\frac{\xi_R}{2} & -i\frac{\xi_R}{2} e^{-i\phi} \\ -\xi_I e^{i\phi} & 0 & i\frac{\xi_R}{2} e^{i\phi} & i\frac{\xi_R}{2} \\ i\frac{\xi_R}{2} & -i\frac{\xi_R}{2} e^{-i\phi} & 0 & \frac{\xi_I}{2} e^{-i\phi} \\ i\frac{\xi_R}{2} e^{i\phi} & -i\frac{\xi_R}{2} & \xi_I e^{i\phi} & 0 \end{pmatrix}. \tag{8.52}$$

Averaging over the circumferential angle ϕ, we obtain the final form of this expression:

$$\hat{H}_{SO} = \begin{pmatrix} 0 & 0 & -i\frac{\xi_R}{2} & 0 \\ 0 & 0 & 0 & i\frac{\xi_R}{2} \\ i\frac{\xi_R}{2} & 0 & 0 & 0 \\ 0 & -i\frac{\xi_R}{2} & 0 & 0 \end{pmatrix}. \tag{8.53}$$

Note that in this final form, $\mathbf{H}_{SO}$ does not depend on ξ_I, and so no longer contains the intrinsic spin-orbit effect.

Part IV

Transport Phenomena

9

Elements of Spintronics

Spintronics, the science and technology of generating, manipulating and detecting spin currents rather than charge currents, is acknowledged as a plausible continuation of electronics [309, 310, 311]. In terms of encoding information, the two states of a spin with reference to a quantization axis, i.e. spin-up or spin-down, may be used to implement a system of binary logic. The technological advantage of spintronics over electronics lies chiefly in significant reductions of energy usage for proper operation. Novel low dissipation devices are envisioned that combine high speed, high density and endurance with non-volatile memory technology. The latter feature is among the major advantages of magnetic storage, involving the continued availability of information when the device is powered down. In the following, we will motivate the strong present interest in spintronics, introduce the basic concepts underlying this recent field of materials studies and discuss various applications of these concepts. By developing the elemental tenets of spintronics, this chapter serves as a preparation for the following one, where we describe actual or potential realizations of spin-based devices by use of carbon nanostructures as material constituents.

Section 9.1 introduces the notion of spin current and the associated causal concept of the spin-resolved quasi-chemical potential. In Section 9.2, essential devices of spintronics circuits are specified: the spin valve, the spin transistor and the spin filter. The rest of this chapter focuses on the crucial processes of spin injection, detection and relaxation. In particular, the mechanisms that limit the lifetime of spin-polarized currents in transmission media are discussed in Section 9.4.

9.1 Spin Current

For initial motivation, we clarify the term *spin current*. Electric current usually refers to charge transport without transfer of net spin polarization. Denoting the overall charge current with I and the spin-resolved alpha (up) and beta (down) contributions to I with $I^{\uparrow}$ and $I^{\downarrow}$, respectively, we have

$$I = I^{\uparrow} + I^{\downarrow}. \tag{9.1}$$

The *net spin currrent* (I_s), in contrast, is found from

$$I_s = \frac{\hbar}{2e}(I^{\uparrow} - I^{\downarrow}). \tag{9.2}$$

In terms of the current density $\mathbf{j}$, this quantity is defined for the charge current as

$$\mathbf{j} = e\Re\{\Psi^*(\mathbf{x})\hat{\mathbf{v}}\Psi(\mathbf{x})\}, \tag{9.3}$$

involving the velocity operator $\hat{\mathbf{v}} = (\hbar/im)\nabla$ with respect to the electron wave function Ψ. The corresponding expression for the spin current density is

$$\mathbf{j}_{s,i} = \frac{\hbar}{2}\Psi^*(\mathbf{x})\frac{\hat{\boldsymbol{\sigma}}_i\hat{\mathbf{v}} + \hat{\mathbf{v}}\hat{\boldsymbol{\sigma}}_i}{2}\Psi(\mathbf{x}), \tag{9.4}$$

with $\boldsymbol{\sigma}_i$ as the Pauli matrices, and $i = x, y, z$. From formulae (9.3, 9.4), one gathers that the charge current changes sign upon time reversal while the spin current does not. This follows from $\hat{\mathbf{v}} \to -\hat{\mathbf{v}}$ and $\hat{\boldsymbol{\sigma}} \to -\boldsymbol{\sigma}$ as $t \to -t$ and implies that the charge current vanishes under conditions of thermal equilibrium. This, however, does not hold for the spin current [315]. Further, charge conservation implies the validity of the continuity equation for the charge current,

$$\frac{\partial n}{\partial t} = -\nabla \cdot \mathbf{j}, \tag{9.5}$$

with n as charge density. There is no direct parallel to this relation for spin currents. While charge does not get lost on its way through a transmission element, an initially polarized spin ensemble may loose spin polarization in the same environment through processes of spin relaxation and dephasing. This is taken into account by a modified continutity equation [309]

$$\frac{d\mathbf{M}}{dt} = -\nabla \cdot \mathbf{j}_s + \mathbf{T}, \tag{9.6}$$

where $\mathbf{M}$ represents the local magnetization, or magnetic moment density, and the term $\mathbf{T}$ accommodates effects that may alter the magnetization, besides the passage of spin current. In the simplest approach, $\mathbf{T}$ may be defined by the nonequilibrium magnetization and an equilibration time constant τ according to

$$\mathbf{T} = -\frac{(\mathbf{M} - \mathbf{M}_0)}{\tau}. \tag{9.7}$$

Spin current may be associated with charge current, or it may be realized as *pure spin current*, defined by the constraint that no net charge is transferred. The four basic modes of charge and spin transfer – (a) $I \neq 0, I_s = 0$, (b) $I \neq 0, I_s \neq 0$, (c) $I \neq 0\ |I_s| = \frac{\hbar}{2e}|I|$, and (d) $I = 0, I_s \neq 0$ – are illustrated in Figure 9.1. Configuration (c) is referred to as *perfectly spin-polarized current*, (d) as *pure spin*

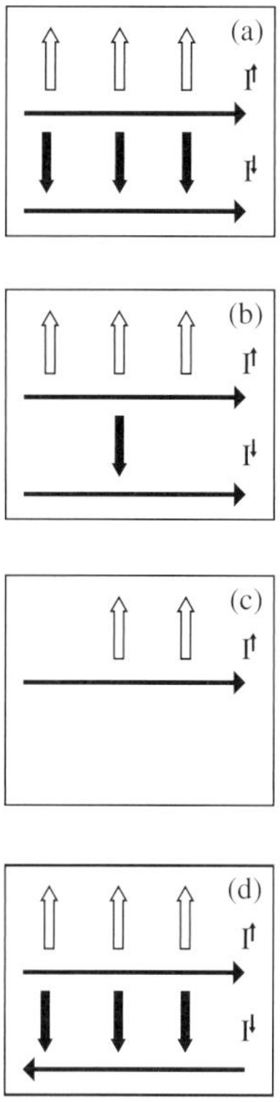

Figure 9.1 Charge and spin transfer: the four basic configurations: (a) charge current without net spin current, (b) charge current with net spin current, (c) perfectly spin-polarized charge current, (d) spin current without net charge current (*pure spin current*).

current. The latter is achieved by letting two currents of equal magnitude but with opposite spin orientations, i.e. $I^{\uparrow}$ and $I^{\downarrow}$, simultaneously flow in opposite directions. Comparing models (a), (b) and (c), the last turns out to be superior in terms power-efficient spin transport. Trivially, no spin-related information is transferred if model (a) is implemented. In case (b), a current of four electrons transmits two spins of size $\frac{\hbar}{2}$. In case (c), however, the same spin information is transported by only two electrons. This difference is tantamount to a reduction of heat losses by 75 percent as one goes from model (b) to model (c), creating a strong incentive to attain the limit of pure spin currents in spintronics devices.

9.1.1 The Quasi-Chemical Potential

We continue this section by introducing some concepts of basic relevance for understanding spin transport. In particular, we emphasize the role of the *spin-resolved chemical potential* in defining the spin current density as well as the spin dynamics within magnetic or non-magnetic media and at interfaces separating them. These are items of crucial importance for the analysis and design of spintronics networks.

To appreciate the relation between the spin chemical potential and the densities of spin-up and spin-down electrons $n_\uparrow$ and $n_\downarrow$, we first inspect the non-polarized electron density n. In conditions of thermodynamic equilibrium, the system is characterized by a well-defined chemical potential μ, and the electron density is found from

$$n_0(\mu) = \int dE\, g(E) F_0(E), \tag{9.8}$$

where F_0 denotes the Fermi–Dirac distribution

$$F_0(E) = (1 + \exp(\frac{E - \mu}{k_B T}))^{-1}. \tag{9.9}$$

In the following, we make reference to metals or degenerate semiconductors, i.e. semiconductors with a high level of doping. In these systems, exposure to electric bias will induce an electric current and thus give rise to a situation of nonequilibrium. The resulting current density is partially dictated by the electric field $\tilde{E}$, partially due to diffusion. We restrict spin transport to a well-defined channel, described by the x coordinate, and write

$$j = \tilde{\sigma} \tilde{E} + eD \frac{\partial n}{\partial x}, \tag{9.10}$$

with $\tilde{\sigma}$ as the conductivity and D as the diffusion constant.

Exercise 9.1 Derive Eq. (9.10) from the diffusion equation and the equation of continuity, in conjunction with the relation between the current density and the electric field, i.e. $j = \tilde{\sigma} \tilde{E}$.

In this nonequilibrium case, the chemical potential is no longer a constant but exhibits some spatial variation. This change is accommodated by the *quasi-chemical potential* $\mu_0(x)$ as a position-dependent addition to the constant μ. The Fermi–Dirac distribution now reads

$$F(E, x) = (1 + \exp(\frac{E - e\phi(x) - \mu - e\mu_0(x)}{k_B T}))^{-1}, \tag{9.11}$$

where ϕ is an electric potential, such that $\tilde{E} = -\frac{\partial \phi}{\partial x}$. The electron density is then obtained from

$$n(x) = n_0(\mu + e\phi + e\mu_0). \tag{9.12}$$

With respect to the current density Eq. (9.10), we conclude from Eq. (9.12),

$$\begin{aligned} j &= -\tilde{\sigma}\frac{\partial \phi}{\partial x} + eD[\frac{\partial n_0}{\partial \mu} e \frac{\partial \phi}{\partial x} + \frac{\partial n_0}{\partial \mu} e \frac{\partial \mu_0}{\partial x}] \\ &= (-\tilde{\sigma} + e^2 D \frac{\partial n_0}{\partial \mu})\frac{\partial \phi}{\partial x} + e^2 D \frac{\partial n_0}{\partial \mu}\frac{\partial \mu_0}{\partial x}. \end{aligned} \tag{9.13}$$

Recognizing that the current density has to vanish for constant chemical potential and thus $\frac{\partial \mu_0}{\partial x} = 0$, we arrive at the *Einstein relation*

$$\tilde{\sigma} = e^2 D \frac{\partial n_0}{\partial \mu}, \tag{9.14}$$

and Eq. (9.13) becomes

$$j = \tilde{\sigma}\frac{\partial \mu_0}{\partial x}. \tag{9.15}$$

This establishes the quasi-chemical potential as the agent that causes net current to flow.

9.1.2 Spin-Resolved Quasi-Chemical Potentials

The ideas outlined in the previous subsection can be readily extended to spin densities and spin current densities, as the quasi-chemical potential is decomposed into a spin-up and a spin-down component. In analogy to Eqs. (9.1) and (9.2), we have [310]

$$\begin{aligned} \mu_0 &\equiv \frac{1}{2}(\mu_\uparrow + \mu_\downarrow) \\ \mu_s &\equiv \frac{1}{2}(\mu_\uparrow - \mu_\downarrow), \end{aligned} \tag{9.16}$$

where the second equation defines the *spin quasi-chemical potential*. The spin-dependent current densities $j_{\uparrow,\downarrow}$, and, consequently, the spin current density $j_s \equiv j_\uparrow - j_\downarrow$, are found from

$$j_{\uparrow,\downarrow} = \tilde{\sigma}\frac{\partial \mu_{\uparrow,\downarrow}}{\partial x}. \tag{9.17}$$

Considering the spin-resolved electron densities, it is of relevance that energy and, through spin-flip processes, particle exchange may take place between the two spin reservoirs. Thus, they are assumed to be at equilibrium and therefore characterized by the same chemical potential μ. This implies for the spin-up and spin-down densities that

$$
\begin{aligned}
n_\uparrow(x) &= n_{\uparrow,0}(\mu + e\mu_\uparrow + e\phi) \approx n_{\uparrow,0} + \frac{\partial n_{\uparrow,0}}{\partial \mu}(e\mu_\uparrow + e\phi) \\
n_\downarrow(x) &= n_{\downarrow,0}(\mu + e\mu_\downarrow + e\phi) \approx n_{\downarrow,0} + \frac{\partial n_{\downarrow,0}}{\partial \mu}(e\mu_\downarrow + e\phi).
\end{aligned}
\tag{9.18}
$$

The expansion in Eq. (9.18) is based on the notion that the chemical potential is much larger than the sum $e\mu_{\uparrow,\downarrow}+e\phi$. This is usually valid in degenerate conductors, where the electrons that contribute to the spin-polarized density tend to be in the vicinity of the Fermi level E_F, and $\mu \approx E_F$. Since $\frac{\partial n_0}{\partial \mu}$ represents the density of states, g, one may express the spin-resolved electron densities by the spin-resolved density-of-state distributions, namely,

$$
\begin{aligned}
n_\uparrow(x) &= n_{\uparrow,0} + g_\uparrow e\mu_\uparrow + g_\uparrow e\phi \\
n_\downarrow(x) &= n_{\downarrow,0} + g_\downarrow e\mu_\downarrow + g_\downarrow e\phi.
\end{aligned}
\tag{9.19}
$$

From the local charge neutrality condition $n_\uparrow + n_\downarrow = n_0$, it follows from Eqs. (9.19) that

$$
g(\mu_0 + \phi) + g_s\mu_s = 0. \tag{9.20}
$$

Here we have used the abbreviations

$$
\begin{aligned}
g &= g_\uparrow + g_\downarrow, \\
g_s &= g_\uparrow - g_\downarrow.
\end{aligned}
\tag{9.21}
$$

Introducing the *electron spin density* $s = n_\uparrow - n_\uparrow$, and subtracting the two equations (9.19) from each other, we find, by use of Eq. (9.20),

$$
s = s_0 + eg_s(\mu_0 + \phi) + eg\mu_s = s_0 + 4e\mu_s\frac{g_\uparrow g_\downarrow}{g}. \tag{9.22}
$$

By virtue of the local charge neutrality premise, there is no charge accumulation. The presence of a finite spin quasi-chemical potential μ_s, however, implies a certain amount of spin accumulation. If $s = s_0 + \delta s$, this nonequilibrium spin density is calculated from

$$
\delta s = 4e\mu_s\frac{g_\uparrow g_\downarrow}{g}. \tag{9.23}
$$

The consequences of these observations for the behavior of spin currents and densities in composite media will be discussed in Section 9.4.

9.2 Spintronics: Basic Devices

In this section we introduce three basic elements of spintronics circuits: the *spin valve*, the *spin transistor* and the *spin filter*. A common feature of these designs is that they subject charge carriers to magnetic interactions in order to generate or manipulate spin currents.

9.2.1 Spin Valve

The most elementary electronic device operating with spin currents is the *spin valve*. It that takes advantage of multilayer structures displaying the *giant magnetoresistance* (GMR) effect. The basic mechanism of spin-polarized transit is illustrated in Figure 9.2. Electrons are assumed to traverse a structure consisting of two ferromagnetic layers that enclose a normal metal. The resistance of this arrangement depends on the relative orientation of the magnetic moments in the ferromagnets, being lower in case of parallel and higher in case of antiparallel alignment. The GMR effect may be explained in the framework of the *two currents model* [312, 313, 314], which treats electrons with up and down polarization as two independent parallel currents, and in the language of *spin-dependent*

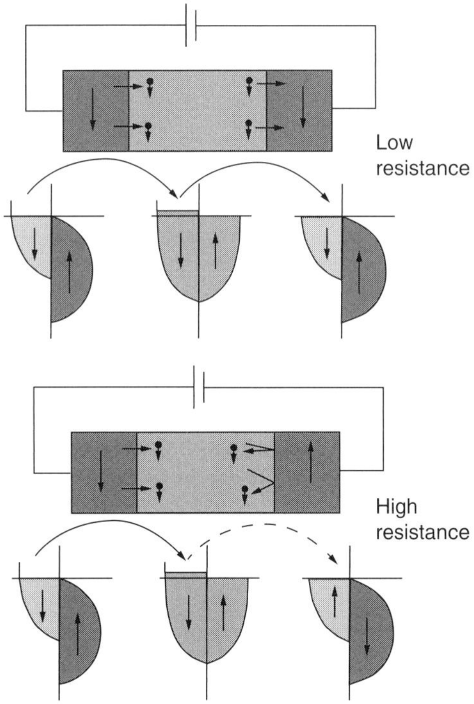

Figure 9.2 Scheme of a spin valve circuit with spin-resolved densities of state for the intermediate unpolarized metal and the enclosing ferromagnetic layers. For the sake of clarity, it has been assumed that one of the bands containing the two spin moities in the ferromagnet is completely filled. An electron whose spin is aligned with the majority orientation at the Fermi energy of the ferromagnet can be transferred into the corresponding majority state of the right layer. Once the magnetization of the latter is reversed, however, no state that matches the electron's spin orientation is available. From [318]. Reprinted with permission from AAAS.

scattering. The comparison between the two magnetic arrangements in terms of resistance is based on the notion that an electron whose spin is antiparallel to the majority spin orientation of its ferromagnetic host at the Fermi energy E_F will be scattered with higher probability than an electron with parallel spin, translating into a shorter mean free path and thus a higher resistance in the former configuration. Further, the traversed metals can be chosen as shown in Figure 9.2, where electrons with majority spin at E_F will hardly experience a change in their number as they move from the ferromagnetic to the nonmagnetic layer. This is in contrast to an electron with the opposite spin direction, as the minority spin density at E_F undergoes a marked change at the interface, making scattering events likely. In summary, if we denote the resistance encountered by an electron in the spin-antiparallel configuration by $R_{\uparrow\downarrow}$ and that for the spin-parallel alternative by $R_{\uparrow\uparrow}$, we find that the overall resistance for an electron with any spin polarization traversing the sandwich structure with the two magnetizations aligned in opposite directions (upper panel in Figure 9.3) is $R_{\uparrow\downarrow} + R_{\uparrow\uparrow}$. For a current with zero spin polarization, this result reduces to $R_{AP} = \frac{R_{\uparrow\downarrow}+R_{\uparrow\uparrow}}{2}$. If however, the alternative with both magnetizations in the same direction (lower panel in Figure 9.3) is implemented, the overall resistance for unpolarized current is $R_P = \frac{2R_{\uparrow\downarrow}R_{\uparrow\uparrow}}{R_{\uparrow\downarrow}+R_{\uparrow\uparrow}}$. Thus, the difference between the resistances in the two cases is seen to be

$$\Delta R = R_{AP} - R_P = \frac{(R_{\uparrow\downarrow} - R_{\uparrow\uparrow})^2}{2(R_{\uparrow\downarrow} + R_{\uparrow\uparrow})}. \tag{9.24}$$

When first measured [316], the *magnetoresistance ratio* $MR \equiv \frac{\Delta R}{R_P}$ [1] turned out to be around 85 percent, greatly exceeding the well-known anisotropic magnetoresistance effect. This value has been further improved by utilizing the phenomenon of *tunnel magnetoresistance* (TMR), where a thin insulating tunnel barrier takes the place of the non-ferromagnetic spacer. Among the virtues of this design is an almost vanishing dependence on temperature, while the GMR was reported to decrease sensitively as the temperature increases [317].

Usually the ferromagnetic materials at the two ends of the spin valve are chosen such that they differ strongly with respect to magnetic hardness, such that the magnetization of one end is easy and that of the other is hard to reverse. In this way it is possible to switch between high and low resistance with respect to spin transmission by changing the orientation of a polarizing external magnetic field.

Most prominent among the applications of the GMR effect is *magnetic recording*. Read heads that extract information stored in magnetic domains on hard disk drives of computers employ the fingerprint feature of this effect, change of resistance with the relative direction of magnetization. Bits, as units of information, are

[1] The magnetoresistance ratio is frequently indicated in terms of conductances. With $G_P \equiv 1/R_P$, and $G_{AP} \equiv 1/R_{AP}$, it holds $MR = (G_P - G_{AP})/G_{AP}$.

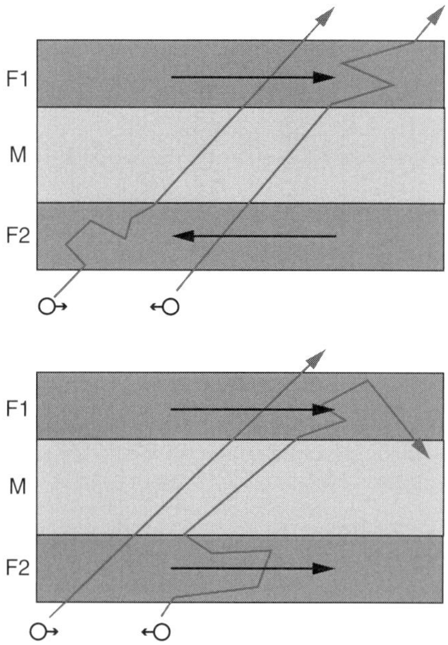

Figure 9.3 Illustration of the two-currents model [312, 313, 314] to account for the GMR effect. Antiparallel alignment between the majority spin orientation of the ferromagnetic host at the Fermi energy and the spin of the electron traversing the structure correlates with higher resistance ($R_{\uparrow\downarrow}$, see text) than parallel alignment ($R_{\uparrow\uparrow}$), implying that configuration (a) has a higher overall resistance than configuration (b). From [318]. Reprinted with permission from AAAS.

encoded on the disk as up (1) and down (0) magnetization within the domains. A GMR read head is a spin valve whose soft layer aligns with the magnetization directions of the traversed domain, as sketched in Figure 9.4. A test current senses the variation of the prevailing magnetization, which induces variations of the spin valve resistance, and makes it possible to map the magnetic orientation of any hard drive segment into an electric signal. This mechanism is expected to provide a storage capacity of 10 gigabits per square inch, a density corresponding to 120 billion bits stored on a regular 3.5-inch disk drive.

9.2.2 Spin Transistor

A spin analog of the field effect transistor (FET) was proposed in 1990 by Datta and Das [319]. Their design is based on a parallel between the spin-valve scheme outlined in the previous subsection and an *electro-optic modulator* used to manipulate the polarization of a light beam passing from an optical polarizer to an optical analyzer. In the arrangement suggested by Datta and Das, the counterparts of the latter elements are provided by the ferromagnetic contacts of a structure similar to a spin valve, with the difference that the magnetizations of both ends are kept

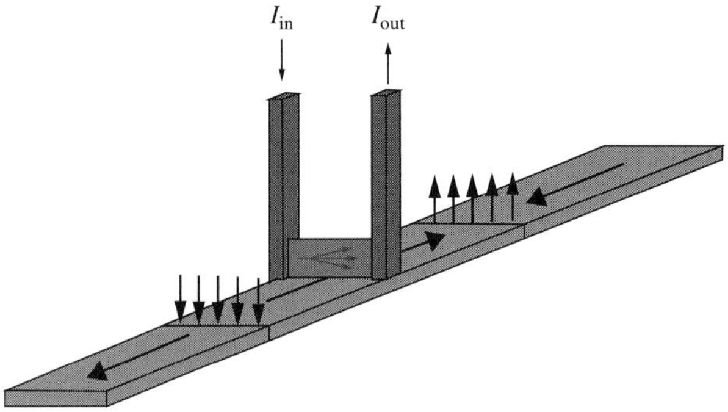

Figure 9.4 Basic mechanism of extracting information magnetically encoded on hard disk drives by use of the GMR effect. The current passing through the read head is sensitive to the magnetization of the traversed magnetic domain, according to the spin-valve principle.

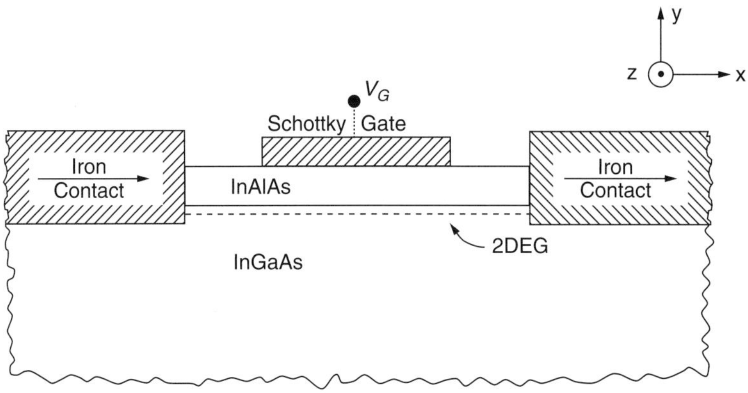

Figure 9.5 Sketch of a spin transistor operating with a semiconductor heterostructure wedged between two ferromagnetic layers. Admission is made for an external bias that gives rise to an electric field perpendicular to the semiconductor components. In the setup shown, electrons move along the positive x axis.

parallel to each other and that the intermediate region consists of a narrow-gap semiconducting heterostructure such as InGaAs/InAlAs. With these premises, a prototypical spin transistor may be conceived as shown in Figure 9.5.

To clarify the principle of the Datta–Das spin transistor in greater detail, we assume that the electrons traversing the composite shown in Figure 9.5 are injected into the heterostructure with spin polarization in the $+x$ direction. With the $+z$ axis chosen as quantization axis, their eigenstate with respect to the $\boldsymbol{\sigma}_x$ operator may be decomposed in the following way:

$$\begin{pmatrix} 1 \\ 1 \end{pmatrix} = \begin{pmatrix} 1 \\ 0 \end{pmatrix} + \begin{pmatrix} 0 \\ 1 \end{pmatrix}. \tag{9.25}$$

As spin-polarized electrons traverse an effective two-dimensional electron gas (2DEG) between the two semiconductor layers, the degeneracy between the $+z$ and $-z$ spin components is broken. This *zero-field spin splitting* is understood as a consequence of an *external spin-orbit effect*, induced by an effective electric field oriented perpendicular to the plane of the 2DEG. An analogous situation has been considered in Section 8.1 where the the Rashba effect, i.e. the impact of Stark interaction on the spin-orbit coupling was discussed (see Eq. (8.16)). For the problem at hand, the *Rashba Hamiltonian*, $\hat{H}_R$, can be cast into a compact form [320],

$$\hat{H}_R = \frac{\alpha}{\hbar}\boldsymbol{\sigma} \times \hat{\mathbf{p}} \cdot \mathbf{e}_y, \tag{9.26}$$

where $\boldsymbol{\sigma}$ denotes the Pauli matrices, and $\mathbf{e}_y$ the unit vector into the y direction perpendicular to the interface between the semiconductor layers (see Figure 9.5), while α, the *Rashba coupling constant*, governs the strength of the interaction. As emphasized in Section 8.1, $\hat{H}_R$ is a model Hamiltonian[2] that originates from the inclusion of both spin-orbit coupling and the Stark effect at the level of second-order perturbation theory. With the geometry chosen as shown in Figure 9.5, and assuming that the electrons may be represented by plane waves, $\hat{H}_R$ simplifies to

$$\hat{H}_R = \alpha(\boldsymbol{\sigma}_z k_x - \boldsymbol{\sigma}_x k_z). \tag{9.27}$$

It is readily shown that this operator gives rise to a shift between the wave numbers $k_\uparrow$ and $k_\downarrow$ of two electrons that enter the interface from the source electrode with the same energy ($E_\uparrow = E_\downarrow$) but opposite spin polarization in z direction. Due to Rashba interaction, the energy of the electrons with spin-up (spin-down) polarization will increase (decrease) by $\alpha k_\uparrow$ ($\alpha k_\downarrow$). For the sake of simplicity, we treat electrons moving into the $+x$ direction ($k_z = 0$). Then it holds,

$$\begin{aligned} E_\uparrow &= \frac{\hbar^2 k_\uparrow^2}{2m_e^*} - \alpha k_\uparrow \\ E_\downarrow &= \frac{\hbar^2 k_\downarrow^2}{2m_e^*} + \alpha k_\downarrow . \end{aligned} \tag{9.28}$$

Since $E_\uparrow = E_\downarrow$, a phase shift $\Delta\theta$ will accumulate between the electrons on their way along an interface of size L. From Eq. (9.28), $\Delta\theta$ is found to be

$$\Delta\theta = (k_\uparrow - k_\downarrow)L = \frac{2m^*\alpha L}{\hbar^2}, \tag{9.29}$$

and thus proportional to the Rashba coupling constant α.

[2] H_R may be viewed as the Dirac Hamiltonian reduced to two dimensions, and with the spin degrees of freedom rotated by $\frac{\pi}{2}$.

For a phase shift of $\Delta\theta = \pi$, the electron spin polarization experiences a full reversal as the 2DEG is traversed.[3] In this case, the current in the drain electrode will be reduced as compared with the reference case of unperturbed traversal. The two electrodes thus act as analogs of the polarizer and the analyzer in electro-optic modulators. Since the size of the Rashba constant, and thus the phase shift undergone by the spins can be varied by manipulating the gate voltage (see Figure 9.5), the Datta–Das model represents a spin-based realization of the field effect transistor (FET) principle.

Early estimates suggested that heterostructures exist that, at low temperature, combine sizeable mean free path lengths in the order of 1 μm with Rashba coupling constants of adequate magnitude. Still, efforts aimed at the practical realization of the Datta–Das model have to contend with low spin injection efficiencies as well as short spin relaxation and dephasing times.

Some prototypes, however, have been implemented. Thus, Koo et al. [321] realized a spin-injected FET based on the design by Datta and Das and reported observation of conductance oscillations in response to varying the gate voltage, thus tuning the strength of the spin-orbit interaction experienced by the spin-polarized electrons, or, equivalently, manipulating the Rashba coupling. The Datta–Das device is not the only spin transistor scheme presently discussed. The *magnetic bi-polar transistor* represents an alternative design [322].

9.2.3 Spin Filters

Like its electro-optical analog, a spin transistor is most efficient if it operates with maximum polarization. This involves the highest achievable degree of light polarization in opto-electronics and of spin polarization in spintronics, defined as $|g_\uparrow(E_F) - g_\downarrow(E_F)|/(g_\uparrow(E_F) + g_\downarrow(E_F))$, where $g_\uparrow(E_F)$, $g_\downarrow(E_F)$ stand for the density-of-states of electrons with up or down spin orientation, respectively, at the Fermi energy. A spin polarization of 1 can, in principle, be accomplished by employing a *half-metal* as the polarizing ferromagnet. Half-metallic materials are ferromagnets with only one occupied spin-polarized sub-band at the Fermi level, implying metallic behavior for electrons with one spin orientation and semiconducting or insulating characteristics for the other, as illustrated in Figure 9.6 in terms of the density of states.

By this criterion, the elementary 3*d* ferromagnets are not half-metals, although some of them, such as cobalt or nickel, have a completely occupied 3*d* $\uparrow$ band, and only 3*d* electrons of the opposite direction contribute to the density-of-states distribution at the Fermi energy. In these cases, the Fermi level also intersects the

[3] Consider that a phase shift by $\Delta\theta = \pi$ between the two eigenstates of $\hat{\sigma}_z$ on the right-hand side of Eq. (9.25) is equivalent to multiplying one of these eigenstates by a factor of -1. This results in a sign reversal of the $\hat{\sigma}_x$ eigenvalue for the new, phase-shifted state.

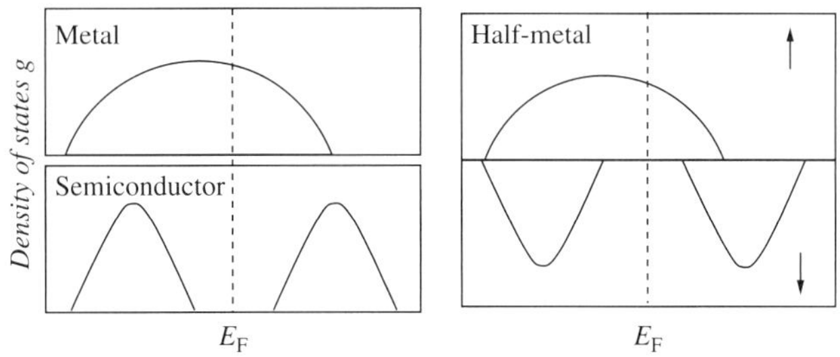

Figure 9.6 Representation of the density of states around the Fermi energy of a half metal, subdivided into spin-up and spin-down components versus the corresponding scheme for metals and semiconductors.

unpolarized 4s band, such that conduction is not limited to one spin orientation alone. Half-metallic ferromagnets were first characterized by de Groot et al. [323] who analyzed the band structure of Mn-based *semi-Heusler alloys* [324]. The spin polarization attainable in a particular half metal can be experimentally examined by recording the spin resolved photoemission spectrum [325]. The Curie temperature of half-metallic compounds is of crucial importance for their potential use as spintronics materials. Complex alloys such as $Co_2Cr_{1-x}Fe_xAl$ display a sensitive dependence of the Curie temperature on the parameter x [326].

We mention also that the spin valve effect described earlier may be used as a spin filtering mechanism [327, 328, 329]. Thus, a spin filter may consist of a three-terminal arrangement, typically composed of a ferromagnetic layer or multilayer between two semiconductors, such that Schottky barriers emerge at the two interfaces. The components of this structure are often referred to in terms of transistor elements: emitter, base and collector. Applying forward bias to the junction between the emitter and the base generates a spin-unpolarized emitter current that consists of *hot electrons*, i.e. electrons of sufficient energy to overcome the barrier at the semiconductor-metal junction. While traversing the ferromagnetic base, electrons with magnetic moments opposite to the local magnetization interact more strongly with the host electrons and thus experience higher resistance than those with the alternative spin orientation, resulting in a longer mean free path for one spin direction than for the other. The spin-filtering effectiveness of this device may be expressed by the *magnetocurrent* MC [311],

$$MC \equiv \frac{(I_{C\uparrow\uparrow} - I_{C\uparrow\downarrow})}{I_{C\uparrow\downarrow}}, \tag{9.30}$$

with $I_{C\uparrow\uparrow}$ ($I_{C\uparrow\downarrow}$) as collector currents with magnetic moments parallel (antiparallel) to the host magnetization. Magnetocurrents exceeding 100 percent by a wide margin have been reported [330]. For a survey of various other spin-filtering techniques, see [311].

Spin filters, as discussed in this section, may be used to produce well-defined net spin populations. For optimal design of spin circuits, problems associated with generating, manipulating and measuring currents with high degrees of spin polarization must be addressed successfully. Further, a major challenge for spintronics arises from the fact that the net spin current in electronic media is not a conserved quantity. Therefore, the spin relaxation and spin dephasing claim particular attention. From these considerations, three basic questions emerge:

1. How can spin-polarized currents be injected into transmission media with high efficiency?
2. What techniques can be used for spin detection?
3. What mechanisms of spin relaxation or spin dephasing are operative in this material?

The remainder of this chapter will deal with these three items.

9.3 Spin Injection

To introduce some basic ideas about the injection of spin currents into materials, we consider the steady flow of an electric current with net spin polarization from a ferromagnetic to a non-magnetic conductor [310, 333]. The objective is to derive the spin current Eq. (9.2) across such a ferromagnet/semiconductor, or F/N, junction. Characteristically, degrees of spin polarization in a ferromagnetic metal amount to 10 to 50 percent. Assuming this margin as initial value $P_{n,0}$ that describes the spin polarization at the interface between the ferromagnet and the non-magnet, we ask for the spatial distribution of P_n across the non-magnet. The motion of the spin-polarized charge carriers may be viewed as a superposition of drift and diffusion. However, if the current is driven by regular electric fields, diffusion will be the dominant effect by far. Therefore, drift will be disregarded in the following development. Introducing the *spin diffusion length* L_s, we write down a diffusion equation for the spin density $s(x) \equiv n_\uparrow - n_\downarrow$ in the non-magnetic conductor:

$$\frac{d^2 s}{dx^2} = \frac{s}{L_s^2}. \tag{9.31}$$

Here we consider, for the sake of simplicity, a one-dimensional model. It is solved by

$$s(x) = s_0 e^{-\frac{x}{L_s}}, \tag{9.32}$$

using $s(x) = 0$ if $x \to \infty$ as a second boundary condition, based on the expectation that the spin ensemble, having undergone a sufficient number of spin-changing interactions with the host, will eventually attain the equilibrium spin polarization of the non-magnet, namely zero. The change of the spin polarization density per length yields the spin current density,

$$j_s(x) = -\bar{D}\frac{ds}{dx}, \tag{9.33}$$

with the spin-averaged diffusion constant $\bar{D}$ as proportionality factor. This constant is related to the spin diffusion length through the *spin relaxation time* τ_s:

$$L_s = \sqrt{\bar{D}\tau_s}. \tag{9.34}$$

We assume that no spin change occurs at the interface between the two media, which is assumed to be infinitely narrow and located at the place $x = 0$. Thus, it holds that $j_s(x = 0) = j_{s0}$, where the latter value refers to the spin current density in the ferromagnet. This premise is realized to a good approximation. It leads to a prediction for the spin polarization density profile in the non-magnetic host:

$$s(x) = j_{s0}\frac{L_s}{\bar{D}}e^{-\frac{x}{L_s}}. \tag{9.35}$$

Integrating this expression in the interval $[0, \infty]$ yields a total spin accumulation of

$$S_{tot} = j_{s0}\frac{L_s^2}{\bar{D}} = j_{s0}\tau_s. \tag{9.36}$$

If the elementary model defined by the preceding equations is to supply useful information about the spin current (9.2) traversing the F/N junction, it must be refined in chiefly two directions. First, it must be acknowledged that spin transport in the ferromagnet is governed by different conductivity constants for the two spin moieties, $\tilde{\sigma}_\uparrow$ and $\tilde{\sigma}_\downarrow$, while the non-magnetic medium is characterized by only one constant, $\tilde{\sigma}_N$. Further, associated with a discontinuity of the chemical potential across the F/N junction [332], spin-dependent contact conductances, labelled $\tilde{\Sigma}_\uparrow$ and $\tilde{\Sigma}_\downarrow$, are to be assigned to spin-up and spin-down charge carriers, respectively, that traverse the interface between the two host materials.[4]

For practical considerations, related to the design of functional spin circuits, questions of *conductivity matching* between the individual segments of the transmission element are of paramount importance. We discuss this issue in terms of the spin chemical potential μ_s that governs the local spin accumulation in spin circuit transmission elements (as introduced in Section 9.1).

[4] A note on dimensions: Based on the relation $j = \tilde{\sigma}\nabla\mu(x)$, where $\mu(x)$ stands for the x-dependent chemical potential, the dimension of $\tilde{\sigma}_{\uparrow,\downarrow}$ and $\tilde{\sigma}_N$ is current density $\times$ length/chemical potential. In the contact zone, however, it holds that $j = \tilde{\Sigma}\Delta\mu$, making the dimension of $\tilde{\Sigma}_{\uparrow,\downarrow}$ equal to current density/chemical potential. Thus, $\tilde{\Sigma}_{\uparrow,\downarrow}$ is a *conductance* rather than a *conductivity*.

Partitioning the conductivity $\tilde{\sigma}$ as $\tilde{\sigma}_\uparrow + \tilde{\sigma}_\downarrow$ and $\tilde{\sigma}_s$ as $\tilde{\sigma}_\uparrow - \tilde{\sigma}_\downarrow$ we write the charge and spin current densities as [310]

$$\begin{aligned} j = j_\uparrow + j_\downarrow &= \tilde{\sigma}\frac{\partial \mu_0}{\partial x} + \tilde{\sigma}_s\frac{\partial \mu_s}{\partial x} \\ j_s = j_\uparrow - j_\downarrow &= \tilde{\sigma}_s\frac{\partial \mu_0}{\partial x} - \tilde{\sigma}\frac{\partial \mu_s}{\partial x}. \end{aligned} \tag{9.37}$$

Combining these two equations, we obtain (compare with Eq. (9.22))

$$j_s = P_{\tilde{\sigma}} j + 4\frac{\partial \mu_s}{\partial x}\frac{\tilde{\sigma}_\uparrow\tilde{\sigma}_\downarrow}{\tilde{\sigma}}, \tag{9.38}$$

where we substituted $P_{\tilde{\sigma}}$ for $\frac{\tilde{\sigma}_s}{\tilde{\sigma}}$, introducing a *conductivity spin polarization*. In what follows, we will apply this formula to the three segments of the transmission element shown in Figure 9.7, with the goal to make a statement about the *current spin polarizations* P_{jF}, P_{jC} and P_{jN} of the ferromagnetic, the contact and the non-magnetic region, respectively.

From Eq. (9.38), the initial value of the quantity $P_{jF} \equiv \frac{j_{sF}}{j_F}$ is given by

$$P_{jF}(0) = P_{F\tilde{\sigma}} + \frac{4}{j}\frac{\tilde{\sigma}_{F\uparrow}\tilde{\sigma}_{F\downarrow}}{\tilde{\sigma}_F}\frac{\partial \mu_{sF}}{\partial x}(0). \tag{9.39}$$

With the approach [310]

$$\mu_{sF}(x) = \mu_{sF}(0)e^{\frac{x}{L_{sF}}}, \tag{9.40}$$

this becomes

$$P_{jF}(0) = P_{F\tilde{\sigma}} + \frac{1}{je}\frac{\mu_{sF}(0)}{R_F}. \tag{9.41}$$

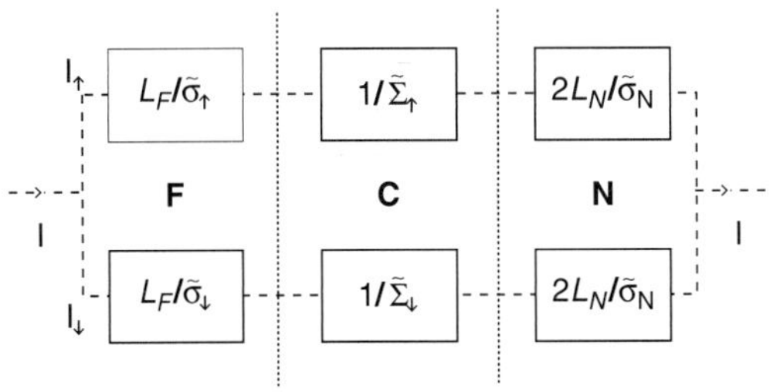

Figure 9.7 Equivalent circuit for spin current traversing a ferromagnet/non-magnet junction. The resistance of the individual system segments is indicated in terms of specific lengths and/or specific conductivities. The lengths of the two extended components are labelled as L_F and L_N, and the symbols $\tilde{\sigma}_{\uparrow,\downarrow}$, $\tilde{\sigma}_N$, and $\tilde{\Sigma}_{\uparrow,\downarrow}$ refer to the conductivities of the ferromagnet as well as the non-magnet, and the conductance of the contact zone between them, respectively, subdivided into spin-up and spin-down orientation. The contact zone is assumed to be infinitely narrow.

The effective resistance of the ferromagnetic zone is defined as

$$R_F \equiv \frac{\tilde{\sigma}_F}{4\tilde{\sigma}_{F\uparrow}\tilde{\sigma}_{F\downarrow}} L_{sF}. \tag{9.42}$$

We point out that R_F is large for half-metallic media (see Section 9.2.3) since in this case the conductivity for one of the spin channels is markedly smaller than that for the other one. For the non-magnetic medium, where $\tilde{\sigma}_{N\uparrow} = \tilde{\sigma}_{N\downarrow} = \tilde{\sigma}_N/2$, analogous considerations [310] lead to

$$P_{jN}(0) = -\frac{1}{j}\frac{\mu_{sN}(0)}{R_N} \tag{9.43}$$

and

$$R_N = \frac{L_{sN}}{\tilde{\sigma}_N}. \tag{9.44}$$

The corresponding treatment of the contact region differs from that of the ferromagnetic and the non-magnetic segments since the former, by assumption, does not have any extension in the transport direction. We describe the current densities through the contact region in analogy to that through the other two system parts, namely

$$j_{\uparrow,\downarrow} = \tilde{\Sigma}_{\uparrow,\downarrow}(\mu_{N\uparrow,\downarrow}(0) - \mu_{F\uparrow,\downarrow}(0)) = \tilde{\Sigma}_{\uparrow,\downarrow}\Delta\mu_{\uparrow,\downarrow}(0). \tag{9.45}$$

With these assignments, we might write expressions for the charge and the spin density, as in Eq. (9.37):

$$\begin{aligned} j &= \tilde{\Sigma}\Delta\mu_0(0) + \tilde{\Sigma}_s\Delta\mu_s(0) \\ j_s &= \tilde{\Sigma}_s\Delta\mu_0(0) - \tilde{\Sigma}\Delta\mu_s(0). \end{aligned} \tag{9.46}$$

The symbols $\tilde{\Sigma}$ and $\tilde{\Sigma}_s$ stand here for the *contact conductance*, $\tilde{\Sigma}_\uparrow + \tilde{\Sigma}_\downarrow$, and the *contact spin conductance*, $\tilde{\Sigma}_\uparrow - \tilde{\Sigma}_\downarrow$, respectively. Repeating the manipulation that led from Eq. (9.37) to Eq. (9.39), we eliminate $\Delta\mu_0(0)$ from this system of equations and arrive at

$$j_s = P_{\tilde{\Sigma}}\, j + \frac{\Delta\mu_s(0)}{R_C}, \tag{9.47}$$

with $P_{\tilde{\sigma}}$ as *conductance spin polarization*,

$$P_{\tilde{\Sigma}} = \frac{\tilde{\Sigma}_s}{\tilde{\Sigma}}, \tag{9.48}$$

and R_C as effective contact resistance,

$$R_C = \frac{\tilde{\Sigma}}{4\tilde{\Sigma}_\uparrow\tilde{\Sigma}_\downarrow}. \tag{9.49}$$

In terms of the contact current spin polarization, Eq. (9.47) assumes the following form:

$$P_{jC} = P_{\tilde{\Sigma}} + \frac{1}{j}\frac{\Delta\mu_s(0)}{R_C}. \tag{9.50}$$

To assess the overall current spin polarization, we solve Eqs. (9.39), (9.43) and (9.50) under the continuity constraint

$$P_j \equiv P_{jF}(0) = P_{jC} = P_{jN}(0). \tag{9.51}$$

This relation is warranted if no spin-flip scattering occurs in the contact. The current spin polarization turns out to be the conductivity polarization average, weighted by the resistances of all three system parts, defining the *spin injection efficiency*:

$$P_j = \frac{R_F P_{\tilde{\sigma} F} + R_C P_{\tilde{\Sigma}}}{R_F + R_C + R_N}. \tag{9.52}$$

This relation makes it possible to assess the problem of *conductivity mismatch*, and to devise schemes that address this challenge. In the case of a purely ohmic contact ($R_C = 0$), the typical condition $R_N \gg R_F$ leads to low spin insertion efficiency. For half-metallic ferromagnets, however, R_F tends to be large, as mentioned in the context of formula (9.42), eliminating or alleviating the conductivity mismatch. From Eq. (9.52), this effect can also be attained if the contact region is designed such that $R_C \geq R_N \gg R_F$. If, specifically, electron transport through the contact region proceeds by tunneling, it holds that

$$R_C \gg R_N,\, R_F, \text{ and } P_j \approx P_{\tilde{\Sigma}}. \tag{9.53}$$

In contrast to *tunnel contacts*, *transparent contacts* are characterized by the condition $R_C << R_N, R_F$. Equations (9.46) to (9.52) summarize the *standard model of spin injection* [310].

We conclude this section with a comment on the density spin polarization in the non-magnetic segment. As implied by Eq. (9.43) in conjunction with Eq. (9.51), we have

$$\mu_{sN}(0) = -jP_j R_N. \tag{9.54}$$

If, therefore, electrons flow from the ferromagnet to the non-magnet ($j < 0$), it holds that $\mu_{sN}(0) > 0$, corresponding to spin injection. The reverse case, in contrast, describes spin extraction. In view of the relation $s = e\mu_s g(E_F)$ that connects the spin density with the number of electronic states in an energy interval $e\mu_s$ at the Fermi level, we conclude from Eq.(9.54) that

$$P_n = \frac{s}{n} = e\mu_{sN}(0)\frac{g_N(E_F)}{n} = -ejR_N\frac{g_N(E_F)}{n}P_j. \tag{9.55}$$

This equation states a proportionality of the density spin polarization and the current spin polarization on one hand and the fraction of electrons in the energy interval ejR_N at the Fermi level on the other.

Spin-Charge Coupling

In the previous section we discussed the insertion of spin-polarized electrons, inducing a spin current in a non-magnetic medium. The resulting non-equilibrium spin distribution is associated with an electromotive force (EMF). Measuring the EMF in open-circuit conditions makes it possible, in principle, to detect a spin signal that is uncontaminated by charge current. To clarify this, we consider once more the F/N junction shown in Figure 9.7, i.e. a sequence of a ferromagnetic and a non-magnetic medium separated by a contact zone. However, we modify this scheme by including a spin source at the right end of the non-magnetic medium, as shown in Figure 9.8. Spins injected at this place will diffuse toward the ferromagnet and the contact, as the spin-polarizing segment of the arrangement. As charge equilibrium is assumed ($j = 0$), this setup generates a pure spin current. The phenomenon of *spin-charge coupling* is manifested by an EMF that may be found from

$$EMF = \mu_{0N}(\infty) - \mu_{0F}(-\infty). \tag{9.56}$$

The charge equilibrium constraint ensures that this difference is equal to the corresponding electric potential difference $\phi_F(\infty) - \phi_N(-\infty)$.

To relate the measurable EMF to the spin quasi-chemical potential, we use the first one of the Eqs. (9.46), which yields

$$\Delta\mu_0 = -P_{\tilde{\Sigma}}\Delta\mu_s, \tag{9.57}$$

relating a change in the quasi-chemical potential to a corresponding change in the spin quasi-chemical potential. Use has been made of the equilibrium condition $j = 0$. Applying this relation to all three system parts, the ferromagnet, the contact and the non-magnet, we arrive at

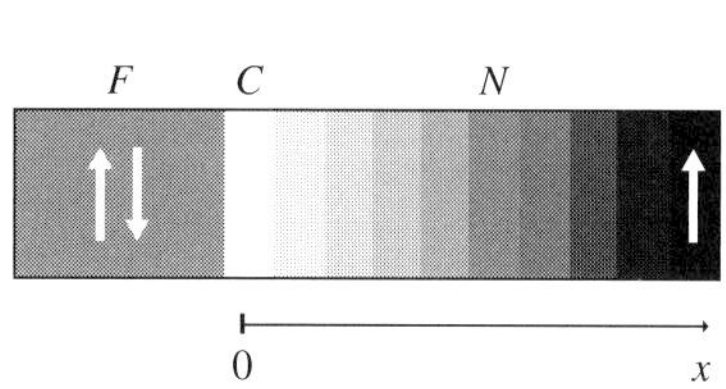

Figure 9.8 Sequence of a ferromagnet, a contact region and a non-magnetic medium, as in Figure 9.7. At the right end of the non-magnet, spins are added to the system. As they diffuse towards the ferromagnet and the contact, they create an electromotive force.

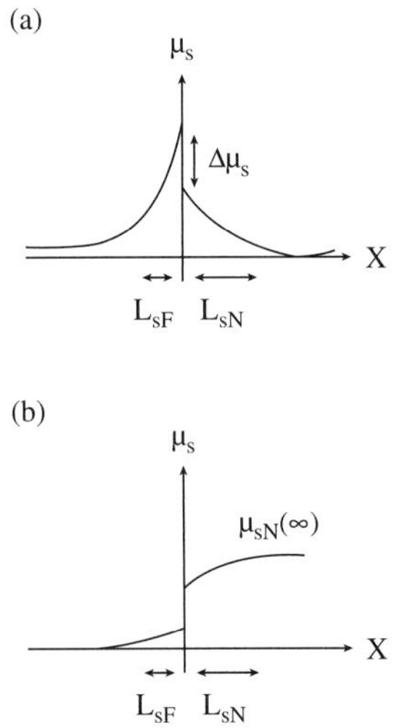

Figure 9.9 Behavior of the spin quasi-chemical potential in the contact zone between the ferromagnetic and the non-magnetic medium for (a) the spin injection and (b) the spin-charge coupling regime. Natural length scales are provided by the characteristic spin diffusion lengths L_{sF} and L_{sN}.

$$
\begin{aligned}
\mu_{0F}(0) - \mu_{0F}(-\infty) &= -P_{\tilde{\sigma}F}\mu_{sF}(0) \\
\mu_{0N}(0) - \mu_{0F}(0) &= -P_{\tilde{\Sigma}}[\mu_{sN}(0) - \mu_{sF}(0)] \\
\mu_{0N}(\infty) - \mu_{0N}(0) &= 0.
\end{aligned} \tag{9.58}
$$

Adding these three equations yields the desired EMF in terms of the spin quasi-chemical potential at the site of the contact:

$$
EMF = \mu_{0N}(\infty) - \mu_{0F}(-\infty) = \mu_{sF}(0)(P_{\tilde{\Sigma}} - P_{\tilde{\sigma}F}) - \mu_{sN}(0)P_{\tilde{\Sigma}}. \tag{9.59}
$$

In both the spin injection regime, as discussed earlier, and the spin-charge coupling regime, as outlined here, $\mu_{sF}(0)$ and $\mu_{sN}(0)$ differ from each other, as the spin quasi-chemical potential is discontinuous in the contact zone. The jump of $\mu_s(x)$ at $x = 0$ is readily derived from Eq. (9.40) and its counterpart for the non-magnetic medium. The crucial difference between the two regimes lies in the asymptotic values of μ_{sN}, which are zero in spin injection and finite in spin-charge coupling conditions. The two modes are depicted in Figure 9.9.

The interface spin currents for the three regions of the setup sketched in Figure 9.8, $j_{sF}(0)$, $j_{sC}(0)$ and $j_{sN}(0)$, may be obtained from Eq. (9.38), in conjunction with the expressions for the three resistances, R_F (Eq. (9.42)), R_N (Eq. (9.44)) and R_C (Eq. (9.49)). We find [333]

$$
\begin{aligned}
j_{sN}(0) &= -\frac{1}{R_N}(\mu_{sN}(0) - \mu_{sN}(\infty)) \\
j_{sF}(0) &= \frac{1}{R_F}\mu_{sF}(0) \\
j_{sC} &= \frac{1}{R_C}(\mu_{sN}(0) - \mu_{sF}(0)).
\end{aligned} \tag{9.60}
$$

Combined with the constraint $j_s(0) \equiv j_{sF}(0) = j_{sC} = j_{sN}(0)$, these equations yield

$$
j_s(0) = \frac{\mu_{sN}(\infty)}{R_F + R_C + R_N}. \tag{9.61}
$$

From Eq. (9.59) and Eqs. (9.60), we derive finally the EMF in terms of $\mu_{sN}(\infty)$, and find

$$
EMF = -(P_{\tilde{\sigma}F} R_F + P_{\tilde{\Sigma}} R_C) j_s(0). \tag{9.62}
$$

This formula in conjunction with Eqs. (9.52) and (9.61) leads to a relation between the EMF and the spin injection efficiency P_j [310]

$$
EMF = -P_j \mu_{sN}(\infty) \tag{9.63}
$$

Exercise 9.2 Verify: Eq. (9.62) is the correct expression for the EMF that manifests spin-charge coupling in an F/N junction.

Assuming that the barrier between the ferromagnet and the non-magnetic medium is a tunnel contact (see Eq. (9.53) and the related commentary), one may obtain a formal description of the EMF (9.63) in terms of the transmission rates for spin-up and spin-down electrons traversing the tunnel barrier, $t_\uparrow$ and $t_\downarrow$, respectively. In particular, the following proportionalities are valid for the current densities across the junction:

$$
\begin{aligned}
j_\uparrow &\propto t_\uparrow g_{F\uparrow}(V + \mu_{sN}) \\
j_\downarrow &\propto t_\downarrow g_{F\downarrow}(V - \mu_{sN}).
\end{aligned} \tag{9.64}
$$

The symbol V stands here for a spin-independent voltage between the ferromagnet and the non-magnet that is modified by the spin-resolved quasi-chemical potential.

Exercise 9.3 Derive from Eq. (9.64) the following expression for the overall current density:

$$
j = j_\uparrow + j_\downarrow \propto (t g_F + t_s g_{sF})V + (t g_{sF} + t_s g_F)\mu_{sN}, \tag{9.65}
$$

with $t_s = \frac{t_\uparrow - t_\downarrow}{2}$, $t = \frac{t_\uparrow + t_\downarrow}{2}$ and $g_s = g_\uparrow - g_\downarrow$, $g = g_\uparrow + g_\downarrow$ (see Eq. (9.21)). Go on to show: if there is no charge current ($j = 0$, or $j_\uparrow = -j_\downarrow$), V decribes, by Eq. (9.65), an EMF between the ferromagnet and the non-magnetic medium. Specifically, it holds that

$$EMF = \frac{P_{gF} + P_t}{1 + P_t P_{gF}} \mu_{sN}, \tag{9.66}$$

with P_{gF} and P_t as the spin polarizations of the density of states and of the transmission probability.[5]

In the following section we will examine the mechanisms that determine the spin relaxation constants τ_s and L_s in Eqs. (9.31–9.32) and (9.34–9.36).

In semiconductors, non-equilibrium spin distributions can be generated through optical excitation, exploiting selection rules for light-induced transitions between spin states. This allows for *optical spin orientation* as a spin polarization technique for electric currents. The reader will find a brief summary of this spin injection procedure at http://faculty.etsu.edu/HAGELBER/comp_chem.html.

9.4 Spin Relaxation

A spin-polarized ensemble of electrons in a non-magnetic electronic system equilibrates by a variety of mechanisms, subdivided into spin-lattice (or longitudinal) relaxation and spin dephasing (or transversal relaxation), and defined by characteristic times T_1 and T_2, respectively. These times govern the evolution of the electronic magnetization $\mathbf{M}$ in a magnetic field $\mathbf{B}$ that is partioned into a *longitudinal* component B_0, conventionally chosen along the $\hat{\mathbf{z}}$ direction, and a *transversal* time-dependent component $\mathbf{B}_1(t)$:

$$\mathbf{B} = B_0\mathbf{z} + \mathbf{B}_1(t). \tag{9.67}$$

The constant T_1 measures the equilibration time for the longitudinal component of the electronic magnetization. It is tantamount to a time constant for a spin-lattice interaction. As the electron spin ensemble relaxes, it loses energy, mostly by emission of phonons that are absorbed by the lattice, where the term lattice stands for a thermal reservoir that is in contact with the spin system. The constant T_2 may be understood as the characteristic time for the phase coherence loss of the magnetization components perpendicular to the longitudinal direction. The latter process

[5] P_{gF} and P_t are defined in accordance with the concept of spin polarization as introduced in Section 9.3: $P_{gF} = (g_{F\uparrow} - g_{F\downarrow})/(g_{F\uparrow} + g_{F\downarrow})$, $P_t = (t_\uparrow - t_\downarrow)/(t_\uparrow + t_\downarrow)$.

is caused by local inhomogeneities as well as the thermal distribution of electron momenta, as both factors give rise to deviations between the precession frequencies of the electrons. The *Bloch–Torrey equations* [345, 346] provide phenomenological relations that summarize these processes:

$$\begin{aligned}\frac{\partial M_x}{\partial t} &= \gamma(\mathbf{M}\times\mathbf{B})_x - \frac{M_x}{T_2} + D\Delta M_x,\\ \frac{\partial M_y}{\partial t} &= \gamma(\mathbf{M}\times\mathbf{B})_y - \frac{M_y}{T_2} + D\Delta M_y,\\ \frac{\partial M_z}{\partial t} &= \gamma(\mathbf{M}\times\mathbf{B})_z - \frac{M_z - M_z^0}{T_1} + D\Delta M_z.\end{aligned} \tag{9.68}$$

Here, γ stands for the electron gyromagnetic ratio, $\mu_B g_s/\hbar$, with μ_B as the Bohr magneton, and D for the diffusion constant, assumed to be scalar. Lastly, M_z^0 is the thermal equilibrium magnetization of the electron ensemble, given by the product of the susceptibility χ and the longitudinal field B_0.

In what follows, we will give a survey of the principal spin relaxation mechansims for conduction electrons in metals and semiconductors. These are, most importantly, Elliott–Yafet and D'yakonov–Perel scattering, further Bir–Aronov–Pikus as well as hyperfine interaction processes.

9.4.1 Elliott–Yafet Scattering

The Elliott–Yafet mechanism involves the effect of impurity, lattice defect or phonon scattering on the spin polarization of a conduction electron [334]. The cause of spin relaxation is here spin-orbit coupling, as described by the Hamiltonian $\hat{H}_{so}$,

$$\hat{H}_{so} = \frac{\hbar}{4m_e^2c^2}(\nabla V\times\hat{\mathbf{p}})\cdot\boldsymbol{\sigma} = \frac{\hbar^2}{2m_e^2c^2}\frac{1}{r}\frac{\partial V}{\partial r}\hat{\mathbf{L}}\cdot\hat{\mathbf{S}}, \tag{9.69}$$

with $\hat{\mathrm{S}}$ as spin operator, and V as spin-independent lattice potential. The presence of spin-orbit coupling due to the host lattice implies that the Bloch wave functions are no longer eigenfunctions of $\hat{S}_z$. Thus, they may be written as

$$\psi_{\mathbf{k}n\uparrow}(\mathbf{x}) = (a_{\mathbf{k}n}(\mathbf{x})|\uparrow\rangle + b_{\mathbf{k}n}(\mathbf{x})|\downarrow\rangle)e^{i\mathbf{k}\cdot\mathbf{x}}, \tag{9.70}$$

where the subscript n denotes a band index. Since spin-orbit interaction is invariant with respect to time inversion, the energy of this state does not change as $t\to -t$. Further, if the lattice is characterized by an inversion center the state is degenerate with

$$\psi_{\mathbf{k}n\downarrow}(\mathbf{x}) = (a^*_{-\mathbf{k}n}(\mathbf{x})|\downarrow\rangle - b^*_{-\mathbf{k}n}(\mathbf{x})|\uparrow\rangle)e^{i\mathbf{k}\cdot\mathbf{x}}, \tag{9.71}$$

as is shown by applying the operations of time and space inversion consecutively on the wave function (9.70).[6] We point out that the perfectly periodic lattice of pristine graphene has a center of inversion symmetry. The down [up] spin admixtures to the states (9.70)[(9.71)] result from the spin-orbit coupling between the conduction band and another, nearby band [335]. For a lattice with inversion symmetry, we conclude from Kramer's theorem that $E(\uparrow, \mathbf{k}) = E(\downarrow, \mathbf{k})$. The spin-orbit interaction cannot split the spin degeneracy within one band, but it can mix states of opposite spin that reside in one band in the presence of another, near-lying band (the coupling mechanism operative in graphene, for instance, involves interaction between the π and the σ band, as outlined in Section 8.1), giving rise to the linear combinations (9.70) and (9.71). Labelling these superpositions in terms of well-defined spins is justified, as the a and b coefficients significantly differ from each other in magnitude, with $a \gg b$, which is rooted in the comparatively small size of the perturbation (9.69). The notations $\psi_{\mathbf{k}n\uparrow}$ and $\psi_{\mathbf{k}n\downarrow}$ thus refer to *quasi-spins*, involving strongly dominant rather than exclusive spin orientations [63].

While spin-orbit interaction defines the quasi-spin states (9.70) and (9.71), it does not provide the essential spin relaxation mechanism. This mechanism consists in elastic scattering of electrons by impurities or lattice defects, leading from a state $\psi_{\mathbf{k}n\uparrow}$ to an isoenergetic state $\psi_{\mathbf{k}n\downarrow}$, or vice versa. Alternatively, the spin relaxation process may be inelastic, associated with the absorption or emission of a phonon with momentum $\mathbf{q}$ and polarization ν by an electron that undergoes a transition between two states differing in quasi-spin and energy. This is the content of the *Elliott–Yafet spin relaxation mechanism.* Intuitively, it is based on the structure of the quasi-spin states (9.70), (9.71), implying a momentum dependence of the spin orientation. The elementary process that determines Elliott–Yafet relaxation is thus a spin-flip induced by momentum scattering. This observation, when combined with Eqs. (9.70) and (9.71), leads to an approximate expression for the spin relaxation rate $1/\tau_s$, namely [336]

$$\frac{1}{\tau_s} \approx \frac{\langle b_{\mathbf{k}n}^2 \rangle}{\tau_m}, \tag{9.72}$$

with τ_m as momentum relaxation time. For the sake of simplicity, we do not distinguish here between the spin relaxation and the spin dephasing time but assume that

$$T_1 = T_2 = \tau_s. \tag{9.73}$$

This relation is valid in isotropic or cubic solids, provided that $\gamma B_0 \ll 1/\tau_c$, with τ_c as the average time that passes between fluctuations of the magnetic field

[6] Apply first the time-reversal operator $-i\sigma_y \hat{C}$, with $\hat{C}$ as complex conjugation operator, and follow with space inversion: $\mathbf{k} \to -\mathbf{k}$.

experienced by a spin. Intuitively, if this field correlation time is small enough, the energy uncertainty corresponding to τ_c exceeds the Zeeman splitting induced by $\mathbf{B}_0$, implying that the longitudinal and transversal spins are equally affected by the dephasing field [336].

The result (9.72) may be derived from comparing spin-changing and spin-conserving momentum-scattering events in the framework of the first Born approximation, assuming a scattering potential V_m:

$$|\langle\psi_{\mathbf{k}n\uparrow}, V_m\psi_{\mathbf{k}n\downarrow}\rangle|^2 \approx |b|^2|\langle\psi_{\mathbf{k}n\uparrow}, V_m\psi_{\mathbf{k}n\uparrow}\rangle|^2. \tag{9.74}$$

Realizing that $1/\tau_m \sim \rho$, with ρ as resistivity, one may rewrite relation (9.72) to obtain a result about the relaxation time T_1, namely

$$\frac{1}{T_1} \sim \langle b_{\mathbf{k}n}^2\rangle\rho, \tag{9.75}$$

a statement known as *Yafet relation* [337]. This formula links the temperature dependence of the resistivity to that of the spin relaxation rate. For low temperatures T, Yafet found a proportionality between $1/\tau_s$ and T^5.

9.4.2 D'yakonov–Perel' Interaction

A great variety of hosts, such as compound semiconductors like GaAs or InAs, do not possess a center of inversion. Also, an externally applied electric field can destroy the symmetry of a crystal with respect to space reflection about a center. In those cases, the condition $E(\uparrow, \mathbf{k}) = E(\downarrow, \mathbf{k})$ that governs Elliott–Yafet scattering reduces to *Kramers degeneracy*,

$$E(\uparrow, \mathbf{k}) = E(\downarrow, -\mathbf{k}), \tag{9.76}$$

provided the Hamiltonian is time-reversal invariant. Removal of inversion symmetry implies that electron states with opposite spin orientations are no longer constrained to be degenerate, and therefore they can be coupled. The essence of the *D'yakonov–Perel' spin relaxation mechanism* lies in coupling different spins states by spin-orbit interaction that can be mapped on a Zeeman Hamiltonian $\hat{H}_{DP}$ with a momentum-dependent effective magnetic field, in analogy to the effect underlying the spin transistor (see Section 9.2.2):

$$\hat{H}_{DP}(\mathbf{k}) = \frac{1}{2}\hbar\boldsymbol{\sigma}\cdot\boldsymbol{\Omega}(\mathbf{k}). \tag{9.77}$$

The symbol $\boldsymbol{\Omega}$ stands for a vector of Larmor frequencies, related to the effective magnetic field $\mathbf{B}_{eff}$ through

$$\boldsymbol{\Omega}(\mathbf{k}) = \frac{e}{m}\mathbf{B}_{eff}(\mathbf{k}). \tag{9.78}$$

Time-reversal symmetry constrains the vector function $\mathbf{\Omega}(\mathbf{k})$ according to [336],

$$\mathbf{\Omega}(\mathbf{k}) = -\mathbf{\Omega}(-\mathbf{k}). \tag{9.79}$$

The protoptype for $\hat{H}_{DP}$ is an effective Hamiltonian that applies to systems with C_{2v} symmetry such as heterostructures of the zinc blende class, with axes along the [110] (=x) and [1$\bar{1}$0] (=y) directions. The zinc blende structure may be described as a diamond lattice composed of two different atom types A and B, such that all next neighbors of type A atoms are of type B. In contrast to the diamond lattice, the zinc blende lattice lacks a center of inversion and thus presents an example for *bulk inversion asymmetry* (BIA). The lowest-order form preserving the specific symmetry of this structure is given by [338]

$$\hat{H} \sim (\alpha x^2 + \beta y^2)z. \tag{9.80}$$

The momentum operator $\hat{p}_x(\hat{p}_y)$ transforms as x, (y) and likewise, $\boldsymbol{\sigma}_x$ $(\boldsymbol{\sigma}_y)$ as $\sim yz$ $(\sim -xz)$. Therefore, the structure of the corresponding D'yakonov–Perel' Hamiltonian is

$$\hat{H}_{DP} \sim \alpha \boldsymbol{\sigma}_y \hat{p}_x + \beta \boldsymbol{\sigma}_x \hat{p}_y, \tag{9.81}$$

or, equivalently,

$$\hat{H}_{DP} \sim \frac{1}{2}(\alpha + \beta)(\boldsymbol{\sigma}_y \hat{p}_x + \boldsymbol{\sigma}_x \hat{p}_y) + \frac{1}{2}(\beta - \alpha)(\boldsymbol{\sigma}_x \hat{p}_y - \boldsymbol{\sigma}_y \hat{p}_x). \tag{9.82}$$

Identifying $\frac{1}{2}(\alpha + \beta)$ with the *Dresselhaus factor* γ_D and $\frac{1}{2}(\beta - \alpha)$ with the *Bychkov–Rashba term* α_{BR}, we obtain the commonly used form

$$\hat{H}_{DP} \sim \gamma_D(\boldsymbol{\sigma}_y \hat{p}_x + \boldsymbol{\sigma}_x \hat{p}_y) + \alpha_{BR}(\boldsymbol{\sigma}_x \hat{p}_y - \boldsymbol{\sigma}_y \hat{p}_x). \tag{9.83}$$

From this expression, one readily derives an associated effective magnetic field, according to the definition (9.78):

$$\mathbf{B}_{\mathit{eff}} \sim \frac{\hbar}{\mu_B} \begin{pmatrix} (\gamma_D + \alpha_{BR})k_y \\ (\gamma_D - \alpha_{BR})k_x \\ 0 \end{pmatrix}. \tag{9.84}$$

We assume here that $\hat{H}_{DP}$ acts on plane-wave states. As $\mathbf{B}_{\mathit{eff}}$ is proportional to $\mathbf{\Omega}$, Eq. (9.84) is consistent with condition (9.79).

If all spins would experience an effective magnetic field of the same strength and the same direction, no dephasing would occur. Thus the essential cause of the phase coherence loss through D'yakonov–Perel' interaction is the momentum-dependence of the effective magnetic field in $\hat{H}_{DP}$, inducing variations in the precession rates, and ultimately the decay of the initial spin polarization. D'yakonov–Perel' and Elliott–Yafet interaction can coincide in systems that lack inversion symmetry. The two modes of spin relaxation may be viewed as complementary. While the latter is operative during scattering events, the former proceeds between these events. Assessing their relative contributions to spin relaxation involves the mean free time between two momentum-changing events, τ_m. While the Elliott–Yafet mechanism becomes more efficient as τ_m decreases, the opposite is true for the D'yakonov–Perel' mechanism. Between two consecutive collisions, the spins precess through an angular difference $\Delta\varphi \sim \bar{\Omega}\tau_m$. For a given τ_m, two regimes can be distinguished, defined by the conditions $\bar{\Omega}\tau_m \geq 1$ and $\bar{\Omega}\tau_m \ll 1$. In the first case, the angular advance of the spin ensemble is in the order of one full precession cycle, and the decoherence rate of a spin ensemble increases in proportion to the width of the distribution of Larmor frequencies, $\Delta\Omega$. D'yakonov–Perel' interaction is usually identified with the second case. Here, the frequently interrupted precession can be modeled as a diffusion process, consisting of a statistical sequence of angular steps. Each spin carries out a *random walk on the Bloch sphere*, and its overall precession angle after a precession time T is given by [339]

$$\varphi^2 = (\Delta\varphi)^2 \frac{T}{\tau_m}. \tag{9.85}$$

For basic information on the concept of the Bloch sphere, see the first special-topic section appended to Chapter 8. Introducing a *spin coherence time* τ_s by the condition $\varphi(t = \tau_s) = 1$, we conclude

$$\frac{1}{\tau_s} \sim \bar{\Omega}^2 \tau_m. \tag{9.86}$$

The rate at which the precessing spin ensemble decoheres is therefore proportional to the time that passes between two successive collisions, implying an inverse dependence of Elliott–Yafet and D'yankonov–Perel' interaction on the electron collision frequency, as confirmed by comparing Eq. (9.86) with Eq. (9.72).

9.4.3 Bir–Aronov–Pikus Mechanism

A third spin relaxation effect, coexisting with the Elliott–Yafet and D'yakonov–Perel' mechanisms, is found in p-doped semiconductors that lack inversion symmetry. In the latter materials, electron spin-flip may be induced by exchange interaction with holes. This process, termed Bir–Aronov–Pikus (BAP) mechanism [340], is modeled as a local interaction according to:

$$\hat{H} = A\hat{\mathbf{S}} \cdot \hat{\mathbf{J}}_H \delta(\mathbf{x}). \tag{9.87}$$

The interaction constant A is here proportional to the exchange integral between the conduction and valence band states. The operators $\hat{\mathbf{S}}$ and $\hat{\mathbf{J}}_H$ refer to the spin of the electron and the angular momentum of the holes, respectively, and $\mathbf{x}$ denotes the relative position between the two. The spin relaxation rate according to the BAP mechanism depends on the properties of the hole. The most frequently applied formula rests on the assumption of a *heavy hole*[7] in a non-degenerate semiconductor [311], i.e a semiconductor at a moderate level of doping:

$$\frac{1}{\tau_s^{BAP}} = \frac{2a_B^3 N_A v_{\mathbf{k}}}{\tau_0 v_B}\left(\frac{n_H}{N_A}|\psi(0)|^4 + \frac{5}{3}\frac{N_A - n_H}{N_A}\right). \tag{9.88}$$

This expression combines the hole density n_H and the acceptor density N_A with the *Sommerfeld factor* $|\psi(0)|^2$ as a measure for the probability of finding the electron at the hole site, and thus for the screening of the hole potential. Further, a_B stands for the exciton Bohr radius, $\frac{\hbar^2\varepsilon}{e^2 m_{ec}}$, and τ_0 is a characteristic time, related to the exchange splitting of the hole-electron configuration. More specifically, $\tau_0 = \frac{64\hbar}{3\pi}\frac{E_B}{\Delta_{ex}^2}$, where E_B is the Bohr exciton energy ($= \frac{\hbar^2}{2m_c a_B^2}$), and Δ_{ex} the exchange splitting of the exciton ground state.

The spin relaxation rate according to Eq. (9.88) depends on temperature mostly through the Sommerfeld factor and the hole density. At sufficiently low temperature and high acceptor density, the BAP process may outweigh the rivaling spin relaxation mechanisms. This is illustrated in Figure 9.10 by a comparison between Bir–Arononv–Pikus and D'yakonov–Perel' interaction in p-doped GaAs.

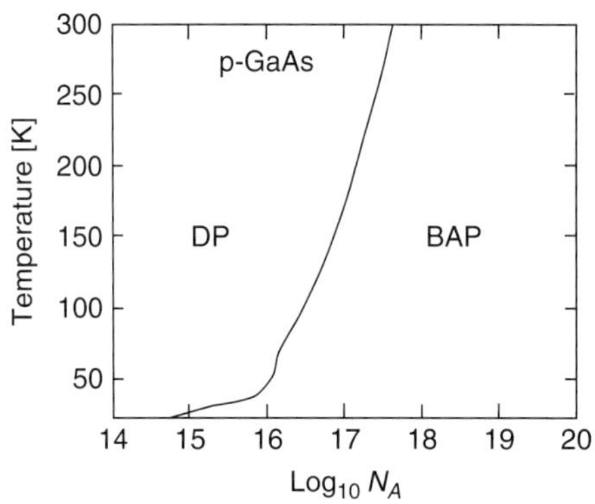

Figure 9.10 Comparison between the Bir–Arononv–Pikus (BAP) and D'yakonov–Perel' (DP) spin relaxation mechanism in terms of temperature and acceptor density [341]. In the regime labeled BAP (DP) the former (latter) process is dominant. The acceptor density N_A is in cm^{-3}. Reprinted with permission from [341]. Copyright (2002) by the American Physical Society.

[7] i.e. $m_{ev} \gg m_{ec}$, with m_v as valence band (hole) mass and m_{ec} as conduction band (electron) mass.

9.4.4 Hyperfine Interaction

Magnetic hyperfine interactions provides a further relaxation pathway for spin ensembles, and also a process for coherence loss of single spins, for instance those of localized electrons trapped in quantum dots [311]. With reference to the latter arrangement, hyperfine interaction has been identified as a significant contribution to the decoherence of a spin qubit in contact with a nuclear spin bath at low temperature [342].

While hyperfine interaction has been confirmed as an important factor for spin relaxation phenomena in GaAs (e.g. [343]), where the nuclei at all lattice sites carry a spin of $\frac{3}{2}\hbar$, it is of substantially smaller impact in Si where the isotope of maximum abundance, ^{28}Si, has no nuclear spin in its ground state and hyperfine effects are largely due to ^{29}Si, whose natural abundance in Si amounts only to 4.7 percent. Likewise, the nuclear spin of ^{12}C, the prevailing isotope in carbon, vanishes, and ^{13}C admixtures contribute a marginal 1 percent to natural carbon. Further, recent measurement on spin relaxation in ^{13}C graphene [344] suggest that hyperfine interactions play a negligible role in these processes, and that, by inference, their relevance for the field of magnetism in carbon nanostructures is low.

10

Spin Transport in Carbon Nanostructures

After the review of spin interactions in carbon nanostructures in Chapters 7 and 8, and the summary of mechanisms and material properties essential for spintronics in the previous chapter, the present chapter will combine these two domains. In the first place, we will attempt to clarify why carbon nanostructures are of major interest as potential carrier materials in spintronics circuits. Further, we will characterize the physical effects that make these structures relevant as elements of spin networks. Since devices based on these effects have been conceived but are still quite far from the manufacturing stage, our account will be about currently discussed models of these devices, and their experimental examination. We focus on two classes of materials: graphene and carbon nanotubes.

10.1 Graphene Spintronics

Graphene is of high interest as nanoelectronic material. Among the features of graphene that motivate this interest are experimentally confirmed electron mobilities as high as 2.00 10^5 $\mathrm{cm^2V^{-1}s^{-1}}$ [347], which is about a hundred times higher than typical mobilities of electrons in silicon.[1] This suggests that graphene-based transistors will function at substantially higher speeds than elements of conventional electronics, while being more efficient in terms of power consumption. Test devices have been shown to be capable of operating at hundreds of gigahertz [348]. Another graphene property of direct relevance for nanotechnological applications is its gate-tunable charge carrier density, resulting from the double-cone structure of the energy surfaces close to the Dirac points. Further, in terms of heat resistance, graphene transistors are distinctly superior to the tools of presently existing

[1] The mobility μ_e is related to the conductivity by the equation

$$\tilde{\sigma} = ne\mu_e,$$

with n as the concentration of charge carriers.

electronics, as they have been shown to perform at temperatures higher by 20–30 °C than their silicon counterparts [349]. The challenge of the semi-metallic character of two-dimensional periodic graphene crystals can be met in various ways. Thus, one may tailor a finite band gap in controlled ways by modifying the graphene sheet with adsorbates. In addition, manufacturing electronic devices from the pristine graphene crystal requires its dimensional reduction, associated with the formaton of a finite band gap (see Section 4.5).

Likewise, graphene holds the promise of major technological innovation in the field of spintronics. Recent research has identified a variety of features that make graphene attractive as a spin channel material [350]. Thus, the weak spin-orbit effects in pristine graphene give rise to the expectation of long spin relaxation times and, related to this, long-distance spin propagation. In terms of spin relaxation mechanisms, hyperfine interaction has been found to be negligible in graphene (see Section 9.4.4). The Bir–Aronov–Pikus process is relevant for semiconductors in the limit of substantial hole doping (see Section 9.4.3), and therefore of minor significance for graphene spin relaxation. This process is currently understood to be governed by Elliott–Yafet and D'yakonov–Perel' scattering (see Sections 9.4.1 and 9.4.2), and the relative importance of these two mechanisms continues to be the subject of a lively debate.

In what follows, we will discuss graphene realizations of elementary spintronics devices and then contrast local and non-local spin valve arrangements involving graphene as a transport medium, where the latter configuration will turn out to allow for a separation between charge and spin currents.

10.1.1 Graphene Spin Valves and Spin Filters

Various proposals have been put forward to implement basic devices of spintronics with graphene as the medium of spin transport. Here we single out two simulations, both based on modeling from first principles, that demonstrate the potential usage of zGNRs as spin valves [351] and spin filters [352].

The computations yielded magnetoresistance ratios MR = $(R_{AP} - R_P)/R_P$, see Section 9.2.1) in the order of 10^6 percent for a zGNR with width $n = 8$. These MR values are higher by factors of several thousand than those for previously suggested spin valve arrangements, consisiting, for instance, of Fe electrodes separated by a non-magnetic MgO layer a few nanometers in length. A rationale for this behavior can be inferred from the discussion of magnetism in zGNRs presented in Section 7.1.1. In the parallel mode (P), the spin density distribution of the zGNR is represented by the image shown in Figure 7.5(a). The localized edge states display uniform spin orientation. In the antiparallel mode (AP), however, electrodes attached to the upper and lower ends of the ribbon induce opposite spin orientations at opposite ends of the zGNR. As Son et al. were able to show, this spin symmetry mismatch is enhanced by an orbital symmetry mismatch in certain regimes of the bias voltage, and most prominently at zero bias. This effect is a reflection of band structure changes caused by the external magnetic fields in the leads. In the AP configuration, the changes in the two leads differ from each other. Thus, energy intervals can be identified where bands of a given spin orientation belong to different orbital symmetries on the left and the right. As a consequence, zero transmission is found in the respective energy regime. The spin valve characteristics of a zGNR are thus based on a cooperative effect involving both spin and orbital symmetry matching. The device, when employed in the AP configuration, is also predicted to act as a spin filter, yielding currents with spin polarizations beyond 99 percent.

A density functional theory study by Son et al. [352] demonstrated that no involvement of ferromagnetic leads is needed to endow graphene with spin filter properties. Again, a zGNR provides the transmission element between two electrodes. Since states of opposite spin orientation in zGNRs are spatially separated, being localized at the two zGNR edges, a properly chosen potential difference across the zGNR width may make the medium conducting for one spin moiety and insulating for the other. Graphene thus exhibits the feature of half-metallicity, as introduced in Section 9.2.3. This is illustrated in Figure 10.1 by the example

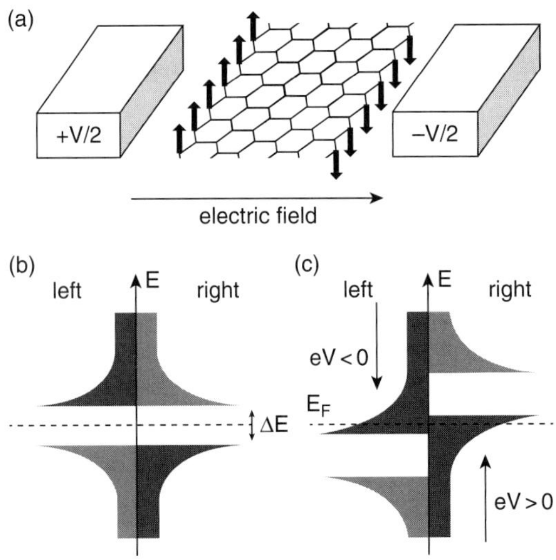

Figure 10.1 The half-metallicity of zGNRs. (a) An electric field is applied along a zGNR, extending between electrodes, in horizontal direction. (b) Scheme of the density-of-states distribution at opposite ends of the zGNR. The spin-up (spin-down) moiety is shaded darkly (lightly). (c) Applying the critical field closes the energy gap ΔE selectively for spin-down electrons. Reprinted from [6], with permission from the author.

of a zGNR used as a transmission element, and assumed to be in its ground state configuration, such that the left and the right edge states exhibit opposite spin polarizations. A critical electric field may then be applied, such that the energy gap ΔE is closed for one spin orientation and enhanced for the other. With reference to Figure 10.1(b), this implies exposing the states localized at the left edge to a positive electric potential of sufficient size, and the corresponding states on the right side to an equal and opposite potential. As a consequence, only spin-down electrons will traverse the junction. The critical field strength is estimated to be $3.0/w$ V/Å, where w is the width of the zGNR, as measured from zigzag edge to zigzag edge. We mention that an analogous half-metallicity effect has been described for zSWCNTs [264].

10.1.2 Non-Local Magnetoresistance Effects in Graphene

Use of graphene as a spin transport medium is predicated on generating, measuring and manipulating spin currents in this medium. In a series of pioneering experiments, Tombros et al. [4] used the *Hanle spin precession effect* to demonstrate spin injection into graphene followed by spin transport. The challenge to separate the spin from the charge current was braved by employing a four-terminal (or *non-local*) arrangement rather than the conventional two-terminal device. In what follows, we sketch the operation principle of the former, emphasizing its superiority over the latter.

The spin valve arrangement introduced in Section 9.2.1 performs in the *local* mode. Electric current flows through a combination of two ferromagnetic layers separated by a non-magnetic buffer, as shown in Figure 9.2. Information about the spin current I_s is derived from the voltage drop across this device, in conjunction with the spin-dependent resistances of its constituents. Likewise, in spin valve resistance measurements the spin signal is contaminated by the charge signal. In Figure 10.2, this drawback is illustrated by signals obtained from a basic two-terminal spin valve circuit. Shown is the experimentally detected local resistance for parallel and antiparallel orientations of the electrode magnetization [4]. In accordance with the expected spin valve behavior, the second case yields a distinctly higher resistance than then former. The data are, however, determined by both the charge current Eq. (9.1) and the spin current Eq. (9.2), and further effort is needed to distinguish the latter. Thus, is there a way of recording the pure spin current, eliminating the background effect of the charge current?

This is achieved by replacing the local with a non-local observation mode, involving the extension of a two-terminal arrangement to a four-terminal setup [4, 354]. A ferromagnet in contact with graphene may act as spin source. In the model illustrated by Figure 10.3, the spin current is due to diffusion, as described

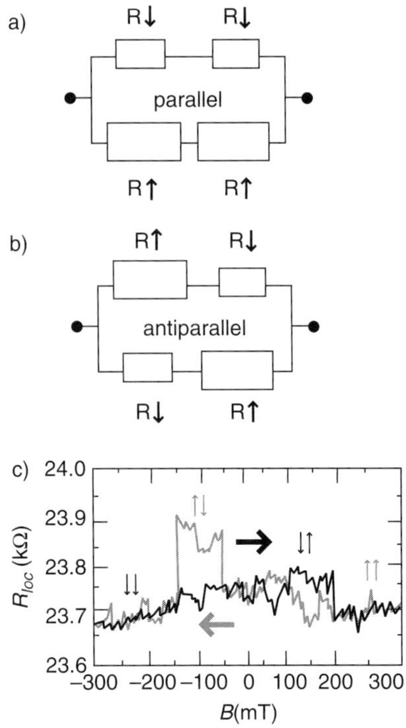

Figure 10.2 The local resistance of a spin valve. Shown is an illustration of the two-current model, as discussed in Section 9.2.1, for parallel (a) and antiparallel (b) magnetization in the two ferromagnetic electrodes (compare with Figure 9.2). Panel (c) displays the local resistance of a two-terminal spin valve arrangement as a function of the external magnetic field. The variable field was applied in two opposite directions, as indicated by black and grey lines, giving rise to configurations of parallel and antiparallel magnetization. Reprinted from [354] with permission from the author.

in Section 9.3. The addition of spin carriers establishes a non-equilibrium state, characterized by a finite spin chemical potential [310]. When the spin carriers traverse the spin relaxation length L_s, the spin chemical potential diminishes and reduces eventually to zero, as the graphene density of states returns to its equilibrium profile. This sequence is represented in panel (a) of Figure 10.3 in terms of the density-of-states (DOS) distribution and in panel (b) in terms of the spin chemical potential.

The salient modification that leads from local to non-local spin transport measurement lies in the separation between a current and a voltage path. With reference to Figure 10.4, both paths are specified by a pair of ferromagnetic electrodes, the current loop by FM3/FM4, and the voltage loop by FM1/FM2 [4, 354]. In the example shown in the figure, the contact FM3 serves as a spin injector that adds spin-up electrons to a graphene sheet placed on a substrate, and covered with a

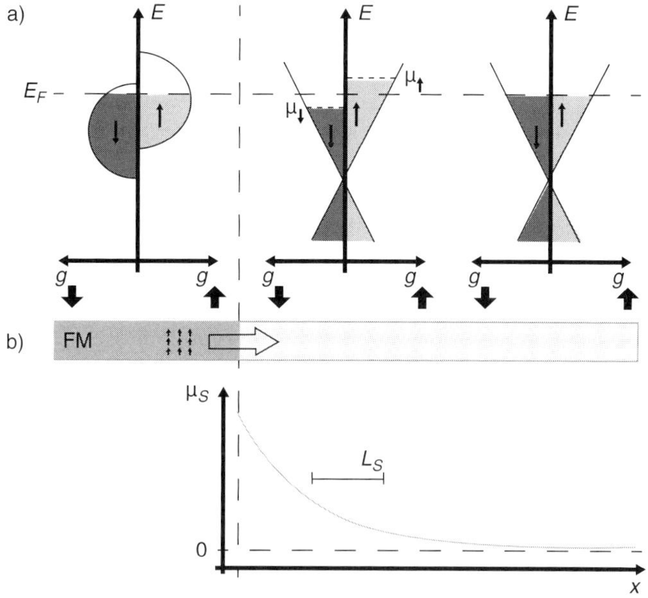

Figure 10.3 Schematic representation of spin injection from a ferromagnetic electrode into graphene [354]. (a) Evolution of the DOS: As spins from the ferromagnet, satisfying $g_\uparrow(E_F) > g_\downarrow(E_F)$, are added to graphene, the spin imbalance of the ferromagnetic source is imposed on the non-magnetic medium. Sufficiently far from the interface, however, equilibrium is restored, corresponding to $E_{F\uparrow} = E_{F\downarrow}$. This process is reflected by exponential reduction of the spin chemical potential from the interface to the bulk (b). Reprinted from [354] with permission from the author.

thin Al_2O_3 layer. As the spins are added to a graphene segment below FM3, they diffuse toward FM2 as well as FM4 where they are extracted, such that the current loop is closed. As the voltage loop is an open circuit, no charge current flows in this loop. The spin diffusion from the contact FM3, however, creates spin-charge coupling conditions, as described in Section 9.3, giving rise to a pure spin current. The different spin quasi-chemical potentials probed by the contacts FM2 and FM1 result in an electromotive force that can be derived with arguments analogous to those presented in Section 9.3.1. The spin current density on account of the change of μ_s along the voltage loop may be evaluated according to the second equation in (9.46) [310].

If the magnetizations of FM3 and FM4 are aligned, this amounts to a surplus of spin-up electrons in graphene at the place below FM3, and a reduction below FM4. This situation is reflected by the behavior of the relative spin quasi-chemical potentials $\mu_\uparrow - \mu_0$ and $\mu_\downarrow - \mu_0$, as shown in panel (b). In accordance with the orientation of their magnetizations, the contacts FM1 and FM2 probe the quasi-chemical potentials of the up-spins. The corresponding potential difference

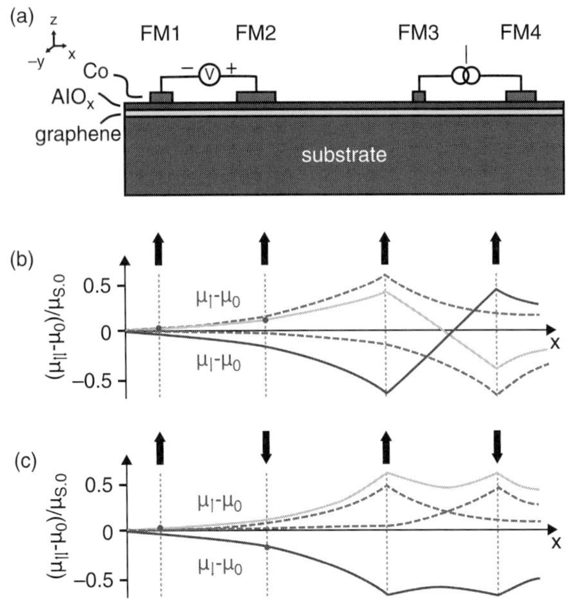

Figure 10.4 Experimental setup for a spin transport measurement in non-local spin valve geometry, involving a lateral design of four ferromagnetic electrodes (a). The pair FM1/FM2 defines the voltage circuit, and the pair FM3/FM4 the charge circuit of the device. Panels (b) and (c) show the differences between the spin quasi-chemical potentials $\mu_\uparrow$, $\mu_\downarrow$ and the chemical potential μ_0 for completely aligned (b) and for antiparallel magnetizations (c) of the ferromagnetic contacts [4, 354]. These quantities are normalized with respect to the spin quasi-chemical potential at the place of injection, FM3. The dashed lines refer to the spin-up contributions due to FM3 and FM4, the light solid line to their sum. The dark solid line describes the overall spin-down contribution. Reprinted from [354] with permission from the author.

$V_{nl} = \mu_\uparrow(2) - \mu_\uparrow(1)$ is combined with the information about the current, I, into the non-local resistance R_{nl}:

$$R_{nl} = \frac{V_{nl}}{I}. \tag{10.1}$$

It should be noted that R_{nl} is a resistance in name only, since the voltage and the current refer to different sections of the four-terminal device. The quantity might be understood as a spin signal normalized to the injection current.

From comparing panels (b) and (c), one infers that the antiparallel arrangement obtained by reverting the magnetizations of FM2 and FM4 results in a sign switch of the spin signal. In this setting, spin-down electrons are extracted at contact FM4, which is equivalent to spin-up injection at this point. The electrode FM2 now detects the spin-down chemical potential. This results in a negative non-local spin signal V_{nl}.

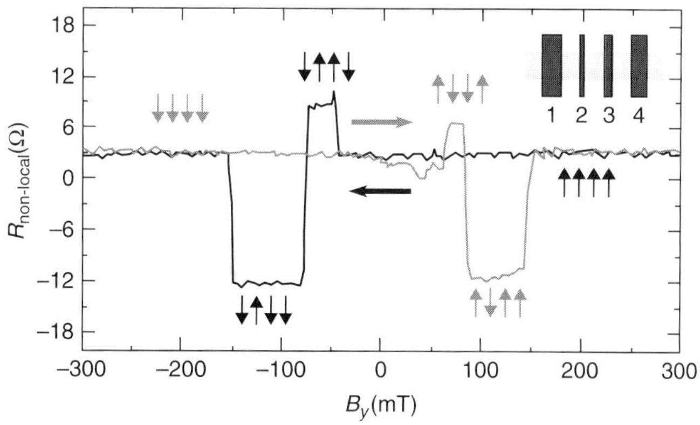

Figure 10.5 The non-local resistance versus the external magnetic field for the experimental setup displayed in Figure 10.4. As in Figure 10.2, the black and grey lines correspond to opposite magnetic sweep directions. The magnetic orientations of the metal electrodes are indicated, and the electrode spacing is shown in the inset. The data were recorded at a temperature of 4.2 K. Reprinted with permission from [4]. Copyright (2007) by Nature Publishing Group.

The non-local observation mode allows for a clear presentation of the spin valve effect, involving a graphene layer, mounted on an SiO_2 substrate, as spin propagation medium. This is documented in Figure 10.5 by non-local resistance data, plotted versus the external magnetic field. In the example shown, the four ferromagnetic contacts have different widths, corresponding to different coercive fields. Thus, the critical field strengths for magnetization reversal vary among the four contacts, which allows them to realize a variety of magnetic configurations. Comparing with the results displayed in Figure 10.2 for the analogous local case, one notes a substantially clearer spin signal that reacts sensitively to the magnetic configuration of the metal contacts. As the noise that compromises measurements of the local resistance is largely eliminated here, the data provide strong evidence for spin injection and detection.

The Al_2O_3 layer between graphene and the ferromagnetic contacts addresses the *conductivity mismatch problem*, as described in the context of spin injection in Section 9.3. As shown there, a contact zone between the ferromagnetic and the non-magnetic medium can be used to tune the overall current spin polarization, achieving high spin injection efficiency even in cases where the resistances of the ferromagnetic and the non-magnetic medium differ substantially from each other [355, 356].

The Spin Signal

To clarify the origin of the of the electromotive force in the voltage loop, we consider a simplified model that may serve as a basic approximation to the F/C/N sequence at contact FM2 in Figure 10.4(a). In accordance with Figures 10.4(b) and (c), we assume that spin injection at FM3 has given rise to non-equilibrium conditions at the site of FM2. This generates spin-dependent potential differences across the the junction. For simplicity, we assume that the barrier between the ferromagnet and graphene is a tunnel contact, and that the proportionalities Eq. (9.64) (see Section 9.3) are valid for the current densities across the junction. The case of pure spin current is realized if there is no charge current: $j = 0$, or $j_\uparrow = -j_\downarrow$, corresponding to open circuit conditions. With reference to the scheme shown in the middle of Figure 10.3, this balance might be realized by a current from the spin-up surplus population in graphene to the ferromagnet versus a spin-down current of equal magnitude, flowing in the opposite direction. By formula (9.65), the constraint of vanishing charge current yields the desired EMF,

$$EMF = \frac{P_{gF} + P_t}{1 + P_t P_{gF}} \mu_{sN}, \tag{10.2}$$

with P_{gF} and P_t as the spin polarizations of the density of states and of the transmission probability. The EMF thus probed yields the non-local resistance (10.1) through

$$R_{nl} = \frac{EMF}{I}. \tag{10.3}$$

An EMF emerges in any junction that involves a conductor in equilibrium, described by P_g, and another conductor with non-equilibrium spin populations, characterized by the spin quasi-chemical potential μ_s. We point out that in the ideal case of perfect spin selectivity, P_{gF} and P_t are both +1 (-1) in case of FM2 and FM3 in parallel (antiparallel) magnetization with respect to each other, implying that the EMF detected by FM2 reduces to $+\mu_{sN}$ $(-\mu_{sN})$.

Exercise 10.1 With reference to Figure 10.4, demonstrate that the spin signal switches its sign as the relative magnetization of the electrodes FM2 and FM3 (parallel or antiparallel) is reversed.

Hanle Precession Measurement of the Spin Relaxation Time in Graphene

From the above discussion, the suitability of a medium for spintronics applications depends critically on the spin diffusion length L_s and the spin relaxation time τ_s. Recent measurements of the latter quantity for graphene as non-magnetic conductor have employed the spin valve technique in the non-local geometry, as

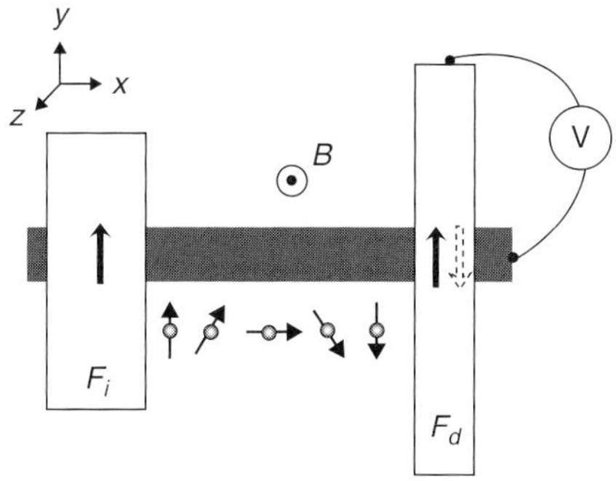

Figure 10.6 Scheme of an experimental arrangement for a Hanle precession measurement. An injection contact is distinguished from a detection contact, as elements of the current and the voltage loops of a non-local spin valve network. Reprinted from [354] with permission from the author.

outlined earlier. Specifically, the Hanle precession effect [4, 357] has been used in these experiments. To generate Hanle spin precession, one applies an external magnetic field in a direction perpendicular to that of the spin injection and to the spin diffusion channel, as sketched in Figure 10.6. The setup shown there provides a reference geometry, with an external magnetic field pointing in the positive z-direction, and a spin density vector oriented initially along the positive y axis: $\boldsymbol{s}(t = 0, x = 0) = s_0\mathbf{e}_y$. Likewise, spin detection proceeds by a ferromagnetic contact placed at distance L from the injection point along the graphene channel.

As we consider diffusive motion, the time needed for the spins to move through the distance L is governed by a probability distribution that depends on the graphene diffusion constant. Further, on their journey from the injector to the detector contact, the spins undergo precession and relaxation. In the model adopted here, we neglect the drift of the electrons due to the effect of electric fields and adopt the following expression to model the evolution of the spin density at the detection point:

$$s_y(L, t, B) \propto \frac{1}{4\pi Dt} \exp(-\frac{L^2}{4Dt}) \cos(\omega_L t) \exp(-\frac{t}{\tau_s}), \tag{10.4}$$

with ω_L as the Larmor frequency, $\omega_L = \frac{eg_s}{2m_e} B$ and D as diffusion constant. The time constant τ_s is here to be understood as a transversal relaxation time, T_2. The dependence of the spin quasi-chemical potential μ_s on time and on the size of the magnetic field is described in an analogous fashion. With the help of the analytic expression for the quantity $\mu_s(L, B)$, one may extract the desired information about the spin relaxation dynamics from measured values of non-local resistance since, by Eqs. (10.2, 10.3),

$$R_{nl} \propto \pm\mu_s(L, B). \tag{10.5}$$

To obtain a prediction for μ_s at $x = L$, we integrate over all possible arrival times and find

$$\mu_s(L, B) \propto \mu_s(0, B) \int_0^\infty dt\, s_y(L, t, B), \tag{10.6}$$

where the plus and the minus signs (see Eq. (10.5)) refer to the cases of detector magnetization parallel and antiparallel to the injector magnetization, respectively. The integral in expression Eq. (10.6) can be evaluated analytically [358], and the solution can be fitted to measured curves of the non-local resistance as a function of the external magnetic field, as illustrated in Figure 10.7. The spin relaxation time may be extracted from these fits.

We point out that at a sufficiently high external magnetic field, the numerical simulation of the non-local resistance must include the interaction between the field and the magnetization of the injector and detector contacts [358].

In the following step we discuss possible mechanisms to account for spin relaxation in graphene.

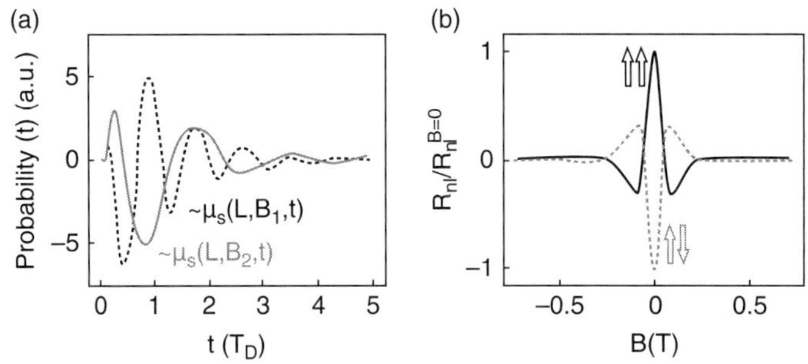

Figure 10.7 Observing Hanle precession of a spin ensemble by use of the non-local resistance in spin valve arrangements. (a) spin density according to Eq. (10.4) at the detection site in the presence of an external magnetic field, oriented perpendicularly to the spin transport channel, as shown in Figure 10.6. The spins are assumed to be injected at a contact F_i. The distribution includes the effects of spin diffusion, relaxation, and precession for magnetic fields of two different magnitudes, where $B_1 > B_2$. The time is indicated in units of the characteristic diffusion time, $\tau_D = \frac{L^2}{2D}$, corresponding to the maximum of the pure diffusion distribution. (b) The non-local spin resistance R_{nl} for parallel (solid curve) and antiparallel (dashed curve) magnetizations of the injecting and detecting ferromagnets. Shown is the resistance normalized to the case of zero magnetic field, i.e. vanishing precession. Reprinted from [354] with permission from the author.

10.2 Estimates of Spin Relaxation Times in Graphene

The discussion about the dominant spin relaxation mechanism in graphene has focused mostly on Elliott–Yafet and D'yakonov–Perel' scattering, as introduced in Sections 9.4.1 and 9.4.2, respectively. While both processes are based on spin-orbit coupling, they are complementary with respect to the underlying spin relaxation mechanisms. The crucial difference between Elliott–Yafet and D'yakonov–Perel' processes is that relaxation occurs during scattering events in the former, and between consecutive events in the latter case. In the following we will comment on theoretical and experimental findings related to spin relaxation in graphene due to both interaction types.

10.2.1 The Elliott–Yafet Mechanism in Graphene

The fingerprint of Elliott–Yafet interaction is a proportionality between the momentum and the Elliott–Yafet spin relaxation times, τ_p and τ_s^{EF} (see Section 9.4.1, formula (9.72)). Inquiring about the spin relaxation time in graphene due to the intrinsic spin-orbit effect, we operate with the corresponding term in the Hamiltonian (8.16) [359],

$$\hat{H} = \hat{H}_0 + \hat{H}_{SO} = i\hbar_{vF\eta} \cdot \boldsymbol{k} - \xi_I \eta_z \boldsymbol{\sigma}_z, \tag{10.7}$$

which refers to the K valley. The eigenfunctions of $\hat{H}_0$ may be written as (see Eq. (4.26))

$$\begin{aligned} \Psi^{\pm}_{\mathbf{k},\rightarrow} &= \frac{1}{\sqrt{2}} \begin{pmatrix} 1 \\ \frac{\varepsilon^{\pm}_{\mathbf{k}}}{|\varepsilon^{\pm}_{\mathbf{k}}|} e^{i\theta_{\mathbf{k}}} \end{pmatrix} |\rightarrow\rangle, \\ \Psi^{\pm}_{\mathbf{k},\leftarrow} &= \frac{1}{\sqrt{2}} \begin{pmatrix} 1 \\ \frac{\varepsilon^{\pm}_{\mathbf{k}}}{|\varepsilon^{\pm}_{\mathbf{k}}|} e^{i\theta_{\mathbf{k}}} \end{pmatrix} |\leftarrow\rangle, \end{aligned} \tag{10.8}$$

where $\varepsilon^{\pm}_{\mathbf{k}} = \pm\hbar v_F|\mathbf{k}|$ denotes the eigenenergies of electron one-particle states, and $\theta_{\mathbf{k}} = \arctan\frac{k_y}{k_x}$. We consider states with surface-parallel spins since the intrinsic spin-orbit Hamiltonian $\hat{H}_{SO}$ commutes with $\boldsymbol{\sigma}_z$ and therefore leaves the spin z component unaltered. Note that the spin functions $|\rightarrow\rangle$ and $|\leftarrow\rangle$ may be represented in terms of up and down states:

$$\begin{aligned} |\rightarrow\rangle &= \frac{1}{\sqrt{2}}(|\uparrow\rangle + |\downarrow\rangle) \\ |\leftarrow\rangle &= \frac{1}{\sqrt{2}}(|\uparrow\rangle - |\downarrow\rangle). \end{aligned} \tag{10.9}$$

The action of $\hat{H}_{SO}$ modifies the solutions in the following way,

$$\Psi^{\pm}_{\mathbf{k}\rightarrow} = \frac{1}{2}[\begin{pmatrix} \sqrt{1+\frac{\xi_I}{\varepsilon_{\mathbf{k}\pm}}} \\ \sqrt{1-\frac{\xi_I}{\varepsilon_{\mathbf{k}\pm}}}e^{i\theta_{\mathbf{k}}} \end{pmatrix} |\uparrow\rangle + \begin{pmatrix} \sqrt{1-\frac{\xi_I}{\varepsilon_{\mathbf{k}\pm}}} \\ \sqrt{1+\frac{\xi_I}{\varepsilon_{\mathbf{k}\pm}}}e^{i\theta_{\mathbf{k}}} \end{pmatrix} |\downarrow\rangle],$$
$$\Psi^{\pm}_{\mathbf{k}\rightarrow} = \frac{1}{2}[\begin{pmatrix} \sqrt{1+\frac{\xi_I}{\varepsilon_{\mathbf{k}\pm}}} \\ \sqrt{1-\frac{\xi_I}{\varepsilon_{\mathbf{k}\pm}}}e^{i\theta_{\mathbf{k}}} \end{pmatrix} |\uparrow\rangle - \begin{pmatrix} \sqrt{1-\frac{\xi_I}{\varepsilon_{\mathbf{k}\pm}}} \\ \sqrt{1+\frac{\xi_I}{\varepsilon_{\mathbf{k}\pm}}}e^{i\theta_{\mathbf{k}}} \end{pmatrix} |\downarrow\rangle], \tag{10.10}$$

and $\varepsilon_{\mathbf{k}\pm} = \pm\sqrt{(\hbar v_F|\mathbf{k}|)^2 + \xi_I^2}$. Using the small fraction $\xi_I/\varepsilon_{\mathbf{k}\pm}$ as perturbation parameter, we expand these solutions up to first order and obtain

$$\Psi^{\pm}_{\mathbf{k}\rightarrow} = \frac{1}{\sqrt{2}}[\begin{pmatrix} 1 \\ e^{i\theta_{\mathbf{k}}} \end{pmatrix} |\rightarrow\rangle + \frac{\xi_I}{2\varepsilon_{\mathbf{k}\pm}} \begin{pmatrix} 1 \\ -e^{i\theta_{\mathbf{k}}} \end{pmatrix} |\leftarrow\rangle]$$
$$\Psi^{\pm}_{\mathbf{k}\leftarrow} = \frac{1}{\sqrt{2}}[\begin{pmatrix} 1 \\ e^{i\theta_{\mathbf{k}}} \end{pmatrix} |\leftarrow\rangle + \frac{\xi_I}{2\varepsilon_{\mathbf{k}\pm}} \begin{pmatrix} 1 \\ -e^{i\theta_{\mathbf{k}}} \end{pmatrix} |\rightarrow\rangle]. \tag{10.11}$$

By the structure of these wave functions, and by comparison with the Elliott–Yafet form (see equations (9.70) and (9.71)), we infer the probability for a spin-flip event to be $|b_{\mathbf{k}}|^2 = \frac{\xi_I^2}{4\varepsilon_{\mathbf{k}_F}^2}$. Applying the relation between the spin relaxation and the momentum relaxation times in Elliott–Yafet scattering (see Eq. (9.72)) on the states (10.10), we conclude

$$\tau_s^{EY} \approx \frac{4\varepsilon_{\mathbf{k}_F}^2}{\xi_I^2}\tau_m, \tag{10.12}$$

with the constant τ_m as the momentum relaxation time, or the characteristic time that passes between two successive collisions. This relation might be used to predict the magnitude of the characteristic time for spin relaxation by Elliott–Yafet scattering in graphene, with intrinsic spin-orbit interaction as elementary mechanism. As a charge carrier density one may adopt $n_i \approx 10^{12}$ cm^{-2}, which is close to the value typically realized in spin relaxation experiments [360]. A conservative assessment of the spin relaxation time is obtained by adopting a value of the electron mobility close to the lower limit of this quantity at the assumed charge carrier density: $\mu_e \approx 3.0\ 10^3$ cm^2 V^{-1} s^{-1}. Using $\mu_e = \frac{e\tau_m}{m_e}$ [361], one obtains a result for τ_m. Further, one derives a numerical estimate of $\varepsilon_{\mathbf{k}_F}$, since $\varepsilon_{\mathbf{k}_F} = \hbar v_F k_F$, and $k_F = \sqrt{2\pi n_i}$. If the upper experimental limit for ξ_I is adopted, 100 μeV, the spin relaxation time according to Eq. (10.12) is in the order of 100 ns. If, instead, an average of the computed spin-orbit coupling constants ($\approx$ 10 μeV) is inserted into formula (10.12), the predicted spin relaxation grows by two orders of magnitude, exceeding a microsecond.

However, expectations of extraordinarily long spin relaxation times in graphene, based on the above assessment, were disappointed by experimental testing. Hanle spin-precession experiments (see earlier, Section 10.1) yielded charge carrier spin lifetimes in the order of 100–200 ps at room temperature [4, 362, 363, 364, 365, 366, 367, 368, 369]. Also, the out-of-plane relaxation rate turned out to be larger than the in-plane rate: $\tau_s^{\perp} < \tau_s^{\parallel}$ [370]. By the above arguments, this finding is not compatible with a spin-orbit interaction mechanism based on the operator (10.7) as perturbation Hamiltonian.

To rationalize the discrepancy between modeling and measurement, one might invoke the effect of randomly distributed impurities and other contaminants, such as adsorbates or topological defects in the graphene lattice, the influence of the dielectric substrate, or deformations of the graphene sheet. All of these factors may cause enhanced spin decoherence. Exploring an extrinsic rather than intrinsic spin-orbit relaxation process, as induced by a ripple with 100 nm as radius of curvature [243], did not yield any fundamental change of the result obtained by Tombros et al. For this situation, a Rashba coupling constant on the order of 10 μeV is reported, leading to times of similar magnitude as found for the case of intrinsic spin-orbit interaction.

Summarily, the experimental spin relaxation times at room temperature were found to be distinctly lower than the theoretical prediction. Nevertheless, they turned out to be on the order of low-temperature values ($T < 100$ K [371]) detected in semiconductors. The corresponding spin relaxation lengths in graphene are in the micrometer range.

In order to ascertain Elliott–Yafet scattering as the mechanism of spin relaxation in graphene, Hanle spin precession measurements of both the spin relaxation time and the diffusion coefficient D were performed for monolayer as well as bilayer graphene samples [357]. Acknowledging the proportionality between D and τ_m [310], the authors reported linear scaling of τ_s with D for temperatures up to about 100 K (see Figure 10.9), confirming the Elliott–Yafet relation (10.12) in the low-temperature regime.

10.2.2 Elliott–Yafet versus D'yakonov-Perel' Mechanism in Graphene

A criterion that distinguishes the D'yakonov–Perel' from the Elliott–Yafet mechanism is the different behavior of these two processes with respect to momentum relaxation, as characterized by the constant τ_m. While the Elliott–Yafet spin relaxation rate is inversely proportional to τ_m (see Eq. (10.12)), the spin relaxation rate by D'yakonov–Perel' interaction is directly proportional to this constant (see Eq. (9.86)).

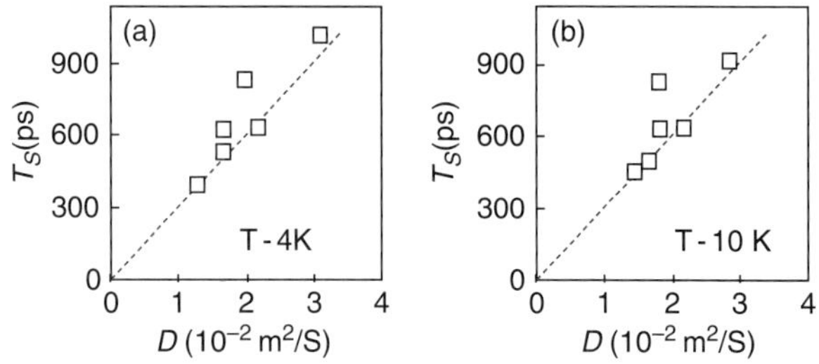

Figure 10.8 Linear scaling of the spin relaxation time τ_s with the diffusion coefficient D for T = 4K (panel (a)) and T = 10K (panel (b)) in single-layer graphene from Hanle spin precession measurement [357]. As D is proportional to the momentum relaxation time τ_m, this finding is in accordance with the Elliott–Yafet scattering relation (10.12). Reprinted with permission from [357]. Copyright (2011) by the American Physical Society.

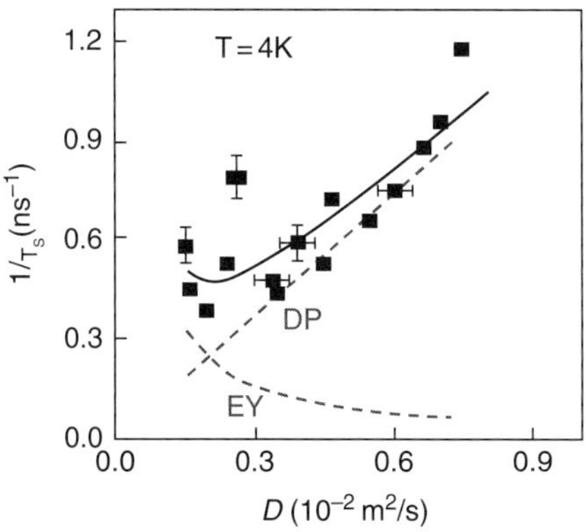

Figure 10.9 Spin relaxation rates versus diffusion coefficients in various samples of bilayer graphene from Hanle spin precession measurement [357] at T = 4 K. For sufficiently large diffusion constant, and thus momentum relaxation time, the correlation between these quantities is consistent with D'yakunov–Perel' interaction as the underlying physical mechanism, i.e. linear. Reprinted with permission from [357]. Copyright (2011) by the American Physical Society.

The quantity $\bar{\Omega}\tau_m$, involving the averaged D'yakonov–Perel frequency $\bar{\Omega}$ (see Eq. (9.86)), provides a measure for the change of the spin polarization between two successive collisions. The characteric frequency may be expressed as $\bar{\Omega} = \xi_{DP}/\hbar$, where the constant ξ_{DP} may include various phenomena capable of inducing spin relaxation, such as the Rashba effect due to both an electric field perpendicular to the graphene sheet and the curvature of the sheet [243]. Between two collisions, the spin polarization change may proceed in two possible directions, both having equal likelihood, and so, a random walk model applies. The average change of the spin polarization after N_{col} collisions thus amounts to $\sqrt{N_{col}}\frac{\xi_{DP}\tau_m}{\hbar}$. We define the D'yakonov–Perel' spin relaxation time τ_s^{DP} by the relation

$$\frac{\xi_{DP}\tau_m}{\hbar}\sqrt{N_{col}} \approx 1 \tag{10.13}$$

in conjunction with the statement that $N_{col} = \frac{\tau_s^{DP}}{\tau_m}$. Therefore,

$$\frac{\xi_{DP}\tau_m}{\hbar}\sqrt{\frac{\tau_s^{DP}}{\tau_m}} \approx 1, \tag{10.14}$$

$$\tau_s^{DP} \approx \frac{\hbar^2}{\xi_{DP}^2 \tau_m}. \tag{10.15}$$

The form of this equation agrees with the D'yakonov–Perel' relation (9.86). As suggested by recent Hanle spin precession experiments [357], the linear scaling between the spin relaxation rate and the momentum relaxation time, as implied by Eq. (10.15), is a feature of bilayer graphene at low temperature. Experimental data that confirm this finding are shown in Figure 10.9. We note that the values of τ_s measured in bilayer graphene at low temperatures are in the nanosecond range.

The relative insignificance of the Elliott–Yafet effect in bilayer graphene when compared with single-layer graphene has been rationalized as a consequence of a higher degree of impurity screening in the former. Other experimental studies led to similar conclusions [372].

10.3 Spin Transport through Few-Electron SWCNT Quantum Dots

Turning to spin transport through carbon nanotubes, we build on the quantum dot arrangement described in Section 5.6. Extension of this setup to a *double quantum dot* system provides the simplest possible frame for studying the interaction of the electron spin with a degree of freedom characteristic for graphene, namely the valley parameter. This interaction leads to selection rules for electron passage through the double quantum dot, and, conversely, for the *Pauli blockade* phenomenon as the spin counterpart of the *Coulomb blockade*, which impedes the transmission of charge.

In more detail, when giving an outline of nanotube quantum dots, we made reference mostly to single dots operated in the mode of low-bias spectroscopy. These systems are governed by a simple alternative: a single electro-chemical potential level may lie within the bias window, or this may not be case. If the first condition is realized, the dot is conducting, while the latter sets up a Coulomb blockade, interrupting the circuit. High-bias spectroscopy allows for a greater variety of arrangements, as more than one electrochemical potential level may lie within the bias window. Still more complex configurations emerge from double quantum dots as transmission elements, in conjunction with high bias. The physical implementation of such as device has been achieved by employing an ultra-clean SWCNT that connects source and drain electrodes while being suspended over two gate electrodes [373, 374, 375, 376]. The two gate voltages, which can be varied independently of each other, determine the charge carrier occupations of the two dots. Thus, through suitable adjustment of the gate voltages, one may populate the dots with holes, or electrons, or a mixture of both charge carrier types.

The basic scheme of a double quantum dot is indicated in Figure 10.10(a), while panel (b) illustrates the interplay of the two dots. For both of them, the bias window encompasses several excited state levels.

Note that the two gate voltages dictate the population of the double quantum dot which can only change if well-defined amounts of energy are added to the individual dots. Thus, the gate voltages have to assume specific values for population change to occur. *Conductance resonances* [377], corresponding to choices of the parameters (V_L, V_R) such that the Coulomb barrier is lifted in both dots simultaneously, are found at triple points, shared by three regions of stable occupation.

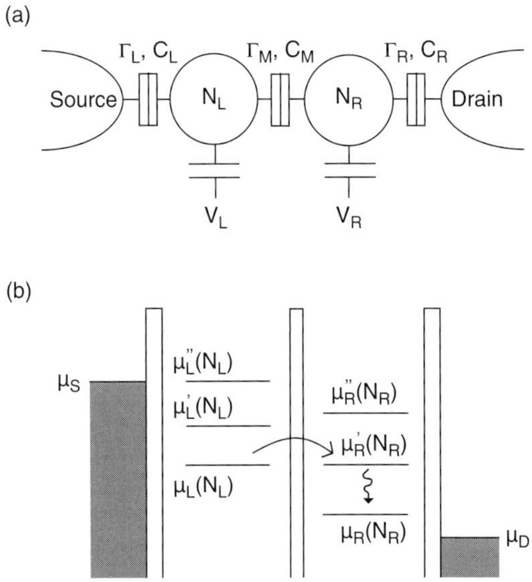

Figure 10.10 Two quantum dots, connected in series. (a) The elementary components of a double quantum dot device. The system is driven by two control parameters, the gate voltages V_L, V_R, and characterized by five capacitances and three tunneling rates (Γ_S, Γ_M, Γ_D). Each one of the five contacts may be modeled by a resistance connected in parallel with a capacitance. (b) Sketch of the double dot region, involving the transition of an electron from the ground state of the left to an excited state level of the right dot. Reprinted with permission from [121]. Copyright (2015) by the American Physical Society.

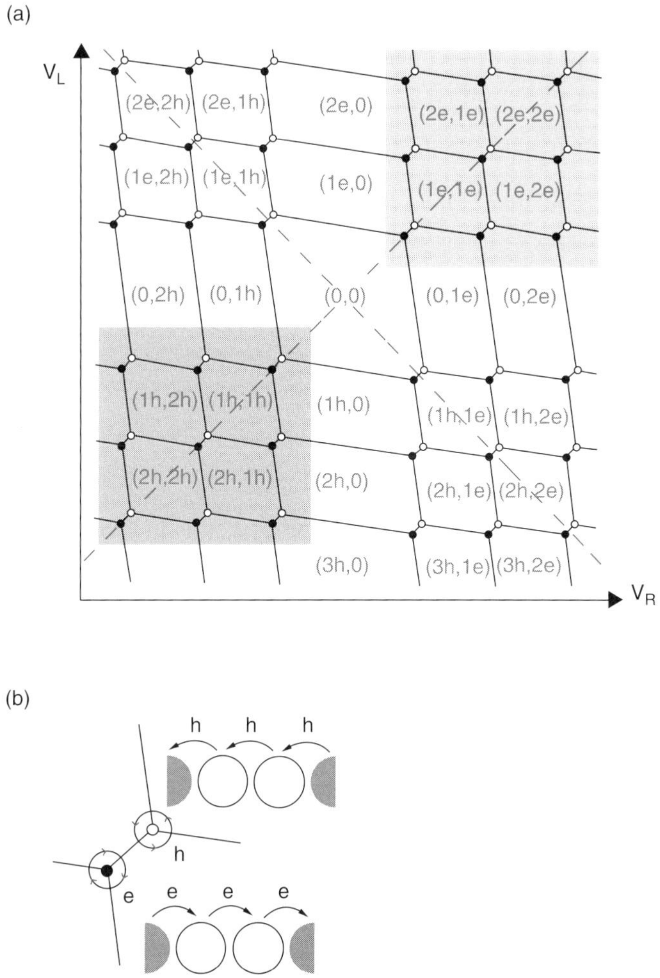

Figure 10.11 Stability diagram in a space spanned by the two gate voltages V_L and V_R. Weak coupling between the two dots is assumed. (a) Scheme of the occupation regimes encompassing up to two electrons and three holes. The triple points are marked by closed or open circles that indicate places of maximum electron or hole conduction, respectively. (b) Electron (hole) conduction through the double dot represented as clockwise (counterclockwise) rotation around the respective triple points. Reprinted with permission from [121]. Copyright (2015) by the American Physical Society.

This feature is best illustrated by a *stability diagram*, as shown in Figure 10.11 for the case of weak interdot coupling. The regions of stable occupation are labeled in terms of their electron and hole numbers, and the network of triple points partitions the $\{V_L, V_R\}$ space into cells that form a distorted honeycomb pattern. Any borderline of any region is indicative of the gate voltage needed to provide the energy difference that separates the two regions divided by this line. Thus, in order to go from [0, 1h] (no electron in the left dot, one hole in the right) to [0, 1e] (no electron in the left dot, one electron in the right), one has to add two times the charging energy E_C (see Section 5.5) and also the amount of the energy gap E_g.

The points distinguished by closed (open) circles refer to places of enhanced electron (hole) conductance. From Figure 10.11(b), when an electron is channeled through the two dots the gate voltages trace out a loop around closed circles in the clockwise sense of rotation, while hole conduction corresponds to surrounding the open circles in the opposite sense. These processes can be summarized by the following schemes:

$$\begin{gathered}\text{Electron conduction:}\\ (N_L, N_R) \rightarrow (N_L + 1, N_R) \rightarrow (N_L, N_R + 1) \rightarrow (N_L, N_R)\\ \text{Hole conduction:}\\ (N_L + 1, N_R + 1) \rightarrow (N_L + 1, N_R) \rightarrow (N_L, N_R + 1) \rightarrow (N_L + 1, N_R + 1).\end{gathered} \tag{10.16}$$

We point out that the interdot capacitance C_M (see Figure 10.10(a)) may be understood as a coupling parameter for the strength of the interaction between the two dots. If C_M vanishes, the double dot reduces to an arrangement of two separated single dots and the honeycomb pattern displayed in Figure 10.11(a) straightens out into a rectangular grid with regions bounded by parallels to the V_L, V_R axes. In the opposite extreme, i.e. with C_M as dominant capacitance, the double dot operates effectively as a single dot with charge $-e(N_L + N_R)$. In this case, the distance between adjacent triple points increases maximally, and the honeycomb network once more simplifies to an ensemble of parallel lines, as there is no strict distinction any more between the configurations (N_L, N_R) and (N_R, N_L).

Whether or not an electron transfer from the left to right dot takes place, as suggested in Figure 10.10(b), depends on selection rules involving the spin and the valley degrees of freedom. An electronic state in a single quantum dot may be specified by a spin and a valley quantum number, and, if high-bias spectroscopy is performed, by a *longitudinal quantum number* in addition to these two. This quantum number labels the states that may be occupied by an electron in either dot, assigning the index 1 to the ground state, 2 to the first excited state, and so on. As admission is made for two dots in contact with each other, many-electron states emerge, and these labels may cease to be good quantum numbers. Further, they are mixed by perturbative effects, such as spin-orbit interactions, hyperfine interactions, and disorder.

In the following, we will consider various two-electron states that can be sustained by the double-dot arrangement and the impact of their properties on the conductance of the dot. In the high-bias case displayed in Figure 10.10 the gate voltages V_L and V_R are adjusted such that the electrochemical potential ground state level of the left dot matches the first excited level of the right. In this situation, a double dot populated by a single electron will be conducting. Two electrons populating the double dot, however, will form multiplets of Fermionic two-particle states, characterized by several combinations of spin and valley degrees of freedom. In this case, the conductance of the device is dictated by selection rules involving these observables. In double quantum dots based on conventional semiconductors, e.g. a two-dimensional electron gas (2DEG) confined by heterostructures made of III-V materials, such as GaAs [123], the electronic states may be specified in terms of spin multiplicities, i.e. spin singlet, triplet etc. The two-electron states in nanotube double dots, however, are determined by their spin and their valley quantum numbers in equal measure. To illustrate this, we assume that the gate voltages are balanced to favor a (0,2) occupation scheme: two electrons confined to the same nanotube dot (see Figure 10.12(a)). The ground state is given here by a *longitudinal singlet* that involves only the single-particle state of lowest energy, with electrochemical potential $\mu(N)$. The singlet based on this state, however, mixes both spin orientations, $\uparrow$ and $\downarrow$, and both Dirac points, K and K'. This state, along with the first two excited states of the (0,2) configuration is made explicit in Table 10.1. From Table 10.2, it may be realized in six different ways of arranging two electrons whose single-particle wave functions differ in at least one quantum number, in accordance with the exclusion principle.

As a second electron is admitted into the double quantum dot, the energy of the system undergoes a uniform shift by the charging energy E_C. The spin-orbit interaction, however, lifts the degeneracy of the six lines of the S(0,2) ground state, as it depends on the relative orientation between the spin and the orbital magnetic moment. For one of the six states, this effect yields energy decrease. It thus singles out a unique (0,2) solution of lowest energy.

Using the schemes given in Table 10.1, the reader may verify that the first excited energy of the (0,2) device is sixteenfold degenerate. This degeneracy is broken by exchange interactions which introduce a splitting $\Delta_{AS,S'}$ between a tenfold antisymmetric and another sixfold symmetric multiplet. If this splitting is distinctly smaller than the spacing of the single-particle levels, the latter provides a good approximation for energy difference between the ground and the first excited state of the (0,2) system.

As the lowest single particle state in the left dot is occupied, rather than the first excited single particle state in the right, we go from the (0,2) to the (1,1) system,

Table 10.1 *The three lowest two-electron states of the type (0,2) in terms of the spin and the valley quantum number. The term* longitudinally symmetric/antisymmetric *refers to the behavior of the wave function upon exchange of the labels for the single-particle states, 1 and 2. Normalization constants have been omitted. The corresponding (1,1) states are generated from those shown in the table by replacing the label 1(2) with* 1_L (1_R)*, indicating that each quantum dot is populated with one electron.*

State label	State composition
$S(0,2)$	$\|1\tau s_z\rangle_1\|1\tau' s_z'\rangle_2 - \|1\tau' s_z'\rangle_1\|1\tau s_z\rangle_2$
$AS(0,2)$	$\|1\tau s_z\rangle_1\|2\tau' s_z'\rangle_2 - \|2\tau s_z\rangle_1\|1\tau' s_z'\rangle_2 + \|1\tau' s_z'\rangle_1\|2\tau s_z\rangle_2 - \|2\tau' s_z'\rangle_1\|1\tau s_z\rangle_2$
$S(0,2)'$	$\|1\tau s_z\rangle_1\|2\tau' s_z'\rangle_2 + \|2\tau s_z\rangle_1\|1\tau' s_z'\rangle_2 - \|1\tau' s_z'\rangle_1\|2\tau s_z\rangle_2 - \|2\tau' s_z'\rangle_1\|1\tau s_z\rangle_2$

Table 10.2 *The ground state of the (0,2) configuration in terms of the quantum numbers* $\tau = \{K, K'\}$ *and* $s_z = \{\uparrow, \downarrow\}$

Label τ, s_z	Energy
$K\downarrow, K'\uparrow$	$-\xi_{SO} + E_C$
$K\downarrow, K\uparrow$	
$K\uparrow, K'\uparrow$	E_C
$K\downarrow, K'\downarrow$	
$K'\uparrow, K'\downarrow$	
$K\uparrow, K'\downarrow$	$\xi_{SO} + E_C$

depicted in Figure 10.12(b). Since the S(1,1) alternative involves electron tunneling between the dots, its energy exceeds that of the S(0,2) solution by a certain margin, the *interdot coupling*. The AS(1,1) state, on the other hand, is distinctly lower in energy than the AS(0,2) state, since it does not comprise a single-particle excitation. If the energy separation $\Delta_{S,AS}$ between the (0,2) ground and first excited states exceeds the prevailing thermal energy, transition from the (1,1) to the (0,2) scheme requires a longitudinal symmetry change of the wave function, and so the AS(1,1) state is long-lived. This is the essence of the *Pauli blockade* phenomenon, as manifested in a double quantum dot device.

This feature is well characterized in conventional semiconductor-based double quantum dots [378] where it can be interpreted in terms of spin singlet versus spin triplet states while the valley degree of freedom complicates the effect in the case of nanotube dots. A semiconductor double dot occupied with one electron localized on the right side and another electron added to the left is only conducting if the two-particle wave function is a spin singlet. If a spin triplet is realized, the passage

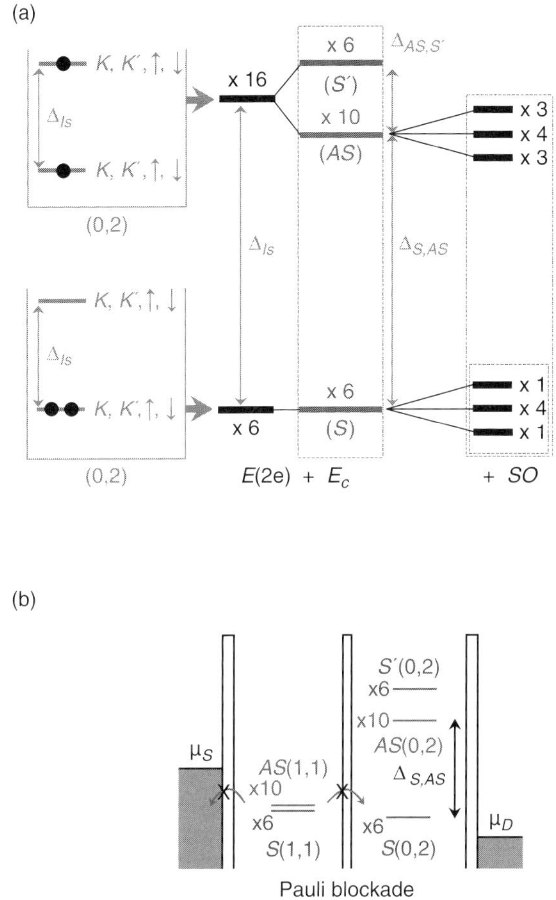

Figure 10.12 Electronic states in a double quantum dot device based on nanotubes. (a) The state (0,2): only the right dot is occupied. It contains two electrons, which form longitudinally symmetric (S) and antisymmetric (AS) states, split by spin-orbit interaction. The multiplicities of the lowest AS state and the two lowest S states are indicated. (b) Conditions of Pauli blockade in the double quantum dot. Formation of a long-lived AS(1,1) state will prevent any transition into a (0,2) state. Reprinted with permission from [121]. Copyright (2015) by the American Physical Society.

of the added electron to the right dot will be blocked by virtue of the exclusion principle.

In both cases, i.e. for double dots built from conventional semiconductors as well as nanotubes, the blockade can be broken by bias reversal, allowing the electron in the left dot to move into the adjacent lead. These systems thus interrupt the current asymmetrically, i.e. in one direction of electron flow, but not in the other, and so act as *Pauli rectifiers*. Beyond bias reversal, there are other various mechanisms that can break the Pauli blockade [121]. These are: (a) dephasing, (b) spin flip, (c) valley flip and (d) a combination of (b) and (c). By *valley flip* we understand the transition $K \to K'$, or vice versa. Commenting on process (a), we point out that the wave function of the blocking state AS(1,1) is constructed in analogy to the wave function indicated in Table 10.1 for AS(0,2), and thus may be represented as a superposition of two determinants that emerge from each other through the operation of time reversal. Replacing in the expression for AS(0,2) the longitudinal label 1(2) with 1_L (1_R), we write

$$\begin{aligned}\psi_{AS(1,1)} &\propto |1_L\tau s_z\rangle_1|1_R\tau' s_z'\rangle_2 - |1_R\tau s_z\rangle_1|1_L\tau' s_z'\rangle_2 \\ &+ e^{\frac{2i}{\hbar}\delta t}(|1_L\tau' s_z'\rangle_1|1_R\tau s_z\rangle_2 - |1_R\tau' s_z'\rangle_1|1_L\tau s_z\rangle_2).\end{aligned} \tag{10.17}$$

The phase factor acknowledges here that the energy differences between the Kramers doublet components $K' \uparrow$ and $K \downarrow$ on the two sides may deviate from each other by a margin δ. The time-dependent phase factor thus induced adds an admixture of the conducting S(1,1)' state to the blocking AS(1,1) state (see Table 10.1). In the limit $e^{2i\delta t} = -1$, the blocking state turns into the conducting state.

Exercise 10.2 Confirm that the state AS(1,1) vanishes as the transition to AS(0,2) is made ($1_L = 1_R = 1$). Demonstrate that this is not true for the state S(1,1)'.

The competing relaxation processes (b), (c) and (d) all depend on the specific composition of the AS(1,1) state. Assuming that it consists of two $K \downarrow$ single-particle states, it reduces to a single 2×2 determinant, namely

$$\psi_{AS(1,1)}[K\downarrow, K\downarrow] \propto |1_L K\downarrow\rangle_1|1_R K\downarrow\rangle_2 - |1_R K\downarrow\rangle_1|1_L K\downarrow\rangle_2. \tag{10.18}$$

A valley flip from $K \downarrow$ to $K' \downarrow$ in the left dot turns this wave function into a superposition of an AS(1,1)[$K \downarrow, K \downarrow$] and an S(1,1)'[$K \downarrow, K \downarrow$] component. Equivalent observations are made for a spin flip with or without a concomitant valley flip in the left dot (pathways (c) and (d)). The presence of valley processes in nanotubes gives rise to a greater number of blockade-breaking mechanisms in nanotube-based double dots than in those made of conventional semiconductors.

10.3.1 Spin and Valley Relaxation by Hyperfine Interaction

Hyperfine interaction is a possible pathway of spin relaxation. When discussing this mechanism in Subsection 9.4.4, we pointed out that the low efficiency of this relaxation channel in carbon-based materials, which is rooted in the predominance of ^{12}C over ^{13}C isotopes, strengthens the case for using carbon nanostructures as transmission elements in spintronics circuits. Performing experimental studies on double quantum dot nanotube systems enriched with ^{13}C, Churchill et al. observed a very marked isotope effect on the conductance in the Pauli blockade regime [379]. The ^{13}C-enhanced device exhibited a distinct leakage current, which was interpreted as a consequence of hyperfine interaction, breaking the Pauli blockade. Examination of this effect as a function of a tube-parallel magnetic field demonstrated that it was strongest at zero field and disppeared at a critical field strength of about 6 mT. This result provided confirmation of the hypothesis that the observed relaxation mechanism was hyperfine interaction. The strong deviation between the magnitudes of the electronic and the nuclear magnetic moment results in Zeeman splittings of markedly differing magnitude, and the hyperfine coupling is expected to diminish rapidly once its size is exceeded by the difference between the electronic and nuclear Zeeman coupling [121, 379].

To account for this and related experimental findings [121, 379], P'alyi et al. considered an effective Hamiltonian that operates on the spin and valley degrees of freedom in quantum dots based on carbon nanotubes [380]:

$$\hat{H}_{hf} = \hat{\boldsymbol{\sigma}} \cdot (\hat{\mathbf{h}}^0 \hat{\tau}_0 + \sum_{i=x,y,z} \hat{\mathbf{h}}^i \hat{\tau}_i). \tag{10.19}$$

The operators $\hat{\mathbf{h}}^k$ $(k = 0, x, y, z)$ are constructed from four tight-binding electronic wave functions that represent the four-dimensional product of the spin and the valley space. Further, they contain the nuclear spin operator. The structure of the Hamiltonian, which couples both the electronic spin and the valley degree of freedom to the spin of the nucleus, implies that hyperfine interactions change both dimensions simultaneously and thus break the Pauli blockade. Allowing for spin-orbit interaction which, in the carbon nanotube quantum dots, is stronger than hyperfine interaction, one finds that the fourfold degenerate electronic ground state, $\{\langle K \uparrow |, \langle K' \downarrow |, \langle K' \uparrow |, \langle K \downarrow |\}$ is split into two Kramers doublets, $\{\langle K \uparrow |, \langle K' \downarrow |, \}$ and $\{\langle K' \uparrow |, \langle K \downarrow |\}$ [380]. While hyperfine coupling will not mix these two doublets, it gives rise to mixing within the doublets and thus causes both spin and valley transitions. Numerical estimates based on hyperfine coupling as a blockade-breaking mechanism underestimate the experimentally recorded leakage effect by about two orders of magnitude. The influence of other processes, such valley-mixing due to short-range disorder [381] on the Pauli blockade in carbon nanotube quantum dots is currently investigated.

11

Magnetotransport

This chapter deals with the transport of charge carriers through carbon nanostructures in the presence of an external magnetic field, with emphasis on carbon nanotubes and on graphene. Specifically, we apply what was stated in Sections 4.4 and 5.4 about the interaction between these two types of nanomaterials with external magnetic fields, to the case of charge carrier transmission. As pointed out earlier, the presence of a magnetic field can have a significant impact on the electronic structure of a carbon nanosystem. Thus, the field may make a semiconducting nanotube metallic or turn a metallic tube into a semiconductor (see Sections 5.4.1 and 8.2). This observation suggests that magnetic fields might provide efficient tools to manipulate nanoelectronic circuits involving carbon elements. The present chapter explores this idea in the context of two fundamental phenomena that have been detected in carbon nanostructures, and subjected to detailed theoretical as well as experimental scrutiny, namely quantum magnetoresistance and the quantum Hall effect.

In Sections 11.1 to 11.3, we discuss the magnetoresistance of carbon nanotubes and of graphene, while the remainder of the chapter deals with the quantum Hall effect in graphene. Appreciating the basic results obtained in these two areas of topical research requires some familiarity with two key concepts of current condensed matter physics, namely Anderson localization and conductance in two-dimensional quantum Hall systems. This background information is provided in Sections 11.1 and 11.4. One might argue that the proper place for the final section of this chapter, 11.5.2, covering the quantum spin Hall effect, is in the previous chapter where spin transport phenomena are reviewed, for the twofold reason that the quantum spin Hall effect is about spin transmission, and that it does not involve any external magnetic field. While this objection is justified, the quantum spin Hall effect in graphene is included in the present chapter for facile comparison with other variants of the quantum Hall effect, as described in Sections 11.4 and 11.5.1.

11.1 Strong and Weak Localization

Our preliminary treatment of quantum transport in carbon nanotubes relied on the ballistic assumption, involving charge carriers that pass through the tube without scattering. This idealization is based on a model of pure nanostructures, i.e. systems that are free of any disorder. Any nanostructure grown in the laboratory, however, exhibits deviations from the ideal structure, be they induced by impurities, adatoms, topological defects or vacancies. In Section 11.3 we will argue that for metallic SWCNTs the ballistic hypothesis is satisfied even in the presence of elastic scatterers. In general, however, structural irregularities affect electronic transport in nanotubes, and the description of transmission has to include scattering processes. These processes are characterized by the momentum relaxation time, τ_m, as well as its spatial counterpart, the *elastic mean free path* L_m, i.e. the average length that an electron traverses between two elastic scattering events. At the Fermi energy, the relation between τ_m and L_m is given by

$$L_m = v_F \tau_m, \tag{11.1}$$

with v_F as Fermi velocity. A further characteristic length in the presence of disorder is L_ϕ, the *dephasing length*. This quantity denotes the length over which the wave function of the electron undergoing transport through the structure retains coherence [68]. With reference to carbon nanotubes, on a tube segment of size L_m, a definite phase can be assigned to the electron wave function. In terms of these concepts, the ballistic regime may be defined by the condition

$$L \ll L_m, L_\phi, \tag{11.2}$$

i.e. the nanotube length is significantly smaller than the other two characteristic lengths. Similarly, we specify *classical transport* by the constraint $L_\phi \ll L_m \ll L$. As L_ϕ is here the smallest among the three length constants, the electrons may be modeled as particles that undergo frequent collisions on their paths between the two electrodes.

A very different transport regime is defined by the condition $L_m \ll L_\phi \ll L$. In this case, wave coherence is maintained through multiple consecutive scattering events. In what follows, we will specify this mode of conduction quantitatively and emphasize the characteristic magnetotransport phenomena that emerge in this regime. The SWCNT is here understood as a one-dimensional transmission element, bounded by electrodes, as shown in Figure 11.1. We distinguish between N_{ch} transmission channels, each of which maps into an SWCNT sub-band with index i whose energy $E_i(k)$ lies in the window set by the bounding chemical potentials ($\mu_D < E_i(k) < \mu_S$), as indicated in Figure 11.1 for the case $N_{ch} = 4$. The ballistic condition is satisfied if $k > 0$ throughout, corresponding to electron flow from

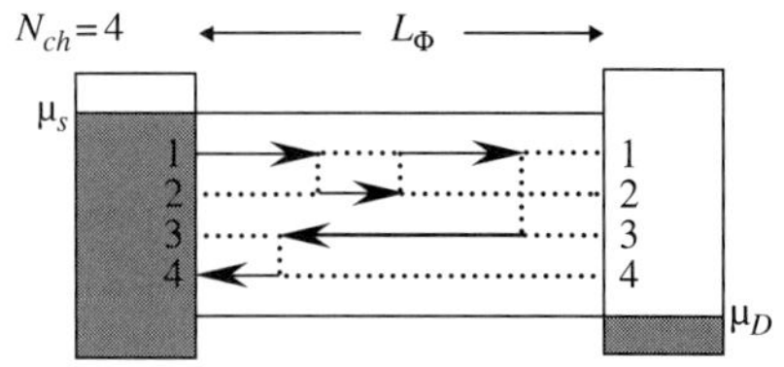

Figure 11.1 Model of an SWCNT used as a transmission element between two electrodes, characterized by chemical potentials μ_S and μ_D. Several transmission channels are indicated, corresponding to sub-bands with energies $E_i(k)$, $i = 1, .., 4$, such that $\mu_D < E_i(k) < \mu_S$. Further, an example is given for transitions that might occur between these channels as an electron is reflected back to the source prior to reaching the drain.

the left to the right electrode and the absence of any scattering events on the way between the two.

To assess the probability for electron transmission through the SWCNT, we consider the transmission probability $\tilde{T}$ through some SWCNT segment. The respective *transmission coefficient* $\tilde{t}$ is given by

$$\tilde{T} = |\tilde{t}|^2. \tag{11.3}$$

Connecting $\tilde{T}$ with the Landauer formula (5.55), we assume that the number of channels available for propagation is one ($N_{ch} = 1$). The quantum conductance may then be approximated as

$$G = \frac{2e^2}{h}\tilde{T}, \tag{11.4}$$

and the corresponding resistance is

$$R = \frac{h}{2e^2}\frac{1}{\tilde{T}}. \tag{11.5}$$

In the simplest approximation, the coefficient for transmission through a nanotube may be modeled as a multi-barrier problem, where each one of N consecutive barriers is characterized by a transmission and a reflection coefficient $(\tilde{t}_n, \tilde{r}_n)$. Considering a situation with two identical barriers, we calculate the overall transmission coefficient $\tilde{t}_{12}$ as a sum over histories that involve multiple pairs of reflection events prior to the final transmission through the second barrier, according to

$$\tilde{t}_{12} = \tilde{t}_1\tilde{t}_2\{1 + \tilde{r}_1\tilde{r}_2e^{i\theta} + \tilde{r}_1^2\tilde{r}_2^2e^{2i\theta} + \ldots\}. \tag{11.6}$$

The premise underlying this formula is that, in view of the relation $L_m \ll L_\phi$, each scattering event involves a phase shift of the electron wave function. The angle θ stands for the phase shift accumulated after a full cycle, consisting of propagation from 1 to 2 and back to 1. As $|\tilde{r}_i| < 1, i = 1, 2$, $\tilde{t}_{12}$ may be expressed as the limit

$$\tilde{t}_{12} = \frac{\tilde{t}_1 \tilde{t}_2}{1 - \tilde{r}_1 \tilde{r}_2 e^{i\theta}}. \tag{11.7}$$

Setting $\tilde{T}_i \equiv |\tilde{t}_i|^2$ and $\tilde{R}_i \equiv |\tilde{r}_i|^2$, $i = 1, 2$, we find

$$\tilde{T}_{12} = \frac{\tilde{T}_1 \tilde{T}_2}{1 - 2\sqrt{\tilde{R}_1 \tilde{R}_2}\cos\theta + \tilde{R}_1 \tilde{R}_2}. \tag{11.8}$$

As we average over θ, the term that depends on $\cos\theta$ disappears. Using Eq. (11.5), we now determine the overall resistance associated with $\tilde{T}_{12}$ as

$$\langle R_{12} \rangle = \frac{h}{2e^2} \frac{1 + \tilde{R}_1 \tilde{R}_2}{\tilde{T}_1 \tilde{T}_2}. \tag{11.9}$$

This relation has a surprising consequence, which is readily clear when one considers two extremely unequal barriers [60]. The first may be a conductor with length L, the second an infinitesimal extension dL of this conductor. In this case, we may set $\tilde{R} \approx 1$ ($\tilde{T} = 1 - \tilde{R} \approx 0$). We allow for elastic backscattering which preserves the phase of the wave function. Assuming $dL \ll L_m$, we set $d\tilde{R} = \frac{dL}{L_m}$ ($d\tilde{T} = 1 - \frac{dL}{L_m}$). In this case, relation (11.9) implies that

$$\langle R + dR \rangle = \frac{h}{2e^2} \frac{1 + \tilde{R} d\tilde{R}}{\tilde{T}(1 - d\tilde{R})} \approx \langle R \rangle (1 + \frac{2dL}{L_m}). \tag{11.10}$$

In the framework of this approach, the behavior of the resistance for large L can be inferred from the differential equation

$$\langle \frac{dR}{dL} \rangle = \langle R \rangle \frac{2}{L_m}, \tag{11.11}$$

which is solved by

$$\langle R(L) \rangle = \frac{h}{2e^2} e^{\frac{2L}{L_m}}. \tag{11.12}$$

This function diverges as L goes towards infinity. Eq. (11.12) predicts that the conductor, if sufficiently long, turns into an insulator. This is a consequence of wave function localization near the Fermi energy. The physical cause of this effect is the presence of disorder in the transmission element, giving rise to scattering from randomly distributed centers, which, in turn, induces electron trapping. This phenomenon is known as *strong localization* [382]. Under conditions of strong localization, conduction proceeds through thermal electron hopping between states localized at different sites.

Confinement emerges here from random potentials acting on one-electron states. This effect is to be distinguished from *Mott localization* [383] which is rooted in interactions between electrons.[1] A general localization criterion can be formulated in terms of the tight-binding approximation, i.e. by use of the Hamiltonian

$$\hat{H}_{TB} = -\sum_{n} E_n \hat{a}_n^{\dagger} \hat{a}_n + V \sum_{\langle n,m \rangle} \hat{a}_n^{\dagger} \hat{a}_m, \tag{11.13}$$

where m, n label neighboring sites. The condition $V = 0$ implies electron confinement, as no transfer of electrons between the sites of the systems will take place in this extreme case. A further conclusion with respect to the extension of electronic states can be drawn from the spectrum $\{E_n\}$. The relevant quantity is here the width W of the distribution of the site energies E_n. If $W = 0$, the case of constant energy is realized. The corresponding eigenstates are Bloch waves, and thus delocalized. As $W > 0$, on the other hand, disorder enters the system, allowing for scattering processes. Thus, the size of the ratio $\frac{W}{V}$ may serve as a criterion for localization. For three-dimensional compounds, a critical ratio $\frac{W}{V}|_c$ can be identified, such that $\frac{W}{V} > \frac{W}{V}|_c$ precludes the existence of extended electron states. For two- and (with few exceptions) one-dimensional systems, an infinitesimal margin of disorder is sufficient to prevent formation of extended states.

Generalizing the result (11.12), one may replace L_m with a *localization length* L_c, which determines the asymptotic behavior of the localized wave function [384], with L_c in the order of L_m. Strong localization occurs when $L_c < L_\phi$, i.e the localization length lies within the range set by the dephasing length. Conversely, the condition $L_c > L_\phi$, in conjunction with $L \geq L_\phi$, defines *weak localization*, a precursor effect of strong localization. This phenomenon does not manifest itself in a qualitative change of the conduction mechanism, but rather in a negative correction of the conductivity. It is of major importance for magnetotransport through non-metallic carbon nanotubes, as it implies *negative magnetoresistance*, involving a diminished resistance in the presence of a magnetic field. In the following we will give an outline of this effect.

11.2 Negative Magnetoresistance

The resistance of the nanotube transmission element depends on the probability for electron reflection, as captured by the reflection coefficient. The reflected wave function may be described as a *sum over histories*, involving alternative propagation pathways from the source towards the drain, and back to the source. The more open channels, i.e. nanotube sub-bands, are available within the limits set by the

[1] Mott localization is frequently defined in terms of a critical on-site Coulomb repulsion energy U_C. The Hubbard model parameter U may be understood as a measure for intra-atomic interaction strength. Mott localization occurs when the on-site Coulomb repulsion between electrons outweighs the dynamic (or hopping) part of the Hamiltonian sufficiently to cause the formation of a band gap. This *Mott transition* from metal to insulator depends on the temperature, the pressure or the composition of the sample.

chemical potentials of the leads, the greater the number of different pathways. Figure 11.1 presents a graphical example for a sequence of channels that an electron may traverse before returning to the source.

At the end point of the round trip, the reflected wave function Ψ_{ref} may then be obtained by summing over all possible pathways [68]:

$$\Psi_{ref}(n \to m) = \sum_p A_p(n \to m). \tag{11.14}$$

Each of the wave function components A_p represents a certain combination of forward or backward propagation segments, where allowance is made for transitions between channels.[2] The ensemble average of the reflection probability associated with the process $m \to n$ equals the square magnitude of $\Psi_{ref}(n \to m)$. In the presence of disorder, i.e. a random distribution of scattering centers, the overall phases of the A_p are expected to fluctuate statistically, implying cancellation of the interference terms,

$$\langle \tilde{R}(m \to n) \rangle = |\Psi_{ref}(n \to m)|^2 = \left\langle \sum_p |A_p|^2 + \sum_{p \neq p'} A_p A_p^* \right\rangle = \left\langle \sum_p |A_p|^2 \right\rangle. \tag{11.15}$$

However, as a characteristic feature of the weak localization regime, we can single out two reflection pathways that undergo constructive interference. This is any process that begins and ends in the same channel and is thus described by $A_p(n \to n)$, as well as its time-reversed counterpart $A_p^{ref}(n \to n)$. In the context of weak localization, the salient point about these two pathways is that the same overall phase is accumulated as they are traversed. As these processes are added coherently, they lead to a marked enhancement of the reflection probability. This is seen from evaluating $\tilde{R}(n \to n)$,

$$\tilde{R}(n \to n) = |(A_1 + A_2 + \ldots.) + (A_1^{ref} + A_2^{ref} + \ldots)|^2 = |A + A^{ref}|^2 = 4|A|^2, \tag{11.16}$$

with $A \equiv A_1 + A_2 + \ldots$ Although these cyclical trajectories, resulting in *enhanced backward scattering*, are rare, they cause a pronounced resistance increase, and a corresponding conductivity reduction [385, 386]. Since quantum interference effects decrease charge carrier mobility, they are a signature phenomenon of weak localization. The result (11.16) is twice the classical finding that is recovered as coherence effects are neglected. We emphasize that this conclusion is a consequence of the non-classical character of electron transport in the regime of weak localization. The described effect hinges on the wave nature of the electron

[2] The following expression exemplifies an amplitude A_p that describes an initial transmission event with j subsequent channel transitions and j segments of electron propagation in a given channel: $A_p(n \to m) = \tilde{t}_{m \to m_1} e^{ik_{m1} L_{m1}} \tilde{r}_{m_1 \to m_2} e^{ik_{m2} L_{m2}} \ldots \tilde{t}_{m_j \to n} e^{ik_n L_n}$.

propagation states. The elementary physical effect encoded by Eq. (11.16) can also be observed in the context of conventional wave phenomena. For instance, shouting into a forest will yield an enhanced echo if the trees are irregularly spaced [386].

Observation of weak localization probes the quantum nature of electrons in the examined medium, since this effect is based on the coherence of electrons that traverse closed paths in opposite directions. This condition requires these paths to be shorter than the dephasing length L_ϕ. Since increasing temperature enhances the probability for inelastic scattering and therefore reduces L_ϕ, the conductivity correction $\Delta\tilde{\sigma}$ due to weak localization exhibits a distinct temperature dependence. For one-dimensional media, $\Delta\tilde{\sigma}$ behaves as $-T^{-\frac{p}{2}}$, with p as a positive constant. In two- and three-dimensional systems the reduction of $\Delta\tilde{\sigma}$ with temperature is described by other functional relationships [386].

The argument that establishes weak localization hinges on the invariance of the overall phase of a component $A_p(n \to n)$ with respect to time reversal. It can thus be examined by applying a magnetic field parallel to the tube axis. Quantum interference arises here from electrons counterpropagating along paths that enclose the same magnetic flux. As a consequence of the Bohm–Aharonov effect, the related electron wave functions will be modified by a magnetic phase that changes its sign upon time reversal. Thus, Eq. (11.16) must be extended according to

$$\tilde{R}(n \to n) = |Ae^{i\alpha_+} + A^{ref} e^{i\alpha_-}|^2, \tag{11.17}$$

where

$$\alpha_\pm = \pm\frac{e}{\hbar}\oint \mathbf{A}\cdot d\mathbf{x} = \pm\frac{2\pi}{\Phi_0}\oint \mathbf{A}\cdot d\mathbf{x}. \tag{11.18}$$

Comparison with Eq. (11.16) clarifies that in this case, the reflection probability is given by

$$\tilde{R}(n \to n) = |A|^2|1 + e^{i(\alpha_+ - \alpha_-)}|^2 = 4|A|^2\cos^2\frac{2\pi\Phi}{\Phi_0}. \tag{11.19}$$

This relation predicts oscillatory behavior of the reflection probability as a function of the magnetic flux threading through the nanotube. In particular, Eq. (11.19) implies that the resistance of the tube varies as a function of the flux with a period of $\frac{\Phi_0}{2}$ [387]. The previous result, $4|A|^2$, is here an upper limit, so quantum interference in the presence of an axial magnetic field leads to an effective reduction of the resistance, thus giving rise to *negative magnetoresistance*. The magnetoresistance (MR) measures the relative influence of an external magnetic field on the resistance of a sample and is thus defined as the ratio $\Delta R/R = (R(\mathbf{B}) - R(\mathbf{B} = 0))/R(\mathbf{B} = 0)$.

This manifestation of the Aharonov–Bohm effect in the weak localization regime has been demonstrated experimentally for multi-walled carbon nanotubes (MWCNTs) as transmission elements [388]. Figure 11.2 shows a comparison

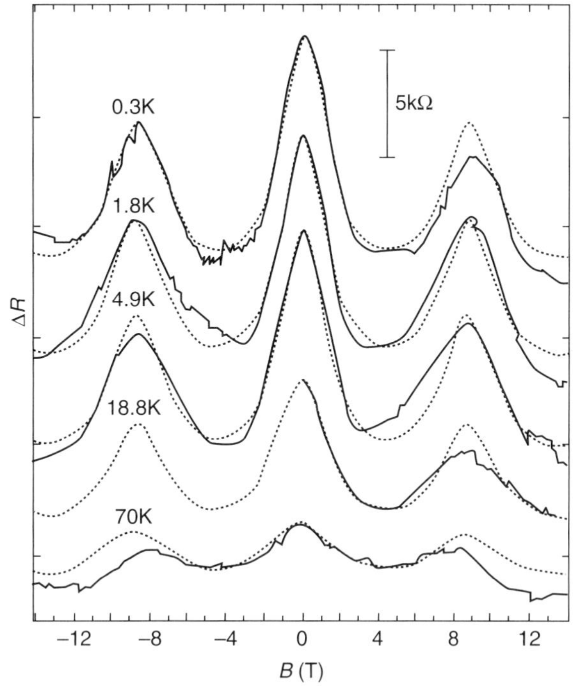

Figure 11.2 Resistance variation of an MWCNT as a function of the strength of a magnetic field applied parallel to the tube axis [388]. Five different temperatures of the sample are indicated as parameters. Aharonov–Bohm oscillations are strongly pronounced. The effect weakens with increasing temperature. The solid (dotted) curve refers to measured (calculated) values. Reprinted with permission from [388]. Copyright (2009) by the Nature Publishing Group.

between experimental and theoretical results for the case of an axis-parallel magnetic field [388]. With modulations in the range of the quantum resistance unit, the predicted periodicity of the resistance with varying field strength is clearly confirmed. The amplitude of the resistance oscillations diminishes markedly as the sample temperature increases.

In the case of a magnetic field applied perpendicularly to the tube axis [389, 390], conductance increase was observed in conjunction with aperiodic fluctuations upon field strength variation.

11.3 Positive Magnetoresistance: the Case of Massless Fermions

Magnetoresistance phenomena in graphene as well as metallic nanotubes present an altogether different scenario than described in the previous section. In what follows, we will make reference to graphene, but the arguments presented carry

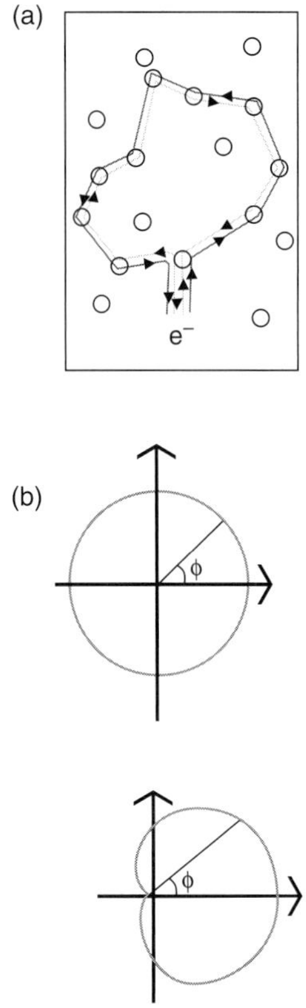

Figure 11.3 Counterpropagating electrons in the graphene plane. (a) Two closed trajectories are indicated. Both of them are assumed to be in the range of the dephasing length L_ϕ, allowing for quantum interference to occur. Both electrons leave the loop with a momentum antiparallel to the entrance momentum. If the scattering processes do not change the helicity of the electrons, they both emerge from the round trip with their pseudospins rotated by π [392]. This leads to mutual cancellation of the two outgoing states. The relative geometric phase accrued in the process of pseudospin rotation leads to destructive interference of the outgoing waves and thus to suppression of backscattering. As illustrated in (b), this results in an anisotropic scattering probability profile, as opposed to the isotropic profile realized in a regular electronic system. Reprinted with permission from [392]. Copyright (2009) by the American Physical Society.

over to metallic nanotubes, as, in both cases, the charge carriers in the Dirac regime have the character of massless Fermions [108, 391].

In any experiment designed to test weak localization in graphene, the probing magnetic field is applied perpendicularly to the graphene sheet, inducing magnetic flux that is enclosed by the paths of two counterpropagating electrons, as shown in Figure 11.3. Once more we make reference to a sample with randomly distributed scattering centers, and once more we assume that the dephasing length distinctly exceeds the elastic mean free path, $L_m \ll L_\phi$. We add the premise that the pseudospin helicity of the electrons does not change during propagation. This assertion, which will be critically examined later, implies *suppression of backscattering*, as a process that reverses the momentum while leaving the pseudospin unchanged, and thus affects the mechanism that causes weak localization. Figure 11.3 contrasts the isotropic scattering probability with an anisotropic profile that describes pseudospinor scattering in graphene.

Quantum interference of chiral charge carriers provides the mechanism for the phenomenon of *weak antilocalization*, as experimentally detected in graphene [392]. As two electrons that initially have the same set of quantum numbers traverse the loop in opposite directions, their pseudospins remain rigidly aligned with their momenta and so rotate through angular intervals of π or $-\pi$ in case of counterclockwise or clockwise propagation through the loop, respectively. At the same time, the pseudospinor wave function of each electron accumulates a geometric phase of magnitude $\frac{\pi}{2}$ as the path shown in Figure 11.3 is traversed.[3] Specifically, the phase is $+\frac{\pi}{2}$ ($-\frac{\pi}{2}$) for the counterclockwise (clockwise) trajectory. The electronic wave functions characteristic of the two return paths thus combine with a relative phase shift of π, resulting in destructive quantum interference [393]. Therefore, backscattering is not possible under the premise that the pseudospin helicity is conserved, as should be expected for long-range potentials that do not distinguish between A and B sites [394]. In this context we point out that the low-energy solutions for opposite helicities, Ψ^+ and Ψ^-, as indicated in Eq. (4.26) (Eq. (4.28)) for the $\mathbf{K}$ ($\mathbf{K}'$) point, describe different molecular states. While Ψ^+ is bonding, corresponding to identical phase factors for A and B sites, Ψ^- is antibonding, involving a sign switch as one goes from the A to the B component.

On the basis of this reasoning, one expects to find a *positive magnetoresistance effect* in graphene. An appropriately calibrated magnetic flux through the loop, however, may increase the relative phase to a whole multiple of 2π, thus establishing constructive interference, and, by the same token, restoring backscattering.

In laboratory graphene samples, several other mechanisms may act toward reinstating weak localization. It is among the unconventional properties of graphene that localization phenomena in this medium do not only depend on inelastic processes (i.e. those that are able to shift the phase), but also on elastic interactions. In fact, several modes of elastic scattering have been identified that are capable of changing the helicity of charge carriers in the vicinity of the Dirac points. Such processes remove the assumption of helicity conservation on which the argument

[3] Note that the electron momentum undergoes a rotation around an angle of π from the initial state of entering the loop to the final state of leaving it, as shown in Figure 11.3.

for antilocalization of massless Dirac Fermions in graphene was resting. In particular, one distinguishes *intervalley* from *intravalley scattering*. This first type of disturbance is typically caused by *sharp* (or short-range) defects, the second type by defects whose size is comparable to the lattice spacing, and further by ripples and dislocations [392, 396]. Intervalley scattering reverses the helicity, since this quantity adopts opposite values in the two valleys (see Section 4.2.1). The impact of intravalley scattering on the helicity, on the other hand, is more indirect. Expanding the tight-binding Hamiltonian (4.5) near the Dirac points up to quadratic order in **k**, one encounters a *warping term* that replaces the isotropy of the Dirac Hamiltonians (4.22) and (4.27) with trigonal symmetry. The added contribution is proportional to $\boldsymbol{\eta}_x(q_x^2 - q_y^2) - 2\boldsymbol{\eta}_y q_x q_y$ [85], giving rise to a threefold arrangement of minima in the landscape around the K/K' point. The energy eigenvalues change accordingly [72],

$$E_{\pm} = \pm\hbar v_F|\mathbf{q}| \mp \frac{3ta_0^2}{8}\sin(3\theta_{\mathbf{q}})|\mathbf{q}|^2, \tag{11.20}$$

and $\theta_{\mathbf{q}} = \tan^{-1}\frac{q_x}{q_y}$. This relation holds for electrons (E_+) and holes (E_-) in the K valley, and one verifies readily that the sign of the warping term switches as one goes to the K' valley. Since, in either case, the valley symmetry is broken by the warping term, a pseudospinor that is eigenstate of the second-order Hamiltonian is not any more eigenstate of the helicity operator, and is thus representable as a superposition of solutions with opposite helicities. Inclusion of quadratic contributions around the Dirac points therefore allows for switches of helicity. From this discussion, consecutive intravalley and intervalley scattering events may restore the original helicity of electrons, thus favoring antilocalization.

A tight-binding analysis of weakly disordered graphene, including trigonal warping [85], arrived at a closed expression for the conductance correction $\Delta\tilde{\sigma}(B)$ due to a magnetic field **B** as a function of both intervalley and intravalley processes, represented by their respective rates, $\tau_{inter}^{-1} + \tau_{intra}^{-1}$,[4]

$$\Delta\tilde{\sigma}(B) = \frac{e^2}{\pi h}[F(\frac{\tau_B^{-1}}{\tau_\phi^{-1}}) - F(\frac{\tau_B^{-1}}{\tau_\phi^{-1} + 2\tau_{inter}^{-1}}) - 2F(\frac{\tau_B^{-1}}{\tau_\phi^{-1} + \tau_{inter}^{-1} + \tau_{intra}^{-1}})]. \tag{11.21}$$

4 This relation is frequently written in terms of length constants characteristic of dephasing (L_ϕ) and of intervalley as well as intravalley scattering (L_{inter} and L_{intra}, respectively). As a function of these lengths, it reads:

$$\Delta\tilde{\sigma}(B) = \frac{e^2}{\pi h}(F(\frac{4l_B^{-2}}{L_\phi^{-2}}) - F(\frac{4l_B^{-2}}{L_\phi^{-2} + 2L_{inter}^{-2}}) - 2F(\frac{4l_B^{-2}}{L_\phi^{-2} + L_{inter}^{-2} + L_{intra}^{-2}})).$$

Here $\tau_B^{-1} = \frac{4eDB}{\hbar}$, where D denotes the diffusion constant, $F(z) = \ln(z) + \psi(0.5 + z^{-1})$, with ψ as the *digamma function*, $\psi(z) = \frac{d\Gamma}{\Gamma}(z)$. This equation predicts diminished conductivity in the extreme case that both rates, τ_{inter}^{-1} and τ_{intra}^{-1}, reduce to zero. This behavior is the hallmark of weak antilocalization. The opposite extreme, involving inter- and intravalley scattering rates that markedly exceed the *dephasing rate* τ_ϕ^{-1}, implies positive $\Delta\tilde{\sigma}(B)$, in accordance with weak localization. Further, if $\tau_{inter}^{-1} \ll \tau_{intra}^{-1}, \tau_\phi^{-1} [\tau_{inter}^{-1} \gg \tau_{intra}^{-1}, \tau_\phi^{-1}]$, the correction is negative [positive].

Experimental evidence for the occurrence of both antagonistic phenomena, weak localization and weak antilocalization, in graphene was obtained by recording changes in the conductance of various graphene samples as a function of an external magnetic field [392]. The anomalous quantum Hall effect (see Section 11.5.1) was found to be operative in each sample, confirming the presence of monolayer graphene. The samples were studied in different regimes of charge carrier concentration, as determined by the applied gate voltage, and at temperatures ranging from 5 K to about 200 K. The measured magnetoconductance profiles prove that regimes of weak localization and antilocalization can be selected by controlled choice of the gate voltage and the sample temperature. Figure 11.4 exemplifies this conclusion, showing the response of $\Delta\tilde{\sigma}$ to the magnetic field strength B, as these two parameters are varied. From Figure 11.4(b), weak localization dominates at low temperature, as demonstrated by the signature conductance dip at $B = 0$. While this fingerprint is present in all three curves, it is most strongly pronounced in the sample with the highest carrier concentration and fades out as the concentration decreases, and thus the metallic character of the sample. At higher temperature (Figure 11.4(c)), a clear transition between the two localization patterns is observed: the dip evolves into a peak, manifesting a crossover to antilocalization as the carrier concentration diminishes and thus the Dirac region is approached. The prevalence of localization over antilocalization at low temperature is rationalized by the relation between the dephasing time, τ_ϕ, and the times characteristic for helicity-changing processes, τ_{intra} and τ_{inter}. The ratio $\tau_\phi/\tau_{intra,inter}$ grows large at low temperature [392], favoring intravalley and intervalley processes. This trend is consistent with a positive conductance correction, and therefore dominance of weak localization.[5]

We mention that, in the presence of disorder, controlled variation of the temperature and the charge carrier density may also give rise to a transition from the weak

[5] Quantitatively, this statement is borne out by the approximation to formula (11.21) in the limit of low magnetic field strengths, the case relevant for the experiment described in [392]:

$$\Delta\tilde{\sigma}(B) \approx \frac{e^2}{24\pi h}\left(\frac{4eDB\tau_\phi}{\hbar}\right)^2\left(1 - \frac{1}{(1 + 2\tau_\phi\tau_{inter}^{-1})^2} - \frac{2}{(1 + \tau_\phi\tau_{inter}^{-1} + \tau_\phi\tau_{intra}^{-1})^2}\right).$$

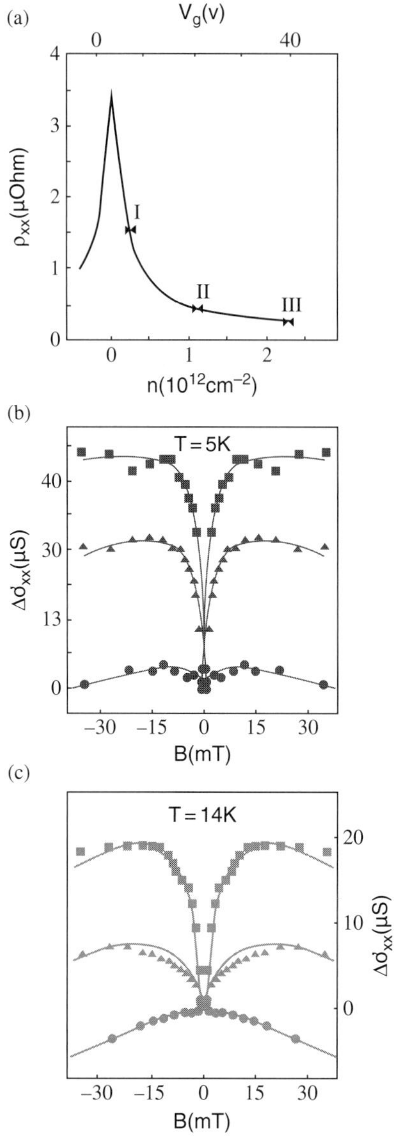

Figure 11.4 Measured magnetoconductances of various graphene samples, differing with respect to charge carrier concentration and temperature [392]. The solid lines represent fits to Eq. (11.21). (a) The resistivity versus the charge carrier concentration. The three bars, each one with a width corresponding to $\Delta V_g = 1V$, demarcate the three gate voltage regimes investigated. (b) Magnetoconductance versus magnetic field strength at $T = 5$ K. The three curves correspond to the three concentrations indicated in panel (a). (c) As (b), for $T = 14$ K. Comparing (b) with (c), one finds that the higher temperature allows for a transition between the two localization schemes, while weak localization prevails at the lower temperature. Reprinted with permission from [392]. Copyright (2009) by the American Physical Society.

to the strong localization regime. For instance, this has been demonstrated by measurements on graphene with ozone impurities, introducing defects of the sp^3 type [397]. This transition is associated with a crossover between the dephasing and the localization length, where $L_c < L_\phi$ defines the regime of strong localization (see Eq. (11.12) and the related commentary.)

In the strongly metallic regime of high electron concentration, far beyond the charge neutrality point, the quantum correction that gives rise to weak antilocalization is expected to become less important. Further, experimental studies on graphene flakes with sizes in the micrometer range [244] did not yield positive magnetoresistance, but rather small or vanishing effects throughout. In many cases, the resistance of the sample remained unaffected by the magnetic field.

This result has been explained by structural imperfections of the investigated graphene devices. These are not globally flat, but contain ripples of varying extension, as confirmed by atomic force microscopy (AFM). The impact of these geometric deformations on the electronic system of graphene is equivalent to that of an irregular arrangement of local magnetic fields, spreading over the sheet. Specifically, the presence of local geometric distortions implies that the three bonds surrounding any carbon atom in the graphene network will, in general, differ from each other. A uniform hopping parameter, as assumed in the tight-binding derivation of Section 4.2 will thus not be sufficient any more to describe the electronic structure of graphene. Replacing the parameter t with the parameters t_1, t_2 and t_3, and retracing the steps that led to the Hamiltonian (4.30), one arrives at the extended expression [9, 398, 399]

$$\mathbf{H} = \boldsymbol{\eta} \cdot (-i\hbar v_F \nabla - \mathbf{A}), \tag{11.22}$$

with

$$\begin{aligned} A_x &= \frac{\sqrt{3}}{2}(t_1 - t_3), \\ A_y &= \frac{1}{2}(t_3 + t_1 - 2t_2). \end{aligned} \tag{11.23}$$

This result holds for the $\mathbf{K}$ valley. For the $\mathbf{K}'$ valley, one obtains $-\mathbf{A}$ rather than $\mathbf{A}$. In both cases, the structural distortions of the graphene sheet are absorbed into a gauge field, giving rise to a *pseudomagnetic effect* [244]. If intervalley scattering can be neglected, the electron subsystems in the vicinity of the two Dirac points are independent of each other. In this case, the geometry-induced gauge field cancels, or strongly reduces, the quantum interference that causes weak localization. Effectively, ripples and dislocations in the graphene sheet enhance the impact of intravalley scattering on the conductivity and thus affect the conductivity correction (11.21) by changing the parameter τ_{intra}^{-1} [392].

Exercise 11.1 Show that admitting local variations in the hopping parameter t gives rise to a vector potential term with components (11.23). Refer to formulae (4.15) and (4.21), as well as Figure 4.2. Associate hopping elements t_i with lattice vectors $\mathbf{R}_i$, $i = 1, 2, 3$.

11.4 The Integer Quantum Hall Effect

The *integer quantum Hall effect* (IQHE), discovered in 1980 by K. von Klitzing, G.Dorda and M.Pepper [400], is a signature phenomeneon of the two-dimensional electron gas (2DEG) in strong, uniform magnetic fields, oriented in a direction perpendicular to its plane. The Hall conductivity turned out to be quantized, being confined to integer multiples of the unit e^2/h, irrespective of the specific properties of the underlying material. Shortly after the initial fabrication of monolayer graphene, a pronounced quantum Hall effect was detected in this novel material [64] as one of the hallmark features of its electronic structure. The quantum Hall effect in graphene, however, was found to be distinguished from that in regular 2DEG systems in various essential respects. To accentuate that difference, we preface a survey of the quantum Hall effects in graphene (Section 11.5) with a summary of some salient features of the integer quantum Hall effect in 2DEG systems.

For a perfectly planar 2DEG sample, the Hall effect is described by a (2×2) magnetoconductivity tensor, connecting the electric field $\mathbf{E}$ and the current density $\mathbf{j}$ in the presence of a perpendicular magnetic field $\mathbf{B}$ according to [60]

$$\mathbf{j} = \tilde{\sigma}\mathbf{E}, \tag{11.24}$$

with

$$\tilde{\sigma} = \begin{pmatrix} \tilde{\sigma}_{xx} & \tilde{\sigma}_{xy} \\ \tilde{\sigma}_{yx} & \tilde{\sigma}_{yy} \end{pmatrix}. \tag{11.25}$$

The x direction is here assigned to longitudinal current, and the y direction to the *Hall current*, resulting from the external magnetic field. Further, $\tilde{\sigma}_{xx} = \tilde{\sigma}_{yy}$, and $\tilde{\sigma}_{yx} = -\tilde{\sigma}_{xy}$. The 2×2 conductivity determines the 2×2 resistivity through the following relations:

$$\begin{aligned} \tilde{\rho}_{xx} &= \frac{\tilde{\sigma}_{xx}}{\tilde{\sigma}_{xx}^2 + \tilde{\sigma}_{xy}^2} \\ \tilde{\rho}_{xy} &= \frac{\tilde{\sigma}_{xy}}{\tilde{\sigma}_{xx}^2 + \tilde{\sigma}_{xy}^2}. \end{aligned} \tag{11.26}$$

The physics of the quantum Hall effect in graphene is encoded in the *Hall conductivity*, i.e. the off-diagonal element $\tilde{\sigma}_{xy}$. We consider the regime of low temperature (not exceeding 4K) and high magnetic fields (≈ 10 T) where thermal transitions between Landau levels can be safely neglected. In the high-field limit, it holds that $\tilde{\sigma}_{xy} = \frac{ne}{B}$.[6] The charge carrier concentration n is determined by the degree of degeneracy of an occupied Landau level. As an overall measure for the occupation of the Landau level hierarchy, one introduces the *filling factor* ν,

$$\nu \equiv \frac{\text{Areal density of electrons}}{\text{Areal density of states in a Landau level}} = \frac{nL_xL_y}{g} = \frac{nh}{Be}, \tag{11.27}$$

where Eq. (4.68) has been used. The definition of ν applies to either of the two Zeeman-split spin orientations. From the dependence of ν on B, one concludes that the filling factor adopts integer values N for selected magnetic field strengths B_N, namely

$$B_N = \frac{nh}{e}\frac{1}{N}. \tag{11.28}$$

If $B = B_N$, a whole number $N = \nu$ of Landau levels are completely filled. Under the premise of high magnetic field strength, it holds that $|\tilde{\sigma}_{xy}| \gg |\tilde{\sigma}_{xx}|$. In this case, one concludes from Eq. (11.26) that

$$\tilde{\rho}_{xx} \approx \frac{\tilde{\sigma}_{xx}}{\tilde{\sigma}_{xy}^2} \approx 0. \tag{11.29}$$

The dominance of the transversal conductivity component over the longitudinal thus implies an extremely small longitudinal resistivity. In case of N completely filled Landau levels, i.e. for $B = B_N$, we conclude from $\tilde{\sigma}_{xy} = \frac{ne}{B}$ and Eq. (11.28) that

$$\tilde{\sigma}_{xy} = N\frac{e^2}{h}. \tag{11.30}$$

From this result, the transversal conductivity is given by integer multiples of $\frac{e^2}{h}$, in accordance with the experimental finding [400]. Relation (11.30) implies thus the existence of a *quantum of conductance*, or, equivalently, a *quantum of resistance*.[7] The *Hall resistance* $\tilde{\rho}_{xy}$ is the inverse of $\tilde{\sigma}_{xy}$, and therefore

$$\tilde{\rho}_{xy} = \frac{1}{N}\frac{h}{e^2}. \tag{11.31}$$

[6] This result is consistent with the classical relation that defines the *magnetoconductance tensor*. In two dimensions, this relation reads [60]:

$$\begin{pmatrix} j_x \\ j_y \end{pmatrix} = \frac{\tilde{\sigma}_0}{1 + (\omega_c\tau_m)^2}\begin{pmatrix} 1 & -\omega_c\tau_m \\ \omega_c\tau_m & 1 \end{pmatrix}\begin{pmatrix} E_x \\ E_y \end{pmatrix}.$$

Here, ω_c stands for the cyclotron frequency, $\frac{eB}{m_e}$, and τ_m for the momentum relaxation time, while $\tilde{\sigma}_0 = \frac{ne^2\tau_m}{m_e}$. In the high-field limit $\omega_c\tau_m \gg 1$, one obtains $\frac{ne}{B}$ for $\tilde{\sigma}_{xy}$.

[7] Note that in two dimensions, the units of conductivity (resistivity) and conductance (resistance) are the same.

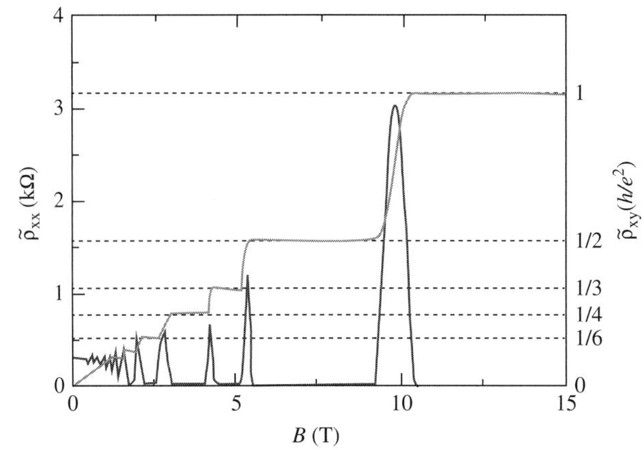

Figure 11.5 Schematic representation of the integer quantum Hall effect in terms of the Hall resistivity versus the external magnetic field for a two-dimensional electron gas. The Hall resistivity exhibits plateaus determined by fractions of the resistance quantum $\frac{h}{e^2}$. Note that the number of occupied Landau levels decreases with increasing magnetic field strength (compare with relation 11.28). The data were taken at 0.3 K in a InGaAs-based heterostructure [401].

Discontinuous changes of the Hall resistance occur at the critical magnetic field strengths $B = B_N$, as displayed in Figure 11.5. At these values, the longitudinal resistivity displays a pronounced maximum.

The reduction of occupied Landau levels with increasing magnetic field strength, as is evident from Figure 11.5, reflects the increase of the Landau level degeneracy with the field, as described by relation (4.68). More specifically, enhancing the field strength adds capacity for accommodating electrons to each Landau level. Thus, electrons from higher levels will undergo transitions to lower levels. As a critical field strength is reached, the highest level is emptied of electrons, while lower levels are completely occupied, and thus the overall number of occupied Landau levels is reduced by one. The sharp quantization of the Hall conductivity and resistivity, as documented in Figure 11.5, appears to follow from the sharp energy quantization of the Landau levels. A physically real sample, however, is marred by disorder, such as structural defects, edges, or atomic impurities, all of them breaking the translational invariance that was used in the derivation of Eq. (11.30). These effects modify the spatial extension of the electronic states and broaden their energies. Thus, changing the gate voltage of the sample gradually should gradually populate the Landau levels and therefore result in a continuous rise of the conductivity, in opposition to a step profile. Contrary to this argument, von Klitzing et al. noticed that the quantization condition Eq. (11.30) was more accurately satisfied if the studied sample contained a higher degree of disorder.

The answer of this puzzle lies in the phenomenon of electron localization, as introduced earlier in this chapter. Localization due to disorder turned out to be of key significance for the explanation of the quantum Hall effect. To appreciate this, we consider a transversally confined sample. At high magnetic field strength, one may readily propose a realistic extension of the Landau spectrum, accommodating the potential energy due to the finite width of the sample:

$$E_N = \pm\hbar\omega_c(N + \frac{1}{2}) + V_c(y). \tag{11.32}$$

This relation introduces a *confinement potential* $V_c(y)$, which reflects deviations from ideal lattice periodicity and so alters the harmonic oscillator potential underlying Eq. (4.56). Figure 11.6(a) displays the impact of confinement on the Landau levels, as a function of the finite coordinate coordinate y (compare with the discussion of Landau levels in Section 4.4). The potential rises steeply at the edges of the sample. Panel (b) adds the influence of randomly distributed defects in the system. At sufficiently high field strength, Landau states (4.64) are narrowly centered around a point y_0, justifying the approximation $V_c(y) \approx V_c(y_0)$. Recalling the equation $y_0 = \frac{\hbar}{eB}k_x$, it turns out that the energy levels according to Eq. (11.32) imply a dispersion relation, as illustrated in Figure 11.6(c) for two Landau levels.

From the foregoing discussion and its illustration in Figure 11.6, disorder in the sense of deviation from ideal periodicity is the cause of Landau level broadening, turning the energy levels into energy bands. This conclusion is reiterated in Figure 11.7 in terms of the density of states. As confirmed by numerical simulations [402], Landau states in a disordered system are extended only within small regions around their ideal energy eigenvalues $E_N = \hbar\omega_c(N + \frac{1}{2})$. These regions are separated by *mobility gaps*. States with energies inside the mobility gaps experience localization due to disorder, as indicated in panel (b) of Figure 11.7. Note that the filling factor adopts integer values at the centers of the extended regimes (the energies E_N) and half-odd-integer values at the centers of the mobility gap.

As the states within a mobility gap are filled consecutively, by variation of the magnetic field or the gate voltage, the number of occupied extended states is left constant, and thus, the conductivity or resistivity of the sample remains unchanged. The well-defined conductivity plateaus in the quantum Hall effect are thus a reflection of localization in the energy regime between consecutive Landau state eigenenergies E_N. Longitudinal conductance is restricted to the window of extended states, as reflected by the maxima of $\tilde{\rho}_{xx}$ in Figure 11.5. This behavior is rooted in the emergence of a conducting state that involves the formation of a transient quantum percolation pathway in the bulk, associated with tunneling between magnetic bound states [403].

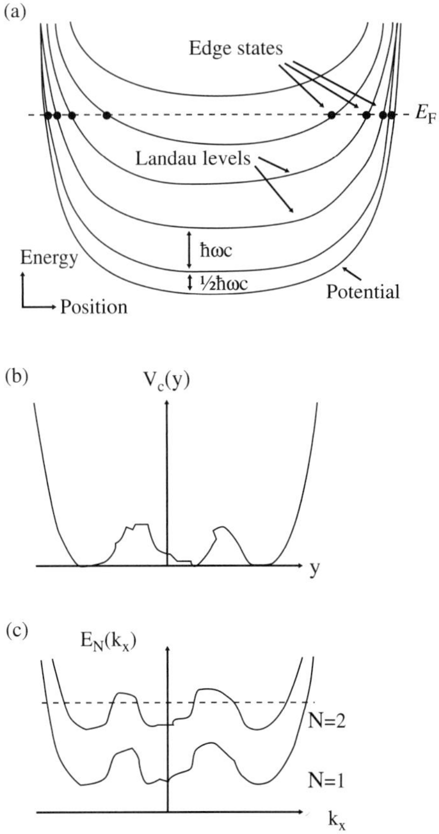

Figure 11.6 Impact of a confinement potential $V_c(y)$ on the Landau levels. (a) Sketch of $V_c(y)$ for a 2DEG system of finite width. Several Landau levels are shown with an added potential energy contribution due to $V_c(y)$ (see Eq. (11.32)). (b) Extension of (a), including the effect of disorder in the sample. (c) Representation of the disperson relation derived from Eq. (11.32). Two Landau levels, augmented by $\tilde{V}_c(k_x) = V_c(y_0)$, are displayed versus k_x.

At first sight, this result is surprising, since mobility gaps are a hallmark of localization in three-dimensional systems, in the precise sense that the localization length L_c diverges at the energies that limit the gap. In two-dimensional systems, on the other hand, all states are localized in case of vanishing or low magnetic field [404].[8] When they are exposed to high magnetic fields, however, their behavior undergoes a crucial qualitative change. To illustrate the impact of a strong magnetic field on the conductivity of a two-dimensional sample, we adopt a model that involves classical cyclotron orbits, as shown in Figure 11.8. As the perpendicular magnetic field imposes circular motion on the charged particles, the

[8] This phenomenon does not preclude situations in which the localization length L_c is large in comparison with the dimensions of the sample. In these cases, conductance does not disappear, as effectively extended charge-carrying states may be identified.

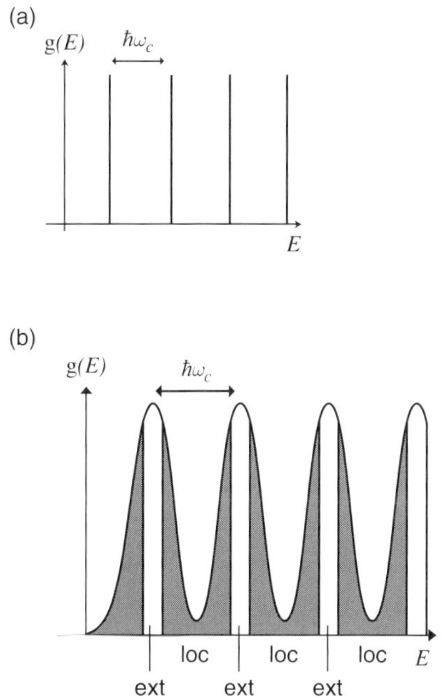

Figure 11.7 Density-of-states distributions for Landau levels in an homogeneous two-dimensional electron gas. (a) The case of an ideally clean sample: the Landau levels (4.56) are represented by sharp lines. (b) The case of a physical sample, containing deficiencies: the Landau levels are broadened by disorder. Localized states are distinguished from extended states. Landau states become localized within *mobility gaps*, separating the zones around the Landau level energies where the extended states reside.

conductance due to the bulk states of the system reduces to zero. Charge carriers located at the edge of the sample, in contrast, perform a significantly different type of motion than bulk particles. Unable to complete a circular orbit, they bounce back from the boundary and continue, under the influence of the magnetic field, to move in segments of circles, as illustrated in Figure 11.8(a). This classical image presents a qualitatively valid model of *finite edge currents* as a result of dimensional reduction, yielding partial restoration of conductance. The scheme of a multiterminal device to study the quantum Hall effect is shown in Figure 11.8(b). The edge currents that account for the conductance of this system are indicated.

The quantized conductivity of 2DEG Hall samples is thus attributed to edge states that are extended at all energies. Each Landau level adds another quantum of conductivity, due to the occupation of a new extended edge state. We point out that this feature is substantiated by the Landau energy bands resulting from a confining potential, as shown in Figure 11.6(c). In particular, the drift velocities at the edges of the sample are found by evaluating the derivatives of the Landau bands $E_n(k_x)$, according to

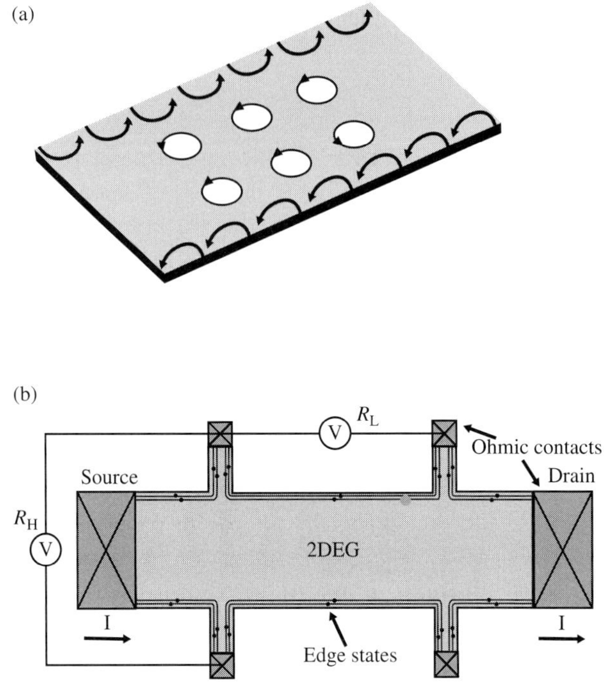

Figure 11.8 Emergence of edge currents in a two-dimensional Hall system. (a) Rectangular sample with charge carriers performing cyclotron motion under the influence of a perpendicular magnetic field. Reflection from the boundary makes the particles travel along the edges of the sample in consecutive jumps. This gives rise to a chiral edge current with well-defined direction of motion and well-defined sense of rotation [405](b) Model of a *Hall bar*. The edge currents are indicated [406].

$$v_x = \frac{1}{\hbar}\frac{\partial E_N(k_x)}{\partial k_x}|_{E=E_F} = \frac{1}{eB}\frac{\partial V_c(y)}{\partial y}|_{E=E_F, y=y_0}. \tag{11.33}$$

When taken together with Figure 11.6(c), this identity states that the currents associated with each Landau band are determined by the two points where the band intersects the Fermi energy. Opposite edges correspond to opposite directions of electron motion. As the direction of the current at a given edge is preserved, backscattering of electrons in edge states is suppressed. The only manifestation of scattering events for these electrons is a phase shift due to forward scattering [407].[9] The absence of backscattering at Hall sample edges is rooted in the *chiral property* of the edge currents that move under the influence of the magnetic field in

[9] Backscattering of charge carriers at the edges of Hall samples may be induced by tunneling between the edges. Such an event, however, is only probable in the extreme case of microscopic separation between the edges.

either a uniquely clockwise or uniquely counterclockwise manner ([408], compare with Figure 11.8(a)).

11.5 Quantum Hall Effects in Graphene

Here we give a brief survey of the *anomalous quantum Hall effect* (AQHE) in graphene (Section 11.5.1) and focus on its consequences on the magnetotransport properties of the graphene sheet. This outline is followed by a sketch of the *spin quantum Hall effect* (SQHE) in graphene (Section 11.5.2).

11.5.1 The Anomalous Quantum Hall Effect in Graphene

As we go from the regular two-dimensional electron gas to graphene, we pass from the integer to the anomalous quantum Hall effect. Once more, we focus on the massless Fermion region in the proximity of the Fermi energy of graphene. This regime is dictated by the Dirac spectrum Eq. (4.55), rather than the harmonic oscillator spectrum. In what follows, we survey the salient features of the quantum Hall effect in graphene and mention related experimental results where available.

Extending the formula for the discrete conductivity (11.30) in the integer quantum Hall effect to the low-energy regime of graphene, one may conjecture that the correct expression is

$$\tilde{\sigma}_{xy} = g_s g_v N \frac{e^2}{h}, \tag{11.34}$$

where g_s (= 2) and g_v (= 2) stand for the degree of degeneracy due to the spin and valley degrees of freedom, and $N = 0, \pm 1, \pm 2, \ldots$ Negative values are admitted to accommmodate hole conduction. Formula (11.34), however, does not include the correct contribution of the Landau level with $E = 0$ to the conductivity. As is evident from the spectrum (4.55) and the related commentary, only one pseudospin projection contributes to the zero level, while both projections are equally represented in other levels. Thus, in the case of $|N| = 0$, the degeneracy is only half of what one finds for $|N| > 0$. Adjusting expression (11.34) accordingly, we arrive at

$$\tilde{\sigma}_{xy} = g_s g_v (N + \frac{1}{2}) \frac{e^2}{h} = g_p g_v (2N + 1) \frac{e^2}{2h}, \tag{11.35}$$

and $N = 0, \pm 1, \pm 2, \ldots$ The conductivity is thus determined by a sequence of half-odd-integer indices. This is an anomalous condition as compared with the regular quantum Hall effect, as observed in two-dimensional electron gases and encoded by formula (11.30). The anomaly of the *anomalous quantum Hall effect* found in graphene reflects the hybrid nature of the Landau level with $E = 0$, which is equally filled with both types of charge carriers, containing only half as many

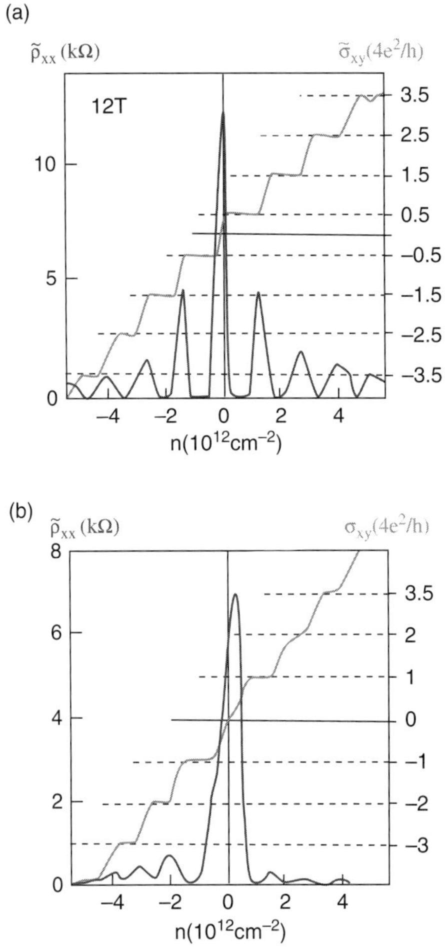

Figure 11.9 Hall conductivity (step profile) and resistivity (oscillatory profile) in graphene, demonstrating the anomalous quantum Hall effect for monolayer graphene (a). While monolayer graphene exhibits conductivity plateaus at half-odd-integer multiples of $\frac{4e^2}{h}$, the plateaus of bilayer graphene are found at integer multiples (b). In both cases, no plateau exists at the Fermi level. Reprinted with permission from [64] (a) and [410] (b). Copyright (2005) and (2006) by Nature Publishing Group.

electrons (holes) as any other Landau level with $E > 0$ ($E < 0$). The validity of the quantization rule (11.35) for graphene has been verified in several experiments [64, 409]. The data represented by the step profile in Figure 11.9 clearly exhibit the characteristic sequence of $\tilde{\sigma}_{xy}$ plateaus [64]. The oscillatory distribution denotes the longitudinal resistivity $\tilde{\rho}_{xx}$, as given by Eq. (11.26). Both are plotted versus the charge carrier concentration in graphene.

The sequence of half-odd-integer indices that governs expression (11.35) may be viewed as a direct consequence of the geometric phase associated with the pseudospinor states in the Dirac regime of graphene (compare with Eqs. (4.26), (4.28)). In more detail, the conductivity as obtained from the Kubo formula at $T = 0$, in the DC limit, and treating the electric field as a perturbation, reads [75]

$$\tilde{\sigma}_{xy} = \frac{e^2}{2\pi h}\Im \sum_m \oint d\mathbf{k}\langle m|\nabla_{\mathbf{k}}|m\rangle. \tag{11.36}$$

The sum extends over all occupied Landau levels and includes the valley and sublattice indices.[10] The Hall conductivity is thus directly related to the geometric phase (4.29). Each contributing state adds the topologically non-trivial phase $g_s g_v \pi$. Thus, taking into account the difference in the sublattice degeneracy of the level with $N = 0$ and the remaining levels yields once more the conductivity formula (11.35).

The $N = 0$ level of *bilayer graphene*, on the other hand, contains twice as many states as that of monolayer graphene. Consequently, an integer quantum Hall effect is observed, as shown in Figure 11.9(b). This is in accordance with the single-valued nature of the massive Dirac Fermions found in bilayer graphene (see Eq. (4.40)). Still, the conductivity quantization in this medium does not conform with the conventional integer quantum Hall effect, as described in the previous section. Thus, there is no plateau at the zero level corresponding to metallic bulk properties for the $N = 0$ state, in marked contrast to the behavior of regular 2DEG systems [410].

For both systems, the relative position of the Fermi levels E_F and the Landau levels E_N can be controlled by the gate voltage, which, in turn, determines the charge carrier density, as, e.g. displayed in Figure 11.9. An alternative way of shifting or manipulating the difference between the Fermi level and a given Landau level is by varying the magnetic field. From formula (4.55), in conjunction with the definition of the Dirac frequency, $\omega_D = v_F\sqrt{\frac{2eB}{\hbar}}$, enhancing the magnetic field strength B increases the Landau level energies while widening the differences between adjacent levels. Steadily increasing B will therefore cause the sequential passage of Landau levels through E_F and thus give rise to marked oscillations in the longitudinal conductivity, so-called *Shubnikov–de Haas oscillations* [411].

Exercise 11.2 From equating E_F with E_N, conclude that $\tilde{\sigma}_{xx}$ is periodic in $1/B$ (ignoring, for simplicity, the confinement potential).

[10] The reader will find further commentary on the Kubo formula at http://faculty.etsu.edu/HAGELBER/comp_chem.html.

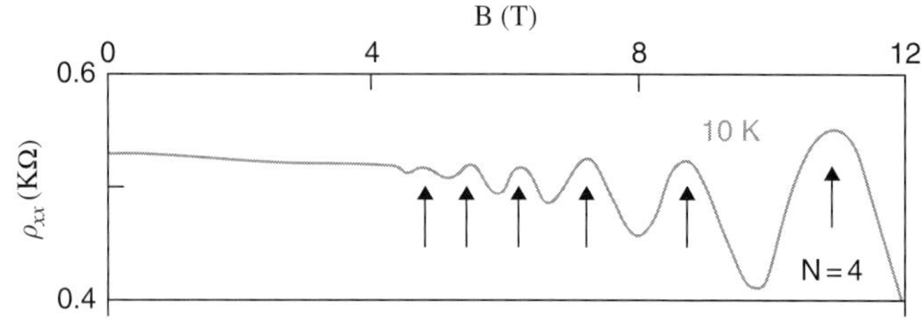

Figure 11.10 Longitudinal resistivity of graphene as a function of the magnetic field, observed at constant gate voltage ($V_g = -60$ V, [64]). Shubnikov–de Haas oscillations are clearly visible. In graphene, only one set of resistivity oscillations is observed for both charge carrier types. Reprinted with permission from [64]. Copyright (2005) by Nature Publishing Group.

For illustration of this effect, we add measured Shubnikov–de Haas oscillation data in Figure 11.10, which shows the longitudinal resistivity ρ_{xx} of graphene as a function of the magnetic field [64]. Due to the $1/B$ periodicity, the difference between adjacent maxima increases with field strength. Note that an alternative procedure to generate these oscillations is to steadily raise the gate voltage. In this mode of representation, the resistivity maxima are evenly spaced, as displayed in Figure 11.10.

By direct extension, periodicity with the reciprocal magnetic field is found also for the electron density of states at the Fermi energy:

$$g(E, B) = g_p g_v g \sum_n \delta(E_F - E_N) = g_p g_v g \sum_n \delta(E_F \mp v_F \sqrt{2eB\hbar} \sqrt{N + \frac{1}{2} \pm \frac{1}{2}}). \tag{11.37}$$

Thus, all quantities that depend on the density of states, such as the pressure, the entropy, or the Helmholtz free energy, change periodically as the Landau levels pass consecutively through E_F. We point out that the Fermi energy itself exhibits the same periodic behavior. As a Landau level rises above the Fermi threshold, it depopulates. The excess electrons occupy the next lower level that is allowed since, by (4.68), the degree of degeneracy of any Landau level increases linearly with the magnetic field. As a consequence, the Fermi energy moves to the next lower level, and the process continues as the magnetic field continues to increase.

Shubknikov–de Haas oscillations are affected by the electron spin, which has not been taken into account in the preceding discussion. This element can be readily included by augmenting the energy expression (4.55) by the Zeeman term, $E_{s,m_s} = \frac{1}{2} g_s m_s B$.

Investigating the quantum Hall effect in samples of high electron mobility and small effective electron mass led to the discovery of the *fractional quantum Hall effect* [412]. Quantum Hall conductance plateaus at $q\frac{e^2}{h}$ were detected with great accuracy where the factor q was found to be a simple fraction, such as $\frac{2}{3}, \frac{1}{5}, \frac{2}{5}, \frac{3}{5}, \frac{4}{5}, \ldots$ This phenomenon is understood as a signature of a condensation of the electronic system into a liquid state. Many-body processes, i.e. interelectronic Coulomb repulsion, play a crucial role in generating this state. The existence of the fractional quantum Hall effect in graphene had been predicted [413] prior to its first experimental observation in 2009, made possible by the availability of sufficiently pure suspended graphene samples [414]. The mechanisms causing the fractional quantum Hall effect in graphene continue to be explored by both experiment and theory.

11.5.2 The Quantum Spin Hall Effect in Graphene

As the charge Hall effect leads to a separation between positive and negative charge in the direction perpendicular to that of the current, the spin Hall effect induces an analogous separation between the up and down spins of the charge carriers. Spin selectivity is already achieved by the anomalous Hall effect [415], as encountered in ferromagnets. Here, the magnetization due to spin polarization causes a spin-dependent deflection of charge carriers, as shown in panel (b) of Figure 11.11. This mechanism gives rise to a Hall voltage, as well as spin accumulation.

Figure 11.11(c) illustrates the pure spin Hall effect, which involves spin accumulution, but no Hall voltage. Arguments for the existence of a quantum spin Hall effect in graphene were first presented by Kane and Mele [8]. Establishing this effect requires demonstration that $\tilde{\sigma}_{xy}$ assumes opposite signs for opposite spin orientation of the charge carriers in the graphene lattice.

Qualitatively, one might understand the quantum spin Hall effect as coincidence of two quantum Hall effects for charge carriers with opposite spins [416]. This implies that, once more, the disorder in the sample is at the root of the phenomenon, and currents are carried by edge channels. The quantum spin Hall effect, however,

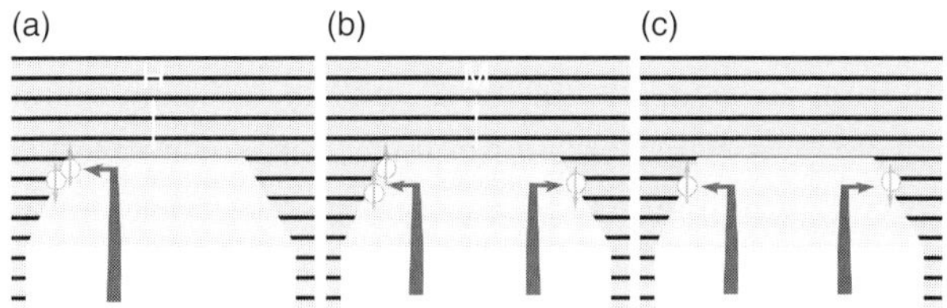

Figure 11.11 Variants of the Hall effect. (a) The regular Hall effect: deflection of charge carriers without spin accumulation, giving rise to a Hall voltage. (b) The anomalous Hall effect: spins separate, while a Hall voltage emerges. (c) The pure quantum spin Hall effect: no Hall voltage, but spin separation. From [417]. Reprinted with permission from AAAS.

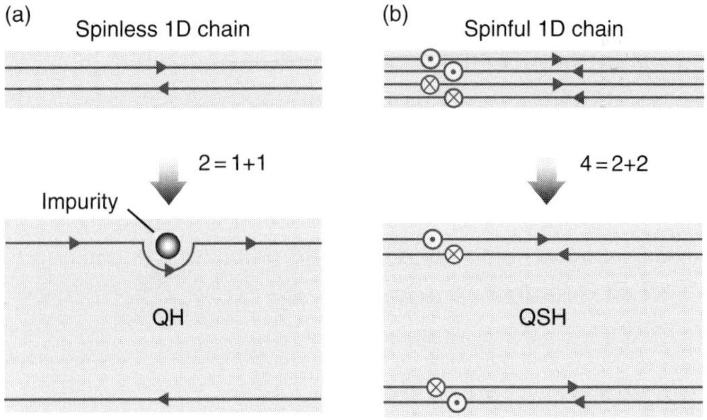

Figure 11.12 Edge channels in the quantum Hall effect (panel (a)) and in the spin quantum Hall effect (panel(b)) [416]. In both cases, spatial separation into edge states with opposite helicity takes place. For the quantum spin Hall effect, however, two states with opposite spin orientation are located at each edge. These states are helical, as the spin orientation of the charge carriers that occupy them correlates with their direction of motion. The helicity switches as one goes from one edge to the other. Spin-up (spin-down) orientation is represented by a dot (a cross) in a circle. Copyright 2011 Annual Reviews.

is characterized by two currents of opposite helicity at opposite edges, correlated with the two spin orientations. The corresponding edge states are termed *helical*, referring to a rigid connection between the direction of motion and the spin orientation. The resulting physical scenario may be viewed as a superposition of two counterpropagating currents with opposing spin polarizations. A comparison between the charge and the spin quantum Hall effects in terms of edge currents is presented in Figure 11.12.

From panel (b) of Figure 11.12, charge carriers in quantum spin Hall systems may move forward or backward at each of the two edges, while one of the two directions is prescribed in quantum charge Hall systems. The consequence of the added degree of freedom in the former case is the emergence of a *pure spin current* at the edges of quantum spin Hall samples. With reference to the right panel of Figure 11.12, the time-reversal operation, $\hat{T}$, causes the simultaneous interchange between the forward and backward directions of the indicated trajectories, and between the two spin orientations. In contrast to the model shown in panel (a), that of panel (b) is thus invariant with respect to $\hat{T}$.

The presence of both directions of motion, forward and backward, at the edges of quantum spin Hall systems seems to allow for backscattering events in these edge channels. In fact, these events are strongly suppressed. This feature is due to mutual cancellation of the two alternative backscattering paths allowed under the constraint of time-reversal invariance.[11] A magnetic impurity that would generate an additional phase shift and thus give rise to backscattering, on the other hand, violates the time reversal symmetry of quantum spin Hall systems. So does an external magnetic field that is an accessory of the quantum charge Hall effects, both the regular (integer) and the anomalous (half-odd-integer), but is not present in their spin counterpart.[12]

In terms of the conductivities as defined by the response equation (11.24): as $\mathbf{j}$ changes its sign under time reversal, but $\mathbf{E}$ does not, the conductivity tensor $\tilde{\boldsymbol{\sigma}}$ must be odd with respect to $\hat{T}$. The equivalent relation for the spin current reads

$$\mathbf{j}_s = \tilde{\boldsymbol{\sigma}}_s \mathbf{E}. \tag{11.38}$$

Since $\hat{T}$ leaves $\mathbf{j}_s$ invariant, the same holds for $\tilde{\boldsymbol{\sigma}}_s$.

We emphasized that neither the interaction with a magnetic impurity nor with an external magnetic field is compatible with the invariance of the quantum spin Hall effect under time reversal. If this variant of the Hall effect proceeds in the absence of a magnetic field, what is its underlying mechanism? Once more, symmetry considerations haven proven to be essential in addressing this question. As pointed out in Chapter 8, the Hamiltonian of the spin-orbit interaction in two-dimensional graphene (see Eq. (8.8)), namely

$$\hat{\mathbf{H}}_{SO} = \xi_I \boldsymbol{\eta}_z \boldsymbol{\tau}_z \boldsymbol{\sigma}_z, \tag{11.39}$$

is time-reversal invariant. This suggests an essential role for spin-orbit coupling in generating the quantum spin Hall effect. The Hamiltonian (11.39) may be obtained from a tight-binding model that confines hopping to equal sublattice sites (see Section 8.1) and correlates spin orientation with sense of rotation, therefore providing a physical process for the separation of up- and down-spin currents, as characteristic for the quantum spin Hall effect.

Kane and Mele [8] thus considered a model that augments the Dirac Hamiltonian for the low-energy regime of graphene by the spin-orbit operator (11.39). As implied by the form of $\hat{\mathbf{H}}_{SO}$, the intrinsic spin-orbit interaction gives rise to spin gaps with opposite signs at the two Dirac points, K and K', i.e. the spin polarization of states above and below the gap switches as one goes from one Dirac point to the other. An electric field will thus induce Hall currents flowing in opposite directions for opposite spins, resulting in a pure spin current.

[11] The allowed backscattering paths surround the impurity in the clockwise or the counterclockwise sense, resulting in a geometric phase difference of π ([416], see Section 11.3).

[12] Haldane [418] demonstrated that the quantum Hall effect, as described in Section 11.4, does not require the presence of a net magnetic field, i.e. it may involve a periodic field that generates an overall vanishing magnetic flux. The spinless model considered by Haldane breaks time-reversal symmetry. This symmetry is restored in spin Hall systems, which may be viewed as realizations of two Haldane models in coincidence, for electrons with opposite spin orientations.

For clarification, we write down the full Hamiltonian $\hat{\mathbf{H}} = \hat{\mathbf{H}}_0 + \hat{\mathbf{H}}_{SO}$, with $\hat{\mathbf{H}}_0$ as Dirac Hamiltonian at the K or the K' point, according to Eq. (4.22) or (4.27), respectively, for opposite spin orientations:

$$\begin{aligned} \hat{\mathbf{H}}_{\pm}(\mathbf{k}) &= \hbar v_F (k_x \eta_x + \pm k_y \eta_y) \pm \xi_I \eta_z \text{ , if } S_z = +1, \\ \hat{\mathbf{H}}_{\pm}(\mathbf{k}) &= \hbar v_F (k_x \eta_x + \pm k_y \eta_y) \mp \xi_I \eta_z \text{ , if } S_z = -1. \end{aligned} \tag{11.40}$$

Here subscript +(−) on the left-hand sides of these equations refers to the K (the K') point. The spin-orbit interaction that accounts for the difference between the cases $S_z = +1$ and $S_z = -1$ adds an effective mass term to the Dirac Hamiltonian. Making admission for a magnetic field, one may determine the eigenstates and eigenvalues of $\hat{\mathbf{H}}_{\pm}(\mathbf{k})$ for both spin orientations in terms of Landau states and Landau levels, taking ultimately the limit of vanishing field ($\mathbf{B} \to 0$) [419]. Applying the Kubo formula (11.36) to determine the Hall conductivity yields for both valleys $\tilde{\sigma}_{xy}^{\uparrow\downarrow} = \pm\frac{e^2}{h}$. This result is compatible with the constraint of invariance with respect to time reversal since the overall Hall conductivity is zero:

$$\tilde{\sigma}_{xy} = \tilde{\sigma}_{xy}^{\uparrow} + \tilde{\sigma}_{xy}^{\downarrow} = \frac{e^2}{h} - \frac{e^2}{h} = 0. \tag{11.41}$$

The *spin current density* is found from the charge current density for the two spin components, $\mathbf{j}^{\uparrow\downarrow}$ through

$$\mathbf{j}_s = \frac{\hbar}{2e}(\mathbf{j}^{\uparrow} - \mathbf{j}^{\downarrow}). \tag{11.42}$$

Combining Eqs. (11.41) and (11.42), we obtain a non-vanishing *spin Hall conductivity*:

$$\tilde{\sigma}_{xy}^{s} = \frac{\hbar}{2e}(\tilde{\sigma}_{xy}^{\uparrow} - \tilde{\sigma}_{xy}^{\downarrow}) = \frac{e}{2\pi}. \tag{11.43}$$

As Kane and Mele [8] showed, making reference to an earlier model by Haldane [418], the spin-orbit operator (11.39) may be understood as a low-energy approximation to a tight-binding Hamiltonian analogous to that given by Eq. (4.5). Hops between sites are restricted to next-nearest instead of nearest neighbors. Kane and Mele applied this model to a finite strip of graphene, specifically to a graphene nanoribbon of the zigzag type (zGNR). Figure 11.13 shows the band structure derived from the model as a function of the periodic $\mathbf{k}$ space coordinate, k_x. Comparing with the zGNR band structure displayed in Figure 7.1 where spin-orbit coupling has been neglected, we find a distinct gap opening at the one-dimensional projections of the Dirac points. These band gaps appear in the bulk of the analyzed nanoribbon. The flat band that accounts for the metallic character of zGNRs within the framework of the tight-binding approximation is deformed into two bands that cross the gap at the Brillouin zone boundary ($k_x = \frac{\pi}{a}$), connecting the K and the

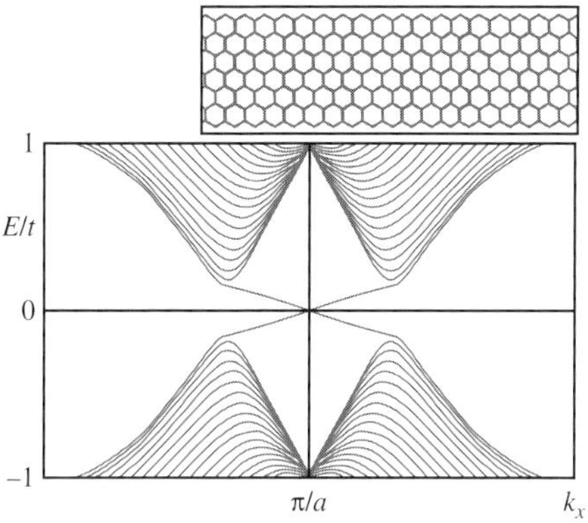

Figure 11.13 Band structure of a zGNR with spin-orbit interaction, as obtained from tight-binding simulation [8]. Spin-orbit coupling induces a gap at the one-dimensional projections of the Dirac points K and K'. These points are connected by two bands of opposite spin orientation, each one occupied with two degenerate spin-polarized edge states, localized at opposite edges of the ribbon. The energy is measured in terms of the tight-binding interaction parameter t. The parameter a stands for the lattice constant. Reprinted with permission from [8]. Copyright (2005) by the American Physical Society.

K' point. These bands turn out to be spatially confined to the edges of the nanoribbon. Further, each band is *spin filtered*, having a unique spin orientation, and each band contains two degenerate spin states, localized at the two edges of the ribbon. Note that the two crossing bands differ from each other with respect to the slope at the crossing point, correlating, by Eq. (11.33), the spin orientation with the crystal momentum.

The quantum Hall effect gives an example for a *topological state of matter*, characterized by an insulating bulk state in conjunction with boundary states that are robust with respect to disorder. These states are characterized by *topological invariants*. In the case of the integer quantum Hall effect, this invariant is provided by the index N that governs the sequence of $\tilde{\sigma}_{xy}$ plateaus (see Eq. (11.35)).[13] Being a discrete quantity related to the topology of the system, this index, and consequently the edge current that it describes, is stable with respect to small perturbations. Further, it does not depend on the geometric shape of the edge, remaining unaffected by edge deformations.

Quantum spin Hall systems are classified as topological insulators, systems that exhibit quantized transport properties that are rooted in topological characteristics of their electronic structure [404, 421]. While for quantum Hall systems an

[13] In terms of topology, the index N is a *Chern number* [420]. It is obtained by integrating over the *local adiabatic curvature* of the ground state solution evaluated on the first Brillouin zone, which has the topology of a torus.

external magnetic field is instrumental for the separation between insulating bulk and conducting edge states, quantum spin Hall systems display an analogous effect without any need for such a field. The process that causes the contrasting features of the bulk and the surface is here internal rather than external, given by spin-orbit coupling instead of interaction with a magnetic field. The physical manifestation of this mechanism is transport of charges with well-defined helicity, corresponding to the emergence of two charge transport channels at the surface of the sample or, in the two-dimensional case, its edge, involving charge carriers with opposite directions of motion and opposite spin orientations. Helicity may here be understood as a *symmetry-protected* property, in accordance with the even behavior of quantum spin Hall systems under the time-reversal operation. As a consequence, backscattering in either channel is excluded (see Figure 11.12(a)), since this process would alter the preserved helicity. As in the case of the quantum Hall effect, where left-moving and righ-moving electrons are confined to two distinct channels, the relevant degrees of freedom are spatially separated from each other. Specifically, states of opposite helicity are relegated to opposite edges of the sample. Helicity conservation correlates reversal of momentum with spin reversal, implying that an electron scattered backward needs to flip its spin. Thus, symmetry protection acts as a mechanism that suppresses backscattering and therefore leads, within the framework of the model defined by the spin-orbit Hamiltonian (11.39) to dissipation-free edge currents. Extending the model (11.39) to weak interelectronic many-body interactions, however, will import a term that does not violate this requirement, yet allows for inelastic backscattering [8].

Making allowance for a Rashba term (see Eq. (8.16)) in the Hamiltonian, Kane and Mele demonstrated the persistence of the quantum spin Hall phase even in situations where space reflection symmetry is broken [422]. These perturbations were shown to preserve gapless states of the type shown in Figure 11.13 for Rashba interactions not too large as compared with spin-orbit coupling. For the quantum spin Hall effect, just as for its charge counterpart, the tenacity of gapless edge or surface states is associated with a topological invariant [416, 422]. This invariant differs from the index N (11.43) which, after all, measures the charge conductivity and thus a quantity that reduces to zero in spin Hall systems. The logic of spin calls for a fundamentally different classification scheme [404, 421]. In qualitative terms: topological invariants that specify quantum spin Hall states belong to the Z_2 *group* that contains two elements only, 0 and 1 ($\nu = 0, 1$). These elements pertain to a simple alternative, related to the notion that an even number of helical edge states may undergo pairwise annihilation, leaving the system in a regular, i.e. topologically trivial insulating state ($\nu = 0$). If, in contrast, initially an odd number of edge states is present, one will remain and preserve the topologically non-trivial character of the system $\nu = 1$.

We emphasize that the quantum spin Hall effect in pristine graphene is, due to the small size of its spin-orbit gap, of systematic rather than practical relevance. Admitting hydrogen or fluorine adatoms, however, one can increase the magnititude of spin-orbit coupling in graphene and, by implication, the spin Hall effect, substantially, as is outlined in Section 12.7. While not being of direct relevance to applications, the analysis of this effect in pristine graphene by Kane, Mele and other authors [8, 421, 422] has greatly advanced the theory of topological insulators. Here, as in many other cases, graphene has been seen to provide a supreme laboratory for exploring basic phenomena in condensed-matter physics.

Part V

Composite Materials

12

Impurities

This chapter deals with magnetic composites consisting of carbon nanostructures and impurity species. The host systems included here are the prototypes fullerene, graphene and carbon nanotubes. In each case discussed, emphasis will be placed on the mechanisms by which the electronic ground state of the aggregate adopts magnetic properties. In many cases, the considered carbon allotrope provides a framework that preserves the magnetism of the guest species, as paradigmatically realized by some metallofullerenes with enclosed lanthanide atoms or clusters. In some systems, however, magnetism is not imported by the guest species but evolves as the carbon allotrope interacts with the externally added moieties. This situation is well exemplified by the much-studied compound tetrakis-dimethylamino-ethylene-C_{60} (TDAE-C_{60}).

In sections 12.1 and 12.2, we refer first to magnetic metallofullerenes, arguably the most traditional among the structures included here. After all, research interest in fullerenes with enclosed metal components emerged soon after the discovery of C_{60} [139]. From metal atoms or metal atom clusters as guest species inside fullerene cages, we turn to single group V atoms as encapsulated components. Systems of the form A@C_{60}, with A = N, P, have proven to be efficient in preserving the spin of the enclosed atom, which makes them interesting as physical realizations of qubits in quantum computing. The following Section (12.3) discusses a variety of magnetic phases that originate from electron transfer to fullerenes. This mechanism is operative in very diverse composites, ranging from fullerides to hybrids of fullerenes and organic molecules.

The remainder of this chapter deals with compounds involving graphene and nanotubes. Specifically, graphene is considered as substrate of two adatom types, hydrogen and fluorine, and computational as well as experimental findings on the magnetic phases of hydrogenated and fluorinated graphene are surveyed. The final section of this chapter summarizes various results on carbon nanotubes in combination with magnetic metal components, ranging from atoms to nanoparticles.

12.1 Magnetic Metallofullerenes

The magnetism of endohedral metallofullerenes with enclosed paramagnetic guest species arises from the interplay of various effects. Thus, the intrinsic magnetic moment of the enclosed metal component may affect the magnetic moment of the unit as a whole. Further, this component may be in a cationic state, as a consequence of electron transfer from the encapsulated atom or cluster to the fullerene enclosure. In cases where the effective number of transferred electrons is odd, as in Er(I)@C_{82} or Er(III)@C_{82}, this process confers paramagnetism to the fullerene cage. Whenever the enclosed species is a metallic cluster, the interaction between its atomic constituents will give rise to an overall magnetic moment that, in general, is of major influence on the magnetic properties of the endohedral complex.

Early experimental as well as theoretical studies on magnetic endohedral metallofullerenes focused on systems of the form Ln@C_{82}, where Ln stands for a lanthanide atom (see Section 6.2.1). Pioneering experiments on the species with Ln = La [423] and Gd [424] involved SQUID measurements on powder samples in the temperature range $3\text{K} \leq T \leq 300\text{K}$. Numerous projects involved the species Gd@C_{82} which we will discuss in the following as an exemplary system for endohedral metallofullerenes with encapsulated magnetic metal atoms.

For Gd@C_{82}, a Weiss temperature of -0.65 K was found, in conjunction with an effective magnetic moment of 6.90 μ_B. Further studies were directed at the magnetization of the species as a function of the ratio between the external magnetic field and the absolute temperature, B/T. The obtained universal curve was fitted with a Brillouin function, defined by the parameters $J = 3.38$ and $g = 2.0$, resulting in an effective magnetic moment of $\mu_{eff} = g\sqrt{J(J+1)}\mu_B = 7.70\mu_B$. A later study [425], once more involving SQUID measurements, narrowed the gap of the magnetic moments derived from Curie–Weiss analysis (μ_{eff} = 6.66 μ_B at T >40 K) and from Brillouin function fitting (μ_{eff} = 6.74 μ_B). In these and other experiments [426], the effective magnetic moment of Gd@C_{82} was determined to be somewhat smaller than that of the free Gd^{3+} ion, for which μ_{eff} is found to be 7.94 μ_B, as $J = S = \frac{7}{2}$, and $g = 2.0$. The reason why embedding Gd^{3+} into the C_{82}^{3-} cage yields a slight reduction of μ_{eff} lies in the magnetic interaction between the cage and the core. In particular, Electron spin resonance (ESR) measurements of Gd@C_{82} in solution [427] shed light on the origin of this decrease. Gd@C_{82} was investigated in two solutions, namely CS_2 and trichlorobenzene (TCB). In the former environment, the metallofullerene exists in a dimerized state, in the latter, however, as a monomer. Dimerization of the cages quenches the cage spin of $S = \frac{1}{2}$. ESR spectra recorded for Gd@C_{82} in CS_2 revealed an enhanced magnetic moment, compatible with that of Gd^{3+}. A smaller value for μ_{eff} was extracted from the spectra obtained in TCB solution, in agreement with the assumption of a total spin $S = 3$. This assignment is in accordance with antiferromagnetic coupling between the metal core and the fullerene shell. The ferromagnetic alternative, corresponding to $S = 4$, was found at a slightly higher energy, with an energy difference of 1.79 meV between the two configurations.

12.1.1 Tri-metallic Nitride Fullerenes

Interest in the magnetism of metallofullerenes with embedded metal clusters is motivated chiefly by the prospect of fabricating fullerene-based species with exceptionally high magnetic moments, surpassing those of monometallofullerenes. Small clusters of transition metal atoms possess magnetic moments often substantially exceeding those of isolated atoms [428, 429]. Thus, electron spin resonance (ESR) research yielded evidence for a high spin ground state of the diatomic molecule Gd_2 [428]. The magnetic moment per atom was found to be 8.82 μ_B, substantially exceeding that of the Gd atom in the gas phase (6.53 μ_B), and the coupling between the two Gd atoms turned out to be ferromagnetic. Containing metal clusters with high magnetic moments in fullerene enclosures is of particular relevance for the design of novel contrast agents to be used in magnetic resonance imaging (MRI), as will be discussed later in fuller detail (see Chapter 14). In this context, various experimental as well as computational efforts have focused on trimetallic nitride fullerenes whose basic features were outlined in Section 6.2.1.

In a pioneering study from 2005, Wolf et al. determined the magnetization of Me_3NC_{80} clusters by SQUID measurements (see Section 3.5), with Me = Tb, Ho, as a function of the temperature [430]. From Brillouin function fitting, the authors determined the total angular momentum quantum number as well as the magnetic moment for both materials, resulting in $J = 12$, $\mu = 18\mu_B$ and $J = 16$, $\mu = 20\mu_B$ for Me = Tb and Ho, respectively. The measured magnetization for Tb_3NC_{80}, normalized to the saturation value, is shown in Figure 12.1 versus the ratio of the applied magnetic field and the temperature, along with the fitted Brillouin curve. The Tb atom g-factor assumed when fitting the curve was that of the free Tb^{3+} cation, $g = 1.50$, and an analogous treatment was applied to Ho_3NC_{80} ($g(Ho^{3+}) = 1.25$). This assumption is based on the charge transfer model that assigns a formal charge of $-6e$ to the C_{80} cage, and of $-3e$ to the central N atom, such that the distribution of charges in the tri-nitride fullerene complex may be described by the formula $(Me^{+3})_3\ N^{3-}C_{80}^{6-}$. Absorption of six electrons by the cage is plausible in view of the icosahedral symmetry of C_{80}. Imposing I_h symmetry on the neutral fullerene gives rise to a fourfold degenerate HOMO that is, however, occupied only by two electrons. Thus, it does not come as a surprise that C_{80} (I_h) is the least stable among the seven C_{80} isomers that obey the isolated pentagon rule [431]. Donation

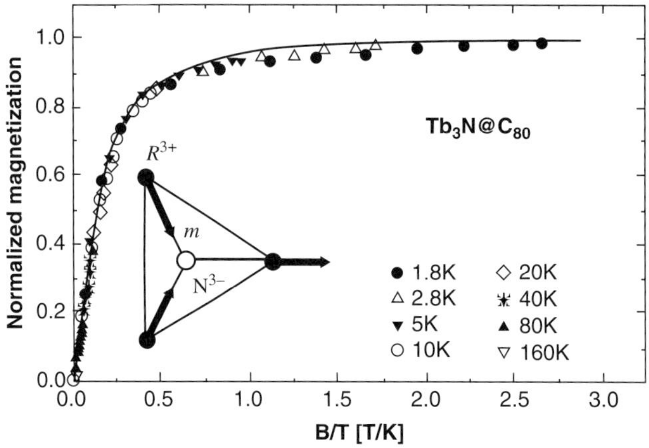

Figure 12.1 Magnetization of Tb_3NC_{80} as a function of the ratio between the external magnetic field and the absolute temperature. The experimental data were obtained by SQUID measurement. The magnetization is normalized to its saturation value. The solid line denotes the fitted Brillouin function. The inset shows a possible configuration of the magnetic moments associated with the three Tb atoms. The value for the total magnetic moment, as extracted from the Brillouin function, is incompatible with a collinear arrangement. Reprinted from [430] with permission from Elsevier.

of six electrons by the encapsulated metal atoms to the cage, however, yields a C_{80}^{6-} (I_h) closed shell configuration. C_{80} as enclosure for tri-metallic nitride clusters therefore prefers icosahedral symmetry.

The charge transfer model is of only approximate validity. A more adequate description includes hybridization effects between the sixfold charged core and cage ions [432, 433, 434].

The total magnetic moments for Me_3NC_{80}, Me = Tb, Ho, derived from the respective Brillouin functions, exclude collinear alignment of the three individual moments. The latter would be $\mu = 27\mu_B$ ($\mu = 30\mu_B$) for Me = Tb (Me = Ho), to be compared with the actual experimental findings $\mu = 18\mu_B$ ($\mu = 20\mu_B$), as indicated earlier. The authors resolved this discrepancy by proposing that the orientation of the atomic moments is determined by Me-N ligand fields, as shown in the inset of Figure 12.1. This model rationalizes the measured values.

Much experimental [435, 436] and computational [437, 432] work has been devoted to the trigadolinium unit Gd_3NC_{80}, motivated by potential applications of this species (Chapter 14) as well as by the intrinsic interest in Gd as an element with a high magnetic moment. Specifically, the magnetic moment of Gd^{3+} is carried solely by the spins of the $4f$ subshell, without any orbital admixtures, making Gd-based clusters enclosed in fullerene cages attractive as a testing ground for models of metallofullerene magnetism.

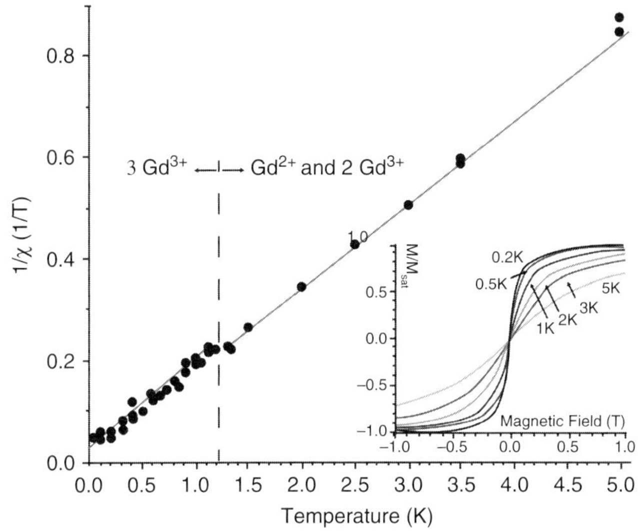

Figure 12.2 Results of SQUID experiments on Gd_3NC_{80} [435], assisted by modeling, using a Heisenberg Hamiltonian. The displayed Curie–Weiss plots show an almost linear relation between the inverse of the susceptibility and the temperature. A disruption of the linear profile at $T = 1.2K$, however, is indicative of a spin transition. The inset shows the normalized magnetization against the magnetic field, for various temperatures in the interval 0.2–5.0 K. The susceptibility is normalized to the saturation magnetization. Reprinted with permission from [435]. Copyright (2011) by the American Institute of Physics.

SQUID measurements on small particles containing a number of about 10^{15} $Gd_3N@C_{80}$ clusters were performed in the temperature interval 0.1–5.0 K [435]. The absence of any hysteresis indicated paramagnetic behavior of the sample. The data were analyzed by use of the Curie–Weiss law. The resulting plot of χ^{-1} as a function of the temperature is shown in Figure 12.2, along with a representation of the normalized magnetization versus the magnetic field, with the temperature as a parameter. At $T \approx 1.2K$, a discontinuity in the χ^{-1} function appears, suggesting critical behavior of the sample at this temperature. What change of the magnetic order is reflected by this irregularity?

In what follows, we summarize the interplay between measurement and modeling that provides specific information about the magnetic phase transition at $T = 1.2K$. Primary susceptibility data were generated from the derivatives of the normalized magnetization with respect to the magnetic field at $B = 0$ at different temperatures. The resulting data set was employed to fit an analytic expression for χ, as obtained within the Heisenberg model. The respective Hamiltonian was written in the following way:

$$\hat{H} = -\sum_i g_i \mu_B \hat{\mathbf{J}}_i \cdot \mathbf{B} + \sum_{i<j} \tilde{J}_{ij}(g_i - 1)(g_j - 1)\hat{\mathbf{J}}_i \cdot \hat{\mathbf{J}}_j. \tag{12.1}$$

The indices i, j run over the three Gd atoms. The second term involves the corresponding Landé factors g_i and g_j. It is based on the identity

$$\mathbf{S}_i = (g_i - 1)\hat{\mathbf{J}}_i. \tag{12.2}$$

Exercise 12.1 Show the validity of formula (12.2), using expression (2.21) for the Landé factor $g_i = g_{i,F}$.

The factor $\tilde{J}_{ij}$ denotes the exchange coupling between the total angular momenta $\hat{\mathbf{J}}_i$ and $\hat{\mathbf{J}}_j$. The Hamiltonian (12.1) is diagonalized in the space of all possible configurations of $\hat{\mathbf{J}}_1$, $\hat{\mathbf{J}}_2$, and $\hat{\mathbf{J}}_3$, which generates an eigenvalue spectrum $\{E_k\}$. Using the relation

$$\mu_k = \frac{\partial E_k}{\partial B_k} \tag{12.3}$$

assigns magnetic moments μ_k to the eigenvalues E_k. For small magnetic field strength, the magnetic susceptibility per cluster can be evaluated by use of the formula

$$\chi = \frac{\beta}{Z} \sum_k \mu_k^2 e^{-\beta E_k}, \tag{12.4}$$

where Z denotes the partition function (2.24).

Exercise 12.2 Establish relation (12.4) by use of Eq. (12.3) as well as the thermally averaged magnetization (2.26), and the definition of the susceptibility, Eq. (2.35). Expand the magnetic energy with respect to the magnetic field.

Fitting calculated and measured χ values selects the dominant magnetic states of the trimetallic nitride fullerene, and determines the exchange coupling factors $\tilde{J}_{ij}$. From this treatment, the two low-temperature regimes appearing in Figure 12.2 are seen to correspond to different spin configurations formed by the three Gd centers. Specifically, for temperatures below $T = 1.2K$, an antiferromagnetic arrangement is obtained, with small $\tilde{J}_{ij}$ constants ($\approx 6.3\ 10^{-4}$ meV). Here, the interaction between three Gd^{3+} cores, each one in its $^8S_{7/2}$ ground state, gives rise to an effective magnetic moment $\mu_{eff} = 7.9\mu_B$. When appplied to the regime $T > 1.2K$, the same evaluation yields a higher magnetic moment of $\mu_{eff} = 9.7\mu_B$. This enhancement reflects the change of one Gd ion from the $Gd^{3+}(^8S_{7/2})$ to the $Gd^{2+}(^7F_6)$ state, as caused by the transfer of one electron from the cage to the enclosed cluster.

Náfrádi et al. [436] reported results of a combined SQUID and electron spin resonance (ESR) study on Gd_3NC_{80} in the temperature regime 2 K–250 K and arrived at ferromagnetic alignment of the three $Gd^{3+}(^8S_{7/2})$ cores for $T \lessapprox 100K$. At high temperature above 150 K, the $S = 7/2$ spins were seen to be uncorrelated.

Computational investigations mostly support the conclusion that $Gd_3N@C_{80}$ at $T \approx 0$ K is antiferromagnetically organized, as suggested by the SQUID experiment of Chen et al. [435]. This result is confirmed by scalar-relativistic DFT calculations [438] and also by computations performed at the DFT+U level [439, 432]. Qian et al. carried out non-collinear DFT+U calculations [437] and obtained a ferromagnetic alignment of the three $^8S_{7/2}$ states, resulting in an overall magnetic moment of $\mu_{eff} = 21\mu_B$, as the most stable configuration. A non-collinear state with $\mu_{eff} = 14.1\mu_B$ turned out to be higher in energy by the small but distinct margin of 10 meV.

Various tri-metallic endohedral single-molecule magnets with mixed metal components have been characterized by experiment and modelling [440, 441, 442]. Among them is the species $Dy_nSc_{3-n}@C_{80}$, with n = 1,2,3. Signature hysteresis profiles have been detected for all three composites by SQUID measurement. In particular, the remanence of the system with $n = 2$ turned out to be markedly larger than for that with $n = 3$. This behavior reflects the different nature of the magnetic interactions operative in the two dysprosium subsystems, Dy_2 and Dy_3. In the former case, demagnetization requires the coordinated reversal of two spins, or transition into a non-magnetic excited state. This process is substantially hindered in comparison with the latter case, where non-collinear ordering of the three Dy spins in conjunction with ferromagnetic coupling, corresponding to negative exchange constants $\tilde{J}_{ij}$, $i, j = 1, 2, 3$[1] (see Eq. (3.33)) gives rise to a sixfold degenerate magnetic ground state, and thus to facile demagnetization in zero-field conditions.

Tables 12.1 and 12.2 summarize experimental results for the effective magnetic moments of various magnetic metallofullerenes, as obtained by SQUID and XMCD measurement, respectively.

12.2 Fullerenes with Encapsulated Group-V Elements

The unit $N@C_{60}$ was first manufactured in 1998 by ion implantation into solid C_{60} [176]. As shown by ESR spectroscopy, the N impurity in the fullerene cage provides a paramagnetic center, where the guest atom retains the electronic structure of the free nitrogen atom [443]. Analogous observations have been made for C_{60} containing atomic phosphorus [444], and for $N@C_{70}$ [443]. As free atoms, both N and P have a half-filled p shell, corresponding to a spin quantum number

[1] An example for *magnetic frustration*, related to an ambiguous magnetic ground state configuration, as, in the simplest scenario, realized by an ensemble of three antiferromagnetically coupled spins.

Table 12.1 *Effective magnetic moments μ_{eff} of magnetic metallofullerenes by SQUID measurment. The values of μ_{eff} are deduced from the Curie–Weiss Law or the Brillouin relation and indicated in units of μ_B. T[K] stands for the temperature or the temperature range in Kelvin. All C_{80} (C_{82}) isomers are of the type $I_h(7)$ ($C_{2v}(9)$).*

System	$\mu_{eff}(CW)$	$T[K](CW)$	$\mu_{eff}(B)$	$T[K](B)$	*Reference*
$La@C_{82}$	0.38	$5-40$			[423]
	3.25	$4-25$			[445]
	1.91^a	$200-300$			[445]
	1.83	>190	0.2	2	[446]
$Ce@C_{82}$	0.67	$2-20$			[445]
	3.25^a	$200-300$			[445]
	3.36	>190	≈ 0.8	2	[446]
			≈ 2.5	>60	[446]
$Gd@C_{82}$	6.90	$5-200$	7.70	$3-200$	[424]
	6.66	>40	6.74	>8	[425]
	6.91	>2	6.95	40	[426]
$Gd_3N@C_{80}^b$	7.9	$0.2-1.2$			[435]
$Tb@C_{82}$	7.51	>20	6.80	55	[426]
$Tb_3N@C_{80}^b$			6.0	$2-160$	[430]
$Dy@C_{82}$	9.25	>20	8.48	75	[426]
$Ho@C_{82}^c$			6.41	$2-5$	[447]
	6.30	>20	5.63	$12-50$	[425]
	6.33	>20	5.55	40	[426]
$Ho_3N@C_{80}^b$			6.7	$2-320$	[430]
$Er@C_{82}$	7.06	>20	6.37	125	[426]
$ErSc_2N@C_{80}^b$	9.8	$4-50$			[448]
$Er_2ScN@C_{80}^b$	7.3	$4-50$			[448]
$Er_3N@C_{80}^b$	6.7	$4-50$			[448]
$Er_3N@C_{80}^b$	3.4	$2-300$	≈ 3.5	2	[449]
$Tm_3N@C_{80}^b$	≈ 9.6	$120-300$			[450]

[a] The measurements are performed for an $M@C_{82} \cdot (CS_2)_{1.5}$ solvate;
[b] The magnetic moments are normalized to the number of atoms;
[c] The measurement is performed on a mixture of endohedral metallofullerenes enclosing Ho, with $Ho@C_{82}$ as dominant component. *CW* stands for Curie-Weiss, *B* for Brillouin.

$S = 3/2$. Thus, $N@C_{60}$ and $P@C_{60}$ may be considered as spin traps, a feature of high potential relevance for the physical realization of *quantum computing architectures*. The atomic spins confined to the fullerene enclosure branch into four equidistant Zeeman levels. These levels are further split by hyperfine interaction with the nuclei of the considered N or P isotopes. In the case of N, this might be ^{14}N

Table 12.2 *Effective magnetic moments μ_{eff} of magnetic metallofullerenes by x-ray magnetic circular dichroism (XMCD) measurement. The values of μ_{eff} are obtained either from the Curie–Weiss Law or from the relation $\mu_{eff} = g_J [J(J+1)]^{\frac{1}{2}}$, with g_J as the Landé factor of Dy^{3+} (4/3), Ho^{3+} (1.25) and Er^{3+} (1.2).*

System	$J(L/S)^a$	μ_{eff}	$T[K]$	*Reference*
$Ce@C_{82} - C_{2v}$	CW	0.82	$6-20$	[451]
	CW	1.00	20-60	[451]
$Ce_2@C_{82} - I_h$	CW	0.75	$6-20$	[451]
	CW	0.51	$20-60$	[451]
$Gd@C_{82} - C_{2v}$	CW	6.8	$10-40$	[452]
$Dy@C_{82} - C_{2v}$	CW	9.5	$10-25$	[452]
	$3.9(2.7/1.2)^b$	5.8	6	[453]
	$3.1(1.9/1.2)$	4.8	4	[454]
	$1.0(0.3/0.7)$	1.9	300	[454]
$Dy@C_{88} - I$	$3.4(2.3/1.1)^b$	5.2	4	[454]
$Dy@C_{88} - II$	$3.3(2.2/1.1)^b$	5.0	4	[454]
$Ho@C_{82} - C_{2v}$	$3.8(2.8/1.0)^b$	5.3	6	[453]
$Er@C_{82} - C_{2v}$	CW	8.5	$10-40$	[455]
$Er_2@C_{82} - C_{2v}$	CW	8.5	$10-40$	[455]
$Er_2C_2@C_{82} - C_{2v}^b$	CW	8.4	$10-40$	[455]
$Er_2C_2@C_{82} - C_s$	CW	7.8	$10-40$	[455]
$ErYC_2@C_{82} - C_s$	CW	7.8	$10-40$	[455]
$Er_2@C_{90}$	$3.5(2.8/0.7)^b$	8.5	6	[453]

[a] J is calculated by use of sum rules from L and S which are obtained by XMCD measurement;
[b] the S values are from *ab initio* calculations.

or ^{15}N, associated with nuclear spin quantum numbers $I = 1$ or $I = 1/2$, respectively. This interaction gives rise to twofold (threefold) splitting of each electronic spin level for ^{15}N (^{14}N). Figure 12.3 shows ESR signals obtained for $^{14}N@C_{60}$ and $^{31}P@C_{60}$, along with an energy level scheme, demonstrating the effect of hyperfine interaction on the Zeeman-split spin levels of $^{14}N@C_{60}$ [444, 456].

The spin Hamiltonian for $N@C_{60}$, including hyperfine interaction, may be written in the following way [457]:

$$\hat{H}_{spin} = \frac{g\mu_B B}{\hbar}\hat{\mathbf{S}}_z + \frac{g_I \mu_n B}{\hbar}\hat{I}_z + A\hat{\mathbf{S}} \cdot \hat{\mathbf{I}}, \tag{12.5}$$

with A as hyperfine coupling constant. A fine structure, or *zero-field splitting* component has been omitted in Eq. (12.5) since it vanishes for spherical cages (the

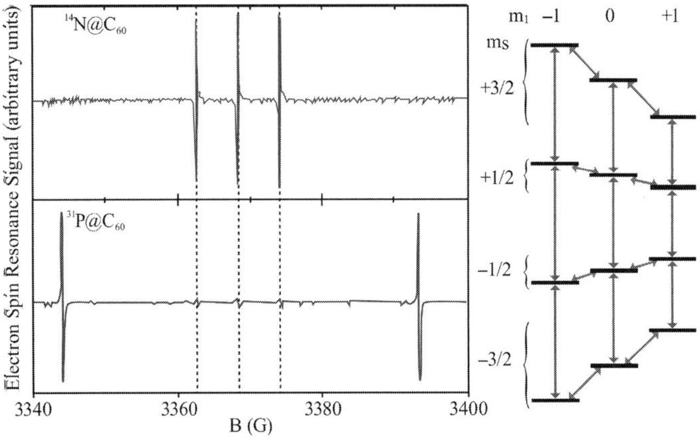

Figure 12.3 Electron spin resonance(ESR) signals obtained at room temperature for dilute $^{14}N@C_{60}$ and $^{31}P@C_{60}$ powders (left panel), and the Zeeman levels of $^{14}N@C_{60}$, modified by hyperfine interaction (right panel) [444, 456]. Reprinted from [444], with permission from Wiley Publishing.

case of $N@C_{60}$) and also for non-spherical cages, such as C_{70}, in rapidly tumbling motion.

12.2.1 Endohedral Fullerenes as Spin-Based Qubits

An intriguing feature of endohedral fullerenes with enclosed group-V atoms is the long lifetime of their electron spins, reflecting the screening effect of the fullerene enclosure. Thus, spin-lattice relaxation times in the order of seconds have been observed for magnetically dilute powder samples of $N@C_{60}$ at low temperature (T = 5K, [444]). Using a pulsed paramagnetic ESR technique, Morton *et al.* reported dephasing times up to about 200 μs for $N@C_{60}$ in a CS_2 solution at $T = 190$ K [458]. Their excellent spin coherence properties make fullerenes encapsulating N or P atoms interesting as candidates for qubits in quantum computing. Numerous proposals for the physical implementation of qubits have been examined, among them quantum dots [459], vacancy centers in diamond [460] and models based on superconductors [461]. In what follows, we will focus on various proposals to exploit the well-isolated group-V atoms in C_{60} as the hardware of a quantum computer. Some of the basic ideas underlying quantum computing with spins are summarized in the special-topic section attached to this chapter (Section 12.9). As outlined there, the elementary logical operations enacted in quantum computation are formally described as unitary transformations applied on a superposition of qubits. An actual quantum computer may consist of an array of spins manipulated by a succession of ESR or NMR pulses.

In particular, Suter and Lim proposed a scalable design involving C_{60} molecules with encapsulated N or P atoms [462]. A scanning tunneling microscope might

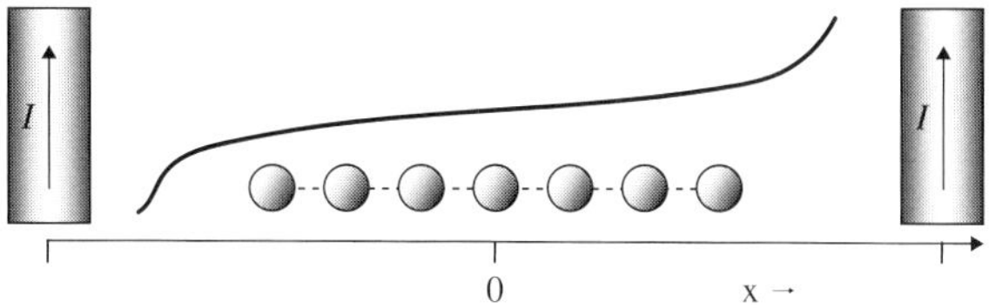

Figure 12.4 Suggested setup for a spin quantum register assembled from fullerenes with encapsulated N or P atoms [444, 462]. Two current-loaded wires create a magnetic field along the fullerene chain. The spatial variation of the field makes it possible to target each element individually by applying suitably tuned ESR or NMR pulses. Reprinted from [444], with permission from Wiley Publishing.

be employed to place a linear array of endohedral fullerenes, each one of them contributing one qubit, on silicon surfaces. Individual qubits might be controlled by use of ESR (NMR) pulses for the electronic (nuclear) subsystem. Addressing a selected qubit may proceed by exposing the fullerene line to a magnetic field gradient, which accomplishes spatial resolution, associating with each endo-fullerene cage specific transition frequencies between Zeeman levels. A suggested arrangement [462] is shown in Figure 12.4. A sequence of equidistant fullerenes extends between two current-loaded wires that are oriented parallel to the supporting surface. The magnetic field felt by a fullerene at position x between the wires is then given by

$$B = \frac{\mu_0}{2\pi} I \left(\frac{1}{x + L/2 + R} + \frac{1}{x - L/2 - R} \right). \tag{12.6}$$

The position coordinate x is here measured from the center between two wires with radius R, whose vertical distance is L. Each of them is traversed by a current I.

The resonance frequency variation inherent in this scheme causes detuning between adjacent electron spins. This feature decreases the weight of spin exchange processes in the dipolar interaction between them [444], effectively reducing the coupling between spins m and $m+1$ to a term proportional to $\hat{S}_{m,z} \cdot \hat{S}_{m+1,z}$. This effect enhances the decoherence time of the electron spin chain.

Nuclear spin relaxation, however, is still a much slower process than electron spin relaxation. Brown et al. employed electron and nuclear spin echo spectroscopy to determine the times T_1 and T_2, characteristic for spin-lattice relaxation and spin dephasing, respectively, in $^{15}N@C_{60}$ [463]. The measurement included both, the electronic and the nuclear subsystem. At low temperature (T = 10 K), the dephasing time for electronic spins (T_{2e}) was found to be 160 μs, while the analogous quantity for the nuclear spins (T_{2n}) turned out to be about three orders of magnitude larger (135 $\pm$ 10 ms). For $T \geqslant 50$ K, T_{2n} was seen to approach the electronic spin-lattice relaxation time, T_{1e}. To put the experimental result for T_{2e} into perspective, one might compare it with estimated times needed for a CNOT operation (see Section 12.9) when implemented by electron spins, about 50 – 100 ns [444]. On the basis of these estimates, the time T_{2e} allows for more than one thousand logical operations, carried out by manipulating the electron subsystem.

Both the magnitudes of the electronic and nuclear relaxation times and their proportions have prompted the proposal to capitalize on both subsystems in the process of quantum computing, using the electron spins as a *working register*, employed for initialization and processing, and the nuclear spins as a long-term memory. Schemes for *coherent transfer* of information between electronic and nuclear degrees of freedom have been designed theoretically [444, 464] and validated experimentally [463].

To circumvent the numerous technical challenges associated with realizing the vision of targeting every element of the register[2], alternative concepts have been developed that employ *cellular automata*. The pivotal move consists here in distributing a single qubit over a sequence of *cells*, each one of them to be realized by a quantum mechanical two-state system. In this manner, a qubit may be implemented by a spin chain rather than an individual spin. Such a chain could consist of an XYXY motif, with X and Y as physically distinguishable cells [465]. A string of alternating $^{15}N@C_{60}$ and $^{31}P@C_{60}$ units has been discussed as possible realization of this design [466]. In terms of the nuclear spin eigenfunctions, the state $|0\rangle$ ($|1\rangle$) might be represented as $|+\frac{1}{2}\rangle|+\frac{1}{2}\rangle|-\frac{1}{2}\rangle|-\frac{1}{2}\rangle$ ($|-\frac{1}{2}\rangle|-\frac{1}{2}\rangle|+\frac{1}{2}\rangle|+\frac{1}{2}\rangle$). The update rules to manipulate the array of qubits then have the form $X_f^{\mathbf{U}}$, and equivalently for the cells of type Y. This is to be understood as a conditional rule of procedure. Each cell of the type X is exposed to a unitary transformation $\mathbf{U}$ if and only if the *field variable* has the value f. The field variable is here the difference between the nearest neighbors in states $|1\rangle$ and $|0\rangle$. This architecture has been shown to be capable of executing all standard quantum algorithms [467]. Operating with global signals that are addressed to all cells of a given type, the cellular automaton approach thus avoids the complications associated with local targeting of distinctive qubits.

In closing this section, we mention recent attempts to generate an ordered array of $N@C_{60}$ molecules by incorporating them into SWCNTs [457, 468]. SWCNTs filled with fullerenes are referred to as *nano-peapods* [469]. Benjamin et al. inserted a mixture of nitrogen-encapsulating C_{60} and the pure C_{60} molecules into an SWCNT at an $N@C_{60}/C_{60}$ concentration of about 1 percent. Other attempts to use SWNCTs as a scaffolding for magnetically functionalized fullerenes include

[2] For instance, the need for a clear separation between the resonance frequencies of adjacent cells in a spin quantum register puts high requirements on the bandwith of the microwave radiation employed for addressing if the number of these cells is high. Also, the magnetic field gradient along the qubit array must be large and stable in time [466].

the work by Simon et al., who formed arrays of the substitutional complex $C_{59}N$ inside SWCNTs by vacuum annealing [470].

12.3 Magnetism from Charge Transfer in Fullerene-Based Compounds

Since Haddon et al. reported the metallic behavior of thin fulleride films, consisting of C_{60} intercalated with alkali metal atoms [471], a wide variety of alkali-doped fullerene compounds has been discovered. In this section, we present a brief overview of some of the most extensively studied magnetic phases of fullerides and their derivatives, such as $NH_3A_3C_{60}$ (A = Rb, Cs). We further highlight the compound TDAE-C_{60} (tetrakis-dimethylamino-ethylene-C_{60}), where magnetism arises from charge transfer between two non-magnetic components. While some of these materials claim interest because of their high T_c superconducting properties, all of them display magnetic phenomena that depend sensitively on the details of their geometric structures and chemical composition. On account of these features, fullerides are acknowledged to be a fertile ground for the exploration of organic magnets.

12.3.1 Magnetic Fullerides

In the following we consider magnetic structures induced in fullerides by electron transfer from the alkali atom moiety to the fullerene cages. A survey on the geometric and electronic properties of fullerides has been given in Section 6.3. Combined with alkali metal or transition metal atoms, C_{60} acts as an electron acceptor. Specifically, in A_nC_{60}, with $n = 1, 3, 6$, electron transfer proceeds from the valence shell of A_n to the threefold degenerate T_{1u} LUMO of C_{60}. Upon transfer of an odd number of electrons into this LUMO level, C_{60} adopts a non-zero spin.

The symmetry of C_{60} as a fulleride constituent depends on the crystallographic order of the fulleride. If this order is of lower than cubic symmetry, the cage will distort from icosahedral to D_{3d} or D_{2h} symmetry. Accordingly, the three components of the T_{1u} LUMO level are split into two lines in trigonal or hexagonal lattices, and into three distinct lines in tetragonal or orthorhombic environments. These effects are summarized in the scheme 12.5 [472].

We mention that the D_{5d} isomer of C_{60} is not realized in fullerides since D_{5d} is not a crystallographic point group. Further symmmetry lowering beyond the effect of the crystal field originates from *Jahn–Teller distortion*, associated with structural deformation in systems with degenerate electronic ground states. This effect may reduce the icosahedral C_{60} symmetry in cubic environments or the D_{3d} symmetry of the C_{60} molecule in trigonal or hexagonal lattices. It may give rise to a D_{5d} C_{60} isomer and further to distortions along various symmetry axes of the molecule if

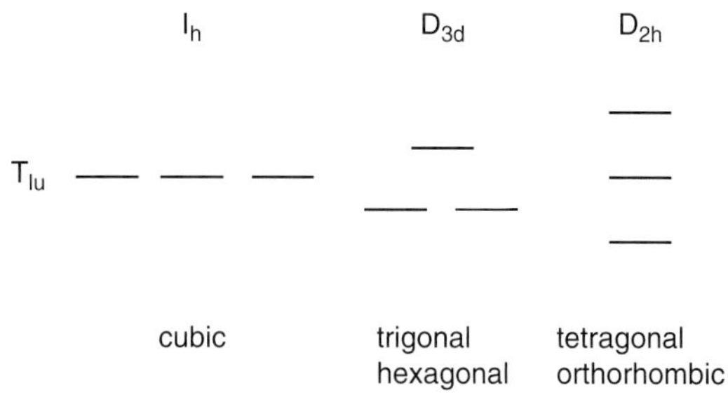

Figure 12.5 Impact of different crystallographic environments on the T_{1u} LUMO level of C_{60} and on the molecular symmetry of C_{60}.

the crystal field is not strong enough to impose translation symmetry on the system. Since the Jahn–Teller effect is of crucial importance for magnetism in fullerene charge-transfer complexes, we provide a brief outline of this phenomenon in the following subsection and discuss its manifestations in fullerides.

Jahn–Teller Distortion of Fullerene Anions

In this section we consider non-magnetic interactions that nevertheless exert a strong influence on magnetic fullerene complexes, causing structural deformations of the fullerene cage. They originate from symmetry-breaking electron-phonon interactions, or, equivalently, those between the electronic and the nuclear degrees of freedom. These processes are captured by the *Jahn-Teller theorem* [473, 474]. Its intuitive content is the following: any polyatomic molecule with an orbital electronic degeneracy is unstable with respect to symmetry-lowering distortion along at least one normal coordinate. This distortion removes the degeneracy. We emphasize that the Jahn–Teller theorem excludes linear molecules as well as cases of Kramers degeneracy, associated with alternative spin states of the same molecular energy.

On the simplest level, the logic of the Jahn–Teller theorem may be illustrated by a twofold degenerate electronic system that interacts with a single vibrational mode. Up to quadratic order, the Hamiltonian for such a system may be written as the following 2×2 matrix:

$$\hat{\mathbf{H}} = -\mathbf{1}\frac{\hbar^2}{2M_*}\frac{\partial^2}{\partial X^2} + \mathbf{1}\frac{k}{2}X^2 + K\begin{pmatrix} -X & 0 \\ 0 & X \end{pmatrix}. \tag{12.7}$$

Here, M_* denotes the reduced mass associated with the normal coordinate X (see, e.g. [33]), and k, K are constants. The Hamiltonian is represented in the electronic basis defined by the two degenerate states. The essential feature of $\hat{\mathbf{H}}$ is that it includes a non-vanishing linear term in the coordinate X. Therefore, the

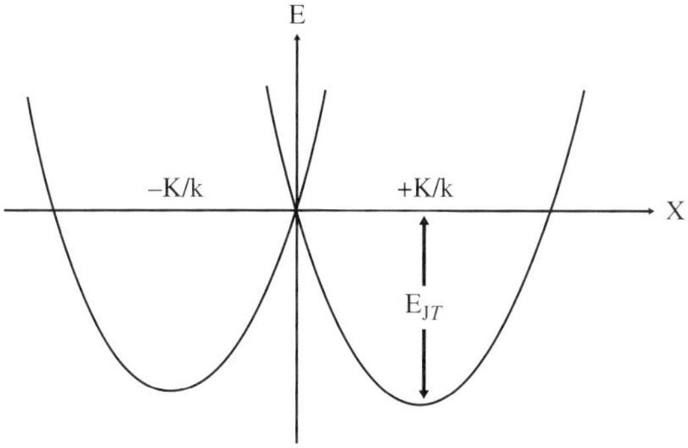

Figure 12.6 Interaction of two degenerate electronic states with a Jahn–Teller active mode, labeled X. The system distorts from $X = 0$, adopting new equilibrium configurations at $X = \pm\frac{K}{k}$. At either position, the initial degeneracy is lifted.

place $X = 0$ is not a point of equilibrium for the considered system. This essential conclusion can be made more tangible by splitting the Hamiltonian into two separate expressions [475]:

$$\hat{H}_{\pm} = -\frac{1}{2M_*}\frac{\partial^2}{\partial X^2} + \frac{k}{2}X^2 \pm KX. \tag{12.8}$$

The operators $\hat{H}_{\pm}$ involve two harmonic oscillator potential energy curves that are shifted with respect to each other by a margin $\Delta X = \frac{2K}{k}$, as shown in Figure 12.6. The energy released as the system distorts from $X = 0$ to $X = \pm K/k$ is termed *Jahn–Teller energy*, E_{JT}.

By the Jahn–Teller theorem, adding an electron to C_{60} will distort the spherical shape of the molecule. A single additional electron will occupy the C_{60} LUMO, a threefold degenerate orbital. The theorem predicts for this case a distortion that takes the molecule from its native I_h symmetry to a lower group. In this situation, the *Jahn–Teller active modes*, i.e. the normal coordinates along which the molecule can distort, may be identified by group theoretical analysis. This search singles out a fivefold representation of the icosahedral group, labeled H_g, in the space of normal vibrational modes as the only nuclear degrees of freedom to which a degenerate orbital with the properties of the C_{60} LUMO, an electronic triplet of T_{1u} symmetry in an icosahedral complex, may couple. Eight modes of the H_g type exist for C_{60} [476].

The quantitative treatment of the problem involves generalizing the approach (12.7) to the case of a 3×3 matrix that describes the interaction of the T_{1u} electronic system with all five components of a given H_g mode. This defines a problem in a multi-dimensional space of vibrational modes. Solving this problem does not consist in finding the minima of potential energy curves, as in the one-dimensional case, but rather of the adiabatic potential energy surface for the ground state of the system. These minima correspond to C_{60}^- geometries at lower than icosahedral symmetry. Specifically, the anion may distort into six isomers of D_{5d} symmetry, and further ten D_{3d}, fourteen D_{2h}, thirty C_{2h} isomers [477, 478]. The configurations within any one of these four groups deviate from each other merely by their orientations and thus have been termed *orientomers* [478]. In the absence of perturbations, all orientomers share the same energy. As they are separated by finite potential energy barriers, tunneling will take place between them.

In terms of possible distortion effects, similar conclusions can be drawn for C_{60}^- and for C_{60}^{2-}. The case of C_{60}^{3-}, however, seems to differ markedly from the C_{60}^- prototype. A plausible occupation scheme for three electrons transferred to C_{60} involves populating the LUMO according to the pattern $[T_{1u,x}^1, T_{1u,y}^1, T_{1u,z}^1]$. This arrangement is orbitally non-degenerate and is thus expected to be stable with respect to Jahn–Teller distortion. For C_{60}^{3-} in A_3C_{60} crystals (A = K, Rb, Cs), however, the assumption that the three excess electrons distribute isotropically over the $T_{1u,i}$, $i = x, y, z$ orbitals is not justified. This was shown explicitly by tight-binding analysis of C_{60}^{3-} in the cubic environment of the A_3C_{60} lattice, where allowance was made for interelectronic repulsion and electronic-vibrational interactions [479]. This computation yielded *orbital disproportionation* of the three T_{1u} occupancies, resulting in the modified scheme $[T_{1u,x}^2, T_{1u,y}^1, T_{1u,z}^0]$. The basic physical reason for this rearrangement is that the energy associated with the interelectronic repulsion in the case of even $T_{1u,i}$ population turns out to be larger than that for disproportionate population combined with Jahn–Teller distortion of C_{60}. This exemplifies an antagonism between the Jahn–Teller theorem and Hund's rule, as the former favors full orbital occupancy, the latter, in contrast, high-spin solutions [480]. The uneven population pattern results in LUMO level splitting into three separate bands whose bandwidth is compatible with the experimental finding of 1.2 eV for K_3C_{60} [481].

It appears tempting to understand the partially occupied fullerene LUMO in alkali doped fullerides as an incompletely filled conduction band. If this intuition is valid, the transferred electrons delocalize, rather than stay attached to a well-defined fullerene site, and fullerides should consequently be metallic. This, however, is not borne out by observation [472]. A more detailed view can be obtained in the framework of the Hubbard model (see Section 3.1). The mobility of an electron depends here on the Coulomb term U. If this term is large, the energy needed to overcome the Coulomb repulsion between an electron and its nearest neighbor is high. In consequence, the electron may be confined to its site. If this is the case, the respective fulleride is appropriately described as an insulator [480]. If the number N of electrons transferred to C_{60} in a fulleride is odd, absorption of an electron by C_{60} increases the Jahn–Teller distortion and thus causes an energy lowering while effectively reducing the Coulomb repulsion U. This decrease, however, is not sufficiently strong to turn the on-site repulsion into an on-site attraction [480].

Within the Hubbard model, electron hopping between neighboring sites is the factor that counteracts localization at a given site. The relation between the two rivaling agents that govern the Hubbard model – confinement versus hopping – is captured by the ratio of the on-site Coulomb repulsion integral and the hopping term, U/t.

A strong trend of electrostatic confinement combined with low inter-site fluctuations, as reflected by large values of U/t, favors the formation of localized magnetic moments. For one particular configuration, the magnetic nature of the ground state according to the Hubbard model may be readily inferred. This is the case of half-filling, as described by the simplified model where a single half-filled orbital is attached to each site. In this case, antiferromagnetic order prevails, as situations of neighboring sites with spin-parallel electrons are disadvantaged. By the Pauli principle, no hopping would occur between these sites which translates into lower energy for the antiferromagnetic solution, and higher energy for the ferromagnetic alternative. At half-filling, the Hubbard model predicts the system to exist in the state of an *antiferromagnetic Mott insulator*.

12.3.2 Magnetic Phases of A_nC_{60}, n = 1,3

A wide range of structural and related magnetic phases have been identified for A_nC_{60} [480]. In this section, we record some of the magnetic signatures of the extensively studied families A_nC_{60}, with n= 1,3 and A = K, Rb, Cs.

A_1C_{60}

Polymeric compounds are realized by $n = 1$ with A = K, Rb, Cs, where the alkali metal atom is located at the octahedral site. At high temperature, these systems stabilize in the FCC structure. Cooling to temperatures below the regime of 350–400 K results in a transition to a quasi-one-dimensional arrangement [275], characterized by C_{60} chains. Compression along the a direction of the FCC crystal with simultaneous extension in the b and c directions leads to the formation of linear polymers, bonded by [2+2]-cycloaddition. Thus, cooling of the high-temperature phase of A_1C_{60} gives rise to a covalently bound one-dimensional substructure, as shown in Figure 12.7. Detailed characterization of the AC_{60} polymer phase succeeded by NMR experiments where the chemical shift of the ^{13}C resonance line was monitored [484]. At high temperature, a single NMR line was detected, providing evidence for a monomer phase where all C atoms are equivalent. This pattern

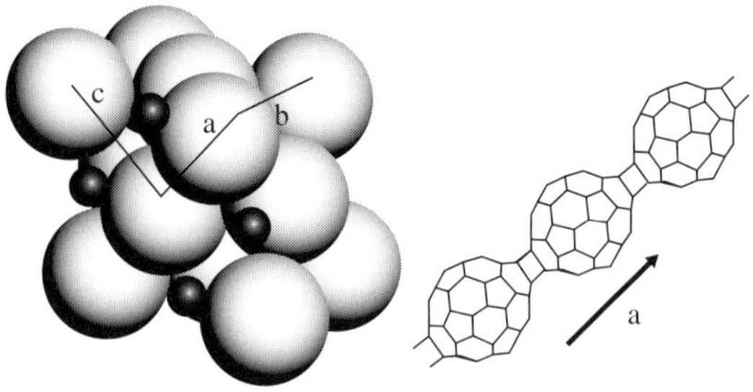

Figure 12.7 Emergence of a quasi-one-dimensional orthorhombic phase of AC_{60} by cooling the high-temperature FCC phase [480]. A small compression along the *a* direction of the face (left) gives rise to the formation of a linear C_{60} polymer, where adjacent fullerenes are covalently bonded by [2+2] cycloaddition (right). Reprinted from [480], with permission from IOP Publishing.

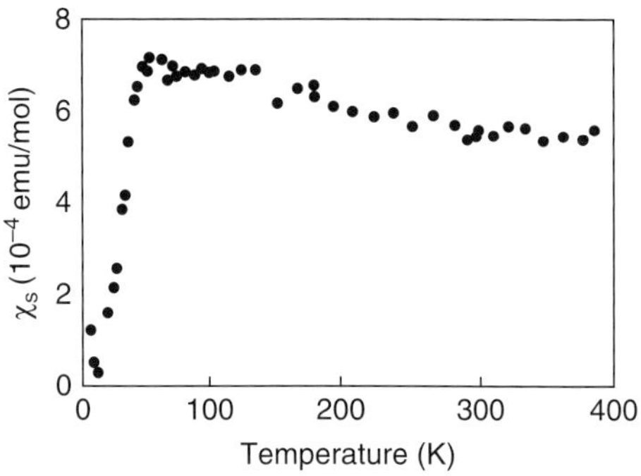

Figure 12.8 The measured spin susceptibility χ_S of RbC_{60} in its quasi-one-dimensional phase versus temperature. Above a critical temperature of about 50 K, χ_S displays, to a good approximation, Pauli characteristics. The marked decrease of χ_S for $T \leq 50K$ is attributed to the Peierls instability of the $(C_{60})_n^-$ chain in RbC_{60}. Reprinted with permission from [275]. Copyright (1994) by the American Physical Society.

changes as the temperature is reduced below 380 K. In the low temperature regime, nine NMR lines become distinguishable, which are indicative of fullerene polymerization. This process results in a deformed cage with nine inequivalent positions. Depending on the intercalating alkali metal species, the structure of the emerging phase may be orthorhombic (A = K) or monoclinic (A = Rb, Cs) [482]. Figure 12.7 displays the transition from the high-temperature FCC lattice to the orthorhombic arrangement.

We now illustrate some salient features of AC_{60} magnetism by the example of RbC_{60} [275]. The spin susceptibility χ_S for the quasi-one-dimensional phase of RbC_{60} was found to be large and consistent with Pauli-like behavior (see Section 2.3.2). Below $T = 50K$, however, it was seen to tend towards zero, as documented by the measured RbC_{60} data shown in Figure 12.8, indicating a further phase transition. The abrupt drop of χ_S at a critical temperature was found to be associated with a *Peierls instability*, turning the rows of C_{60} molecules into chains of a dimers, and inducing a metal-to-insulator transition. This transition leads to a magnetically ordered phase which displays the properties of a spin wave state. The resulting system has been characterized as an antiferromagnetically coordinated insulator [275]. Similar conclusions have been drawn with respect to CsC_{60} and KC_{60} [483].

A_3C_{60}

Having commented on AC_{60}, we focus on the complex magnetic features of the fulleride family A_3C_{60}, where A = K, Rb, and Cs. In contrast to most other fulleride compounds, these trivalent salts tend to be metallic. Since 1991 when these materials were first fabricated, they have attracted major interest, chiefly because of their High-T_c superconducting properties [486]. While there are various fulleride superconductors (see e.g. [487]), the most thoroughly investigated and understood are those of composition A_3C_{60}. In these systems, the superconducting state has been found to be restricted to simple cubic structures in conjunction with the trivalent configuration of C_{60}^{3-}, corresponding to a half-filled T_{1u} conduction band (compare with the scheme 12.5). In an orthorhombic lattice, these fullerides may exist also as antiferromagnetic insulators, rather than superconductors, as documented by Figure 12.9 which shows the transition temperatures T_N and T_c for the former and the latter materials, respectively. We mention here that the antiferromagnetic systems included in the figure are of composition $NH_3A_3C_{60}$. Later, we will focus more closely on these compounds which are obtained by adding an ammonia molecule to each octahedral site of A_3C_{60}.

As inferred from Figure 12.9, fullerene anions occupy a larger volume in the insulating than in the superconducting phase, which is equivalent to stating that their average distance from each other is greater in the former than in the latter phase. This feature is in accordance with the finding that not only temperature but also pressure variation can induce a change between the two phases, a *Mott–Hubbard* transition [488]. On which side of this metal-insulator transition is a given A_3C_{60} compound located? This question can be decided by inspecting the critical ratio between the on-site Coulomb repulsion energy U and the T_{1u} bandwidth W. For given U, the condition for the insulating state, $\frac{U}{W} > \frac{U}{W}|_c$, may be realized by narrowing the bandwidth W. This in turn is accomplished by reducing the interaction strength between adjacent C_{60}^{3-} sites. To achieve this goal, one might decrease the pressure on the considered system, thus enhancing the inter-fullerene distances. This mechanism was confirmed by examining the compound Cs_3C_{60} [488, 490] which, at ambient pressure, is an antiferromagnetic insulator. Under hydrostatic

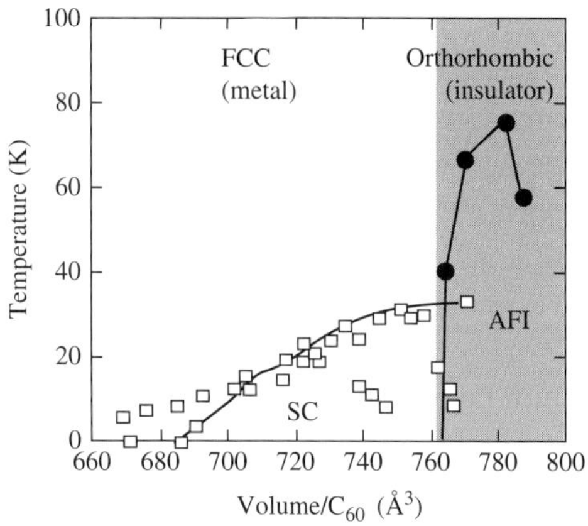

Figure 12.9 Critical temperatures for the onset of superconductivity (T_c, open squares) and antiferromagntism (T_N, filled circles) in trivalent fullerides [489] versus the C_{60}^{3-} volume. The labels SC and and AFI stand for superconductor and antiferromagnetic insulator, respectively. The shaded region indicates the orthorhombic insulator phase. Reprinted from [489], with permission from IOP Publishing.

pressure of 7 kbar, however, the sample changed from antiferromagnetic to superconducting. This behavior is specific to Cs_3C_{60}, since the spatial extension of Cs metal atoms exceeds that of K or Rb atoms, giving rise to a maximum interfullerene distance for A = Cs within the A_3C_{60} family.

Less evidently, the insulating condition may also be established by reducing the orbital degeneracy of the C_{60}^{3-} centers [491, 492]. We assume here a half-filled system. If, in this case, there is no orbital degeneracy, i.e. only one channel is open for an electron that hops between neighboring sites, $\frac{U}{W}|_c \approx 1$.[3]

In this case, $U > W$ favors the insulating state. If the degree of orbital degeneracy is N instead of 1, however, the critical ratio is approximated by $\frac{U}{W}|_c \approx \sqrt{N}$

[3] For half-filled bipartite lattices, and for $U \gg W$, Gunnarsson *et al.* derive the approximation $\Delta E \approx U - \sqrt{N}W$ [487], with ΔE as the band gap of the system, and N as the degree of orbital degeneracy of any given site. This estimate is based on the expression for ΔE in the assumed case of half-filling, namely

$$\Delta E = E(NM + 1) + E(MN - 1) - 2E(MN),$$

with M as the number of lattice sites (ΔE may be understood as the difference between the vertical electron affinity and the vertical ionization energy of the system). From the multiplicity of available states in conjunction with a normalization factor $1/\sqrt{N}$ for the wave function that describes the orbital system at any one of the M sites, one obtains within the Hubbard model: $\langle i|\hat{H}|a\rangle = \sqrt{N}t$, where $\langle i|$ and $\langle a|$ stand for states localized at adjacent sites. From

$$E(MN + 1) \approx E(NM) + NU - \sqrt{N}t$$
$$E(MN - 1) \approx E(NM) - (N - 1)U - \sqrt{N}t,$$

one concludes, relating the hopping parameter to the band width W($W \approx 2t$, see, for instance, [115]), that $U - \sqrt{N}W$ is an approximate expression for ΔE. Thus, the gap closes if $U = \sqrt{N}W$.

[491], yielding for A_3C_{60} the more stringent inequality $U > \sqrt{3}W$ as criterion for the Mott–Hubbard insulator state. This correction is related to an increase of the hopping rate if admission is made for an orbital degeneracy in the assumed half-filled systems. Elementary arguments suggest that this rate scales with the square root of N, the degree of degeneracy [491]. This reasoning suggests that the metallic state is more likely to be adopted by A_3C_{60} then by A_1C_{60}.

This trend is exemplified by the FCC structure of the A_3C_{60} lattice for A = Rb, Cs. The electronic state of C_{60}^{3-} in A_3C_{60} is preferentially trivalent rather than monovalent, which in turn favors cubic symmetry for the lattice structures of these compounds. From the scheme 12.5, an orthorhombic crystal field gives rise to a monovalent state. In this electronic configuration, an on-site Coulomb repulsion energy slightly larger than the bandwidth may already induce Mott localization, and thus lead to an insulating ground state.

12.3.3 $NH_3A_3C_{60}$ with A = Rb, Cs

Both features – symmetry reduction and increase of the inter-fullerene distance – are realized for compounds of the composition $NH_3A_3C_{60}$, with A = K, Rb [493, 494, 495, 496]. Originally, the guiding interest in adding an ammonia moiety to the fulleride network was fabricating a material with enhanced inter-fullerene separation, and so to enhance T_c, the transition temperature for the onset of superconductivity [495]. This expectation was based on the BCS theory of superconductivity [497] which predicts a monotonic increase of T_c with the electronic density of states at the Fermi level, $g(E_F)$. As the A_3C_{60} lattice is expanded, the width of the T_{1u} band shrinks, and $g(E_F)$ increases. When synthesized in the laboratory, however, $NH_3A_3C_{60}$ turned out to be a *Mott insulator* with antiferromagnetic coordination.

As ammonia molecules are introduced into the fullerides A_3C_{60}, a structural phase transition takes place that lowers the face-centered cubic (FCC) symmetry of the pure fulleride to face-centered orthorhombic (FCO) symmetry. For K_3C_{60}, this transition is found to occur at $T = 150$ K [493]. It leads to an expansion of the fulleride frame, associated with doubling the lattice constants of the high-temperature structure along all three crystal axes. From x-ray diffraction measurement [493], this change is associated with the formation of an ordered pattern of K_o-NH_3 pairs, where K_o stands for the K atom at the octahedral site (see Section 6.2.1). The resulting ferroelectric arrangement, as displayed in Figure 12.10, minimizes the electrostatic energy associated with these pairs: they form chains that are uniformly aligned in the direction of the electric dipoles, while the dipoles in the nearest chains point in the opposite direction.

In contrast to the A_3C_{60} matrix, $NH_3K_3C_{60}$ does not exhibit high-T_c superconductivity [498]. On the basis of NMR investigations, Tou et al. [494]

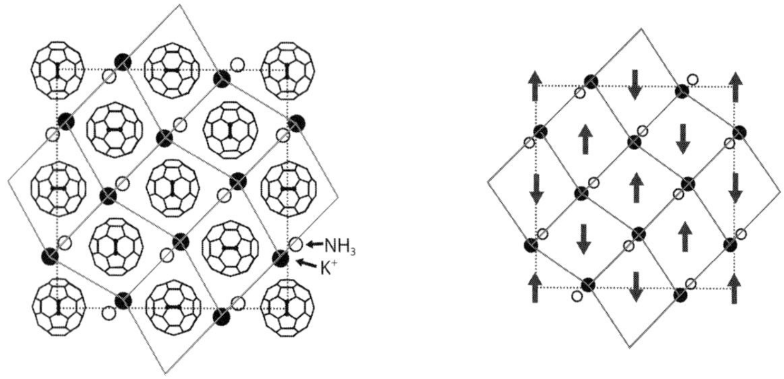

Figure 12.10 Coincidence of antiferromagnetic and ferromagnetic order in $NH_3K_3C_{60}$. K_o-NH_3 pairs are indicated. (a) Parallel to the K_o-NH_3 chains, the fullerenes display uniform orientation, corresponding to antiferromagnetic coordination. In the direction perpendicular to these chains, however, the orientations of the fullerenes alternate between nearest neighbors, which is the signature of ferromagnetic coordination. (b) The array of magnetic moments resulting from the orientational ordering scheme indicated in panel (a). Reprinted from [489], with permission from IOP Publishing.

characterized the compound as a localized electron system, where each fullerene carries a spin of $S = 1/2$, over the investigated temperature range [4.2, 300 K]. Rather than a metal, the system proved to be a Mott–Hubbard insulator [494, 499]. A transition to a three-dimensional long-range antiferromagnetic order was found at $T_N \approx 45K$. Below this Neél temperature, a distinct correlation between orientational and magnetic order emerges. The antiferromagnetic structure is due to alternate stacking of C_{60} sheets [496], each of which displays an antiferromagnetic and a ferromagnetic motif in coexistence with each other. This is demonstrated by the projection of the FCO structure of $NH_3K_3C_{60}$ shown in Figure 12.10. Along the direction defined by the K_o-NH_3 pairs ([110]), the orientation of the C_{60} molecules is uniform. From Figure 12.10 (b), the spins located on these centers are coupled antiferromagnetically. The converse holds for the C_{60} chains that run, along the $[1\bar{1}0]$ direction, perpendicularly to the K_o-NH_3 pairs. Their orientation alternates, and their spins are ordered ferromagnetically. The origin of this remarkable correspondence between spin and spatial degrees of freedom will be further discussed in the following section, dealing with TDAE-C_{60}.

From ^{13}C NMR measurement, a critical temperature of $T_s = 150$ K has been inferred for the transition of $NH_3K_3C_{60}$ from FCC to FCO order [489]. Above this temperature, the distorted fullerenes carry out a hindered rotation that may be understood as a rapid passage between several stable orientations, to freeze at T_s into the order that is conducive to the magnetic phase attained at T_N, as shown in Figure 12.10 (b). It is instructive to compare this sequence of a structural and

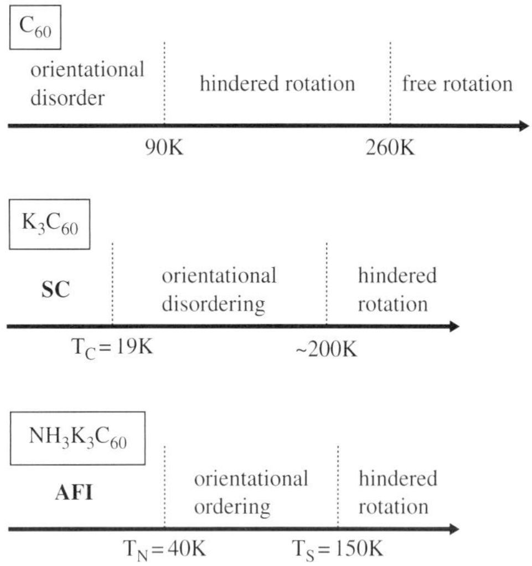

Figure 12.11 Structural and electronic transitions in C_{60}, K_3C_{60} and $NH_3K_3C_{60}$. The last, in contrast to the other two systems, maintains orientational order in its low-temperature structure. This condition is essential for the emergence of antiferromagnetism (*SC*: superconductor, *AFI*: antiferromagnetic insulator). Reprinted from [489], with permission from IOP Publishing.

an electronic phase transition with the parallel behavior of K_3C_{60} and C_{60}. This is done in Figure 12.11. The fullerene molecules in the C_{60} compound undergo rapid rotations above T_s, carry out a hindered rotation below this threshold and finally adopt a phase of orientational disorder. As C_{60} is doped with alkali metal atoms, fullerene rotation takes place, but it is substantially hindered and ceases for strongly doped systems, such as K_6C_{60}. The low-temperature structure of A_3C_{60} is orientationally disordered. This condition, nevertheless, does not suppress the onset of superconductivity.

12.3.4 TDAE-C_{60}

While the C_{60} molecule *per se* is non-magnetic, the compound tetrakis-dimethylamino-ethylene-C_{60} (TDAE-C_{60}), discovered in 1991 [500], has been shown to exist in different modes of magnetic ordering at low temperature. The TDAE molecule ($C_2N_2(CH_3)_8$) is known to be a strong electron donor. As C_{60} is a strong electron acceptor, electron transfer from the TDAE molecule to the fullerene is among the defining properties of TDAE-C_{60}. Two phases of TDAE-C_{60} have been detected, labeled α and α'. The former phase exhibits a structural transition at $T = 170K$ [501], leading from the c-centered monoclinic structure *C2/c* to a primitive monoclinic lattice ($P2_1/n$). Figure 12.12(b) indicates the small

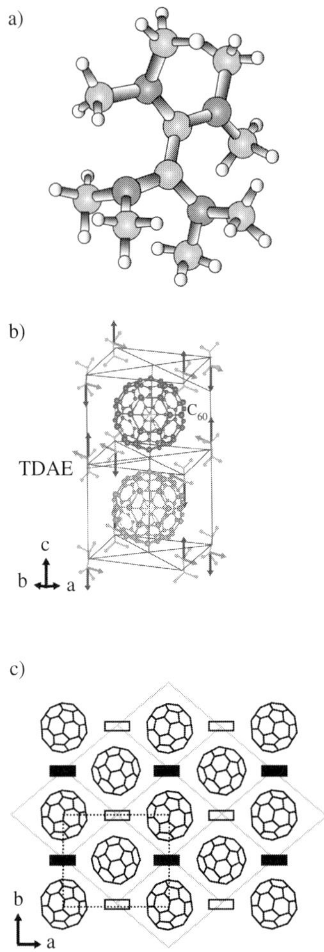

Figure 12.12 The crystal structure of TDAE-C_{60} in the α phase at low temperature ($T < 170K$) from x-ray diffusion measurement [501]. Panel (a) shows the TDAE molecule (C atoms: light grey, N: dark grey, H: white circles). Panel (b): The structure of α-TDAE-C_{60} at room temperature. The displacements of the TDAE molecules that lead from the room temperature to the low-temperature lattice, involving shifts into the crystallographic b and c directions, are indicated by arrows. Every fullerene is surrounded by eight TDAE molecules, and vice versa. Panel (c): view along the c direction. Open (closed) rectangles stand for TDAE molecules above (below) the fullerene plane. Reprinted with permission from [502] (a,b) and [501] (c). Copyright (2014) by Nature Publishing Group (a,b) and (2005) by the American Physical Society (c).

displacements (in the order of 0.07 $\AA$ [501]) that deform the room temperature into the low-temperature phase.

The α phase is ferromagnetic, with a Curie temperature of 16 K [500], which is exceptionally high for organic ferromagnets. The unit cell at low

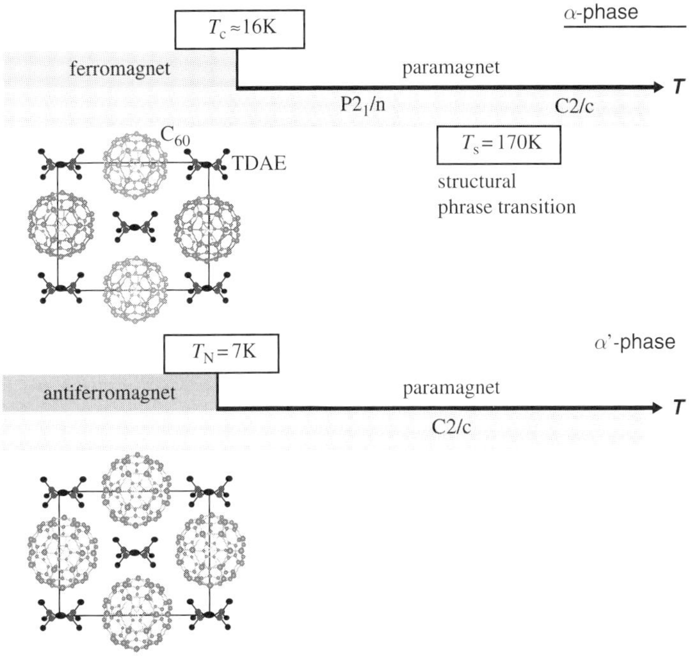

Figure 12.13 Structural and magnetic order of TDAE-C_{60} in the α (upper panel) and the α' (lower panel) phase (from [502]). The former compound undergoes a transition to a structure of lowered symmetry at $T = 170K$. The low temperature structures of the α (the α') phase involve two hexagons (a hexagon and a pentagon) of adjacent C_{60} molecules facing each other. These two structures correlate with different types of orbital ordering. The former scheme leads to ferromagnetism, the latter to antiferromagnetism in the ground states of α and α' TDAE-C_{60}, respectively. Reprinted with permission from [502]. Copyright (2014) by Nature Publishing Group.

temperature includes two inequivalent fullerene molecules, oriented in such a way that hexagons of adjacent molecules face each other [502]. The α' phase is paramagnetic at room temperature, changing to antiferromagnetic coordination at a Neél temperature of 7 K ([503], see Figure 12.13). No structural phase transition is observed in this case. The unit cell of this phase contains only one C_{60} molecule. The contact between neighboring fullerenes involves a pentagon and a hexagon facing each other. The α' phase transforms into the α phase [501, 504] upon thermal annealing and is thus metastable.

Recent computational-experimental studies, involving soft x-ray photoelectron spectroscopy measurements on an α-TDAE-C_{60} single crystal [505], confirmed the charge transfer model for the electronic structure of the α phase, corresponding to the effective charge states TDAE$^+$ and C_{60}^-. This result supported conclusions drawn from earlier magnetization [506] and NMR [507] experiments. Dimer formation among the weakly bonded $TDAE^+$ ions along the c axis of the crystal was proposed [505].

The electron transferred from the TDAE moiety to the fullerene endows C_{60} with a spin of one-half. It occupies one of the three degenerate T_{u1} orbitals of C_{60} in icosahedral (I_h) symmetry (see Section 6.3). This gives rise to *charge-transfer induced intramolecular Jahn–Teller* (CTJT) distortion, reducing the I_h symmetry of neutral C_{60} to a lower group, and splitting the LUMO triplet T_{1u} into a singlet and a doublet, $A_{2u} + E_u$.

The different magnetic behavior of the two TDAE-C_{60} phases is the consequence of a cooperative effect that involves spin and orbital degrees of freedom, as well as the lattice geometry. The key for understanding this effect lies in a correlation between the relative orientation of occupied $LUMO_i$, $i = x, y, z$, orbitals at adjacent sites and their magnetic interaction. The notation $LUMO_i$, $i = x, y, z$, refers here to the split orbitals that emerge from the T_{1u} state of C_{60} in a non-cubic environment. We assume that the ground state configuration of C_{60}^- involves a filled $LUMO_z$ orbital of C_{60}. For illustration, Figure 12.14 shows two nearest-neighbor C_{60}^- orientomers, formed by Jahn–Teller distortion due to electron transfer into one of the $LUMO_i$, $i = x, y, z$ orbitals. In panel (a), both orientomers are aligned, while in panel (b), their axes are perpendicular with respect to each other. Configuration (a) may be treated in the framework of the single-band Hubbard model at half-filling which is known to yield an antiferromagnetic ground state (see Section 3.1.1). In the alternative situation, shown in (b), intermolecular electron transfer proceeds still preferentially between corresponding $LUMO_i$, $i = x, y, z$ components, for instance between the $LUMO_z$ orbitals at neighboring sites. This case, however, involves a transition between a ground state and an excited state, resulting in a final configuration with two electrons that inhabit different $LUMO_i$ orbitals, both attached to the same site. This condition favors a spin triplet over a singlet state. Since no spin flip occurs during the transition between neighboring fullerenes, ferromagnetism is a plausible ordering scheme for the case of perpendicular orientomers. This reasoning is borne out by a quantitative treatment that applies the one-dimensional multi-band Hubbard formalism to the problem depicted in Figure 12.14.

From these considerations, the charge transfer-induced intermolecular Jahn–Teller effect imposes a certain orbital structure on TDAE-C_{60} in its low-temperature α form; i.e. its ferromagnetic phase. The constraint that nearest-neighbor orientomers are perpendular with respect to each other may be satisfied by two different models, shown in Figure 12.15. In the first (I), the axes of elongation alternate between alignment along the c axis and in the $a - b$ plane. In the second (II), all axes lie within the $a - b$ plane. Which one of these structural versions is realized preferentially?

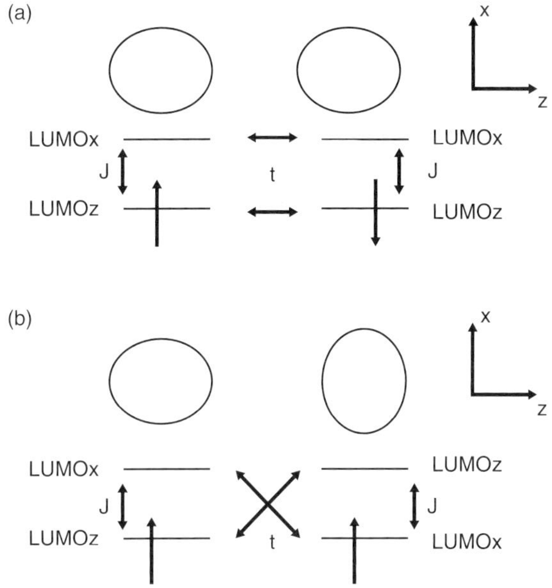

Figure 12.14 Relation between relative alignment of singly occupied C_{60} orbitals $LUMO_i$, $i = x, z$ at adjacent centers, and the resulting magnetic order. For the sake of clarity, only x and z orbitals are included. (a) $LUMO_i$ orbitals of the same type are occupied on both fullerenes. The axes of both are parallel to each other. Intermolecular electron transfer proceeds between the two ground states, but not between the ground state of one molecule and the excited state of the other. This configuration gives rise to antiferromagnetic order. (b) $LUMO_i$ orbitals of different orientation are occupied. The fullerene axes are now perpendicular to each other. Intermolecular electron transfer proceeds between the ground and the excited states of the two centers. This configuration favors ferromagnetic order. Reprinted from [508] with permission from Elsevier.

Electron spin resonance (ESR) experiments [509, 510] where uniaxial pressure was exerted on the TDAE-C_{60} crystal supplied a helpful hint for settling this question. The authors reported a marked decrease of the ferromagnetic transition temperature T_c as a function of pressure directed along the crystal axis. In particular, T_c responded sensitively to pressure applied along the c axis which suggests the validity of model (I). Also, these observations led to the discovery of a further TDAE-C_{60} phase, termed β-TDAE-C_{60}. Ferromagnetism was found to vanish at a critical pressure of about 9 kbar.

For greater pressure, and at 300 K, the onset of a new phenomenon is detected, namely the formation of 1-dimensional C_{60} polymer chains, where the individual molecules connect by [2+2] cycloaddition [511]. This phase was shown to persist after pressure release. This β phase of TDAE-C_{60} remained stable up to a temperature of 520 K, where irreversible depolymerisation was seen to occur, inducing a

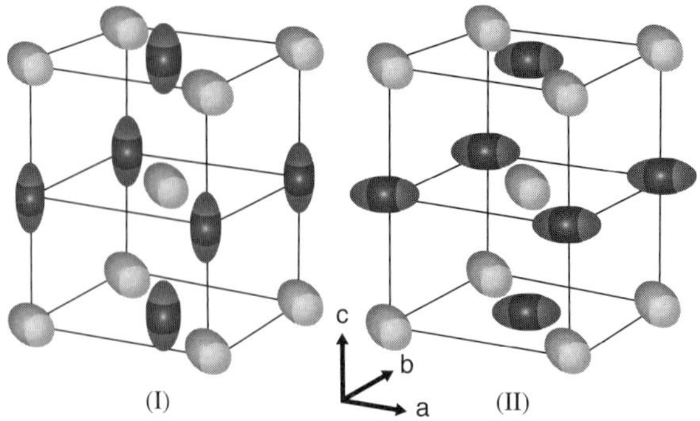

Figure 12.15 Alternative models for orientational ordering of the TDAE-C_{60} α phase at low temperature. The constraint of ferromagnetism implies that the elongation axes at nearest-neighbor sites are perpendicular with respect to each other. For simplicity, the TDEA molecules have been omitted from this scheme. Reprinted with permission from [501]. Copyright (2005) by the American Physical Society.

new magnetic configuration [512]. This high-temperature phase is paramagnetic. As documented by magnetic susceptibility measurement, twice as many spins contribute to the magnetism of β as to that of α TDAE-C_{60}. This finding has been attributed to the TDAE moiety whose spins are quenched in the α and α' phases but are present in the β phase.

We mention that linear and non-linear *AC* susceptibility (i.e. a susceptibility that arises in response to a small AC field applied to the sample) data, obtained for a single-crystal sample of α-TDAE-C_{60} [513], suggest the presence of intrinsic randomness in this system. This interpretation was supported by an analysis of the critical behavior close to the ferromagnetic phase transition [514]. As a reflection of this feature, a spin-glass transition at $T = 7$ K was identified.

A simpler magnetic ordering scheme than found in α-TDAE-C_{60} is realized for the α' phase below the Neél temperature, in accordance with the symmetry considerations illustrated in Figure 12.14 (a). As the unit cell of the crystal contains here only one C_{60} molecule, the system consists of identical orientomers with antiferromagnetic coordination, as shown in Figure 12.16. From ESR experiments [503], this system exhibits a magnetic anisotropy, with the largest axial g-factor along the *c* axis.

12.4 Adatoms on Graphene

We focus in this section on the two most extensively investigated graphene-adatom systems, namely hydrogen and fluorene in contact with graphene sheets. Local magnetism induced by hydrogen chemisorbed on graphene was already outlined

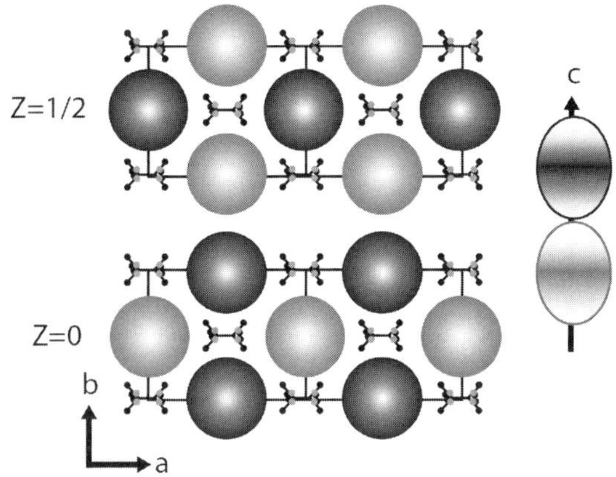

Figure 12.16 Antiferromagnetic ordering scheme of TDAE-C_{60} in the α' phase. Dark grey and light grey signify opposite spin orientations. Reprinted with permission from [503]. Copyright (2007) by the American Physical Society.

in the context of the discussion of graphene defects, presented in Section 7.2. Here we will characterize magnetic effects due to layers of adatoms, as realized by the hydrogen adsorbate systems graphane and graphone, as well as their fluorine counterparts. We will emphasize strategies for tailoring the magnetic features encountered in graphene (summarized in Chapter 7) that emerge from computational modeling of these composites as well as experimental efforts to confirm the predicted phenomena. Most important in this context are ways to enhance the size of the magnetic effects.

12.5 Hydrogen on Graphene

The simplest and most widely studied adsorbate species on graphene is hydrogen. Hydrogenated graphene was first fabricated in 2009 [515]. This was achieved by exposing suspended graphene to an argon-hydrogen plasma at low temperature (4–160 K) and pressure (10 Pa). A novel route to manufacturing a large-area, high-quality monolayer of graphene with adsorbed H atoms proceeds by use of plasma-enhanced chemical vapor deposition [516]. Hydrogenated graphene at maximum adatom coverage is named *graphane*.

Attaching an H atom to a C atom in the graphene sheet causes a small protrusion of the hydrogenated C atom. Likewise, the three neighboring C atoms are slightly displaced, and the resulting local electronic structure is determined by sp^3 rather than the sp^2 bonding that prevails in pure graphene (see Figure 12.17). The most widely investigated structural prototypes of graphane involve a *chair-like* and a *boat-like* arrangement of hydrogen adatoms on the graphene sheet,

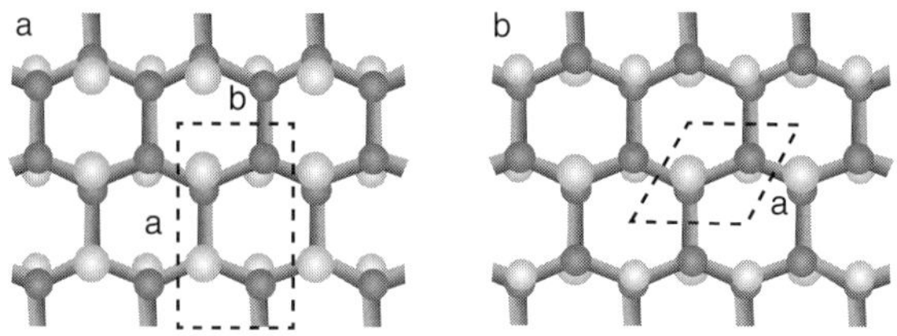

Figure 12.17 The boat (a) and the chair (b) arrangement: the two fundamental hydrogen adsorption structures for graphane. First-principles computation [518, 519] as well as scanning tunneling microscopy (STM, [523]) favor the chair model. From [517], with permission by the authors.

as illustrated by Figure 12.17. The former pattern involves a regular alternation between neighboring adsorption centers, as H atoms attaching to different sublattices are located on different sides of the graphene plane. The boat structure, in contrast, allows for H atom attachment to the same side of the plane at adjacent centers. From first-principles computation, the chair-like structure is most stable among the competing configurations [518, 519]. This conformation is stabilized by diamond-like bonds throughout, corresponding to uniform sp^3 hybridization.

Between the extremes of graphene and graphane lies partially hydrogenated graphene, which has has been termed *hydro graphene* [520]. Reaction with H atoms turns the semi-metal graphene into a semiconductor or insulator. For completely saturated graphene, i.e in the graphane limit, electronic structure theory at the level of the GW approximation[4] [521] predicts large band gaps, amounting to 5.4 eV and 4.9 eV for the chair-type and the boat-type conformation, respectively. Scanning tunneling spectroscopy (STS) studies on various hydrogenated graphene samples yielded band gap results ranging from 0.6 to 1.4 eV [523]. This finding was rationalized by density functional theory simulations of the respective adatom configurations, as observed by STM. From these calculations, the band gap is seen to depend sensitively on the level of H atom coverage. Thus, for hydrographene with an H atom coverage of only 17 percent, the minimum of 0.6 eV was obtained, matched by the computed value. Tuning the size of the gap by controlled variation of the level of H-atom coverage, and thus defining the sample in terms of semimetal, semiconductor, or insulator, is among the main incentives for exploring hydrographene.

In Section 7.2 we commented on the magnetic effect due to a single H atom adsorbed to graphene. As allowance is made for one H atom per unit cell, each chemisorption site contributes a magnetic moment of 1 μ_B. This has been established by density functional theory for various superstructures and corresponding

[4] The GW approximation is a Green's function scheme based on many-body perturbation theory [522].

distances between adjacent adatom sites, namely $d_{H-H} = 3(1+k)a_0$, with $k = 1-5$ [234]. When hydrogen dimers are considered, a magnetic moment of 2 μ_B is obtained in cases where both H atoms are adsorbed to the same sublattice, and thus to the same side of the graphene sheet [524]. Conversely, if the two H atoms are arranged in ortho- or para-dimer configurations[5], their combined magnetic moment vanishes. This behavior is in agreement with the rules based on Lieb's theorem, relating the overall magnetic moment of a graphene derivative to the sublattice imbalance (see Section 7.2.1). Accordingly, all magnetic moments due to H atoms adsorbed to the same side of a graphene sublattice are expected to be parallel to each other. This scheme might result in ferromagnetic ordering of hydrographene with only one hydrogenated side, a material referred to as *graphone* [525].

Graphone at maximum H atom coverage involves unilateral hydrogenation of all graphene C atoms within a certain sublattice. From computational simulation of this system, the presence of H adsorbates ruptures the π electron network, unpairing and localizing the p_z orbitals of the C atoms adjacent to a hydrogenated site [525]. The resulting magnetic moments are thus located at the H-atom vacancies in the hydrogen layer covering the graphene sheet. Ideally, each one of these vacancies contributes a magnetic moment of 1 μ_B to the overall magnetization of the system. Adopting, however, the analogous case of graphene magnetism due to defects as a guideline, one will expect to find non-magnetic behavior beyond a certain threshold of the adsorbate density. Berashevich and Chakraborty investigated the corresponding phenomenon for H atoms on graphene [526] by use of a hybrid DFT procedure. Specifically, they studied graphane with H atom vacancies and monitored the preference of ferromagnetically ordered versus spin singlet states, where H atom vacancies were admitted on both sides of the graphene sheet. Formation of vacancies on one side, corresponding to sites of the same graphene sublattice (the *AA scheme*), turned out to be preferred to vacancy distribution over both sides, mixing the sublattice sites (the *AB scheme*). Accordingly, ferromagnetism was found to prevail over antiferromagnetic order. While the energy difference between these alternatives was found to increase as the number of vacancies was enhanced from very dilute to intermediate levels it dropped towards zero in the limit of high vacancy concentration.

The evolution of magnetic substructures on graphene has been discussed in the context of H atom diffusion [527]. More specifically, DFT computation was used to determine energy barriers between alternative patterns of H atom attachment, starting from configurations with high magnetic moments. A hydrogen dimer with the two H atoms chemisorbed to adjacent sites within the same sublattice was found to be substantially less stable than with the atoms at next-neighbor positions, and thus

[5] Two adatoms attached to an hexagonal ring form an ortho- (para-) dimer if they attach to corners of minimum (maximum) separation within the hexagon.

attached to different sublattices. The computation reported in [527] yielded a margin of 1.15 eV for the energy difference between these alternatives. The transition state connecting the two turned out to be higher in energy by 0.65 eV than the less stable state. By thermal activation or tunneling, an initially magnetic H-dimer state is thus expected to decay into a non-magnetic final state. Similar conclusions have been drawn for H-atom trimers. Here, an initial arrangement with $\mu = 3\mu_B$ was seen to end up with the minimal possible magnetic moment of $\mu = 1\mu_B$. Strictly analogous observations were made for supercells, comprising substantially larger clusters of H adatoms. The arrangement of maximum stability proved to be non-magnetic throughout if the number of adsorbates was even, and it was found to adopt the minimal possible magnetic moment, $\mu = 1\mu_B$, in case of odd adsorbate numbers.

This scheme corresponds to a choice between two possible values of the nullity η (see Eq. (7.9)), namely 0 and 1, for even- or odd-numbered sets of H adatoms, respectively. This finding is plausible, considering that η may be understood as a label that counts the number of unbound states at the Fermi level.

Naturally, supported rather than suspended graphene has been used in experimental realizations of hydrographene with unilaterally attached H atoms [528, 529, 530]. The resulting adsorption structures are strongly affected by the choice of the substrate. Effectively, the preferred positions of H atom attachment are determined by the superlattice formed by graphene and the underlying material, i.e. the *Moiré pattern* that emerges from the juxtapposition of two lattices with different periodicities. An experimental study of H atom adsorption on graphene deposited on an Ni(111) substrate resulted in chemisorption on the vacuum side of the graphene layer [530]. By means of x-photoemission sprectroscopy (XPS) as well as thermal desorption spectroscopy (TPD) in conjunction with DFT simulation, it was observed that the H atoms attached preferentially to C atoms above FCC hollow sites of Ni(111). This *meta arranged* hydrogen pattern involved covalent bonding of an H atom to every second C center of graphene, amounting to graphone at a coverage of one monolayer. We mention that the prospect of high hydrogen uptake in combination with facile ways of controlling dehydrogenation makes the composite graphene/Ni(111) interesting for hydrogen storage technology [530].

Other hydrogen adsorption structures are detected for different transition metals as substrates of graphene, such as iridium or platinum [529]. In these cases, lattice incommensurability gives rise to slight graphene corrugation. H adatoms tend to locate in the valleys of the corrugated graphene layer. Since the hydrogenation of a C atom breaks the π bonds between this atom and its neigbors, it causes enhanced bonding between the neighbor atoms and the transition metal atoms of the substrate. This holds specifically for the valley regions of the graphene sheet where the distance between the C centers and the substrate atoms is smallest.

Theoretical predictions of magnetic effects in hydrogenated graphene outnumber experimental observations of these effects by a wide margin. Non-local spin transport measurements performed by McCreary et al. [531], however, supplied experimental evidence for magnetism induced in graphene by H atoms adsorbates. The four-electrode setup described in Section 10.1 was used, involving a non-local spin valve geometry that allows to separate a current from a voltage path. The device displayed in Figure 10.4 was modified, as two Co electrodes were placed between two Au electrodes, reducing the number of ferromagnetic contacts to two. Non-local spin signals were recorded, similar to what is shown in Figure 10.5, with the difference that only two relative magnetizations were employed, those provided by the two interior electrodes. The non-local resistance R_{nl} was recorded with graphene in the pristine state as well as with hydrogen adsorbates at different levels of coverage. A comparison between the spin signal obtained for graphene in the pristine state and after exposure to H atoms at $T = 15$ K and a chamber pressure $P = 1 \times 10^{-6}$ torr is shown in Figure 12.18. Note that positive (negative) R_{nl} signifies parallel (antiparallel) alignment. Clearly, the difference $\Delta R_{nl} = R_{nl}^{\uparrow\uparrow}$-$R_{nl}^{\uparrow\downarrow}$ depends sensitively on the level of H atom coverage of graphene, diminishing from 8.8 Ω to 1.4 Ω within eight seconds of exposure. Further, as the concentration of H atom adsorbates increases, a dip in the non-local resistance around $B_{ext} = 0$ emerges, where the two Co electrodes are in a state of parallel magnetization. The inverse feature is found for antiparallel magnetization, as demonstrated by a minor loop scan (see Figure 12.18 (c)).[6] In both cases, spin signals of markedly reduced magnitude are observed in the regime of low external field.

This finding was rationalized by interactions between conduction electron spins and the localized spins associated with the presence of H adatoms [531]. On the most elementary level of description, the coupling of the spins to each other, and to the external field, are captured by the magnetic Hamiltonian:

$$\hat{H}_{mag} = A_{ex}\hat{\mathbf{S}}_e \cdot \hat{\mathbf{S}}_M + g_e\mu_B\hat{\mathbf{S}}_e \cdot \mathbf{B}_{ext} + g_M\mu_B\hat{\mathbf{S}}_M \cdot \mathbf{B}_{ext}. \tag{12.9}$$

The constant A_{ex} denotes the strength of the exchange between the conduction electron spin $\hat{\mathbf{S}}_e$ and the localized spin $\hat{\mathbf{S}}_M$. The coupling between the two spins opens an additional *spin dephasing channel* for the conduction electrons. For high field strength, however, the Zeeman interaction between the spins and the external field prevails, decoupling the spins from each other. This mechanism motivates the experimentally established appearance of a spin signal reduction at low field strength, which may be viewed as a direct manifestation of magnetic moments induced by the hydrogen adsorbates in the graphene layer.

[6] *Minor loop* refers to a hysteresis circuit with maximum fields that lie below the saturation field strengths.

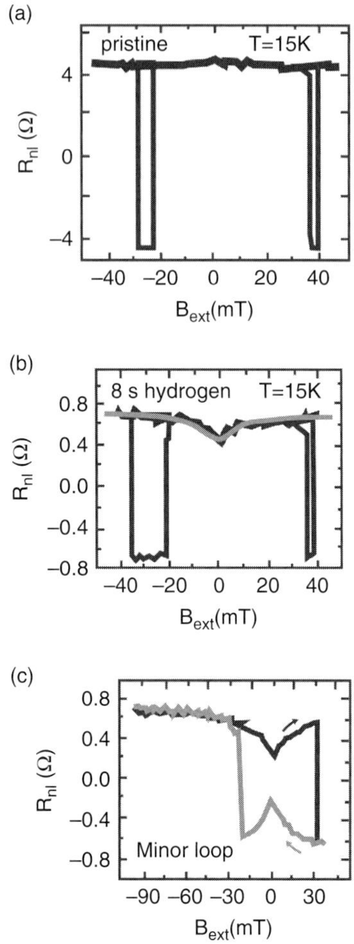

Figure 12.18 The effect of H atom adsorption on spin transport in graphene [531]. The device involves two ferromagnetic contacts. (a) The non-local resistance R_{nl} versus the external magnetic field for pristine graphene. As for the examples discussed in Chapter 10 (see Figures 10.2(c) and 10.5), R_{nl} is displayed for both sweeping directions of the field. (b) The same as in (a), after exposure of the sample for eight seconds to atomic hydrogen at a chamber pressure of 1×10^{-6} torr. With increasing H atom coverage, a dip develops at $B_{ext} = 0$, i.e. a reduction of the spin signal in case of parallel electrode magnetization at low external field strength. The light grey curve is added to emphasize this trend. (c) Results of a minor loop scan, showing that the decrease of R_{nl} corresponds to enhanced R_{nl} for antiparallel electrode magnetization. Reprinted with permission from [531]. Copyright (2005) by the American Physical Society.

This interpretation was further confirmed by Hanle precession measurements (see Section 10.1) which yielded a significantly narrowed distribution upon H atom deposition (compare with formulas (10.2)–(10.4) and Figure 10.7(b)). The fit of the experimental Hanle curve resulted in a greatly increased conduction electron g factor $\tilde{g}$ ($\tilde{g} = 7.13$) as a consequence of the spin exchange interaction due to the H adatoms.

12.6 Paramagnetism in Fluorinated Graphene – or: Graphene Paramagnetism Induced by Adatoms?

Several procedures have been used for manufacturing fluorinated graphene, the most direct among them being the mechanical cleavage of the commercially available three-dimensional compound graphite fluoride [532]. The fluorinated-graphene monolayers generated in this mode turned out to be highly fragile [535]. Following a different production pathway, however, made it possible to fabricate large samples of fluorinated graphene, namely by exposure of graphene to xenon difluoride (XeF_2). Atomic fluorine, released in the process of XeF_2 decomposition, attaches to the graphene sheet. Among the advantages of this deposition technique is that it allows to vary the concentration of F atoms on graphene in a controlled manner. Alternatively, graphene fluorination has been achieved by exposing graphene sheets to CF_4 plasma [533, 534]. The gradual fluorination of graphene may be monitored and quantitatively analyzed by means of Raman and XPS spectroscopy [535]. The label *fluorographene* (or *graphene fluoride*) refers to the limiting stoichiometry of C:F = 1, corresponding to the formula $(CF)_n$[7]. As suggested in Figure 12.19, adsorbates attached to neighboring centers are located on opposite sides of the the graphene plane.

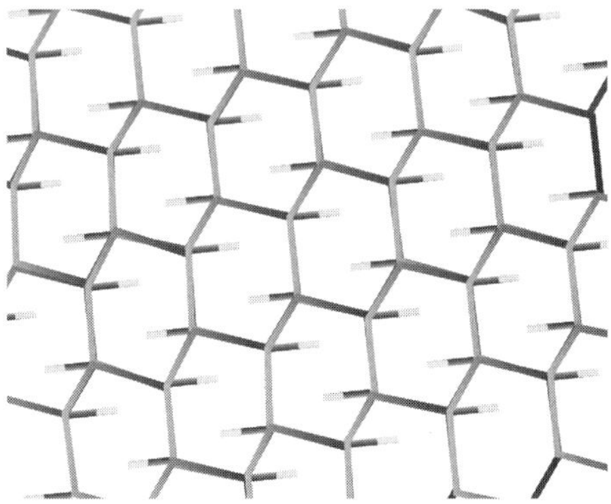

Figure 12.19 Fluorographene: adsorption pattern for fluorinated graphene, CF_x, with a limiting F/C ratio of one.

[7] This composition invites comparison with Teflon (polytetrafluoroethylene), which is formed by carbon chains, where each C atom binds to two F atoms: $(CF_2)_n$.

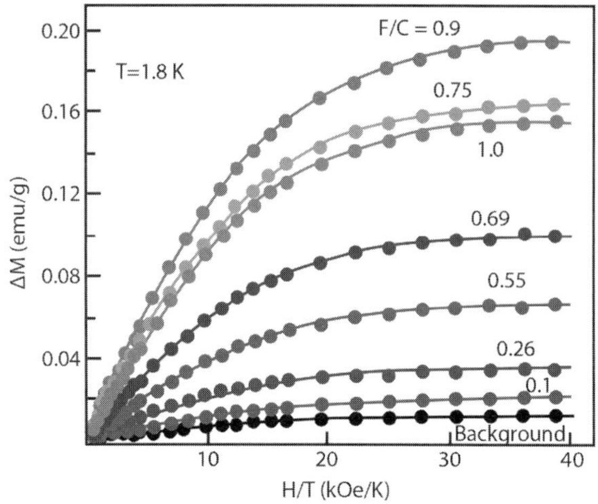

Figure 12.20 Brillouin function analysis of fluorinated graphene for several choices of the fluorine concentration x, at $T = 1.8$ K. The maximum saturation magnetization is reached at $x \approx 0.9$. All curves are fitted setting $J = \frac{1}{2}$ in the Brillouin function. Reprinted from [535], with permission from Wiley Publishing.

While fluorographene is an insulator with an energy gap larger than 3 eV [536], semifluorinated graphene, the fluorine analog of graphone, is known to be metallic. For various values of the fluorine adsorbate concentration, the magnetization of fluorinated graphene samples was determined as a function of the temperature and the strength of the external magnetic field [535]. In all cases considered, the results were found to follow precisely the Brillouin function profile (Eq. (2.31)), giving evidence for paramagnetic behavior. This is documented in Figure 12.20 at $T = 1.8K$. The saturation magnetization is seen to increase with the fluorine concentration until a maximum is attained at $x \approx 0.9$. For the stoichiometric compound fluorographene ($x = 1$), this maximum is reduced by about 20 percent.

For all Brillouin curves, excellent agreement with the data was achieved for $J = S = 1/2$. The overall magnetic effect, however, was found to be weak. Out of about 1,000 fluorine adatoms, only one turned out to contribute to the paramagnetism of the sample. This observation may be rationalized by quenching mechanisms, leading to mutual cancellations of adatom moments. As shown by density functional theory modeling [536, 537], the energy barriers towards fluorine migration on the graphene surface are low, favoring the formation of F atom clusters. This effect is further assisted by the presence of ripples, corrugating the surface [538], and supporting aggregation of F adatoms into clusters. In the interior of these structures, the net magnetization vanishes. Thus, the magnetism of a cluster of fluorine adsorbates is due to edge adatoms with uncompensated magnetic moments.

Clustering occurs preferentially on the hill segments of the corrugated graphene sheet [538]. In regions of high curvature, the sp^2 bonds in graphene adopt sp^3 character associated with the formation of a dangling bond that may be saturated by bonding to an adsorbate atom. From density functional theory computations, the chemisorption energy for hydrogen at a hill site of the rippled graphene landscape may be lower by as much as 2.5 eV in comparison with bonding to the flat surface [539]. Adsorption to valley regions, in contrast, is disadvantaged when compared with the flat case. The tendency of seeking the hills and shunning the valleys affects the evolution of the magnetization due to fluorine adatoms on graphene. From the saturation values that may be inferred from Figure 12.20, the magnetic moment increases linearly for sufficiently low population, while for $x > 0.5$, the rate of change becomes distinctly nonlinear. This trend has been explained with the existence of some critical adatom population, associated with occupation of all hill regions with fluorine clusters. Once this population threhold has been crossed, the adsorption dynamics changes in favor of single adatoms or small clusters, both of which are associated with higher magnetic moments per adatom than extended clusters.

As an aside, we point out that the experimental observation of magnetic moments due to F atom adsorption on graphene is in conflict with density functional theory (DFT) calculations at the generalized gradient approximation level (GGA, see Section 3.2.2) level, as the latter predicted a non-magnetic ground state for fluorinated graphene [540, 541]. This failure of computational modeling from first principles has been ascribed to the tendency of DFT to favor delocalized over localized electronic states [541]. This feature, in turn, resides in a native shortcoming of DFT methodology, namely the self interaction error, as described in Section 3.2.2. This spurious self-repulsion favors electron delocalization, compromising the capacity of DFT to describe spatially confined states. From STM, on the other hand, a single F adatom deposited on the graphene sheet adopts a quasilocalized state at the Fermi energy, extending over a few nanometers around the site of attachment [534]. To capture the magnetic ground state of fluorine adsorbed on graphene, a tight-binding approach has been used [540]. Further, SIE-corrected versions of DFT have been shown to be likewise capable of describing the ground state magnetism of fluorinated graphene. In particular, the hybrid Heyd–Scuseria–Ernzerhof functional [542] turned out to yield a spin-polarized result as the most stable solution. This functional contains *ab initio* admixtures, mitigating the SIE problem.

Hong et al. reported experimental evidence for localized magnetic moments in fluorinated graphene at dilute coverage from observing strong negative magnetoresistance in these materials [534, 543]. The authors determined the conductivity of graphene samples with different levels of fluorine concentration, with and without a magnetic field, in all cases oriented perpendicularly to the graphene plane. They

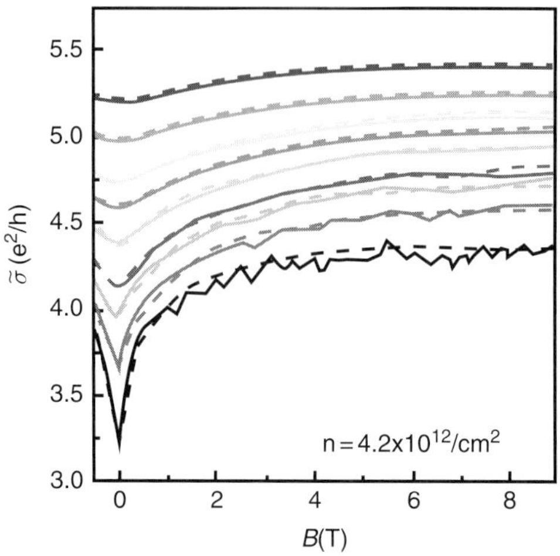

Figure 12.21 Measured conductivities of a fluorinated graphene sample at dilute coverage versus the magnetic field strength, and with the temperature as parameter. From bottom to top, the temperature values are T = 1.5, 10, 20, 30, 50, 75, 100, 150 and 200 K. The charge carrier density of the sample is $n = 4.2 \times 10^{12}$ /cm^2. The measured data are compared with predictions based on formula (11.21), where the characteristic lengths L_ϕ, L_{inter} and L_{intra} are treated as fitting parameters. The distinct negative magnetoresistance effect is indicative of weak localization. Reprinted with permission from [543]. Copyright (2005) by the American Physical Society.

further varied the gate voltage and thus the charge carrier density, the temperature and the field strength. In all cases examined, the magnetic field was found to cause a marked conductivity increase. This effect is documented in Figure 12.21, which displays the observed conductivity $\tilde{\sigma}$ for one of the fluorinated samples as a function of the field strength and for a wide range of temperatures (1.5 K $\leq T \leq$ 200K). The consistent rise of $\tilde{\sigma}$ with the magnetic field strength has been understood as a manifestation of weak localization. To substantiate this hypothesis, the authors fitted the measured conductivity curves with expression (11.21). In each case, extremely good agreement between the observed data and the weak localization model was achieved. The fit yielded the temperature dependence of the characteristic lengths L_ϕ, L_{inter} and L_{intra}.

In all cases considered, the dephasing length turned out to be smaller than that observed in pristine graphene by at least a factor of ten [544, 545], saturating, in the regime of sufficiently low temperature (see Figure 12.21), at $L_\Phi \approx 80 - 400nm$. To exclude the possibility that this reduction was caused by lattice defects or other impurities than F atoms, the authors inspected a defluorinated control sample that was, in all other respects, prepared in the same way as the samples with finite F atom coverage. Extrapolation to zero temperature yielded a vanishing dephasing rate for the control system, but a finite rate for the fluorinated systems. Further, graphene sheets with different F adatom densities were compared, and the dephasing rate was seen to increase with the level of fluorine coverage. In the presence of F atom adsorbates, not only the dephasing length, but also the inter- and intravalley scattering lengths turned out to be significantly smaller than their counterparts in pristine graphene. These observations, in conjunction with arguments based on the physics of the Kondo effect, led to the conclusion that the F adatoms give rise to the formation of magnetic moments in graphene, generating magnetic scattering centers that induce spin flip processes. The decoherence caused by this mechanism counteracts the weak localization effects in graphene. A fingerprint of this behavior is seen in the measured and calculated curves displayed in Figure 12.21, showing the conductivity versus the magnetic field. At high enough temperature, these profiles flatten out rather than exhibiting the increase characteristic for the weak localization regime.

12.7 Spin-Orbit Coupling in Hydrogenated or Fluorinated Graphene

When describing spin relaxation mechanisms in graphene (see Section 10.2), we pointed out that computational predictions of the spin life time of charge carriers in graphene overestimate the experimental finding by several orders of magnitude. Most theoretical estimates of the spin life time are based on the premise of pristine graphene. From the foregoing discussion, however, impurities such as hydrogen or fluorine may have significant impact on the magnetic interactions experienced by an electron on its way through the graphene sheet. Specifically, the previous assessment of spin-orbit interactions must be revised in the light of these findings, acknowledging the presence of defects in any graphene sample. It is intuitively plausible that the change from the sp^2 hybridization of an ideal graphene layer to the sp^3 scheme realized for hydrogenated or fluorinated graphene will influence the size of the spin-orbit effect. Also, the experimentally confirmed presence of localized magnetic moments in graphene with adatoms, is likely to be of consequence for spin-orbit scattering.

For a closer understanding of the spin-orbit effect in graphene functionalized with H or F atoms, it is essential that adsorption reduces the D_{6h} symmetry of pristine graphene [546, 547]. As adsorbates attach to the pure system, its space inversion symmetry is broken. This symmetry lowering occurs irrespective of the ultimate level of coverage; it applies to graphane as much as graphone formation, or to species with less-regular adsorption patterns. Further, pseudospin inversion symmetry, as introduced in Section 8.1, is violated. For both graphane and graphone, the Wigner–Seitz cell of pristine graphene, a regular hexagon, transforms into a puckered ring. The resulting point group is C_{3v}. As the Hamiltonian of graphene with adatoms is no longer invariant with respect to space or pseudospin inversion, spin-orbit coupling is not merely an effect of second order in this system. In particular, its space inversion asymmetry acts in analogy to an externally applied electric field, applied in a direction perpendicular to the graphene layer, giving rise to Bychkov–Rashba-like terms. An expansion of the graphone Hamiltonian in terms of constants that measure the strengths of the space and pseudospin asymmetry effects is found in reference [546]. Use of this Hamiltonian to evaluate the spin-orbit coupling in graphone led to the prediction of a strongly enhanced spin-orbit gap ξ_{SO}, exceeding that of pristine graphene by about a factor of hundred [546]. These results were borne out by electronic structure calculations based on DFT.

An additional strong enhancement of the spin-orbit interaction, amounting to roughly a factor of ten, was found when inspecting fluorinated graphene along similar lines of computational modeling [547]. The authors considered two main sources of this significant effect. In the first approach, it may be interpreted in analogy to the case of hydrogenation, that is, as the result of sp^3 versus sp^2 hybridization. Given the sizable spin-orbit coupling of the F atom, however, the adsorbate itself may contribute significantly to the observed phenomenon. As it turned out, both factors cooperate to induce the substantial spin-orbit splitting in fluorinated graphene, with the native atomic effect due to the fluorine atom as the dominant cause. This was confirmed by a simulation involving F adatoms with artificially quenched spin-orbit coupling. This manipulation led to ξ_{SO} values whose magnitude did not deviate much from typical results for the hydrogenated system.

Among the advantages of a greatly increased spin-orbit interaction in graphene is the associated magnification of the quantum spin Hall effect (see Section 11.5.2), which is not observable in pure graphene on account of its minute intrinsic spin-orbit coupling. Adsorption of foreign atoms may, in contrast, induce a clearly detectable *extrinsic spin Hall effect* [548]. Figure 12.22 provides a schematic view of an experimental setup designed to detect the spin Hall effect in graphene by measuring a non-local resistance R_{nl}. As a charge current I_c traverses the graphene sample from source to drain, electrons interact with hydrogen adsorbates. Bychkov–Rashba interaction correlates the direction of motion of the scattered charge carriers with their spin orientation, as is characteristic for the spin Hall effect. As a consequence, spin-polarized electrons diffuse through the graphene sheet, undergoing further interaction with the adsorbates. This results in a net transverse charge current that may be detected by recording the induced voltage V_{nl} between a second pair of electrodes. This process, involving a spin current that generates a voltage due to the net flow of charges, has been referred to as *reverse spin Hall effect* [550].

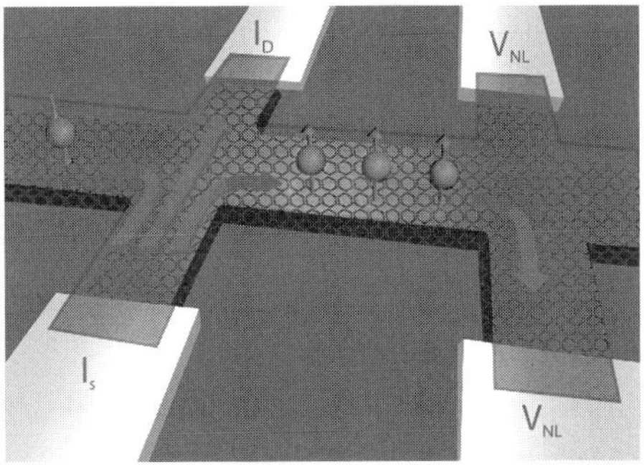

Figure 12.22 Principle of an experimental design to detect the spin Hall effect by measuring the non-local resistance $R_{nl} = V_{nl}/I_c$. The voltage V_{nl} is due to spin-dependent scattering of a spin-polarized current. As graphene is functionalized with adatoms, no external magnetic field is needed to induce this effect. Reprinted with permission from [549]. Copyright (2005) by the American Physical Society.

While increased spin-orbit coupling and, by extension, an increased spin Hall effect in graphene are desirable, these benefits come at the price of functionalizing graphene. Adding hydrogen or fluorene to the graphene sheet changes the nature of the material and thus alters its resistance, introduces disorder etc. This raises the question for the minimal levels of adatom coverage that are required to produce a marked enhancement of the spin-orbit coupling. An experiment conducted by Balakrishnan et al. [549] addressed this question. Specifically, the non-local resistance of an hydrogenated graphene sample was recorded at low levels of coverage in the order of 0.05 percent. The non-local resistance R_{nl} is described as a function of the elementary structural and physical parameters of the experiment in the following way [550]:

$$R_{nl} = \frac{1}{2}\gamma^2 \frac{W}{\tilde{\sigma} L_s} e^{-\frac{L}{L_s}}. \tag{12.10}$$

The symbol γ stands here for the *spin Hall coefficient*, defined as the ratio of the (transversal) spin Hall conductivity $\tilde{\sigma}^s_{xy}$ and the longitudinal charge conductivity $\tilde{\sigma}_{xx}$ [550]:

$$\gamma \equiv \frac{\tilde{\sigma}^s_{xy}}{\tilde{\sigma}_{xx}}. \tag{12.11}$$

The conductivity $\tilde{\sigma}$ can be obtained for various values of the charge carrier density n by recording the local resistivity $\tilde{\rho}$ as a function of n. This was done for both pristine and hydrogenated graphene samples. Relation (12.10) then was used to attain the spin Hall coefficient as well as the spin relaxation length L_s by determining the non-local resistance for different lengths of the sample, L. From this measurement, spin relaxation lengths in the order of 1.0 μm were found at the charge neutrality point and at a charge carrier density of $1 \times 10^{12}\text{cm}^{-2}$.

From the measured conductivity of the samples, their diffusion coefficient D can be derived by use of the relation $D = k_B T \mu_e$, in conjunction with the definition of the charge carrier mobility μ_e ($\tilde{\sigma} = e\mu_e n$), and for a given value of n. Thus, once L_s has been inferred from experiment, it is possible to make a statement about the spin relaxation time τ_s, acknowledging that

$$\tau_s = \frac{L_s^2}{D}. \tag{12.12}$$

In Section 10.2.2, we mentioned experimental evidence for Elliott–Yafet scattering as the prevailing spin relaxation mechanism in graphene. Adopting this process as the physical effect that establishes the spin relaxation time in hydrogenated graphene, one may employ formula (10.12) to relate τ_s to the spin-orbit splitting in the sample, as $\tau_s \approx \frac{\varepsilon_{\mathbf{k}_F}^2}{\xi_{SO}^2}\tau_m$. Here, the momentum relaxation time, τ_m, can be obtained from the proportionality between the τ_m and D [310], and $\varepsilon_{\mathbf{k}_F}$ stands for the eigenenergies of electron one-particle states at the Fermi level, $\varepsilon_{\mathbf{k}_F} = \hbar v_F k_F$.[8] A spin-orbit splitting of $\xi_{SO} = 2.5 \pm 0.2$ meV was extracted, exceeding the predicted effect in pristine graphene by three orders of magnitude, and demonstrating not only that the expected strong enhancement of spin-orbit interaction in hydrogenated graphene is observable in the laboratory, but also that it is observable at minimally obtrusive levels of H atom coverage.

12.8 Nanotubes with Magnetic Components

We focus on two classes of nanotubes, namely systems containing atomic impurities and those with embedded metallic nanoparticles. While units belonging to the former group might provide novel prototypes for highly effective elements of spintronics circuits, the latter group encompasses a large diversity of members that vary widely with respect to structural and magnetic properties, corresponding to a multitude of potential applications.

[8] The relation between the Fermi energy and the gate voltage is given by [551]

$$\varepsilon_{\mathbf{k}_F} = sgn(V_g)\hbar v_F \sqrt{\alpha\pi |V_g|},$$

where $\alpha = 7.2\ 10^{10}\ \text{cm}^{-2}\ \text{V}^{-1}$ is the gate capacitance according to the relation $n = \alpha V_g$. Reference is made here to the standard case of exfoliated graphene on an SiO_2 substrate with thickness 300 nm. On the other hand, the gate voltage determines the charge carrier density, according to [552]

$$n[10^{10}\text{cm}^{-2}] = 7.2\frac{t[\text{nm}]}{300}\frac{\kappa}{3.9}V_g[V].$$

Here, t denotes the distance between the gate and the graphene layer (“the thickness of the dielectric”), and κ is the dielectric constant of the insulating substrate of the graphene sheet.

12.8.1 Nanotubes with Atomic Impurities

Various proposals for implementing the spin valve principle with low-dimensional elements have focused on carbon nanotubes [553, 554]. Thus, one might exploit the ground state edge magnetism of zigzag tubes (zSWCNTs, see Section 7.2) to realize spintronic transmission elements with parallel or antiparallel magnetic moments at the ends of the tube. Among the major challenges related to the realization of this concept is the problem of embedding the zSWCNTs into a nanoelectronic circuit while leaving their edge magnetism intact [554]. Likewise, attaching the tube to magnetic electrodes is precarious in view of their limited capacity of injecting spin-polarized electrons into the tube. This deficiency is associated with the resistivity mismatch between the electrodes and the carbon nanosystem they attach to, as described in greater detail in Section 9.3.

This situation suggests exploring the potential of ferromagnetic adsorbates or substitutional moieties for generating the spin valve effect [555, 556]. If magnetized electrodes are employed, magnetic impurities in the transmission element induce, in general, unwanted relaxation effects. A simulation of an electrode-free configuration, in contrast, involving a carbon nanotube with randomly dispersed substitutional Ni atoms [556], demonstrated that magnetic impurities in nanotubes may cause magnetoresistance effects of substantial size. The authors used a Green's function procedure to investigate the indirect magnetic coupling $\tilde{J}_{km}$ between Ni atoms at substitutional sites labeled k and m and found that $\tilde{J}_{km}$ was positive (negative), corresponding to parallel (antiparallel) spin orientation (see Eq. (2.70)) if the impurities substituted for atoms in the same sublattice (in different sublattices). The same pattern has been established for magnetic centers embedded into graphene layers [557].

While the RKKY theory of indirect exchange is based on a perturbation approach at second order, the coupling between magnetic centers embedded into nanotubes described in reference [556] was obtained by a non-perturbative procedure [555] that includes the multiple-scattering terms associated with introducing a magnetic impurity into the tube lattice. The coupling between adjacent magnetic impurities as a function of their distance, D, as shown in Figure 12.23, displays a long-range decrease that can be approximated by a $1/D$ law. The same result has been found when adatoms rather than substitutional atoms were investigated, provided they occupied an on-top position, adsorbing to carbon atoms. A much more rapid decline (proportional to $1/D^5$), on the other hand, was found for adatoms placed on *plaquette sites*, i.e. the centers of the hexagonal rings. These trends prevail in metallic SWCNTs. In semiconducting zigzag tubes, however, no extended states at the Fermi level are available for mediating between the distant magnetic

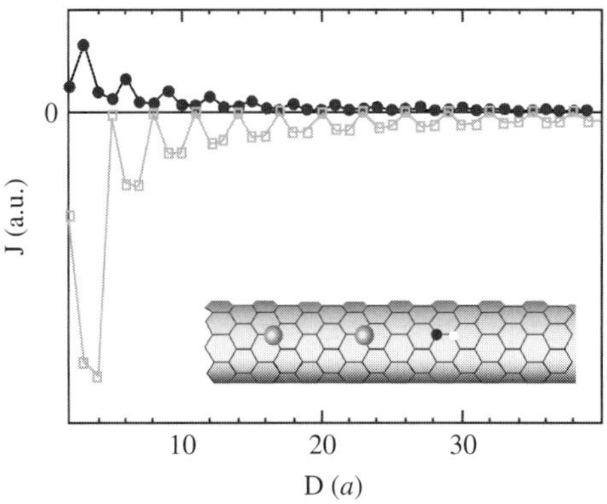

Figure 12.23 The magnetic coupling between substitutional Ni impurities in a carbon nanotube, evaluated by use of a Green's function method [555, 556], versus the distance D between two magnetic centers, in units of the lattice parameter a = 2.45 Å. The sign of the magnetic interaction is seen to depend on the relative location of the impurities: substituting two Ni atoms for C atoms in the same sublattice (A or B) yields a positive coupling constant and thus parallel spin orientation. Correspondingly, antiparallel orientation is found for impurities located in different sublattices. A marked dependence on the distance between the two impurities is a further characteristic feature of this magnetic two-center interaction. Reprinted from [556], with permission from Elsevier.

sites. In this case, no long-range interaction is found, and the magnetic coupling drops off exponentially [555]. A limiting case for the distance behavior of the coupling between substitutional impurities can be inferred by comparison with graphene, where a decrease proportional to $1/D^3$ has been obtained [558]. This is the expected asymptotic tendency for metallic nanotubes as their radius grows large.

Zanolli et al. studied the spin polarization induced in carbon nanotubes due to exohedrally connected magnetic metal nanoclusters, namely Ni_{13} and Pt_{13} [559]. As established by DFT computation, coupling between the π orbitals of the nanotube and the d orbitals of the metal cluster strongly affects the band structure of the tube and confers a net magnetic moment to the tube. The induced magnetic properties were found to depend critically on the chemical environment of the composite of nanotube and nanocluster. This effect might be of interest for the use of these materials as chemical sensing devices.

12.8.2 Nanotube-Nanoparticle Composites

Incorporating magnetic metal components into carbon nanotubes has been a subject of lively experimentation since the early years of nanotube research (e.g.

[560]). Among the motivations for these studies is the potential for carbon nanotubes to provide chemical as well as mechanical protection to the species enclosed, keeping them from oxidation as well as structural deterioration. An equally important motif is the interest in creating novel magnetic materials, characterized by low dimensionality and large geometric aspect ratio [561]. The spectrum of practical applications for these materials is wide, comprising purposes as diverse as sensing for scanning force microscopy [562] and drug delivery in medical therapy [563], to be described in further detail later in this text (see chapter 14). In the following paragraphs, we mention some recent efforts to fabricate novel hybrids, consisting of magnetic nanoparticles in the interior of carbon nanotubes.

The *floating catalyst chemical vapor deposition method* provides a versatile tool for manufacturing various mixtures between carbon nanotubes and magnetic nanoparticles [564]. The chemical vapor deposition (CVD) procedure of producing carbon nanotubes relies on the use of a catalyst, such as a 3d transition metal, and thus involves much lower temperatures than the arc discharge or laser ablation methods. In the floating catalyst method, a separate catalytic production step in the fabrication of carbon nanotubes is circumvented, as the carbon and the metal that acts as catalyst are obtained from the decomposition of the same component [565]. This production technique has been employed to generate both SWCNTs and MWCNTs. The raw material may be a combination of a hydrocarbon, e.g. benzene, methane or xylane, and ferrocene, $Fe(C_5H_5)_2$.

Prischepa et al. utilized a xylane/ ferrocene mixture to create MWCNTs filled with Fe nanoparticles [564]. The concentration of ferrocene (measured in weight percent) turned out to be a parameter that allows to steer the nanoparticle composition and the architecture of the resulting nanosystem. High-resolution transmission electron microscopy (HRTEM), assisted by Mössbauer spectroscopy, yielded a composite of two phases, namely cementite (Fe_3C) and ferrite (or $\alpha-$Fe, an allotrope of iron that crystallizes as BCC lattice and contains carbon admixtures). The cementite phase was shown to increase markedly with the the ferrocene concentration. Also, higher ferrocene concentration favored the emergence of small ferromagnetic nanowires, coexisting with sphere-like particles inside the tube. The latter species was also detected in exohedral coordination, intercalating between the tubes.

Turning attention to the decomposition process of ferrocene as floating catalyst for MWCNTs containing iron nanoparticles, Dillon et al. investigated the impact of ferrocene pyrolysis on the magnetic parameters of the product species [561]. The reaction proceeded in two stages. A first step, involving sublimation of ferrocene from the solid to the gas phase, was followed by the heat disintegration of gaseous ferrocene, where the temperature of the first stage was lower than that of the second. Both, the sublimation and the pyrolysis temperature turned out to be variables that exert considerable influence on the magnetic features of the final complex. Thus, higher pyrolysis temperature led to lower coercivity, while optimizing the sublimation temperature resulted in a substantially enhanced saturation magnetization of the Fe@MWCNT product.

In another CVD experiment aimed at generating novel phases consisting of mixtures of MWCNTs and magnetic nanoparticles, Ritter et al. used a catalytic route to produce carbon nanotubes with Co nanoparticles [566]. From SEM and HRTECM observation, the resulting MWCNTs had an outer diameter of 30–40 nm, and contained metal particles. X-ray diffraction (XRD) spectroscopy revealed the coexistence of BCC and FCC Co structures. SQUID measurements added data about the magnetization of the Co nanoparticles. Two types were distinguished. Larger particles, with diameters ranging in the nanometer regime, not exceeding about 10 nm, turned out to be ferromagnetic, as manifested by distinct hysteresis profiles. A second class of Co particles, with sizes in the atomic regime, displayed paramagetic behavior at low temperature. These findings document the coexistence of different structural as well as magnetic phases within the same nanotube.

12.9 Special Topic: Some Basics of Quantum Computing

This section is intended to provide some background for the discussion presented in Section 12.2, where spins enclosed in fullerene cages were considered as potential realizations of qubits. Classical computers store information in the form of *bits*, physical units that can adopt the values zero or one. At any time, an individual element realizes one of these two alternatives. In a quantum computer, bits are replaced with *qubits*, states defined on the Bloch sphere, as described in Section 8.4. For the following discussion, we replace the spin-one-half states $\{|-\frac{1}{2}\rangle, |+\frac{1}{2}\rangle\}$ specified in Section 8.4 with a general two-state system, labeled $\{|0\rangle, |1\rangle\}$. A single qubit may the be written as

$$|\psi\rangle = \alpha|0\rangle + \beta|1\rangle, \tag{12.13}$$

with $|\alpha|^2 + |\beta|^2 = 1$, and $\alpha = \cos\frac{\theta}{2}$, $\beta = \sin\frac{\theta}{2}e^{i\phi}$. Thus, qubits are understood as quantum superpositions of both binary states. Analogously, a two- qubit system is represented in a four-dimensional basis of product states, according to

$$|\psi\rangle = A_{00}e^{i\phi_{00}}|00\rangle + A_{10}e^{i\phi_{10}}|10\rangle + A_{01}e^{i\phi_{01}}|01\rangle + A_{11}e^{i\phi_{11}}|11\rangle. \tag{12.14}$$

The square magnitudes of the coefficients $A_{i,j}$, $i, j = 0, 1$ yield the probability of finding the system in the respective basis state. Likewise any N-qubit state may be expanded into a linear combination of 2^N basis terms. In any case, reading the result of a quantum computation requires making a measurement on the N-qubit system and so to collapse its wave function into one of the superposed alternatives.

In this way, the product of the quantum computation is a classical distribution. The operation of a quantum computer, however, differs essentially from that of a probabilistic classical computer, i.e. a randomized classical device. While, in the latter case, the system develops by applying stochastic matrices, the evolution of the former system proceeds by unitary transformations, such that the phase coherence of the evolving state is maintained.

A quantum computation thus consists in executing a sequence of unitary transformations on the initial state. Some of these operations have the status of *quantum gates*. The simplest example for these may be given by the *NOT* gate applied on a single qubit. With reference to the representation (12.13), this operation maps $|1\rangle$ into $|0\rangle$, and vice versa. This exchange is achieved by the Pauli matrix $\boldsymbol{\sigma}_x$:

$$\begin{pmatrix} 0 & 1 \\ 1 & 0 \end{pmatrix} \begin{pmatrix} |1\rangle \\ |0\rangle \end{pmatrix} = \begin{pmatrix} |0\rangle \\ |1\rangle . \end{pmatrix} \tag{12.15}$$

We note that applying the NOT operation on a qubit may be represented in terms of a *qubit rotation*. For any matrix $\mathbf{A}$ that satifies $\mathbf{A}^2 = \mathbf{1}$, it holds that $e^{i\frac{\theta}{2}\mathbf{A}} = \cos\frac{\theta}{2}\mathbf{I} + i\sin\frac{\theta}{2}\mathbf{A}$. The choice of $\mathbf{A} = \boldsymbol{\sigma}_x$ yields, therefore,

$$e^{-i\frac{\theta}{2}\sigma_x} = \begin{pmatrix} \cos\frac{\theta}{2} & -i\sin\frac{\theta}{2} \\ -i\sin\frac{\theta}{2} & \cos\frac{\theta}{2} \end{pmatrix} . \tag{12.16}$$

If $\theta = \pi$, this rotation matrix in the single-qubit space turns into $-i\boldsymbol{\sigma}_x$, i.e. the Pauli matrix that describes the NOT gate, multiplied by a phase factor.

Mathematically, *all quantum gates* may be understood as rotations applied to one or multiple qubits. In practical terms, Eq. (12.16) provides the basis for developing qubit manipulation techniques. In the case of a spin ensemble, for instance, this succeeds by exposing the spins to suitably shaped magnetic signals. If we denote by $B_0 \cdot \mathbf{n}$ the magnetic field acting on the magnetic moment $\boldsymbol{\mu}$ associated with a spin, may it be that of an electron, the related time evolution operator is

$$\hat{U}(t) = e^{-\frac{i}{\hbar}\hat{\mathbf{H}}t} = e^{\frac{i}{\hbar}B_0\mathbf{n}\cdot\hat{\boldsymbol{\mu}}t} = e^{-\frac{i}{2\hbar}B_0\mu_B\mathbf{n}\cdot\boldsymbol{\sigma}t}, \tag{12.17}$$

with $\mathbf{n}$ as a unit vector. Comparing this relation with Eq. (12.16) yields $\theta = \frac{1}{\hbar}B_0\mu_B t$, where the rotation axis is determined by the choice of $\mathbf{n}$.

The *Hadamard gate* H mixes the components of a single qubit[9], according to

$$\mathsf{H}\begin{pmatrix} |0\rangle \\ |1\rangle \end{pmatrix} = \frac{1}{\sqrt{2}} \begin{pmatrix} 1 & 1 \\ 1 & -1 \end{pmatrix} \begin{pmatrix} |0\rangle \\ |1\rangle \end{pmatrix} = \frac{1}{\sqrt{2}} \begin{pmatrix} |0\rangle + |1\rangle \\ |0\rangle - |1\rangle . \end{pmatrix} \tag{12.18}$$

[9] We refer to the Hadamard gate by use of the notation H, to distinguish it from a Hamilton operator.

Table 12.3 *Truth table for the controlled-NOT (CNOT) operation, acting on the sequence of a control and a target qubit. The latter changes its state if and only if the former is in the state* $\langle 1|_c$. *The symbol* $\oplus$ *denotes the exclusive OR; c and t stand for* control *and* target, *respectively.*

Input c, t	*Output c,* $[c \oplus t]$
0, 0	0, 0
0, 1	0, 1
1, 0	1, 1
1, 1	1, 0

Exercise 12.3 Show: The Hadamard gate may be understood as composed of rotations around the x and the z axes, according to

$$\mathsf{H} = e^{i\frac{\pi}{2}} e^{i\frac{\pi}{4}\sigma_x} e^{i\frac{\pi}{4}\sigma_z} e^{i\frac{\pi}{4}\sigma_x}. \tag{12.19}$$

As an example for a logical operation applied to a system of two qubits, we employ the *controlled NOT* (CNOT) gate, which is of major importance in quantum computation. Here, the quantum register is composed of two qubits, a *control qubit* followed by a *target qubit*. If the control qubit is in the state $\langle 0|_c$, the target qubit does not change. If it is in the state $\langle 1|_c$, the target qubit reverses, i.e. is subjected to the NOT operation. This scheme is summarized by Table 12.3.

Consequently, the matrix representation of the CNOT operation in the space $\{\langle 0|_c\langle 0|_t, \langle 0|_c\langle 1|_t, \langle 1|_c\langle 0|_t, \langle 1|_c\langle 1|_t\}$ is given by

$$\begin{pmatrix} 1 & 0 & 0 & 0 \\ 0 & 1 & 0 & 0 \\ 0 & 0 & 0 & 1 \\ 0 & 0 & 1 & 0. \end{pmatrix} \tag{12.20}$$

In the subspace characterized by $\psi_c = \langle 1|$, the mapping reduces to a NOT gate.

Based on quantum superpositions, quantum computers operate naturally in the parallel mode. For many (albeit not all) tasks their powers exceed those of parallel classical computers. This statement is often substantiated by the inverse of a regular phone book search. Looking up a person's phone number in a phone book provides an example for a *structured search*, as names in phone books are listed in alphabetic order. Reversing the process, however, i.e. looking for the name associated with a given phone number, poses an *unstructured search* problem. Quantum computing networks have turned out to be far better suited to brave this challenge than classical devices. Schematically, the list of phone numbers may be encoded by a superposition of N qubit states, extending the two-qubit expansion of Eq. (12.14). It can be generated by initialzing $\log_2 N$ qubits in the state $|0\rangle$ and then apply the Hadamard gate on each of them, resulting in set that contains all states, from $|00\ldots0\rangle$ to $|11\ldots1\rangle$ with the same amplitude factor, $N^{-\frac{1}{2}}$. These basis states correspond, in binary coding, to all numbers in the range from 0 to $N-1$. Note that in actual search situations, the entries will not be ordered according to the numbers they represent. L. Grover proposed a search strategy that singles out the targeted number by employing an *amplitude amplification operator* of the form [567]

$$\hat{Q} = -\hat{U} \cdot \hat{I}_s \cdot \hat{U}^{-1} \cdot \hat{I}_t. \tag{12.21}$$

The index t stands here for *target*, and s for any other entry than the targeted one. The map $\hat{I}_s$ is defined by $\hat{I} - 2|s\rangle\langle s|$, and $\hat{I}_t$ as $\hat{I} - 2|t\rangle\langle t|$. In the initial move, the Grover algorithm acts on the state $\hat{U}|s\rangle$, where $|s\rangle$ might be chosen as $|00\ldots0\rangle$. While $\hat{U}$ is an arbitrary unitary operator, the N-qubit Hadamard gate is a favorable choice, as it yields a superposition that contains all alternative states with equal amplitudes. Thus, the target state $|t\rangle\langle t|$ enters the superposition initially with the same weight as all competing states. The mapping $\hat{Q}$, however, is constructed such that its repeated application moves the probability of detecting the target entry into the vicinity of one, reducing the probabilities of all alternative choices accordingly. The optimum number of steps has been shown to be $\frac{\pi}{4}\sqrt{N}$. Note that the procedure pursued by a classical device, namely a direct scan of the phone number list, requires an average of $N/2$ steps. Thus, quantum computing yields a quadratic speedup where unstructured search problems are to be solved.

The condition that the answer provided by the Grover algorithm is probabilistic may be seen as a drawback of the method. Arriving at the wrong answer is not systematically excluded. A different methodology is envisaged for *ensemble quantum computing* which involves the cooperation of multiple quantum computers rather than a single one. The result of an ensemble quantum computation is an expectation value rather than the outcome of an individual measurement [568].

13

Networks of Carbon Clusters

This chapter highlights magnetic phenomena in complex extended carbon nanosystems beyond the prototypes fullerenes, carbon nanotubes, and graphene. We begin this survey with carbon frameworks based on structural motives with hyperbolic geometry, as introduced in Section 7.4.1, where elementary units with negative Gaussian curvature were discussed. These nanostructures, termed *carbon nanofoams* [569], combine extreme lightness with a high degree of stability. Among their numerous astonishing materials properties is paramagnetic, and, at sufficiently low temperature, ferromagnetic behavior. The relation between the distinct geometric features of carbon nanofoam and its magnetic characteristics is the subject of Section 13.1.

Carbon nanospheres, as considered in Section 13.2, may be understood, in some sense, as the geometric complements of carbon nanofoams, as they consist of closed carbon surfaces and thus derive from fullerenes. We focus here on nanosphere aggregates which are composed of nanospheres nested into each other, *carbon nano-onions*. The magnetism of these architectures derives from magnetic units embedded in their interior. These impurities range from clusters of transition metal atoms to magnetic nanocrystals.

Sections 13.3 and 13.4 deal with various forms of nanodiamond and nanographite. The latter has been demonstrated to display magnetic features that differ dramatically from bulk graphite. In this context, we introduce *activated carbon fibers* (ACFs), extended microporous structures composed of graphite nanoparticles. Their magnetism is governed by the simultaneous presence of different exchange mechanisms whose strengths deviate substantially from each other. ACFs are also capable of accommodating metal clusters with finite magnetic moments. As host systems, they determine both the spatial distribution and the mutual magnetic coupling of these species.

Comparison is made with another porous carbon allotrope, *zeolite-templated carbon* (ZTC). ZTCs are obtained as negative replicas of zeolite compounds. The different design of the ACF and ZTC matrices gives rise to markedly different magnetic lattices in the two materials.

In the final section of this chapter we comment on magnetism in amorphous carbon.

13.1 Nanoporous Carbon Magnets

Carbon nanofoam is a cluster-assembled allotrope of carbon. It was first fabricated by laser ablation of glassy carbon in an argon atmosphere at a pressure of about 1–100 Torr [569, 570, 574].[1] The porous material has been characterized as an aerogel. Its surface area of 300–400 m^2/g is comparable to that of zeolites, and its density of 2–10 mg/cm^3 is exceptionally low. From electron diffraction [569, 574], nanofoams contain hyperbolic carbon layers. The diffraction spectra indicate the absence of long-range three-dimensional geometric order. Inspection by low-resolution transmission electron microscopy reveals assemblies of connected clusters whose diameters range from 4 to 9 nm and contain several thousand carbon atoms. The cluster web extends into dendritic forms of which the scanning electron micrograph shown in Figure 13.1 gives an impression.

Figure 13.1 Scanning electron microscopy image of carbon nanofoam. Carbon clusters with average diameters of about 6 nm associate to form dendritic structures. Reprinted with permission from [573]. Copyright (2004) by the American Physical Society.

[1] Glassy, or vitreous, carbon is a material of extremely low density. Consisting of pure carbon, it combines properties of both glasses and ceramics. As suggested by recent research, it is composed of fullerene-like structural elements [571]. Nanofoam is obtained from glassy carbon as parent material by ultrafast laser ablation, with a repetition rate in the MHz regime [572]. The high pulse rate provides temperatures and densities sufficiently high to break and reconfigure the bonds between C atoms. The ambient gas (argon) acts as a moderator in this process, tempering the expansion of the ablated carbon gas.

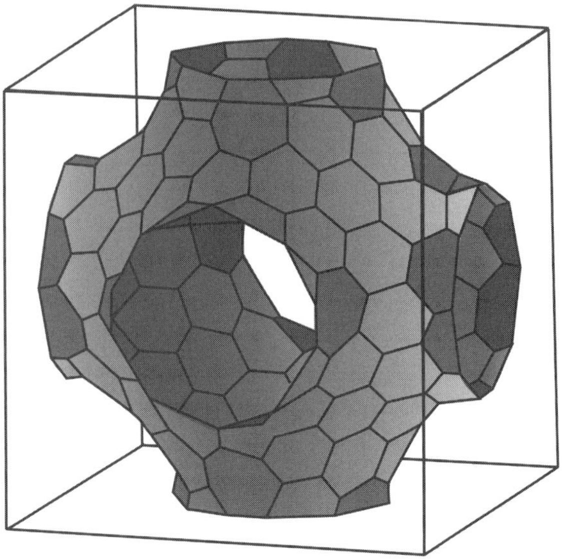

Figure 13.2 Unit cell of a periodic minimal surface of genus 3 (P-surface). This is the geometric model for the unit cell of a Schwarzite with capacity for 216 C atoms.

The elementary structural components of nanofoam have been identified as schwarzites, which were mentioned in this text first as exemplary carbon clusters with negative Gaussian curvature, such as carbon tetrapods (see Section 7.4.1). The term *Schwarzites* acknowledges the pathbreaking contributions of the German mathematician H. A. Schwarz to the theory of *periodic minimal surfaces* [575]. The latter are understood as periodic, negatively curved surfaces that divide space into an outer and an inner region. Each periodic minimal surface is specified by a topological *genus*, i.e. a number of characteristic channels within the unit cell of the surface, and further, how these channels are connected [576]. An example is provided by the "plumber's nightmare" motif shown in Figure 13.2. In this case, the genus is three. Three channels intersect at right angles in the center of the unit cell, which thus has exactly one joint. *D-surfaces* display a tetrahedral arrangement of channels. Similarly, *G-surfaces* possess channels along the (100) and (111) directions of the unit cell.[2] Repeating in all three dimensions of space, these three models represent *triply periodic minimal surfaces*. In contrast to the fullerenes, which form finite ellipsoids, Schwarzites form open structures that extend through three-dimensional space [577].

Vanderbilt and Tersoff discuss a Schwarzite of the D-type whose primitive cell has the striking property of being the *hyperbolic twin* of C_{60} [578]. Besides

[2] The labels P, D and G stand for *primitive*, *diamond-like* and *gyroid*.

hexagons, this fullerene analog contains heptagons, but no pentagons. In spite of its negative curvature, this unit shares salient geometric features with C_{60}. Thus, every C atom is a vertex of two hexagons and one heptagon, every heptagon borders on hexagons only, every hexagon is surrounded by heptagons and hexagons in regular alternation. These local topological signatures in addition to the constraint of preserving the mirror symmetry of the diamond lattice yields a unit cell consisting of 168 C atoms. The computed formation energy of this cluster turned out to be considerably lower than that of its fullerene counterpart since, in the former case the ideal sp^2 bond angle of 120° is approximated more closely than in the latter.

Interconnected Schwarzites are structural elements of carbon nanofoam. In accordance with the diversity of their geometric structures, the electronic structure of nanofoams cannot be uniquely characterized in terms of a prevailing bonding scheme. While sp^2 hybridization dominates, substantial admixtures of sp^3 hybridization have been detected [569]. Thus, the ratio $sp^3/(sp^2 + sp^3)$ has been found to be around 35 percent on average, while reaching up to 60 percent in specific regions. This information was provided by a variety of experimental techniques, among them electron energy loss spectrometry (EELS), which is an expedient tool for distinguishing the two predominant hybridization types in carbon nanostructures [579].[3] From spatially resolved EELS spectra, it was inferred that sp^2 bonding dominates in the cluster cores while the clusters are linked by sp^3 bonds. The latter hybridization scheme is seen to prevail not only on the cluster boundaries but also on the surface of the nanofoam compound as a whole. This finding rationalizes the high resistivity of this material, on the order of 10^7 Ωm [573] at room temperature. Careful examination by use of the *elastic recoil detection* method [580], employing Au atom beams at an energy of 200 MeV [581] yielded 100 ppm as the upper limit for the hydrogen content of the sample. This result eliminates the presence of hydrogen adsorbates as explanation for the experimentally secured sp^3 hybridization.

Arguably, the most remarkable feature of carbon nanofoam is its native magnetism. More specifically, measurement of the magnetic response of the foam yielded a magnetization M that increased with decreasing temperature, as indicative of paramagnetic behavior [573, 581]. The saturation value of M was found to be equivalent to 9.0×10^{-4} μ_B per carbon atom. At temperatures as low as 1.8 K, a slight hysteresis was found, and thus a signature property of ferromagnetism. The Curie temperature was reported to be ≈ 92 K. From low-temperature electron spin resonance (ESR) measurement, the average spin density of carbon nanofoam turned out to be 1.8×10^{20} per gram, corresponding to one unpaired

[3] The main signature of *graphite-like* (sp^2-coordinated) as opposed to *diamond-like* (sp^3-coordinated) carbon in an EELS spectrum is the appearance of peaks due to exciting both a σ^* and a π^* state, at 290 and 285 eV, respectively, in the K-edge regime. Only the σ^* peak is observed in case of sp^3-coordination.

spin in approximately 300 C atoms. Rutherford backscattering (RBS) and mass spectroscopy (MS) analysis ascertained that the magnetic signals did not arise from traces of magnetic metal atoms [573]. We add that carbon nanofoam is a metastable material and looses its room-temperature ferromagnetism within about one day.

All experimentally observed magnetic features of carbon nanofoam support a model of intrinsic magnetic moments, generated by unpaired spins in the clusters that define the lowest geometric scale of the nanofoam structure. This type of intrinsic magnetism differs qualitatively from that associated with terminations or ruptures of the graphene network, as is typical for the edge regions of nanotubes and nanoribbons, or defects in the hexagonal lattice, respectively. Here, finite magnetic moments are induced by the negative curvature of the carbon sheet, giving rise to the formation of heptagons or even larger rings, embedded into the hexagonal web. These geometric irregularities perturb the sp^2 bonding pattern, thus creating unpaired π states, as discussed in Section 7.4.1. Thus, cross-linked tetrapod structures have been proposed as the fundamental units of nanofoam [269]. Naturally these units are highly reactive, which raises the question for the reason of their persistent magnetism. A possible explanation for this phenomenon is provided by the principle of *steric protection*: the tight folding of the hyperbolically curved carbon layers hinders the diffusion of foreign species through the interior of the foams [269, 582].

Mathioudakis and Kelires [582] performed large-scale Monte Carlo (MC) simulations to study the emergence of carbon nanofoam and, by the same token, its magnetic properties. Specifically, the interaction and reconfiguration of representative Schwarzite units, as shown in Figure 13.3, was studied as a function of temperature. These calculations involved cubic supercells with up to 1600 atoms, interacting through the Tersoff empirical potential [583], where periodic boundary conditions were imposed. A vapor of Schwarzite clusters was annealed at 2000 K using the isothermal-isobaric (N, P, T) ensemble at various values of the pressure. Subsequently, the system underwent volume/density relaxation at 300 K. Lastly, a tight-binding formalism [294] was employed to obtain the physical properties of the resulting structure. Besides the fundamental C_{168} unit, this simulation employed primitive and gyroid Schwarzites composed of hexagon-octagon networks (P8 and G8).

Extended porous structures emerged from this investigation. The original Schwarzites change significantly as they conglomerate to form the foam. Closer analysis reveals multiple fragmentations and deformations, as manifested by the appearance of large C atom polygons as well as hybridized structures that involve deviations from the sp^2 scheme. In the interfaces between these subunits, sp^3 sites were seen to be assisted by sp^1 sites that typically occur in pairs or chains.

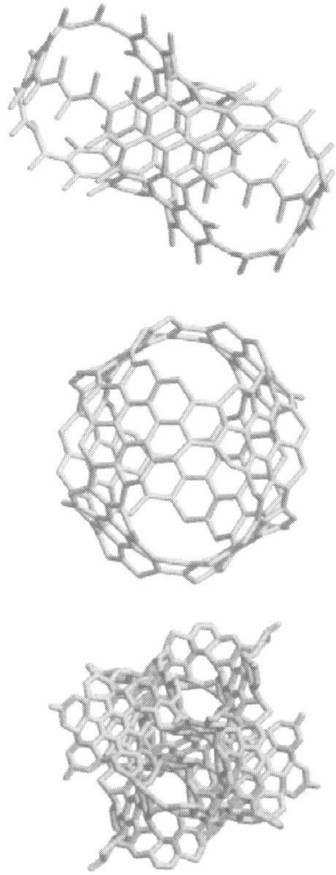

Figure 13.3 Representative carbon clusters with Schwarzite geometries. These units were used as basic materials for large-scale Monte-Carlo simulations, yielding nanofoam structures. Reprinted with permission from [582]. Copyright (2013) by the American Physical Society.

Eight- and nine-membered rings were found, arising mostly from broken and re-configured Schwarzite polygons.

While the component clusters, as shown in Figure 13.3, are semiconducting, the optimized nanofoam structures turned out to be metallic with high conductivities. This feature was attributed to the large polygons that emerge in the process of carbon nanofoam aggregation. Giving rise to locally planar geometries, they are largely sp^2 bonded and are seen to contribute substantially to the density of states at the Fermi energy of the complex.

13.2 Carbon Nanospheres

While the paradigm of a spheroidal carbon nanostructure is given by the fullerenes, a variety of related carbon shell systems have been manufactured [584]. Clusters found in soot, i.e. a mixture of carbon particles that results from the incomplete combustion of hydrocarbons, are typically spherical. As early as 1980, Iijima

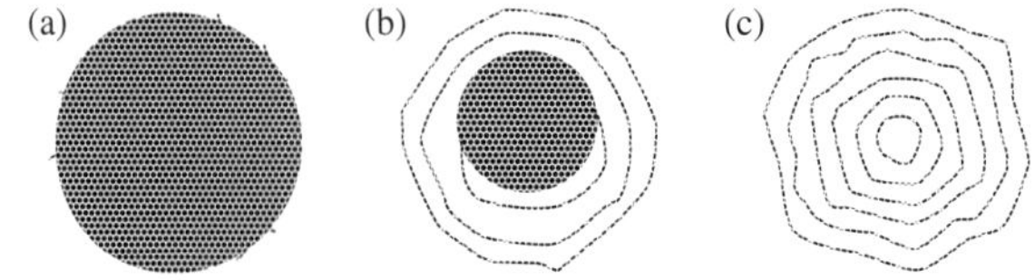

Figure 13.4 Simulated annealing of a 2-nm nanodiamond particle by use of a molecular dynamics procedure [588], leading from pristine nanodiamond (a) to carbon nano-onions. The annealing temperatures are 1673 K (b) and 2273 K (c). Reprinted from [589]. Reproduced with permission from the Electrochemical Society.

[585] reported the discovery of a spherical particle, 3–7 nm diameter, and composed of graphitized carbon. Inspection by high-resolution transmission electron microscopy (HRTEM) revealed a layered morphology. Exposing amorphous carbon to electron bombardment, Ugarte [586] later elucidated the formation process of the novel nanospecies. Under the impact of the electron beam, the sample was found to graphitize, and a distinctive architecture of numerous concentric graphitic shells was found to emerge. Prototypically, these *carbon nano-onions* (CNOs) are realized by sequences of fullerenes nested into each other. In larger quantities, CNOs are produced by annealing a nanodiamond precursor [587] in vacuum or in atmospheres of inert gases. The resulting materials are often inadequately described in terms of ideal fullerenes, but still consist of closed layers of concentric carbon nanoshells. Figure 13.4 shows stages of the transition from a nanodiamond to a carbon nano-onion structure, as modeled by simulated annealing on the basis of molecular dynamics (MD).

CNOs have been used as vessels for magnetic metal clusters [590, 591, 592]. As indicated in Section 12.1, these clusters are prone to oxidation in air, and preparing oxide forms of these clusters compromises their magnetic properties. Surrounding fullerene layers, however, shield pure metal clusters efficiently from the detrimental effects of their chemical environment. A direct method of manufacturing CNOs with endohedral metal clusters by decomposing organometallic compounds is outlined in [591]. Specifically, synthesis of these composites proceeded by thermal dissociation of metallocenes, units of the form $M(C_5H_5)_2$, involving a metal atom (M) sandwiched by two C_5H_5 rings. In the experiments described in [591], M was chosen as Co or Ni. Metallocene samples were vaporized in an evacuated reaction chamber, and the vapor was decomposed by resistive heating. After removal of the hydrogen moiety, the reaction products were cooled to room temperature. As secured by a combination of HRTEM and x-ray diffraction measurement (XRD), this procedure yielded two types of endohedral systems, namely (1) nanotubes and (2) CNOs, both containing clusters of Co or Ni atoms. Nanotubes were more abundant than CNOs, with 70–80 percent versus 20–30 percent, respectively. From

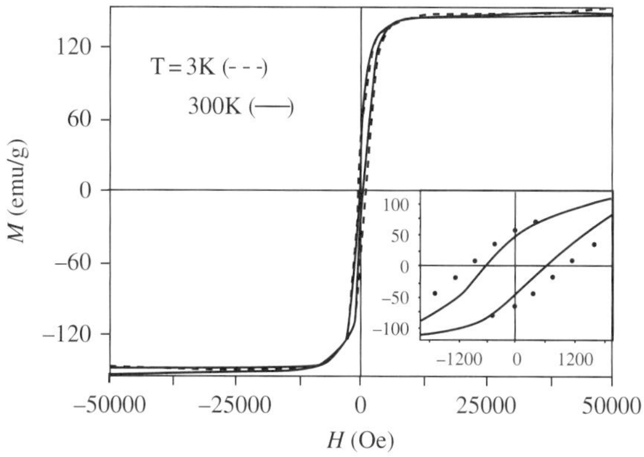

Figure 13.5 Hysteresis loop for Co nanocrystals enclosed in CNOs [591]. The inset shows the central portion of the plot in a scaled representation, indicating the reduction of the coercivity as one raises the temperature from 3 K (dotted line) to 300 K (solid line). Reprinted from [591] with permission from Elsevier.

HRTEM analysis, the encapsulated clusters were seen to range in size from ≈ 20 (Co) to ≈ 40 (Ni). Further, XRD spectra made it possible to assign lattice symmetries to the metal clusters, which thus can be characterized as nanocrystals. The x-ray signatures were consistent with pure FCC structures in the case of Ni, and FCC structures with small HPC admixtures in the case of Co. SQUID technology was used to obtain the magnetic properties of the enclosed species. Figure 13.5 shows the results for the Co containing CNO. The plot of the magnetization versus the inducing magnetic field displays a clearly defined hysteresis loop, giving evidence for ferromagnetism. The coercivity is high, with about 1000 Oe at 3 K and about 600 Oe at room temperature.

It is instructive to compare these observations with another set of experiments that involve carbon arc methods to manufacture magnetic nanocrystals enclosed by carbon shells [593, 594, 595]. Employing a Krätschmer–Huffman electrode arrangement, Ruoff et al. were the first to extract novel metal-carbide species wrapped into carbon coats [593]. Graphite rods doped with La_2O_3 were used as electrodes. Transmission electron microscopy (TEM) assisted by energy dispersive x-ray spectroscopy (EDS) for chemical analysis [596] gave evidence for a single-domain nanocrystal of the type α-LaC_2 embedded in the carbon enclosure. Including a magnetic-field-gradient separator in the experimental setup, Majetich et al. [594] extended the carbon arc technique to fabricating CNOs with internal nanocrystals composed of rare earth carbides. Specifically, they reported detection of Gd_2C_3. A paramagnetic response obtained by SQUID measurement was associated with the presence of Gd^{3+} ions. Applying this procedure to cobalt as guest species in CNO [595] yielded pure Co nanocrystals, with a carbide moiety as minority phase. For not too high external magnetic fields and at temperatures above 160 K, the magnetization of the encapsulated Co nanosystems displayed scaling behavior when plotted with respect to H/T, a fingerprint of superparamagnetism.

The production of carbon-coated magnetic nanocrystals by metallocene dissociation or by various carbon arc techniques is rivaled by a diversity of alternative procedures. Among them are tungsten arc methods [597], the use of combustion, or detonation, processes [598], magnetron and ion beam sputtering [599] and catalytic chemical vapor condensation [600]. A procedural drawback of these strategies consists in their reliance on vacuum equipment. A recently pioneered pathway of synthesizing transition metal nanocrystals inside carbon onion shells, however, avoids this inconvenience [601, 602]. This method involves *ultrasonic cavitation*. In the experiment described in [602], Fe electrodes were immersed into toluene, which was exposed to a source of ultrasound signals. Cluster formation proceeds through *hot spots*, sharply localized zones of high temperature and pressure. These spots develop when ultrasound-induced bubbles implode. At hot spots between the electrodes, toluene molecules dissociate into free radicals and ions, thus turning into a plasma that locally erodes the electrode metal. Rapid quenching leads to formation of nanoparticles. X-ray measurements confirmed the presence of elemental Fe nanocrystals in CNO enclosures. As an approach that circumvents the vacuum requirement, the use of ultrasonic cells may turn out to open a way towards cost-effective manufacturing of magnetic nanocrystals in carbon spheres. This progress will be of likely relevance for medical therapy in various areas, and will be further outlined in Chapter 14.

13.3 Nanodiamond and Nanographite

The first fabrication of nanoscale diamonds proceeded through detonation of explosives, which was pioneered in the early 1960s [603]. *Detonation diamond* arises as a product of oxygen-deficient mixtures of TNT/RDX in closed chambers. Other fabrication techniques employ laser ablation [604], plasma-assisted chemical vapor deposition (CVD) [605], ion irradiation of graphite [606], electron irradiation of CNOs [607], ultrasound cavitation [608] and more. The large multiplicity of nanodiamond manufacturing pathways that have been explored through the past decades is indicative for the high interest in this species, related to a wide spectrum of applications.

The typical length scale of nanodiamonds is 4 to 5 nm. As finite cluster systems, they are characterized by a large surface-to-volume ratio. The surfaces of pristine nanodiamond particles are covered with localized sp^3 dangling bonds, and saturation of these bonds is essential for stabilizing nanodiamonds. This may proceed through surface functionalization by terminating groups or surface reconstruction by *graphitization*. For small systems with sizes below 1.5 nm (*diamondoids*), hydrogenation has turned out to be energetically more favorable than the reconstruction alternative [609, 610]. For larger systems, functionalization competes with graphitization, i.e conversion of surface patches or faces from the sp^3 to and sp^2 bonding scheme.

As diamond is heated above $\approx$ 1000 °C, it converts into graphite. Accordingly, nanodiamond has been shown to graphitize when exposed to high temperatures. Andersson et al. reported heat treatment of nanodiamond powder in an argon atmosphere at various temperatures in the interval between 1600 and 2750 °C [611]. This procedure gave rise to the formation of *graphite nanoparticles*, or *nanographite*. At the lower end of the temperature range, a minority phase of nanodiamond remnants was detected. Some large graphitic particles, $\approx$ 200–500 nm in size, were found to appear at more elevated temperature and to grow into a major component at the high end of the range. Graphite nanoparticles were seen to be present at any temperature, with sizes independent on the heat treatment temperature. Structural information about these units was obtained by XRD, Raman-scattering and transmission electron microscopy (TEM). They turned out to be onion-structured complexes of graphitic elements, a hollow interior enclosed by sequences of graphene sheets stacked above each other, forming nested polyhedra with hexagonal cross-sections, as shown in the Figure 13.6.

The faces of these structures typically include 3–6 graphene layers whose in-plane extension amounts to 7–8 nm. The vertical distance between adjacent layers was determined to be 0.353 nm, which exceeds the corresponding distance in bulk graphite by about 5 percent, indicating a distinct reduction of the interlayer cohesion as one goes from the extended structure to the nanomaterial. Detailed experimental studies revealed the physical processes that underlie the conversion from nanodiamond to nanographite [612]. This conversion was shown to commence at a heat treatment temperature of 1200 °C. Functional groups that saturate the dangling bonds at the nanodiamond surface are removed at sufficiently

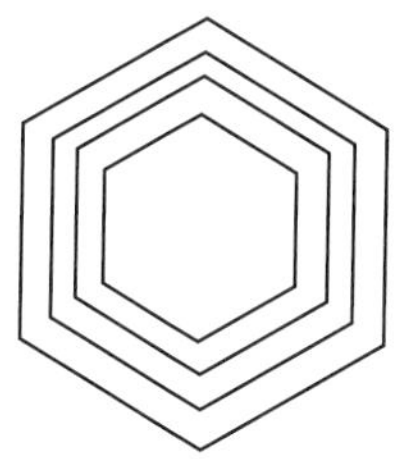

Figure 13.6 Cross-sectional view of graphite nanoparticles, as generated by heat treatment of nanodiamond powder [611]. Planar graphene sheets assemble to form nested polyhedral subunits.

high temperature, and the remaining defects turn into nucleation centers for the reconstruction of diamond (111) planes into graphitic (001) planes. This transformation begins at the nanodiamond surface and continues into the bulk as the heat treatment temperature increases. In the course of this geometric reorganization, the ratio between sp^3 and sp^2 bonds shifts gradually from the diamond- to the graphite-dominated extreme, with more and more sp^3 converted into sp^2 bonds. This development which is completed at 1600 °C has been monitored by means of Raman spectroscopy [612].

Likewise, this development may be traced by recording the evolution of key magnetic quantities during the transition. From magnetization and magnetic susceptibility measurements, nanographite exhibits intrinsic magnetism. At low temperature, and in the absence of any inducing external field, its magnetization was shown to vanish, indicating that the magnetic effects cannot be ascribed to ferromagnetic impurities. Susceptibility studies yielded a pronounced contribution due to Pauli paramagnetism (see Section 2.3.2), related to a shift between the spin densities at the Fermi energy of the system. By relation (2.101), this enhancement corresponds to a marked increase of the density of states at the Fermi energy. Since the σ bonds at the edges of the nanographite substructures can be expected to be saturated by adsorbates, or by neighboring edges within the nested polyhedra, the surplus density at the Fermi energy is due to electrons in nonbonding π states, localized at the ends of the layers. As the layers are confined by both armchair and zigzag edges (compare with Figure 4.8), the additional π states are readily associated with unpaired electrons, localizing at zigzag boundaries within the nanoparticle. The system as a whole, however, differs from a collection of carbon nanoribbons, as described in Section 4.5. Cross-linking bonds between edge states of neighboring graphene layers in nanographite, affecting both the electronic and magnetic features of the nanoparticle, have no counterparts in nanoribbons.

Nanographite thus displays magnetic features that differ dramatically from regular bulk graphite, a strong diamagnet. The distinct paramagnetic Pauli susceptibility of nanographite, χ_P, that reflects the magnetic impact of π edge states counteracts the diamagnetic contributions related to the carbon core electrons, χ_{core}, as well as the effect due to the cyclotron motion of the itinerant electrons, χ_{orb} [611]. It supports the Curie contribution to the overall susceptibility, χ_{Curie}, which measures the magnetic response of localized unpaired electrons. Among these components, χ_P is the one that deviates most from the graphite reference values, (χ_P(graphite) $= 2 \times 10^{-8}$), exceeding it by two orders of magnitude.

13.3.1 Activated Carbon Fibers

Graphite nanoparticles may aggregate to form extended microporous carbon materials with enormous specific surface areas, activated carbon fibers (ACFs)

[222, 251], a three-dimensional assembly of randomly distributed nanographite domains. Carrying out SQUID measurements of the magnetic susceptibility of ACFs as a function of temperature, Shibayama et al. traced the origin of the recorded magnetic effects back to unpaired π electrons at the edges of the graphite nanoparticles [222]. When the sample was treated at a temperature of 1473 °C, it underwent a transition from insulating to metallic behavior. This behavior has been associated with the temperature-induced loss of functional groups at the nanographite edges, opening coherent electron paths between the nanographene sheets. In fact, the transition temperature can be related to a percolation threshold for the graphene π electrons in the fractal network composed of the nanographitic segments. Beyond the transition temperature, not only the charge transport, but also the magnetic properties of ACFs become more graphite-like. The susceptibility is here strongly dominated by its diamagnetic components and shows very little variation as a function of temperature [251].

A disordered spin-glass-analogous magnetic structure was shown to arise in the vicinity of the insulator-metal transition, i.e. a heat treatment temperature around 1200 °C [222]. This phenomenon, occurring at the conductivity threshold, is indicative of the coexistence of edge-localized and itinerant π electrons in nanographite. Typically, the spin glass structure emerges wherever the exchange interaction strengths within a sample undergo statistical variations in space [251, 613]. With reference to a single nanographene sheet, two exchange constants may be discerned, a constant J_0 that determines the magnitude of the ferromagnetic coupling within a zigzag edge, and J_1, defining the interaction strength between separate edges, as shown in Figure 13.7. The mechanism underlying the latter type of exchange is transfer by conducting π electrons. Thus, the onset of metallic features is essential for the appearance of the exchange between different edges. The ratio J_1/J_0 is 10^{-1} to 10^{-2} [223]. Still weaker exchange processes occur between different sheets within one nanographite unit, and between different nanographites. The interplay between the processes described by J_0 and J_1 tends towards an effectively *ferrimagnetic* situation within an individual layer of nanographene [223].

Recent ESR studies on activated carbon fibers [223] yielded, below a transition temperature of 20 K, an *inhomogeneous phase*, specified by a static distribution of ferrimagnetic moments which originate from individual nanographene sheets, arranged in a disordered network. This phase is understood as a spin-glass state, arising from the simultaneous presence of spin coupling mechanisms of differing strengths with random spatial distributions in the neighborhood of the insulator-to-metal transition. At 25 K, motional narrowing is observed, reflecting enhanced electron mobility, and thus fast hopping of electrons between nanographene sheets, causing a transition to a homogeneous spin system. Heat treatment that takes ACFs markedly beyond 1500 °C induces their graphitization.

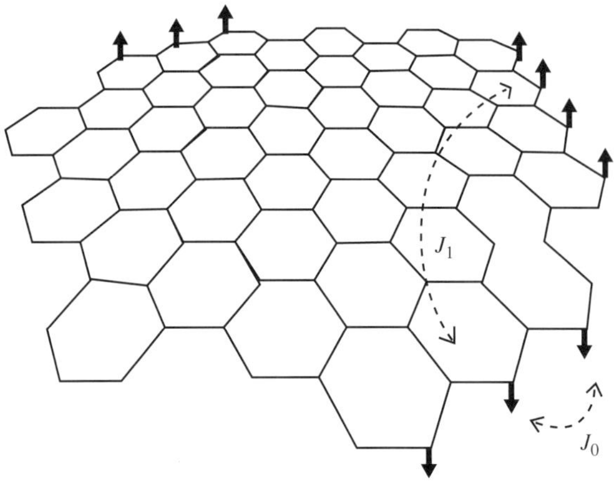

Figure 13.7 Magnetic structure of a representative nanographene sheet, irregularly confined by zigzag as well as armchair edges, which may be understood as a constituent of activated carbon fibers (ACFs). Strong ferromagnetic coupling within the edges is to be distinguished from weaker interaction between moments at separated zigzag edges, with characteristic constants, J_0 and J_1, respectively. Coupling between edges with opposite spin orientations gives rise to a state with effective ferrimagnetic coupling. Reprinted with permission from [223]. Copyright (2009) by the American Physical Society.

13.4 Host-Guest Systems Based on Nanographite

Graphite intercalation compounds (GICs) have been studied due to their intrinsic interest as well as their wide-ranging spectrum of applications [614]. They consist of graphite with guest species localizing in the interlayer spaces. Superconducting and catalytic properties, efficiency as electrode material for novel batteries, and many more technologically relevant features, make these materials attractive targets for research and development. ACF analogs of these compounds may be synthesized by inserting guest species into the hollow regions of the nanographite constituents. This manipulation can give rise to nanomagnetic materials, as was demonstrated for composites of ACFs with endohedral potassium atom impurities (K-ACFs [615]).

In the potassium crystal, the 4*s* electrons of the potassium atom shell are delocalized. In the ACF host, however, the potassium subsystem exists as a cluster. The tendency toward electron delocalization decreases in finite, as opposed to periodic, systems. In the framework of the Hubbard model, this trend may be described in terms of a competition between the hopping and the on-site Coulomb term, as the former prevails in the periodic case and the relative weight of the latter increases as the size of the system shrinks. Electron localization is further promoted by the impact of the nanographene sheets, providing a disordered environment for the potassium clusters. These factors in conjunction with the half-filled 4*s* valence

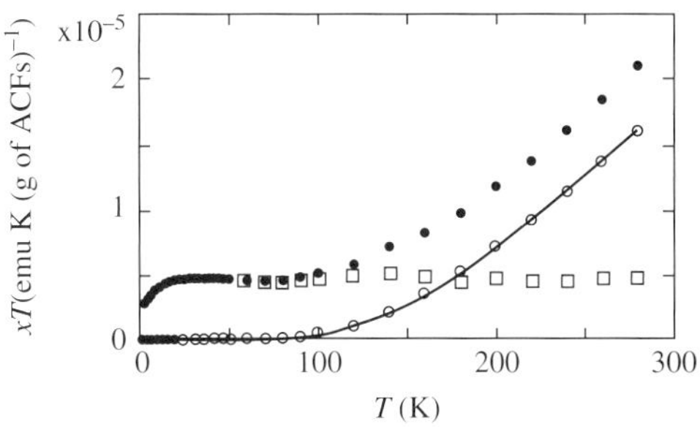

Figure 13.8 The thermal variation of the susceptibility of ACFs with embedded potassium atom clusters. The filled black symbols refer to the measured data, the open squares and the open circles to a decomposition of these data with respect to the nanographite and the potassium cluster contributions, respectively. The observation that the susceptibility product $\chi \times T$ for pure ACFs stays approximately constant in the temperature regime $T \gtrsim 80K$ allows to correct the measured values for the contribution of the ACFs and to infer the susceptibility of the metal clusters. The solid line is a fit based on expression (13.1). Reprinted with permission from [615]. Copyright (2007) by the American Physical Society.

shell of the potassium atoms yield favorable conditions for the formation of a Mott insulator with an antiferromagnetic ground state. In consequence, the alkali metal moiety in K-ACFs stabilizes as a collection of magnetic nanoparticles.

Figure 13.8 contains information about the temperature behavior of pure ACFs and of K-ACFs. Specifically, for pure ACFs, the product of the susceptibility and the temperature, $\chi \times T$, is seen to rise steeply as a function of T in the low-temperature regime $T \lesssim 80K$, followed by almost constant behavior for $T \gtrsim 80K$. This profile changes as potassium atom impurities are added. As the atomic ratio of the number of K and the number of C atoms exceeds 0.2, a marked rise of $\chi \times T$ is recorded if $T \gtrsim 80K$. Ascribing this increase to the effect of the potassium component, one may glean from the data information about the susceptibility of the endohedral potassium clusters as a function of temperature. To a good approximation, this leads to an exponential curve, according to

$$\chi_K = \frac{C_K}{T} \exp(-\frac{\Delta E}{k_B T}). \tag{13.1}$$

The subscript K refers here to potassium, with C_K as the Curie constant of the potassium moiety. Relation (13.1) characterizes the thermal behavior of an insulator with a well-defined spin gap ΔE.

We point out that positive charges are present in the potassium clusters, as consequence of a small margin of electron transfer from these clusters to their nanographite environment. While this feature counteracts the electron localization in the clusters it does preserve their characteristics of an antiferromagnetic insulator.

These findings raise the question for the extent to which the carbon superstructure determines the magnetic order of the guest species. In this context it is interesting to draw a comparison between activated carbon fibers (ACFs) and *zeolite-templated carbons* (ZTCs) as host networks. ZTCs are generated by *carbonizing organic species in microporous host materials* [616]. Thus, *zeolites*, aluminosilicate arrays with high porosity, have been used as templates for tailoring extended carbon structures in a process that involves chemical vapor deposition (CVD). This procedure gives rise to ZTCs, defined as carbon structures which emerge in the channels and caverns of zeolites as their negative replicas, and whose morphology is thus dictated by that of the host. Once the carbon network has formed the host is dissolved by means of an acid treatment.

ZTCs display, in analogy to ACFs, a three-dimensional web of pores that may be inhabited by atomic or molecular clusters. As zeolite replica, however, they form regular arrays, setting them apart from the disordered ACFs. Ideally, they are composed of fullerene-like cages which are connected by carbon nanotubes. The region intermediate between the cages and the nanotube channels contain higher carbon polygons, such as heptagons, introducing negative curvature into the structure as a whole (compare with Figure 13.2). In reality, however, the positively curved elements tend to be fullerene fragments, so-called *buckybowls*, and the system is invaded by foreign atoms, such as hydrogen and oxygen. Figure 13.9 gives an impression of the resulting geometry.

As much as pure ACFs, pure ZTCs are magnetic [617]. In contrast to ACFs, magnetism is here not predominantly an electronic edge effect. To a large extent, it emerges from sterically protected spins located at the saddle points of the sites with negative curvature. Thus, the same mechanism that preserves unpaired spins in carbon nanofoam is operative in ZTCs. Subjecting pure ZTC samples to annealing, and investigating them with respect to the temperature dependence of their susceptibilities resulted a small negative Weiss temperature, less than 1 K in magnitude. This finding indicates weak antiferromagnetism. Estimating the spin density in the samples yielded a magnetic lattice with a single spin in each porous structural unit of the ZTC frame.

Indeed, potassium clusters within ZTCs turned out to adopt an essentially different magnetic order than those in ACFs. This conclusion was drawn from SQUID susceptibility measurements in conjunction with ESR X-band spectrometry applied to CVD-generated ZTCs with incorporated K atoms, termed K-ZTCs [617, 618]. Here the potassium-carbon mixture was prepared by use of a vapor transfer method. Studying the susceptibility in the temperature range $T \leq 300$ K yielded a feature not present in K-ACFs, namely a marked peak in the $\chi \times T$ profile at 16 K, as shown

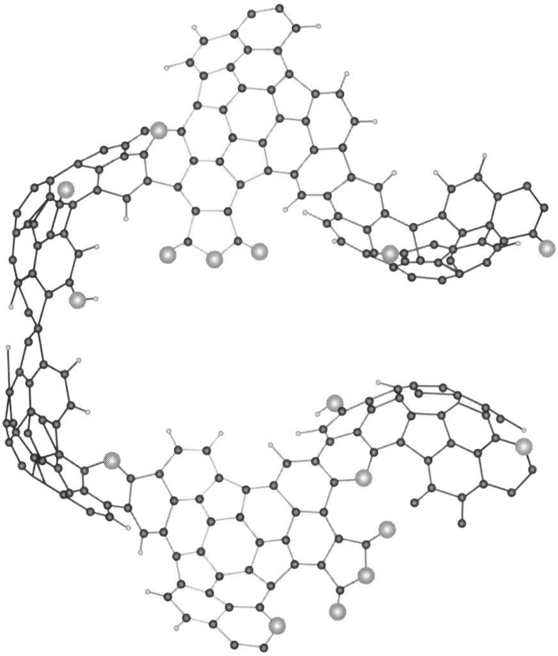

Figure 13.9 The architecture of zeolite-templated carbon (ZTC) [617]. Segments of positive curvature, containing pentagons, are connected by elements that incorporate higher than six-membered rings, inducing negative curvature. Small black spheres refer to C atoms with sp^2 coordination, large and small gray spheres to oxygen and hydrogen adsorbates, respectively. Reprinted with permission from [617]. Copyright (2010) by the American Physical Society.

in Figure 13.10. This trait was interpreted as a reflection of ferromagnetic order. The temperature dependence of the potassium cluster magnetization was found to support this model. Upon cooling the sample, a distinct rise of this quantity was detected, with an onset temperature around $T = 24$ K. Comparison with the magnetic signals from the pure ZTC made it possible to exclude that a host rather than a guest feature was observed. Further, from known magnetic characteristics of potassium-oxygen compounds, oxygen contamination could be ruled out as the cause of the effect. Tracing the magnetization as a function of the external magnetic field at various temperatures yielded, at $T = 2$ K, the definitive fingerprint of ferromagnetic behavior, namely a hysteresis curve. From the observation of vanishing remnant magnetization, however, it was concluded that the ferromagnetism induced in the endohedral potassium clusters was weak.

This behavior is analogous to that of potassium clusters accommodated in nanopores of the zeolite from which the ZTC is derived [619, 620]. These clusters have been labeled as *superatoms*. The optical spectra of these systems may be understood in terms of transitions between states of the 4*s* electron ensemble that is assumed to move in the well-shaped potential formed by the K^+ cores. The resulting shell structure is characterized by a succession of $1s$, $1p$, $1d$, $2s$ orbitals. Accordingly, zeolite cages hosting these clusters have been interpreted as *supercrystals* [619]. Ferromagnetic ordering was reported for selected cluster sizes and thus numbers of $4s$ electrons per cluster. The geometric structure of the porous host turned out to be of defining influence on the magnetic order among the alkali metal atom clusters. Thus, embedding clusters of varying alkali metal atom species into sodalites, variants of the zeolite class with modified aluminosilicate networks [621], yielded stable antiferromagnetism in all cases considered [623].

In summary, the disordered environment of the ACFs gives rise to Mott antiferromagnetism in the embedded potassium ensembles. In the ordered enclosure provided by ZTCs, on the other hand, these units adopt weak ferromagnetism.

Replacing the alkali species potassium with a halogen element, bromine, leads to a range of essentially different host-guest interactions than observed for potassium [617, 622]. Bromine moieties in ACFs react with the carbon matrix through electron absorption into the halogen shell, formation of covalent bonds and physisorption into the interstitial space between the nanographite domains [622]. On comparison with potassium impurities, bromine reverses the direction of charge transfer between the guest and the host. Electron absorption by bromine proceeds from the π electron reservoir at the edges of the nanographite sheet, reducing the orbital diamagnetism of the host. Covalent bonding between carbon and bromine takes place in the form of substitutional reactions at these edges, as C atoms are replaced with Br atoms. Lastly, the Br component assembled in the ACF nanopores exerts pressure on the nanographite units, thus varying their interlayer distances. This temperature dependent effect may be used to control the antiferromagnetic coupling between the layers [622].

Bromine as guest species in ZTCs has been shown to diminish the spin density of the pure species by approximately 60 percent [618], as reflected by the markedly lowered values of the product $\chi \times T$ for Br-ZTC when compared with K-ZTC (see Figure 13.10). Further, this effect has been discussed in the context of diffraction and Raman spectroscopy [617] which demonstrated that bromine adsorption destroys the ZTC superstructure, flattening the curved graphene network. Thus, the radical spins localized at the saddle points of the structural segments with negative curvature vanish as they are not sterically protected any longer. The remaining magnetic response stems from the edges of the flattened graphene sheets.

13.5 Amorphous Carbon

Samples of amorphous carbon with distinct magnetic properties have been obtained through pyrolysis of various organic substances [624, 625]. This process involves

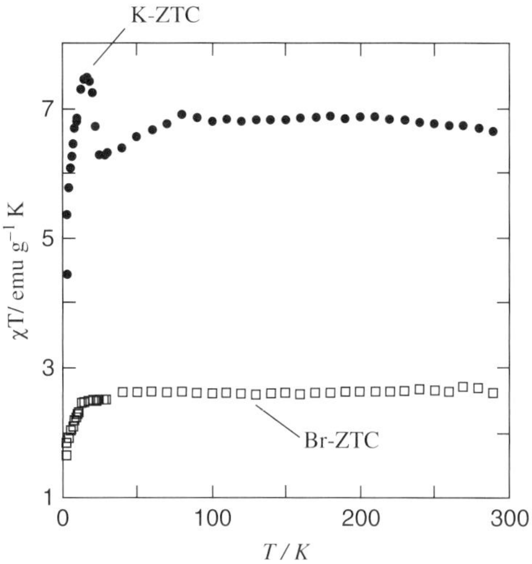

Figure 13.10 The product of the spin susceptibility and the temperature for K-ZTC (zeolite-templated carbon with embedded potassium clusters) and Br-ZTC (zeolite-templated carbon with embedded bromine clusters) as a function of the temperature, by SQUID magnetometer measurement. Reprinted from [618] with permission from Elsevier.

the thermochemical decomposition of the organic species at high temperature in an oxygen and halogen depleted environment. Thus, Murata and Ushijma fabricated amorphous-like carbon by subjecting trialkylboranes to pyrolysis [624]. No long-range order was observable by powder x-ray diffraction or transmission electron microscope (TEM) analysis.[4] Portions of the carbon produced by this procedure turned out to be magnetically active, as secured through ESR spectroscopy assisted by SQUID and vibrating sample magnetometry (VSM).

Two types of magnetic responses are distinguished, both characterized by features in the ESR spectra, as well as differing g factors. The first type, associated with a narrow ESR peak, is compatible with conventional paramagnetism, as induced by π states in a distorted sp^2 network. The temperature dependence of this component conforms with Curie behavior. The second component, on the other hand, deviates from this behavior. It corresponds to irregular, and substantially broader peaks in the ESR spectra. This second magnetic phase was attributed to regions of mixed sp^2 and sp^3 bonding, embedded into an sp^2 background. The appearance of sp^3 centers was rationalized by assuming the presence of impurity atoms in the carbon fabric, namely remnants of the trialkylborane matrix. The authors propose the model sketched in Figure 13.11 to account for the coexistence of both types of magnetism in the amorphous material.

[4] The term *amorphous-like* refers to systems that do not exhibit a long-range order while a short-range order can still be identified.

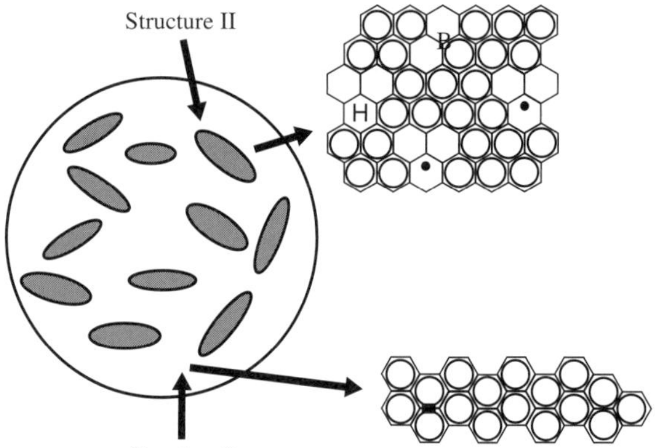

Figure 13.11 The magnetic composition of the amorphous carbon samples manufactured by pyrolysis of trialkylboranes. Integrated into a background that consists primarily of carbon in the sp^2 configuration (*Structure I*) are magnetic islands where sp^2 and sp^3 bonding coexist (*Structure II*). The sp^3 subunits are ascribed to impurities stemming from the original material, trialkylborane. Reprinted with permission from [624]. Copyright (1996) by American Institute of Physics.

Investigating the magnetically active region of samples prepared at a certain pyrolysis temperature (1083 K) yielded a distinct hysteresis effect. Further, the temperature dependence of the measured magnetization was found to be consistent with superparamagnetic relaxation theory. These observations point at superparamagnetic interaction between the active zones. From fitting the observed magnetization to a Brillouin function for the superparamagnetic case, the magnetic moment of a spin cluster was seen to be approximately 500 μ_B.

Enhancing the pyrolysis temperature to 1223 K made the broad, irregular line shapes in the ESR spectrum disappear. Only the narrow feature remained which was understood as a magnetic signal from spins of π electrons integrated into an sp^2 environment. Preparing the sample at an intermediate temperature (1123 K) yielded a superposition of both signals. Further, the saturation magnetization M_s of the sample at room temperature turned out to be sensitively correlated with the pyrolysis temperature T_p. Using triethylborane as starting material, M_s was found to be $\approx$ 32 emu/g for $T_p = 1083$ K, $\approx$ 7 emu/g for $T_p = 1123$ K, and zero for $T_p > 1223$ K. The pyrolysis temperature thus provides a means to control the magnetic texture of the sample. As a procedural rule for the synthesis of carbonaceous materials through pyrolysis of organic compounds, full graphitization of the sample is to be avoided. In this sense, the amorphous-like shape of these materials is essentially linked to their pronounced magnetism.

Similar observations as for boranes as precursors were made for pyrolytic fabrication of magnetic carbon from amines [625, 626, 627]. For instance, Araki et al. applied pyrolysis at temperatures between 800 and 1300 K on a mixture of phenylendiamine ($C_6H_8N_3$) and triazine ($C_3H_3N_2$) and generated a carbonaceous substance with ferromagnetic properties [627]. A marked isomer effect was observed. Three conformations of phenylendiamine were used, termed ortho(o)-, para(p)-, and meta(m)- phenylendiamine. They differ from each other with respect to the attachment sites of the two NH_2 radicals. Using m-phenylendiamine turned out to yield a significantly higher ferromagnetic effect than the competing o- and p-conformations of this organic substance, exhibiting, when pyrolized at 873 K, a saturation magnetization of $M_s = 6.24 \times 10^{-1}$ emu/g and a coercive force $H_c =$ 125 Oe. This finding is in accord with the argument that the possible products of the pyrolytic reaction between triazine and m-phenylendiamine are higher in correlated spins carried by nitrogen radicals than the analogous products obtained from the o- or the p-conformation of phenylendiamine.

Ferromagnatic carbon has further been generated from pyrolysis of hydrocarbons and, more generally, of precursors that were rich in hydrogen [628, 629, 630]. In general, high hydrogen content in the starting materials is conducive to high saturation magnetization [624].

In all experiments that examine the magnetism of carbonaceous compounds, great care must be taken to eliminate traces of magnetic metals as possible sources of the detected effects. Saito et al.[625], for example, analyzed the contents of transition metal elements in ferromagnetic carbon samples obtained from pyrolysis of polyacrylonitrile by use of electrothermal atomic absorption spectrometry. This yielded less than 16, 2 and 1 ppm for remnants of Fe, Co and Ni, respectively. The magnetization that could be attributed to the iron content was estimated to be maximally 3.5 10^{-3} emu/g, and thus by two to three magnitudes lower than the typically observed magnetizations of the carbon samples.

14

Medical Applications

Carbon nanostructures, especially carbon nanotubes, are of high interest for bionanotechnology, and especially medical applications. They have been considered as carrier systems, transporting drugs or other medical devices, such as contrast agents for improved diagnosis, to spatially well-defined areas of the body. Low-weight, high-chemical, mechanical, and thermal stability properties combined with a large surface area are but a few properties that make many carbon nanostructures pertinent to the purposes of biomedicine. Before employing a certain class of carbon nanostructures for clinical treatment, however, issues of its biocompatibility have to be resolved. In more detail, carbon allotropes tend to be hydrophobic, implying low solubility in an aqueous environment. As non-polar materials, pristine carbon nanostructures display only weak interactions with strongly polar molecules such as H_2O. This feature has consequences for the metabolic pathway of these materials, as it promotes their aggregation, which in turn may prevent them from being excreted by the body and thus cause them to accumulate in vital organs. Laboratory studies of CNTs point further at the risk of carcinogenesis, associated with ingestion of long, rigid nanotubes that cannot be eliminated by macrophage cells and may thus act like toxic fibers, in analogy to asbestos [631]. Another concern about administering carbon nanostructures *in vivo* is that they may harbor metal impurities that can be hostile to the body [632].

An expedient way to enhance the solubility of carbon nanostructures is modifying their surfaces by depositing hydrophilic ligand molecules on them. The toxicity of adequateley functionalized CNTs, for instance, has been found to be very low [633]. Surface modification may proceed through covalent or non-covalent bonding of ligand species [634]. In the former case, functional groups may bind to carbon atoms, or they may attach to oxygen groups that are already bound to the carbon surface.[1] The covalent connections have a significant effect on the electronic

[1] Bonding ligands to oxygen groups may be exemplified by graphene oxide, a graphene sheet with randomly distributed sp^2 and sp^3 carbon atoms, where hydroxyl, epoxy, carbonyl and carboxyl are attached to the latter.

structure of graphitic surfaces, as they change sp^2 into sp^3 configurations. Non-covalent interactions, on the other hand, leave the sp^2 network intact. They involve electrostatic, dispersive or inductive forces.

In addition, the surface of carbon nanoparticles may be functionalized with species that have capabilities of recognizing the targeted tissue and thus discriminate between natural and diseased environments. For instance, many types of cancer lead to an overexpression of the *folate receptor*, a membrane protein that binds extracellular folic acid with high selectivity and channels it into the interior of the cell [635]. By functionalizing MWCNTs used as drug carriers with folic acid molecules, this mechanism can be exploited for steering the drug towards the cancerous cells [636] and injecting it into these targets.

Carbon nanostructures may be loaded with magnetic particles in two ways: the particles may be enclosed in a carbon cage (a) or attached to the surface of the carbonaceous system (b). The general concern associated with alternative (a) is the one metioned earlier in the context of pure carbon nanostructures. Carbon-encapsulated magnetic nanoparticles (CEMNPs) tend to be insoluble in an aqueous environment, and thus in biological organisms [637]. However, the insolubility problem and, by the same token, the tendency of the particles to aggregate can be overcome by proper functionalization. Insolubility is not the chief concern about alternative (b). Many carbon nanostructures decorated with magnetic particles have turned out to be soluble in aqueous media [638], which is related to the presence of anchor molecules needed to tie the magnetic units to the carbonaceous substrate. Yang et al., for example, experimented with iron oxide particles (Fe_3O_4) on a graphene oxide substrate [638]. In this case, the particles were immobilized by COOH groups attaching to the graphene sheet. Evidently, this mode of engineering does not make use of the protective capabilities of cage-like carbon nanostructures, and thus lacks the crucial advantage of CEMNPs. However, different magnetic species are associated with health concerns of different severity. While precautionary measures, such as enclosure in carbon cages, have to be in place for the lanthanide complexes involved in MRI diagnosis, no such protection is needed from the iron oxide particles that typically provide the magnetic component of a magnetically guided drug delivery system, as they are biocompatible. This feature is related to the fact that iron oxide dissolves under acidic conditions [639].

In what follows, we focus on three areas of high current activity in research on medically relevant magnetic carbon nanostructures and their development into clinical devices: diagnosis by magnetic resonance imaging (MRI), controllable drug delivery and thermotherapy.

14.1 Diagnosis

Contrast agents are designed to increase the efficiency of imaging techniques and are routinely used in medical diagnostics. Thus, one of the most essential tools of present-day medical diagnostics, magnetic resonance imaging (MRI), employs contrast-enhancing substances to generate images of improved resolution. MRI has found a large number of applications in clinical and laboratory environments [640] because of its ability to provide detailed images of a sample by detecting nuclear magnetic resonance (NMR) signals of the ^{1}H nuclei in water. Additionally, as a non-invasive technique that does not involve any potentially harmful ionizing radiation, MRI is considered to be safer than than other modes of imaging, for instance those employing x-rays. The great versatility of MRI is manifested by a broad spectrum of clinical applications that range from detecting physiological anomalies to observing blood flow, or assessing the general status of the cardiovascular system.

As nuclear magnetic resonance (NMR) spectroscopy is at the very core of the MRI procedure, the MRI signals are determined by nuclear relaxation, specifically that of ^{1}H water nuclei, i.e. protons in H_2O molecules. MRI devices detect and amplify inductive signals that stem from two types of nuclear magnetic relaxation processes:

- Restoration of the ground state magnetization following a resonant high-frequency pulse. The organic sample is placed into a static magnetic field $\mathbf{B}_0$, resulting in a magnetized ^{1}H ensemble, with preferential orientation of the nuclear magnetic moments along the direction of $\mathbf{B}_0$, conventionally chosen as the positive z direction. The equilibrium ground state is disturbed by electromagnetic radiation at the Larmor frequency $\omega_L = g_p \mu_p B_0/\hbar$, with $g_p = 5.59$ as the g factor of the proton and μ_p as the *nuclear magneton* (= 3.15 10^{-8} eV/T). The ensemble is left in an excited state, with an increased fraction of nuclear magnetic moments oriented opposite to $\mathbf{B}_0$. Undergoing *free induction decay*, the ensemble returns to the ground state configuration. This process is referred to as spin-lattice or *longitudinal relaxation* and described by a characteristic time constant, T_1.
- Energy exchange within the proton ensemble. While this process is not associated with a net energy loss from the ensemble, it leads to dephasing of the spins oriented in the directions transverse to that of the static field. Inhomogeneity of the static field has the same effect, as it broadens the distribution of nuclear precession frequencies. The dephasing of the ensemble is known as *spin-spin relaxation*. The time scale of this mechanism is determined by the *transversal relaxation* time constant, T_2.

Both processes were introduced in Section 9.4 to account for relaxation of an electron spin current. As in the case of electrons, a classical model of relaxation dynamics is provided by the Bloch–Torrey equations (9.68). For a more detailed presentation of the physical principles underlying NMR spectroscopy, the reader may consult one of the comprehensive monographs on this topic, for instance [641].

Substances with strong paramagnetism, involving metal atoms or ions with a high number of unpaired electrons, such as lanthanide atoms with incompletely filled f shells, provide large fluctuating local fields, and thus enhance the response of the observed ^{1}H nuclei. More specifically, these *contrast agents* shorten the characteristic relaxation times of the ^{1}H nuclear ensemble. Accordingly, the effect of paramagnetic agents is to manipulate the times T_1 and T_2, and thus the corresponding relaxation rates, $1/T_1$ and $1/T_2$. In this way, differences between relaxation rates in different sections of the imaged sample may be increased, resulting in improved contrast between the MRI signals received from these sections. The effectiveness of a contrast agent in shortening either of the relaxation times is captured by its *relaxivity* r, defined by

$$R_{1,2} \equiv \frac{1}{T_{1,2}} = R^0_{1,2} + r_{1,2}C, \tag{14.1}$$

where R stands for the overall relaxation rate and R^0 for that in the absence of contrast agents, while C denotes the contrast agent concentration. The specific relaxation processes induced by a contrast agent are frequently studied by *nuclear magnetic relaxation dispersion* (NMRD) spectroscopy. This technique involves recording the spin-lattice relaxation rate as a function of the resonance frequency and thus the strength of the static field.

The basic operation of paramagnetic contrast agents can be illustrated by the example of components specifically designed to reduce the spin-lattice time, T_1. This is accomplished by dipolar or contact interaction between the spin magnetic moments of the nuclei and those of unpaired electrons contained in the agent, the latter being larger than the former by about a factor of 660.[2] The rare earth ion Gd^{3+} has turned out to be particularly suitable for MRI contrast enhancement. This is rooted in its magnetic structure which involves an exactly half-filled $4f$-orbital, corresponding to a total of seven unpaired electrons and thus the maximum $4f$ magnetic moment.

Any effect that reduces T_1 will likewise reduce the dephasing time, T_2. In most medical applications of MRI organic tissue is examined. In this environment, it

[2] Since the magnitude of the spin magnetic moment is given by $|\boldsymbol{\mu}_S| = g_e \frac{e\hbar}{2m}\sqrt{S(S+1)}$, the disparity between electrons and protons is due to the large mass ratio of $\frac{m_p}{m_e} \approx 1836$, and the ratio of g factors, $\frac{g_p}{g_e} = 5.59/2.00$.

typically holds that $R_1^0 \ll R_2^0$. As a rule, contrast agents containing Gd ions increase the relaxation rates R_1 and R_2 by approximately equal amounts, implying, by Eq. (14.1), a much greater relative increase of longitudinal than of transversal relaxation. The latter, however, is more sensitive to paramagnetic agents that introduce magnetic inhomogeneities into the static field, such as superparamagnetic iron oxide nanoparticles (SPIONs) [642, 643]. Reduction of T_1 increases the MRI signal intensity, while reduction of T_2 has the opposite effect.

Selective uptake of contrast agents by different types of tissue makes it possible to enhance the difference between the signals from a targeted region and its environment. Contrast agents are therefore assessed not only in terms of their relaxivities, but also their specificities for different tissue types.

Further, acknowledging that a typical MRI diagnosis is performed *in vivo*, contrast agents have to be biocompatible. Specifically, they need to be non-toxic. In particular, they must not be accumulated in the organism, but discharged a short time after administration. Hydrated complexes of Gd are generally toxic. Therefore, the paramagnetic metal is introduced into the MRI-examined tissue as part of a compound (chelate) in which the metal ion is surrounded by detoxifying ligands (chelators) that bind tightly to it. A qualitative distinction is made between two basic relaxation processes due to the presence of paramagnetic metals. The *inner sphere mechanism* involves temporary attachment of a water molecule to the paramagnetic center. In this case, relaxation is achieved by rapid exchange between the water molecules of the first coordination sphere of the metal ion and the surrounding solvent. The *outer sphere mechanism* is related to H_2O interaction with a coordination shell beyond the first, or diffusion of the water molecule past the chelate.

Commercially available Gd-containing MRI contrast agents include the frequently used Gadopentetate dimeglumine (Gd-DTPA), a renally excreted chelate, better known by its brand name, Magnevist. Gadolinium-based contrast agents, however, have been found responsible for a toxic reaction, nephrogenic systemic fibrosis (NSF) [644, 645] in patients with kidney problems. In addition, the currently administered contrast agents suffer from a lack of selectivity in targeting specific organs, and, by the standard of what is theoretically achievable, from low proton relaxivities.

Much research is presently devoted to designing improved MRI contrast agents that are preferable to the standard ones in terms of the relevant optimization parameters, namely safety of application, specificity and relaxation efficiency. Novel nanostructures offer an alternative to the prevailing chelate technology, as they may be used for developing contrast agents that are in most respects superior to those presently available. In this context two groups of carbon-based nanomaterials, namely fullerenes and ultra-short carbon nanotubes, have turned out to be of major interest [143, 646, 647, 648]. A feature shared by these two types of nanosystems is the cage structure that can be utilized to contain heavy metal species, as detailed in Section 12.1. By enclosing the heavy metal atoms within fullerenes or carbon nanotubes with endohedral magnetic impurities, one may lock them safely into the interior of a highly stable molecular cage, rendering them innocuous for organisms without compromising their magnetic effects.

14.1.1 Metallofullerenes

The efficiency of magnetic metallofullerenes for application in MRI has been examined by the standards of traditional MRI contrast agents. Early experiments in this field involved the unit Gd@C_{82}. Instead of contributing to chemical bonding, the $4f$ electrons of the Gd atom shell are expected to generate strongly localized magnetic moments. This feature was confirmed by ESR measurements [649] that examined the magnetization of Gd@C_{82} as a function of both the temperature and the applied magnetic field. A Brillouin function profile was obtained by assuming a total electronic angular momentum $J = 7/2\hbar$, as is found in the free Gd^{3+} ion, suggesting that the fullerene enclosure preserves the magnetic moment of the free ionic metal species. The benchmark measurement of Mikawa et al. [647] provides a reference for the comparison of a metallofullerene-based MRI contrast agent with Magnevist. The authors conducted *in vitro* MRI studies, comparing the proton relaxation enhancement accomplished by polyhydroxylated Gd@C_{82} and Magnevist. The relaxivities obtained for Gd@C_{82} were found to be about twenty times higher than those due to Magnevist at three different external magnetic field strengths. In subsequent *in vivo* studies involving the imaging of lung, liver, spleen and kidney of mice strong signals were observed at doses amounting to 5 percent of the typical clinical dose of Magnevist.

While these results were encouraging, the observed biodistribution behavior of Gd@$C_{82}(OH)_x$ as well as comparable polyhydroxylated complexes, such as $^{166}Ho_n$@$C_{82}(OH)_x$ with $n = 1, 2, 3$, calls into question the suitability of these units for clinical purposes. Thus, significant uptake of the polyhydroxylated metallofullerenes by the mononuclear phagocyte system (MPS), a part of the immune system, was observed. For some time, progress in the field was hindered by the difficulty of separating the Gd@C_n (n = 60, 70, 74, 82) species generated simultaneously by use of the arc discharge method. While Gd@C_{82} clusters only accounted for 10 percent of the synthesized gadofullerenes, they were the only Gd@C_n moiety that could be selectively extracted from the carbon soot [637]. With the advent of more refined methods of fabrication and purification, however, this obstacle was overcome. Bolskar et al. reported the synthesis and subsequent *in vivo* studies of the water-soluble and air-resistant C_{60} derivative

$Gd@C_{60}[C(COOH)_2]_{10}$ [143]. These complexes revealed much more favorable biodistribution features than the previously assessed C_{82} based units, as they started to undergo excretion within one hour after injection. The relaxivity of the C_{60} based MRI contrast agent was determined to be about five to twenty times greater than that of conventional clinical agents [650]. NMRD measurement yielded an outer sphere mechanism as the cause of the observed relaxation enhancement.

As expected, agents based on trimetallic nitride fullerenes (see Section 6.2) with Gd as metal species yielded significantly higher relaxivities than found for monometallofullerenes involving $Gd@C_n$. *In vitro* as well as *in vivo* studies of $Gd_3N@C_{80}$, which was functionalized with poly[ethylene glycol] (PEG) and subsequently hydroxylated ($Gd_3N@C_{80}[DiPEG5000(OH)_x]$) [651, 652], resulted in r_1 enhancement by up to a factor of 37 when compared with commercial Gd-based agents. Likewise, the biodistribution features of these complexes were found to be favorable. Variation of the molecular weight of the PEG derivatives provided enhancement factors as large as about 60 at selected strengths of the static magnetic field, all chosen within the clinical range [653]. From these findings, $Gd_3N@C_{80}$ appears extremely well suited as a metallofullerene core for future MRI contrast agents. The low yield of this species, however, motivates the ongoing search for alternatives, materials that can be produced in higher quantities while being equally advantageous in terms of relaxation rates and biocompatibility. Wang et al. experimented with water-soluble hydroxylated derivatives of trimetallic nitride clusters that mix Gd with Sc, such as $Sc_xGd_{3-x}N@C_{80}O_m(OH)_n$ ($x = 1,2$; $m \approx 12$; $n \approx 26$) [654]. While sizeably lower than those of the trigadolinium clusters, the relaxivities of these units turned out to exceed those of traditional agents by factors of four to five.

Recent developments on refining gadofullerene-based MRI contrast agents include the improvement of their specificty [655, 656]. These efforts are guided by the idea of *molecular imaging*, involving the chemical recognition of selected proteins, cells or parts of cells by the agents [656]. This principle is exemplified by *bio shuttle* systems linked to metallofullerenes with Gd components that have been fabricated and successfully tested *in vitro* [655, 656]. The composition of such a unit is shown in Figure 14.1. Three segments are distinguished: the head component is a cell-penetrating peptide that acts like a key, facilitating the transfer of the system through the membrane of the biological unit of interest. A disulfide bridge links this part to the *address module* of the complex, including a peptide nuclear acid (PNA) that targets c-myc mRNA-expressing cells.[3] Lastly, the trimetallic nitride fullerene $Sc_2GdN@C_{80}$ is covalently attached to the address module. The complex as a whole is termed *Gd-cluster@BioShuttle*.

[3] *c-myc* is a regulator gene. Mutated forms of this gene are present in many types of cancer where they are associated with unregulated gene expression which results in cell proliferation.

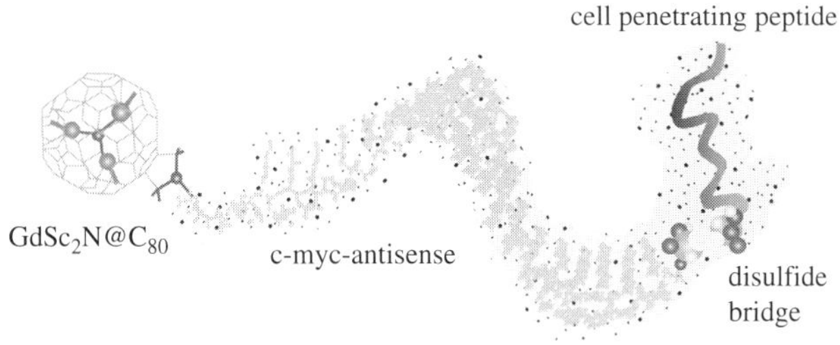

Figure 14.1 Structure of Sc$_2$GdN@C$_{80}$@BioShuttle, consisting of three functional units. A peptide, opens up the membrane of the targeted cell for penetration by the composite. The c-myc-antisense mid-section of the complex hybridizes with the c-myc mRNA, a template for the myc protein, which is implicated in the violent growth of cancer cells. The hybrid remains in the imaged cell. The third module is gado fullerene employed as MRI contrast agent. It is covalently bonded to the central module through a lysine spacer. A fluorescent dye, connected to the lysine spacer, allows one to track the transport of the complex into its destination cell. Reprinted from [656], with permission from the authors.

Fluorescence spectroscopy was used to track the motion of the imaging complex into the cancerous cell, and to confirm its eventual intracellular residence. To make this possible, a fluorescent dye was coupled to a lysine spacer that separated the address module from the imaging module (see Figure 14.1). Tracking the composite by use of an He-Ne laser in regular time intervals after incubation provided evidence for its migration into the target cell and its localization in the cytoplasm. Substantially reduced Gd concentration in conjunction with very strongly increased MRI sensitivity, corresponding to r_1 enhancement factors of 500 and more [655], make Gd-cluster@BioShuttle carrier systems a focus for further research and development. A proof of principle for the effectiveness of the concept in the study of human brain tumor cells was given recently [657].

14.1.2 Metallonanotubes

Ultra-short nanotubes, not exceeding 20 to 200 nm in length, have turned out to be interesting candidates for novel MRI contrast agents [648, 658]. Specifically, it has been shown that single-walled nanotubes can be functionalized for biocompatibility by attaching external groups to their surface [659], have tolerable toxicity if properly functionalized [660, 661] and are largely bioinert, being discharged intact by mammals [662]. The small extension of ultra-short SWCNTs implies that a large number of these units can assemble within a targeted cell, resulting in an increased response, and consequently improved image resolution. The minute size of these tubes also provides a key argument for their biocompatibility, as renal elimination proceeds more readily for smaller than for larger particles [663].

High stability in conjunction with a hollow cylindrical shape makes ultra-short nanotubes potential delivery media for magnetic heavy metal atoms, in precise analogy to fullerenes. Sitharaman and coworkers [648] reported internal loading of ultra-short SWCNTs with aqueous $GdCl_3$ molecules. High resolution transmission electron microscopy (HRTEM) demonstrates the formation of Gd_n^{3+} clusters with varying number of constituents inside the SWCNT cavity. Gadolinium clustering is expected to be catalyzed by sidewall defects of the ultra-short tube [664], created in the course of the SWCNT size reduction procedure. By use of NMRD, the resulting gadonanotubes were investigated with respect to their proton relaxivities. Extremely high effects were found, exceeding the relaxivities of conventional MRI contrast agents by factors on the order of 40 to 90, and thus surpassing those of trimetallic nitride fullerene endohedral fullerenes with encapsulated Gd_3N. As an explanatory model to account for this effect, an inner sphere mechanism has been suggested, involving interaction between the superparamagnetic Gd_n^{3+} clusters and the ^1H nuclei inside the SWCNT void. Given the high observed relaxivities, it is all the more remarkable that a chemical composition analysis of the metal-loaded SWCNTs yielded a very low Gd content, amounting to a relative weight of less than 3.0 percent. Further, the highest relaxation effects were detected at substantially lower magnetic field strength than customarily applied for Gd-based contrast agents. As another interesting feature of gadonanotube agents, these systems displayed a sensitve dependence on the pH value of their environment [658], a feature that may be used to optimize the selectivity of these devices.

14.2 Therapy

Besides the promise of magnetic carbon nanostructures for improved MRI diagnosis, their great potential for medical therapy has been recognized, as manifested by significant experimental efforts in this field. These efforts focus on refining two types of clinical technology, drug delivery and thermal treatment of tumors. In the following, we will give brief surveys on the recent progress in both areas of current development.

14.2.1 Drug Delivery

A well-known problem in chemotherapy is finding a compromise between the desired effects and undesired side effects of the delivered drug. Thus, techniques of site-specific delivery are a subject of high current interest and rapid development, as they promise to limit the damage to healthy cells while targeting the afflicted sites of the body. The concept of *magnetic delivery* involves designing drug carrier systems with magnetic components, and therefore making drug transport sensitive to magnetic interactions. An external magnetic field can then, in principle, be used to guide the carrier systems to the target sites [665]. Once their destination

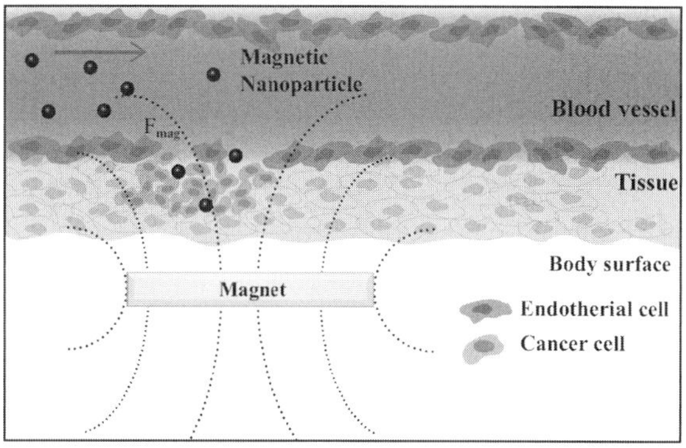

Figure 14.2 The principle of magnetic delivery: magnetic nanoparticles serve as drug carrying systems, as they are loaded with a medical agent. Channeled through arteries, they are directed toward diseased cells by a guiding external magnetic field. This field may further confine them to the target site for a period required to release the drugs. Reprinted from [666], with permission from Elsevier.

is reached, they may be held in place by action of the magnetic force for a time sufficiently long to release the drug. The scheme is illustrated in Figure 14.2.

In view of the rapid decrease of the magnetic dipole field (2.60) with the distance R ($\propto R^{-3}$), the direct implementation of the arrangement shown in the figure is confined to cases where the application site is close to the patient's skin. This deficiency of the method has been addressed by placing the guiding magnets into the interior of the body, near the tissue to be treated. The proof of principle for this extension of the method has been given in animal experiments [667].

Carbon nanostructures have been identified as suitable for magnetic drug delivery [668]. This is owed largely to the cage structure shared by many of these species. As much as the magnetic component, the medical agent may be transported in the interior of the cage [669], or it may be attached to the large cage surface of carbonaceous structures with shell geometry. In what follows, we make reference to the work of Sherlock et al. [670], who used the latter strategy to design a CEMNP for the purpose of delivering drugs to cancerous cells. Figure 14.3 shows the composition of the nanoscopic drug vehicle fabricated and assessed by the authors in terms of its capability to inhibit cell growth. The inner region of the shown particle consists of a magnetic FeCo nanocrystal coated by a graphitic carbon shell (GC). This FeCo/CG core was synthesized by use of a chemical vapor deposition (CVD) method [671]. To make the composite water-soluble, it was functionalized by polyethylene glycol (PEG) derivatives. The resulting nanoparticle was loaded with the antibiotic *Doxorubicin*^{TM} (DOX), which is used in cancer therapy. This drug is attached to the nanoparticle surface by $\pi - \pi$ stacking, a noncovalent interaction.

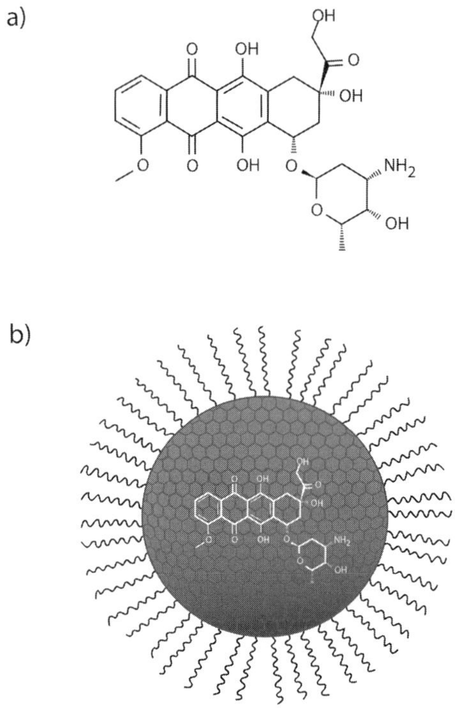

Figure 14.3 Anatomy of a nanoparticle designed to deliver the drug Doxorubicin^{TM} (DOX) to cancerous cells. (a) Structural formula of DOX. The six-membered ring joined to the rest of the molecule by an oxygen bridge is referred to as a daunosamine group. (b) DOX attached to the surface of a FeCo nanocrystal in a graphitic carbon encasement. The complex is functionalized by polyethylene glycol (PEG) derivatives. The typical size of these particles is 4 to 5 nm in diameter. Reprinted from [670]. Copyright (2011) by the American Chemical Society.

Drug delivery by means of nanoparticle transport requires the definition of loading and release mechanisms. Various strategies for implementing both processes have been described and tested in the laboratory [637]. The work of Sherlock et al. involved studies of the role played by the chemical environment of the CEMNP in both stages. This research yielded a sensitive dependence on the pH value of the medium surrounding the CEMNP. In particular, the loading efficiency was found to be distinctly improved under basic conditions (high pH), while acidic conditions (low pH) gave rise to enhanced drug release. This behavior is plausible, since a more acidic environment favors protonation of the daunosamine group of the DOX molecule, and thus leads to increased water solubility [672]. The pH value of the medium may thus be understood as a control parameter for activating drug release. This observation is of particular clinical importance, since cancerous tissue is characterized by elevated levels of acidity.

Further, drug release may be controlled by thermal effects. More specifically, many carbon nanostructures, in particular SWCNTs, have displayed strong absorption of electromagnetic radiation in the near-infrared (NIR) regime [673, 674]. Irradiation in this regime thus gives rise to heat generation at the nanoparticle site. Sherlock et al. explored the consequences of this effect on the cellular uptake of the FeCo/GC-DOX complex and, in particular, on the toxicity of this effect. Photothermally induced heating was found to be highly efficient at inhibiting cell growth in the presence of the CEMNP, leading to a growth reduction of about 65 percent for the irradiated samples as compared with about 30 percent for the samples not irradiated. On the other hand, irradiation did not have any significant impact on cell proliferation when the CEMNP was replaced with the pure drug, demonstrating that the carbon nanostructure is instrumental for the heating effect. Coupling their test of therapeutic effectiveness with an MRI measurement, the authors were able to detect a substantially enhanced cellular uptake of the CEMNP at elevated temperature. Specifically, the uptake doubled at 43 °C as compared with 37 °C, where photothermal therapy was administered for twenty minutes. Heating alone could not account for the inhibition of cell growth, and administering the drug turned out to be crucial for attaining this result. The focus of the following section, in contrast, will be on pure temperature enhancement as a way to destroy cancerous cells.

A challenge addressed by several groups working in the field of CEMNP-assisted drug delivery is the time scale of the drug release [636, 675, 676]. Often, sustained release is preferable over rapid discharge, and the so-called burst effect, i.e. the unloading of a large volume from its nano-vehicle to the tissue in a short span of time, is to be avoided [677]. This situation calls for nanoparticle technologies that allow for control over the rate of drug release.

In response to this need, carrier particles have been tailored specifically to lower the rate of release. Chen et al. encased a complex combined of DOX and magnetite (Fe_3O_4) particles into a porous silicon shell made soluble by addition of PEG ligands [675]. In this experiment, the silicon coat, while providing a protective screen for the drug and the metal components, was seen to act as a barrier that reduced the rate of drug loss from the carrier.

Conversely, drug release from nanoparticles can be accelerated by NIR irradiation [676], as demonstrated in Figure 14.4. Shown are results from an *in vitro* experiment where the antibody Ap[4] was grafted on the outer surface of SWCNTs. Further, DOX was loaded externally, and the amount of drugs released was determined as a function of incubation time in the presence and the absence of NIR. Evidently, the irradiation speeds up the loss of DOX from the carrier by a very significant margin.

[4] Ap stands for *antibody of P-glycoprotein*, where P-glycoprotein denotes a biological agent found in a broad variety of cell lines that display drug resistance.

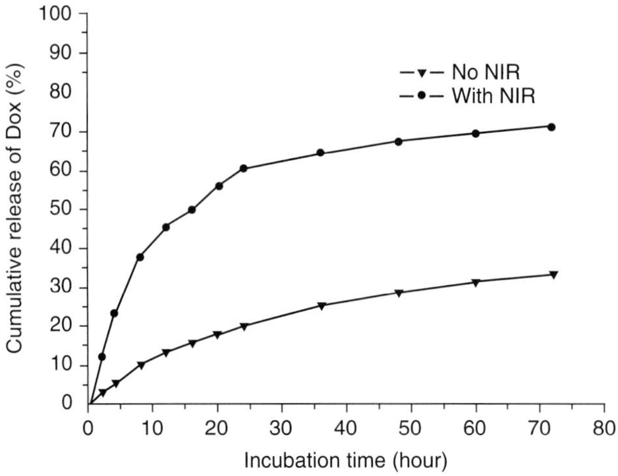

Figure 14.4 *In vitro* release profiles of DOX from SWCNTs functionalized with Ap. The lower curve refers to spontaneous release as a function of time; the upper curve, to release assisted by near infrared radiation (NIR). Reprinted from [676]. Copyright (2010) by the American Chemical Society.

14.2.2 Magnetic Thermotherapy

Magnetic thermotherapy uses magnetic agents to induce elevated temperatures at selected sites of the body by transforming electromagnetic energy into heat. In particular, this method is applied in thermal cancer treatment, where the temperature sensitivity of tumor cells is exploited. In fact, tumor cells are more readily destroyed by high temperatures than healthy cells [678]. While some thermotherapy methods in current use involve heating the whole body and thus affect healthy as much as diseased tissue, the use of *regional hyperthermia* is much preferable. In this case, temperature increase is confined to selected portions of the tissue, ideally to the tumor cells alone. This goal could be realized by channeling appropriately functionalized magnetic metal nanoparticles into the target cells. Increase of the cell temperature is then accomplished by applying an oscillating magnetic field whose energy is absorbed by the nanoparticle and subsequently converted into heat. Pioneering experiments conducted in 1957 first established the efficiency of inducing hyperthermia by use of magnetic nanoparticles [679]. These early studies involved γ-Fe_2O_3 particles 20–100 nm in size. Much research has since been directed at developing and testing alternative nanoparticles as magnetic centers, and also at novel procedures of magnetic heating. A. Jordan et al. demonstrated the heating effect of a superparamagnetic crystal suspension on biological tissue [680]. This concept has been developed into a method of cancer treatment in humans [681].

The heating mechanism by which magnetic nanoparticles give rise to hyperthermia in the surrounding tissue depends on their magnetic properties. Thus, hysteresis losses are employed in case of ferromagnetic nanoparticles, relaxation losses in case of single-domain, superparamagnetic agents. Considering the small size of the particles and also the low AC frequencies used for magnetic thermotherapy, eddy currents are a minor and usually negligible source of heating. The crucial parameter for evaluating the heat release of a nanoparticle, and for comparing different particles in terms of their heating efficiency, is the internal energy that the particle gains and, at equilibrium, dissipates during one cycle. This quantity determines the *specific loss power* (SLP), defined as the rate of energy absorption normalized to the absorbing mass [637]. In medical literature, this parameter is often referred to as *specific absorption rate* (SAR). For ferromagnetic particles, the rate of heat loss to their environment is described by the frequency f of the time-dependent external magnetic field and the area enclosed by the hysteresis loop, according to

$$P = \mu_0 f \oint H dM, \tag{14.2}$$

assuming a system of constant density and unit volume, and with **H** and **M** as collinear vectors. Superparamagnetic materials, i.e single-domain magnetic particles, do ideally not display any hysteresis behavior. Their magnetization, however, does not adjust instantly to the varying external field, but lags the field with a certain delay for which the rotational relaxation time τ_{SPM} provides a measure [682]. This effect is captured by a complex-valued susceptibility,[5]

$$\chi = \frac{\chi_0}{1 + i\omega\tau_{SPM}}, \tag{14.3}$$

which may be written as $\chi \equiv \chi' - i\chi''$. The real and the imaginary parts of the quantity are then given by:

$$\Re(\chi) = \chi' = \chi_0 \frac{1}{1 + (\omega\tau_{SPM})^2} \tag{14.4a}$$

$$\Im(\chi) = \chi'' = \chi_0 \frac{\omega\tau_{SPM}}{1 + (\omega\tau_{SPM})^2}. \tag{14.4b}$$

[5] This relation follows readily from

$$\frac{\partial \mathbf{M}}{\partial t} = \frac{1}{\tau_{SPM}} (\mathbf{M}_0(t) - \mathbf{M}(t)),$$

if one admits complex valued solutions for the magnetization and stipulates that

$$\mathbf{M}_0 = \chi_0 \mathbf{H}_0 \cos \omega t = \Re(\chi_0 \mathbf{H}_0 e^{i\omega t})$$

and

$$\mathbf{M} = \Re(\chi \mathbf{H}_0 e^{i\omega t}).$$

Exercise 14.1 Derive expression (14.3) for the complex valued susceptibility, assuming

$$\frac{\partial \mathbf{M}}{\partial t} = \frac{1}{\tau_{SPM}}(\mathbf{M}_0(t) - \mathbf{M}(t))$$

(see footnote 5).

As we will see, this approach introduces a dissipative admixture into the interaction between the superparamagnetic system and the external field. A general formula for the change in internal energy U undergone by the magnetic material per cycle and unit volume is obtained from the first law of thermodynamics, namely,

$$\Delta U = -\mu_0 \oint M dH. \tag{14.5}$$

With $H(t) = \Re(\mathbf{H}_0 e^{i\omega t})$, where the amplitude $\mathbf{H}_0$ is assumed to be small, the magnetization is

$$M(t) = H_0(\chi' \cos \omega t + \chi'' \sin \omega t), \tag{14.6}$$

combining an *in-phase* with an *out-of-phase* component. Inserting the expression (14.6) into Equation (14.5), one finds that, for each cycle of the external field, the internal energy increases by the following margin,

$$\Delta U = \mu_0 \omega H_0^2 \chi'' \int_0^T \sin^2 \omega t dt, \tag{14.7}$$

and $T = 2\pi/\omega$. Eventually, the rate of heat dissipated per unit volume is seen to be

$$P = f \Delta U = \mu_0 \pi f H_0^2 \chi'', \tag{14.8}$$

in analogy to the result for the ferromagnetic case, Eq. (14.2).

Krupskaya et al. conducted experimental studies on the heating effect due to Fe particles embedded in MWCNTs [683]. These composites were synthesized by use of a CVD procedure applied to ferrocene solvated in hydrocarbon. The size of the endohedral Fe particles was found to range between 5 and 20 nm. While the Fe-MWCNT samples exhibited distinct hysteresis effects when they were frozen or prepared as a powder, they displayed superparamagnetic behavior in liquid dispersion. Accordingly, measuring the inductive heating due to the Fe-MWCNTs in a liquid environment yielded a quadratic dependence of the power output on the amplitude of the AC magnetic field, as predicted by formula (14.8). Significant heating effects were recorded. Thus, the specific loss power at the highest field

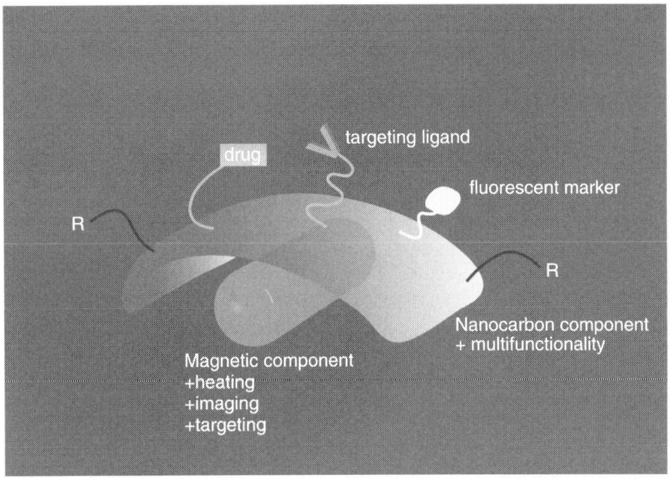

Figure 14.5 Composition of a multifunctional magnetic carbon nanoparticle used as medical device. Reprinted from [670]. Copyright (2011) by the American Chemical Society.

applied, H_0 = 100 kA/m, was determined to be 1879 W/m_{Fe}, with m_{Fe} as mass of the iron component per nanoparticle.

We note that local heating of tissue by use of carbonaceus compounds does not necessarily have to proceed through magnetic interactions. Alternatively, it is possible to exploit strong absorption in the near-infrared regime of the electromagnetic spectrum, as displayed by various carbon allotropes, among them CNTs. The AC fields associated with inductive heating, however, are of lower impact on healthy biological tissue than near-infrared radiation [637]. Further, carbon shells enclosing magnetic species may be used as multi-functional agents, serving as therapeutic devices and as sensors as well. The versatility of magnetic carbon nanostructures used in clinical diagnosis and therapy is graphically illustrated in Figure 14.5.

In closing this chapter, we emphasize the wide variety of health-related applications of magnetic carbon nanostructures. A range that includes using the magnetic component as a tool for imaging, as a device for targeting the region of clinical interest, and as source of heat.

Afterword

This survey of major themes, essential results, and open questions in the field of carbon nanostructure magnetism focused on three main areas of current research activity: the intrinsic magnetism of carbon-based nanomaterials, magnetism caused by impurities embedded into these materials, and their interest as media for spin transport. Discoveries continue to be made in each of these three domains. Far from presenting a closed body of knowledge, the arena of carbon nanostructure magnetism is wide open in many directions, and its future course is difficult to chart. As it reaches into a multiplicity of very diverse disciplines, such as biomedicine, spintronics or quantum information processing, its locus within nanoscience keeps changing.

Various results in this field have given rise to extensive debate and, in some cases, ongoing controversy. Thus, carbon-based magnetism in the sense of intrinsic magnetic effects is an embattled zone of current materials science. These effects involve magnetism due to unsaturated or incompletely saturated edge structures of carbon nano systems, or due to vacancies and voids in the carbon network, or topological defects, as discussed in Chapter 7. In monolayer graphene, but also in other systems derived from this prototype, Lieb's theorem provides the conceptual frame for the quantitative assessment of the resulting magnetic order. Theoretical and computational predictions of these phenomena abound, and some recent experimental findings are in accordance with these predictions, as pointed out in Chapter 7. However, none of the presently available experimental findings related to intrinsic carbon magnetism have been greeted with unanimous acceptance by the scientific community. Even if the existence of ordered magnetic substructures due to edge effects would be experimentally validated beyond doubt, its use in technology would set very high standards of materials engineering. Specifically, it would necessitate designing carbon nanostructures with atomic precision, which is not feasible by use of current routine fabrication modes.

Further, purity of the carbon sample has turned out to be a factor of preeminent importance for any experimental exploration of magnetism in carbon-based systems, imposing a constraint that may be hard to satisfy. This problem was sharply accentuated by the debate on high-T_C ferromagnetism in polymeric phases of C_{60}, as described in Section 7.5. Accurate analysis of the carbon sample with respect to its transition metal content is since acknowledged as imperative for any experimental investigation on carbon magnetism.

Several groups, on the other hand, in recent years presented experimental confirmation of magnetic order in graphene-based systems. For instance, direct observation of zigzag edge magnetism in zGNRs by ESR was reported [221], and magnetic effects due to nanopore boundaries [224], or associated with nano-holes in a graphene sheet [237], were recorded by use of SQUID technology. In all cases, the purity of the samples, and in particular the absence of contaminating metal admixtures, was subjected to thorough scrutiny.

Less contentious is the situation with regard to carbon phases beyond the prototypes graphene, carbon nanotubes and fullerenes. In particular, some novel phases assembled from carbon clusters have been found to be intrinsically magnetic. Carbon nanofoam may be viewed as representative for these materials. As outlined in Section 13.1, it is thought to arise from sterically protected spins nesting in the hyperbolic segments of the nanofoam network. Relatively high Curie temperatures around 90 K have been reported for carbon nanofoam manufactured by laser ablation of glassy carbon in a noble-gas atmosphere. Activated carbon fibers exemplify a different organization principle of magnetic carbon networks, as magnetism originates here from spin coupling within and between nanographitic constituents, and thus from the cooperation of various interactions involving localized π electrons. While many details of the mechanisms that cause magnetism in these complex structures are still to be elucidated, their magnetic nature has been secured by copious experimental evidence.

Magnetism in carbon nanostructures may arise from importing foreign species into the structure. A wide range of such magnetic hybrids has been investigated, with magnetic endohedral fullerenes as the most traditional among them. Various applications have been identified for these mixed species. We emphasized their potential use in medical diagnosis and therapy (Chapter 14), as well as their possible role in quantum computing, where the highly stable fullerene shell could be employed to screen an internal spin from environmental disturbances (Chapter 12). In many cases, the guest moiety is magnetic itself, but by far not in all. Magnetic order as a consequence of electron transfer between two non-magnetic substructures was exemplified by the C_{60}-TDAE and related systems. Investigating them in detail is mostly motivated by interest in the mechanisms that give rise to a variety of magnetic phases in these materials. Technical application of these mixed species, however, does not appear favorable in view of their typically deep Curie temperatures, for instance 16 K for the α phase of C_{60}-TDAE. This feature, a consequence of weak coupling between electron spins, introduces a low-temperature constraint, which creates an impediment toward the practical use of these hybrids.

From both computational modeling and experimental testing, many authors expect a future career of carbon nanostructures as elements of spintronic circuits. Most of the respective proposals exploit the unusual spin transport properties of graphene or species derived from it. The finding that electric fields can induce half-metallicity in zGNRs and zSWCNTs opens the possibility of electrically controlled spin filtering devices based on carbon. Further, zGNRs are expected to provide magnetoresistive junctions with exceptionally high magnetoresistance ratios. Importantly, graphene is an excellent medium for spin transmission, on account of its weak spin-orbit and hyperfine interaction effects. The spin relaxation times obtained by Hanle spin precession measurements in monolayer graphene at room temperature stayed below theoretical estimates but turned out to be in the order of low-temperature values ($T < 100$ K [371]) observed in semiconductors. In bilayer graphene, however, substantially longer spin relaxation times were found, extending into the nanosecond range. Spin relaxation lengths were reported to be in the micrometer regime. Further, measuring the non-local magnetoresistance makes it possible to distill a pure spin signal and thus exclude contaminating charge admixtures from the spin current. What puts graphene on the center stage of spintronic innovation is the prospect of modifying its spin transport properties by external agents. Thus, the half-metallicity of graphene makes it possible to control the spin polarization of the charge carriers by applying an external electric field, aligned parallel to the nanostructure. Likewise, a perpendicular electric field may be utilized to tune the Rashba coupling experienced by the spin carriers.

These possibilities relate to the main present challenge of graphene spintronics: using the insight afforded by basic research to replace the elements of electronics by spintronic devices, and ultimately build operative and efficient spintronic circuits containing graphene-based transmission units. In this context, a major question concerns viable techniques for spin manipulation. The eventual development of a workable graphene spin transistor will depend on the installation of practical spin gating mechanisms. These considerations emphasize, once more, the importance of the Rashba effect, which provides a handle for regulating the spin current, as described in Section 9.2.2.

The principles that determine spin transmission through carbon nanostructures can be most sensitively probed by experiments on nanotube-based double quantum dots. Passage versus blockade of charge carriers on their way through the tube is governed by the interplay between the spin and the valley degree of freedom, defining a transport mechanism characteristic for graphene and graphene-derived structures. In terms of potential applications of this system, we point out that it offers several possibilities to realize qubits. These may be implemented by up or down spins, as a well-defined axial magnetic field may be used to create conditions of degeneracy between two electron states with opposite spin orientations but identical valley quantum number, as illustrated in Figure 8.6. Conversely, a pure valley qubit can be built by occupying degenerate or near-degenerate states of the type $\{\langle K \uparrow |, \langle K' \uparrow |\}$. Time-dependent electric and magnetic fields may be used to manipulate the qubits, exploiting Coulomb and Pauli blockade effects.

The valley degree of freedom that governs the low-energy regime of graphene is currently explored in carbon allotropes beyond graphene. The challenge of generating, transmitting and detecting *valley-polarized* rather than spin-polarized charge carriers has engendered the novel discipline of *valleytronics* [684, 685]. Control of a graphene-based electronic device in terms of *valley filters* and *valley valves* was suggested [684], as counterparts of the corresponding elements of spintronics. Recently, additional media have been identified as potentially suitable for implementations of valleytronics. This emergent discipline operates with materials that have two or more degenerate conduction band valleys positioned at different sites in **k** space. This condition is met not only by graphene, but also by silicon and diamond. Simulating the transport of valley-polarized electrons in bulk diamond yielded long relaxation times on the order of 300 ns at 77 K [685].

The exploration of this non-magnetic, but spin-analogous, dimension is inspired by the electronic structure of graphene with its three determining binary variables: the spin, the valley and the sublattice degrees of freedom. Comprehensive understanding of magnetism in carbon nanostructures involves many of the unique features associated with the graphene paradigm, i.e. phenomena rooted in relativistic quantum physics, or novel topological phases, as exemplified by the quantum spin Hall effect in graphene. At the same time, it opens avenues toward technological innovation in disciplines as different as data processing and biomedical therapy, making carbon nanostructure magnetism a place of intersection for widely varying research interests, fundamental and practical alike.

References

[1] E. Tretkoff, Nanofoam exhibits surprising magnetic properties, APS News 13, 7 (2004)

[2] R. Singh, Unexpected magnetism in nanomaterials, J. Mag. Mag. Mat. 346, 58 (2013)

[3] A.L. Kuzemsky, Unconventional and exotic magnetism in carbon-based structures and related materials, Int. J. Mod. Phys. B, 27, 1330007 (2013)

[4] N. Tombros, C. Jozsa, M. Popinciuc, H.T. Jonkman, B.J. van Wees, Electronic spin transport and spin precession in single graphene layers at room temperature, Nature 448, 571 (2007)

[5] F. Kuemmeth, S. Ilani, D.C. Ralph, P.L. McEuen, Coupling of spin and orbital motion of electrons in carbon nanotubes, Nature 453, 448 (2008)

[6] O.V. Yazyev, Emergence of magnetism in graphene materials and nanostructures, Rep. Prog. Phys. 73, 056501 (2010)

[7] M. Sepioni, R.R. Nair, S. Rablen, J. Narayanan, F. Tuna, R. Winpenny, A.K. Geim, I.V. Grigorieva, Limits on intrinsic magnetism in graphene, Phys. Rev. Lett. 105, 207205

[8] C.L. Kane, E.J. Mele, Quantum spin Hall effect in graphene, Phys. Rev. Lett., 95, 226801 (2005)

[9] M.I. Katsnelson, K.S. Novoselov, Graphene: new bridge between condensed matter physics and quantum electrodynamics, Solid State Comm. 143, 3 (2007)

[10] e.g. M. Getzlaff, *Fundamentals of Magnetism*, Springer 2008

[11] J.C. Slater, *Quantum Theory of the Atomic Structure, Vol. II.* McGraw-Hill, New York, 1960

[12] M.E. Rose, *Elementary Theory of Angular Momentum*, John Wiley, 1957

[13] C. Elster, *Advanced Quantum Mechanics*, *www.phy.ohiou.edu/ elster/phys735/ index.html*

[14] L.H. Thomas, The motion of the spinning electron, Nature 117, 514 (1926)

[15] I.N. Levine, *Quantum Chemistry*, Prentice-Hall, 4th ed., 1991

[16] M.A. Ruderman, C. Kittel, Indirect exchange coupling of nuclear magnetic moments by conduction electrons, Phys. Rev. 96, 99 (1954)

[17] T. Kasuya, A theory of metallic ferro- and antiferromagnetism on Zener's model, Prog. Theor. Phys. 16, 45 (1956)

[18] K. Yosida, Magnetic properties of Cu-Mn alloys, Phys. Rev. 106, 893 (1957)

[19] H. Zhigao, C. Huang, L. Hen, Z. Minyoug, GMR from RKKY coupling in multilayered films, Jour. Mag. Soc. Japan 23, 129 (1999)

[20] R.M. White, *Quantum Theory of Magnetism*, 3rd ed., Springer 2006
[21] J. Kondo, Resistance minimum in dilute magnetic alloys, Progr. Theo. Phys. 32, 37 (1964)
[22] C. Zener, Interaction between the d-shells in the transition metals. II. Ferromagnetic compounds of manganese with Perovskite structure, Phys. Rev. 82, 403 (1951)
[23] P.-G. de Gennes, Effects of double exchange in magnetic crystals, Phys. Rev. 118, 141 (1960)
[24] e.g. S. Blundell, *Magnetism in Condensed Matter*, Oxford University Press 2001
[25] R. Schumann, D. Zwicker, The Hubbard model extended by nearest-neighbor Coulomb and exchange interaction on a cubic cluster rigorous and exact results, Annalen der Physik 522, 419 (2010)
[26] J. Keeling, *Quantum Magnetism*, *www.tcm.phy.cam.ac.uk/ jmjk2/qm//*
[27] E. Ising, Beitrag zur Theorie des Ferromagnetismus, Z. Phys. 31: 253 (1925)
[28] K. Huang, *Statistical Mechanics*, 2nd ed., John Wiley 1987
[29] F. London, Quantum theory of interatomic currents in aromatic compounds, J. Phys. Radium 8, 397 (1937)
[30] M. Springborg, *Methods for Electronic Structure Calculations*, John Wiley, New York 2000
[31] K. Hirao, ed., *Recent Advances in Multireference Methods*, World Scientific 1999
[32] D.B. Cook, *Handbook of Computational Quantum Chemistry*, Dover Publications, Mineola, NY 2004
[33] F. Hagelberg, *Electron Dynamics in Molecular Interactions*, Imperial College Press, London 2013
[34] e.g. G.A. Webb, ed., *Modern Magnetic Resonance*, 1st ed., Springer 2006
[35] B.J. Hammond, W.A. Lester, P.J. Reynolds, *Monte Carlo Methods in Ab Initio Quantum Chemistry*, Singapore, World Scientific 1994
[36] L.H. Thomas, The calculation of atomic fields, Proc. Cambridge Phil. Soc. 23, 542 (1926)
[37] E. Fermi, Un metodo statistico per la determinazione di alcune proprieta dell'atomo, Rend. Acad. Lincei 6, 602 (1927)
[38] J.C. Slater, A simplification of the Hartree–Fock method, Phys. Rev. 81, 385 (1951)
[39] R.Gáspár, Über eine Approximation des Hartree–Fockschen Potentials Durch eine Universelle Potentialfunktion, Acta Phys. Acad. Sci. Hung. 3, 263 (1954)
[40] P. Hohenberg, W. Kohn, Inhomogeneous electron gas, Phys. Rev. 136, B864 (1956)
[41] W. Kohn, L.J. Sham, Self-consistent equations including exchange and correlation effects, Phys. Rev. 140 A1133 (1965)
[42] G. Vignale, M. Rasolt, Current- and spin-density-functional theory for inhomogeneous electronic systems in strong magnetic fields, Phys. Rev. B 37, 10685 (1988)
[43] G. Vignale, W. Kohn, Current-dependent exchange-correlation potential for dynamical linear response theory, Phys. Rev. Lett. 77, 2037 (1996)
[44] S.H. Vosko, L. Wilk, M. Nusair, Accurate spin-dependent electron liquid correlation energies for local spin density calculations: a critical analysis, Can. J. Phys. 58, 1200 (1980)
[45] J.P. Perdew, A. Zunger, Self-interaction correction to density-functional approximations for many-electron systems, Phys. Rev. B 23, 5048 (1981)
[46] L.A. Cole and J.P. Perdew, Calculated electron affinities of the elements, Phys. Rev. A 25, 1265 (1982)
[47] John P. Perdew, Y. Wang, Accurate and simple analytic representation of the electron-gas correlation energy, Phys. Rev. B 45, 13244 (1992)

[48] R.G. Parr, W. Yang, *Density Functional Theory of Atoms and Molecules*, The international series of monographs on chemistry, 1st ed., Oxford University Press 1989
[49] A.D. Becke, Density-functional exchange-energy approximation with correct asymptotic behavior, Phys. Rev. A 38, 3098 (1988)
[50] J.P. Perdew, in *Electronic Structure of Solids '91* ed. P. Ziesche, H. Eschrig, Akademie Verlag, Berlin 1991
[51] J.P. Perdew, Y. Wang, Accurate and simple density functional for the electronic exchange energy: Generalized gradient approximation, Phys. Rev. B 33, 8800 (1986)
[52] A. Becke, Density functional thermochemistry. III. The role of exact exchange, Jour. Chem. Phys. 98, 5648 (1993)
[53] C. Lee, W. Yang, R.G. Parr, Development of the Colle–Salvetti correlation-energy formula into a functional of the electron density, Phys. Rev. B 37, 785 (1988)
[54] J.A. Weil, J.R. Bolton, *Electron Paramagnetic Resonance*, 2nd ed., John Wiley & Sons 2007
[55] A. Carrington, A.D. McLachlan, *Introduction to Magnetic Resonance with Applications to Chemistry and Chemical Physics*, Chapman & Hall, London 1967
[56] *www.chm.bris.ac.uk/emr/Phil/Phil*$_1$
[57] E.J. Zavoisky, Relaxation of liquid solutions in perpendicular fields, J. Phys. (USSR) 9, 211 (1945)
[58] *www.tf.uni-kiel.de/servicezentrum/neutral/praktika/anleitungen/m106*
[59] *www.wmi.badw.de/teaching/Lecturenotes/AS/AS_Chapter4.pdf*
[60] C. Kittel, *Introduction to Solid State Physics*, 8th ed., John Wiley & Sons, Inc.
[61] R. Gross, A. Marx, Applied superconductivity: The Josephson effect and superconducting electronics, *www.wmi.badw.de/teaching/Lecturenotes*
[62] T. Holstein, H. Primakoff, Field Dependence of the Intrinsic Domain Magnetization of a Ferromagnet, Phys. Rev. 58, 1098 (1940)
[63] C. Kittel, *Quantum Theory of Solids*, John Wiley 1963
[64] K.S. Novoselov, A.K. Geim, S.V. Morozov, D. Jiang, M.I. Katsnelson, I.V. Grigorieva, S.V. Dubonos, A.A. Firsov, Two-dimensional gas of massless Dirac fermions in graphene, Nature 438, 197 (2005)
[65] K.S.Novoselov, A.K. Geim, S.V. Morozov, D. Jiang, Y. Zhang, S.V. Dubonos, I.V. Grigorieva, A.A. Firsov, Electric field effect in atomically thin carbon films, Science 306, 666 (2004)
[66] Y. Zhang, L. Zhang, C. Zhou, Review of chemical vapor deposition of graphene and related applications, Accts Chem. Res. 46, 2329 (2013)
[67] A. Geim, *Random Walk to Graphene*, Nobel Lecture, December 8, 2010
[68] R. Saito, G. Dresselhaus, M.S. Dresselhaus, *Physical Properties of Carbon Nanotubes*, Imperial College Press, London 1998
[69] H.S. Philip Wong, D. Akinwande, *Carbon Nanotube and Graphene Device Physics*, Cambridge University Press, Cambridge 2011
[70] S. Reich, J. Maultzch, C. Thomsen, P. Ordejon, Tight binding description of graphene, Phys. Rev. B 66, 35412 (2002)
[71] D.W. Boukhvalov, M.I. Katsnelson, A.I.Lichtenstein, Hydrogen on graphene: Electronic structure, total energy, structural distortions and magnetism from first-principles calculations, Phys. Rev. B 77, 035427 (2008)
[72] A.H.Castro-Neto, F. Guinea, N.M.R. Peres, K.S. Novoselov, A.K. Geim, The electronic properties of graphene, Rev. Mod. Phys. 81, 109 (2009)
[73] M.V. Berry, Quantal phase factors accompanying adiabatic changes, Proc. Royal Soc. A 392, 45 (1984)

[74] G.W. Semenoff, Condensed-matter simulation of a three-dimensional anomaly, Phys. Rev. Lett. 53, 2449 (1984)

[75] M. Katsnelson, *Graphene – Carbon in Two Dimensions*, Cambridge University Press, Cambridge 2012

[76] Y. Sui, J. Appenzeller, Screening and interlayer coupling in multilayer graphene field-effect transistors, Nano Lett. 9, 2973 (2009)

[77] E. McCann, M. Koshino, The electronic properties of bilayer graphene, Rep. Prog. Phys. 76, 056503 (2013)

[78] T. Ohta, A. Bostowick, T. Seyller, K. Horn, E. Rotenberg, Controlling the electronic structure of bilayer graphene, Science 313, 951 (2006)

[79] E.V. Castro, K.S. Novoselov, S.V. Morozov, N.M.R. Peres, J.M.B. Lopes dos Santos, J. Nilsson, F. Guinea, A.K. Geim, A.H. Castro Neto, Biased bilayer graphene: Semiconductor with a gap tunable by the electric field effect, Phys. Rev. Lett. 99, 216802 (2007)

[80] J.W. McClure, Diamagnetism of graphite, Phys. Rev. 104, 666 (1956)

[81] J.W. McClure, Theory of diamagnetism of graphite, Phys. Rev. 119, 606 (1960)

[82] H. Fukuyama, Theory of orbital magnetism of Bloch electrons: Coulomb interactions, Progr. Theor. Phys. 45, 704 (1971)

[83] L. Jiao, L. Zhang, X. Wang, G. Diankov, H. Dai, Narrow graphene nanoribbons from carbon nanotubes, Nature 458, 877 (2009)

[84] M.Y. Han, B. Özyilmaz, Y. Zhang, P. Kim, Energy band gap engineering of graphene nanoribbons, Phys. Rev. Lett. 98, 206805 (2007)

[85] E. McCann, V.I. Falko, Landau-level degeneracy and quantum Hall effect in a graphite bilayer, Phys. Rev. Lett. 96, 086805 (2006)

[86] L.D. Landau, Paramagnetism of metals, Z. Phys. 64 629 (1930)

[87] *www.physics.rutgers.edu/ aluican/research.html*

[88] J.N. Fuchs, F. Piéchon, M.O. Görbig and G. Montambaux, Topological Berry phase and semiclassical quantization of cyclotron orbits for two-dimensional electrons in coupled band models, The Europ. Phys. Jour. B 77, 351 (2010)

[89] M. Brack, R.K. Bhaduri, *Semiclassical Physics*, Westview Press, Boulder, CO 2003

[90] A. Young, Y. Zhang, P. Kim, Experimental manifestation of Berry phase in graphene, in: H. Aoki, M.S. Dresselhaus (eds.), *Physics of graphene*, Springer 2014

[91] R.S. Deacon, K.C. Chuang, R.J. Nicholas, K.S. Novoselov, A.K. Geim, Cyclotron resonance study of the electron and hole velocity in graphene monolayers, Phys. Rev. B 76, 81406 (2007)

[92] G. Li, E.Y. Andrei, Observation of Landau levels of Dirac fermions in graphite. Nature Physics 3, 623 (2007)

[93] Z. Chen, Y.-M. Lin, M.G. Rooks, P. Avouris, Graphene nano-ribbon electronics, Physica E 40, 228 (2007)

[94] C. Tao, L. Jiao, O.V. Yazyev, Y.-C. Chen, J. Feng, X. Zhang, R.B. Capaz, J.M. Tour, A. Zettl, S.G. Louie, H. Dai, M.F. Crommie, Spatially resolving edge states of chiral graphene nanoribbons, Nature Physics 7, 616 (2011)

[95] L. Jiao, X. Wang, G. Diankov, H. Wang, H. Dai, Facile synthesis of high-quality graphene nanoribbons, Nature Nanotech. 5, 321 (2010)

[96] D.V. Kosynkin1, A.L. Higginbotham, A. Sinitskii, J.R. Lomeda, A. Dimiev, B.K. Price, J.M. Tour, Longitudinal unzipping of carbon nanotubes to form graphene nanoribbons, Nature 458, 872 (2009)

[97] P. Han, K. Akagi, F.F. Canova, H. Mutoh, S. Shiraki, K. Iwaya, P.S. Weiss, N. Asao, T. Hitosugi, Bottom-up graphene- nanoribbon fabrication reveals chiral edges and enantioselectivity, ACS Nano 8, 9181 (2014)

[98] D.A. Areshkin, D. Gunlycke, C.T. White, Ballistic transport in graphene nanostrips in the presence of disorder: Importance of edge effects, Nano Letters 7, 204 (2007)

[99] D. Gunlycke, D.A. Areshkin, C.T. White, Semiconducting graphene nanostrips with edge disorder, Appl. Phys. Lett. 90, 142104 (2007)

[100] F. Sols, F. Guinea, A.H. Castro-Neto, Coulomb blockade in graphene nanoribbons, Phys. Rev. Lett. 99, 166803 (2007)

[101] A.H. Castro-Neto, F. Guinea, N.M.R. Peres, Edge and surface states in the quantum Hall effect in graphene, Phys. Rev. B 73 205408 (2006)

[102] A.K. Geim, K.S. Novoselov, The rise of graphene, Nat. Mater. 6, 183 (2007)

[103] Nanotube modeler, copyright JCrystalSoft 2005–2016, *www.jcrystal.com/products/wincnt/*

[104] S. Reich, J. Janina, C. Thomsen, *Carbon Nanotubes: Basic Concepts and Physical Properties*, Wiley, New York 2004

[105] R. Saito, G. Dresselhaus, M.S. Dresselhaus, Trigonal warping effect of carbon nanotubes, Phys. Rev. B 61, 2981 (2000)

[106] J.W. Mintmire, C.T. White, Universal Density of States for Carbon Nanotubes Phys. Rev. Lett. 81, 2506 (1998)

[107] T. Ando, S. Seri, Quantum transport in a carbon nanotube in magnetic fields, J. Phys. Soc. Jpn. 66, 3558 (1997)

[108] T. Ando, Theory of electronic states and transport in carbon nanotubes, J. Phys. Soc. Jpn. 74, 777 (2005)

[109] J.P. Lu, Novel magnetic properties of carbon nanotubes, Phys. Rev. Lett. 74, 1123 (1995)

[110] H. Ajiki, T. Ando, Energy bands of carbon nanotubes in magnetic fields, J. Phys. Soc. Jpn. 65, 505(1995)

[111] T.A. Searles, Y. Imanaka, T. Takamasu, H. Ajiki, J.A. Fagan, E.K. Hobbie, J. Kono, Large anisotropy in the magnetic susceptibility of metallic carbon nanotubes, Phys. Rev. Lett. 105, 017403 (2010)

[112] M. Bockrath, D.H. Cobden, P.L. McEuen, N.G. Chopra, A. Zettl, A. Thess, R.E. Smalley, Single-electron transport in ropes of carbon nanotubes, Science 275, 1922 (1997)

[113] S.J. Tans, M.H. Devoret, H. Dai, A. Thess, R.E. Smalley, L.J. Geerligs, C. Dekker, Individual single-wall carbon nanotubes as quantum wires, Nature 386, 474 (1997)

[114] R.de Picciotto, H.L. Stormer, L.N. Pfeiffer, K.W. Baldwin, K.W. Wes, Four-terminal resistance of a ballistic quantum wire, Nature 411, 51 (2001)

[115] S. Datta, *Quantum Transport: Atom to Transistor*, Cambridge University Press, Cambridge 2005

[116] M.Büttiker, Y. Imry, R. Landauer, S. Pinhas, Generalized many-channel conductance formula with application to small rings, Phys. Rev. B 31, 6207 (1985)

[117] Y. Imry, R. Landauer, Conductance viewed as transmission, Rev. Mod. Phys. 71, S305 (1999), and references therein

[118] J.-C. Charlier, X. Blase, S. Roche, Electronic and transport properties of nanotubes, Rev. Mod. Phys. 79, 677 (2007)

[119] S. Heinze, J. Tersoff, R. Martel, V. Derycke, J. Appenzeller, Ph.Avouris, Carbon nanotubes as Schottky barrier transistors, Phys. Rev. Lett. 89, 106801 (2002)

[120] M. Biercuk, S. Ilani, C. Marcus, P. McEuen, Electrical transport in single-walled carbon nanotubes, in: A. Jorio, G. Dresselhaus, M.S. Dresselhaus (eds.), *Carbon Nanotubes: Advanced Topics in the Synthesis, Structure, Properties and Applications – Topics in Applied Physics* Vol. III, 455, Springer 2008

[121] E.A. Laird, F. Kuemmeth, G.A. Steele, K. Grove-Rasmussen, J. Nygrd, K. Flensberg, L.P. Kouwenhoven, Quantum transport in carbon nanotubes, Rev. Mod. Phys. 87, 703 (2015)
[122] L.P. Kouwenhoven, D.G. Austing, S. Tarucha, Few-electron quantum dots, Rep. Prog. Phys. 64, 701 (2001)
[123] R. Hanson, L.P. Kouwenhoven, J.R. Petta, S. Tarucha, L.M.K. Vandersypen, Spins in few-electron quantum dots, Rev. Mod. Phys. 79, 1217 (2007)
[124] R. Taylor (ed.), *The Chemistry of Fullerenes*, Advanced series in fullerenes, Vol. 4, World Scientific 1995
[125] M.S. Dresselhaus, G. Dresselhaus, P.C. Eklund, Fullerenes, J. Mater. Res. 8, 2054 (1993)
[126] P.W. Fowler, J.I. Steer, J. Chem.Soc., The leapfrog principle: a rule for electron counts of carbon clusters, Chem.Commun., 1403 (1987)
[127] Y.Z. Tan, R.T. Chen, Z.J. Liao, J. Li, F. Zhu, X. Lu, S.Y. Xie, J. Li, R.B. Huang, L.S. Zheng, Y.Z. Tan, R.T. Chen, Z.J. Liao, J. Li, F. Zhu, X. Lu, S.Y. Xie, J. Li, R.B. Huang, and L.S. Zheng, Carbon arc production of heptagon-containing fullerene, Nat. Commun., Aug. 9; 2, 420 (2011)
[128] A. Sorkin, B. Tay, H. Su, Three-stage transformation pathway from nanodiamonds to fullerenes, J. Phys. Chem. A 115, 8327 (2011)
[129] Y.Z. Tan, S.Y. Xie, R.B. Huang, L.S. Zheng, Y.Z. Tan, S.Y. Xie, R.B. Huang and L.S. Zheng, The stabilization of fused-pentagon fullerene molecules, Nat. Chem. Sept; 1(6), 450 (2009)
[130] S. Stevenson, P.W. Fowler, T. Heine, J.C. Ducham, G. Rice, T. Glass, K. Harich, F. Hadju, R. Bible, H.C. Dorn, A stable non-classical metallofullerene family, Nature 408, 427 (2000)
[131] S.S. Park, D. Liu, F. Hagelberg, Comparative investigation on non-IPR C_{68} and IPR C_{78} fullerenes encaging Sc_3N molecules, J. Phys. Chem. A 109, 8865 (2005)
[132] M.S. Dresselhaus, G. Dresselhaus, Fullerenes and fullerene derived solids as electronic materials, Annu. Rev. Mat. Sci. 25, 487 (1995)
[133] W.I.F. David, R.M. Ibberson, J.C. Matthewman, K. Prassides, T.J.S. Dennis, J.P. Hare, H.W. Kroto, R. Taylor, D.R.M. Walton, Crystal structure and bonding of ordered C_{60}, Nature 353, 156 (1991)
[134] R.C. Haddon, L.F. Schneemeyer, J.V. Waszczak, S.H. Glarum, R. Tycko, G. Dabbagh, A.R. Kortan, A.J. Muller, A.M. Mujsce, M.J. Rosseinsky, S.M. Zahurak, A.V. Makhija, F.A. Thiel, K. Raghavachari, E. Cockayne, V. Elser, Experimental and theoretical determination of the magnetic susceptibility of C_{60} and C_{70}, Nature 350, 46 (1991)
[135] *www.users.csbsju.edu/ frioux/c60/BondingC60.pdf*
[136] R. Taylor, The pattern of additions to fullerenes, Phil. Trans. R. Soc. Lond. A, 343, 87 (1993)
[137] R.S. Ruoff, D. Beach, J. Cuomo, T. McGuire, R.L. Whetten, F. Diederich, Conformation of a vanishingly small ring-current magnetic susceptibility of icosahedral C_{60}, J. Phys. Chem. 95, 3457 (1991)
[138] H. Tanaka, K. Takeuchi, Diameter determination of C_{60} and C_{70} monomers in the gas phase using a differential mobility analyzer, Appl. Phys. A 80, 759 (2005)
[139] M.S. Dresselhaus, G. Dresselhaus, P. C. Eklund, *Science of Fullerenes and Carbon Nanotubes*, Academic Press, New York 1996
[140] P.W. Fowler, D.E. Manolopoulos, *An Atlas of Fullerenes*, Clarendon Press 1995
[141] W. Krätschmer, L.D. Lamb, K. Fostiropoulos, D.R. Huffman, Solid C60: a new form of carbon, Nature, 347, 354 (1990)

[142] A.A. Popov, S. Yang, L. Dunsch, Endohedral fullerenes, Chem. Rev. 113, 5989 (2013)
[143] R.D. Bolskar, J.M. Alford, A.F. Benedetto, L.O. Husebo, R.E. Price, E.F. Jackson, S. Wallace, L.J. Wilson, First soluble M@C60 derivatives provide enhanced access to metallofullerenes and permit *in vivo* evaluation of Gd@$C_{60}[C(COOH)_2]_{10}$ as a MRI contrast agent, J. Am. Chem. Soc. 125, 5471 (2003)
[144] C. Ju, D. Suter, J. Du, An endohedral fullerene-based nuclear spin quantum computer, Phys. Lett. A 375, 1441 (2011)
[145] J.R. Heath, S.C. O'Brien, Q. Zhang, Y. Liu, R.F. Curl, K.K. Tittel, R.E. Smalley, Lanthanum complexes of spheroidal carbon shells, J. Am. Chem. Soc 107, 7779 (1985)
[146] Y. Chai, T. Guo, C. Jin, R.E. Haufler, L.P.F. Chibante, J. Fure, L. Wang, J.M. Alford, R.E. Smalley, Fullerenes with metals inside, J. Phys. Chem. 95, 7564 (1991)
[147] M. Takata, B. Umeda, E. Nishibori, M. Sakata, Y. Saito, M. Ohno, H. Shinohara, Confirmation by x-ray diffraction of the endohedral nature of the metallofullereneY@C_{82}, Nature 377, 46 (1995)
[148] K. Kobayashi, S. Nagase, Structures and electronic states of M@C_{82} (M=Sc, Y, La and lanthanides), Chem. Phys. Lett. 282, 325 (1998)
[149] E. Nishibori, M. Takata, M. Sakata, M. Inakuma, H. Shinohara, Determination of the cage structure of Sc@C_{82} by synchrotron powder diffraction, Chem. Phys. Lett. 298, 79 (1998)
[150] R.D. Johnson, M.S. Devries, J. Salem, D.S. Bethune, C.S. Yannoni, Electron paramagnetic resonance studies of lanthanum-containing C_{82}, Nature 355, 239 (1992)
[151] H. Shinohara, H. Sato, Y. Saito, M. Ohkohchi, Y. Ando, Mass spectroscopic and ESR characterization of soluble yttrium-containing metallofullerenes YC_{82} and Y_2C_{82}, J. Phys. Chem. 96, 3571 (1992)
[152] H. Shinohara, Endohedral metallofullerenes, Rep. Prog. Phys. 63, 843 (2000)
[153] B. Kessler, A. Bringer, S. Cramm, C. Schlebusch, W. Eberhardt, S. Suzuki, Y. Achiba, F. Esch, M. Barnaba, D. Cocco, Evidence for incomplete charge transfer and La-derived states in the valence bands of endohedrally doped La@C_{82}, Phys. Rev. Lett. 79, 2289 (1997)
[154] J. Lu, X. Zhang, X. Zhao, S. Nagase, K. Kobayashi, Strong metal-cage hybridization in endohedral La@C_{82}, Y@C_{82} and Sc@C_{82}, Chem. Phys. Lett. 332, 219 (2000)
[155] D. Liu, F. Hagelberg, Impact of internal electron transfer on the structure of C_{74} encapsulating Sc and La metal atom impurities, Int. J. Quant. Chem. 107, 2253 (2007)
[156] K. Shibata, Y. Rikiishi, T. Hosokawa, Y. Haruyama, Y. Kubozono, S. Kashino, T. Uruga, A. Fujiwara, H. Kitagawa, T. Takano, Y. Iwasa, Structural and electronic properties of Ce@C_{82}. Phys. Rev. B 68, 094104 (2003)
[157] L. Mora, R.S. Ruoff, C.H. Becker, D.C. Lorents, R. Malhotra, Studies of metallofullerene primary soots by laser and thermal desorption mass spectrometry, J. Phys. Chem. 97, 6801 (1993)
[158] H. Funasaka, K. Sugiyama, K. Yamamoto, T. Takahashi, Magnetic properties of rare-earth metallofullerenes. J. Phys. Chem. 99, 1826 (1995)
[159] K. Laasonen, W. Andreoni, M. Parrinello, Structural and electronic properties of La@C_{82}, Science 258, 1916 (1992)
[160] J. Ding, L.T. Weng S. Yang, Electronic structure of Ce@C_{82}: An Experimental study, J. Phys. Chem. 100, 11120 (1996)

[161] T. Pichler, M.S. Golden, M. Knupfer, J. Fink, U. Kirbach, P. Kuran and L. Dunsch, Monometallofullerene Tm@C_{82}: proof of an encapsulated divalent Tm ion by high-energy spectroscopy, Phys. Rev. Lett. 79, 3026 (1997)
[162] J. Liu, X. Zhang, X. Zhao, Metal-cage hybridization in endohedral La@C_{60}, Y@C_{60} and Sc@C_{60}, Chem. Phys. Lett. 332, 51 (2000)
[163] M.M. Alvarez, E.G. Gillan, K. Holczer, R.B. Kaner, K.S. Min, R.L. Whetten, La_2@C_{80}: a soluble dimetallofullerene, J. Phys. Chem. 95, 10561 (1991)
[164] H. Shinohara, H. Yamaguchi, N. Hayashi, H. Sato, M. Ohkohchi, Y. Ando, Y. Saito, Isolation and spectroscopic properties of scandium fullerenes (Sc_2@C_{74}, Sc_2@C_{82}, and Sc_2@C_{84}, J. Phys. Chem. 97, 4259 (1993)
[165] E. Nishibori, M. Takata, M. Sakata, A. Taninaka, H. Shinohara, Pentagonal-dodecahedral La_2 charge density in [80-Ih]fullerene: La_2@C_{80}, Angew. Chem. Int. Ed. 40, 2998 (2001)
[166] T. Akasaka, S. Nagase, K. Kobayashi, M. Waelchli, K. Yamamoto, H. Funasaka, M. Kako, T. Hushino, T. Erata, ^{13}C and ^{139}La NMR studies of La_2@C_{80}: first evidence for circular motion of metal atoms in endohedral dimetallofullerene, Angew. Chem. Int. Ed. Engl. 36, 1643 (1997)
[167] H. Shinohara, H. Sato, M. Ohchochi, Y. Ando, T. Kodama, T. Shida, T. Kato, Y. Saito, Encapsulation of a scandium trimer in C_{82}, Nature, 357, 52 (1992)
[168] M. Takata, E. Nishibori, B. Umeda, M. Sakata, M. Inakuma, Y. Yamamoto, H. Shinohara, Triangle scandium cluster imprisoned in a fullerene cage, Phys. Rev. Lett, 83 2214 (1999)
[169] J. Ungerer, T. Hughbanks, The electronic structure of Sc_3@C_{82}, J. Am. Chem. Soc. 115, 2054 (1993)
[170] S. Stevenson, G. Rice, T. Glass, K. Harich, F. Cromer, M.R. Jordan, J. Craft, E. Hadju, R. Bible, M.M. Olmstead, K. Maitra, A.J. Fisher, A.L. Balch, H.C. Dorn, Small-bandgap endohedral metallofullerenes in high yield and purity, Nature 401, 55 (1999)
[171] E.B. Iezzi, J.C. Duchamp, K.R. Fletcher, T.E. Glass, H.C. Dorn, Lutetium-based trimetallic nitride endohedral metallofullerenes: New contrast agents, Nano Lett. 2, 1187 (2002)
[172] M.M. Olmstead, A.de Bettencourt-Dias, J.C. Duchamp, S. Stevenson, H.C. Dorn, A.L. Balch, Isolation and crystallographic characterization of $ErSc_2N$@C_{80}? an endohedral fullerene which crystallizes with remarkable internal order, J. Am. Chem. Soc 122, 12220 (2000)
[173] R.M. Macfarlane, D.S. Bethune, S. Stevenson, H.C. Dorn, Fluorescence spectroscopy and emission lifetimes of Er^{3+} in Er_xSc_3-xN@C_{80} (x=13), Chem. Phys. Lett. 343, 229 (2002)
[174] I.N. Ioffe, A.S. Ievlev, O.V. Boltalina, L.N. Sidorov, H.C. Dorn, S. Stevenson, G. Rice, Electron affinity of some trimetallic nitride and conventional metallofullerenes, Int. J. Mass. Spec. 213, 183 (2002)
[175] LUNA nanoMaterials (*www.lunananomaterials.com*)
[176] T. Almeida Murphy, T. Pawlik, A. Weidinger, M. Hohne, R. Alcala, J.M. Spaeth, Observation of atomlike nitrogen in nitrogen-implanted solid C_{60}, Phys. Rev. Lett. 77, 1075 (1996)
[177] M. Saunders, H.A. Jimenez Vazquez, R.J. Cross, R.J. Poreda, Stable compounds of helium and neon - He@C_{60} and He@C_{60}, Science 259, 1428 (1993)
[178] K. Komatsu, M. Murata, Y. Murata, Encapsulation of molecular hydrogen in fullerene C_{60} by organic synthesis, Science 307, 238 (2005)

[179] E. Dietel, A. Hirsch, B. Pietzak, M. Waiblinger, K. Lips, A. Weidinger, A. Gruss, K.P. Dinse, Atomic nitrogen encapsulated in fullerenes: Effects of cage variations, J. Am. Chem. Soc. 121, 2432 (1999)

[180] K. Kurotobi, Y. Murata, A single molecule of water encapsulated in fullerene C_{60}, Science 333, 613 (2011)

[181] T.Rõõm, L. Peedu, Min Ge, D.Hüvonen, U. Nagel, S. Ye, M. Xu, Z. Bačić, S. Mamone, M.H. Levitt, M. Carravetta, J.Y.-C. Chen, X. Lei, N.J. Turro, Y. Murata, K. Komatsu, Infrared spectroscopy of small-molecule endo fullerenes, Phil. Trans. R. Soc. A 371, 20110631 (2013)

[182] T. Liu, A. Troisi, What makes fullerene acceptors special as electron acceptors in organic solar cells and how to replace them, Adv. Mat. 25, 1038 (2013)

[183] L.S. Wang, J. Conceicao, C. Jin, R.E. Smalley, Threshold photodetachment of cold C_{60}, Chem. Phys. Lett. 182, 5 (1991)

[184] T.P. Martin, N. Malinowski, U. Zimmermann, U.Näher, H. Schaber, Metal coated fullerene molecules and clusters, J. Chem. Phys. 99, 4210 (1994)

[185] U. Zimmermann, N. Malinowski, U.Näher, S. Frank, T.P. Martin, Multilayer metal coverage of fullerene molecules, Phys. Rev. Lett. 72, 3542 (1994)

[186] M.S. Dresselhaus, Future directions in carbon science, Annu. Rev. Mater. Sci. 27, 1 (1997)

[187] A.R. Koran, N. Kopylov, S. Glarum, E.M. Gyorgy, A.P. Ramirez, R.M. Fleming, F.A. Thiel, R.C. Haddon, Superconductivity at 8.4 K in calcium-doped C_{60}, Nature 355, 529 (1992)

[188] K. Kikuchi, S. Suzuki, K. Saito, H. Shiramaru, I. Ikemoto, Y. Achiba, A.A. Zhakidov, A. Ugawa, K. Imaeda, H. Inokuchi, K. Yakushi, Structure and superconductivity of single crystalline C_{60}, Physica C, 185, 415 (1991)

[189] R. Tycko, R.C. Haddon, C. Dabbagh, S.H. Glarun, D.C. Douglass, A.M. Mujsce, Solid-state magnetic resonance spectroscopy of fullerenes, J. Phys. Chem. 95, 518 (1991)

[190] R. Tycko, G. Dabbagh, R.M. Fleming, R.C. Haddon, A.V. Makhija, S.M. Zahurak, Molecular dynamics and the phase transition in solid C_{60}, Phys. Rev. Lett. 67, 1886 (1991)

[191] W.I.F. David, R.M. Ibbersen, T.J.S. Dennis, J.P. Hare, A. Prassides, Structural phase transitions in the fullerene C_{60}, Europhys. Lett. 18, 219 (1992)

[192] E. Koch, *The Doped Fullerenes*, Max Planck Institut fuer Festkörperforschung (2003)

[193] K. Thirunavukkuarasu, V.C. Long, J.L. Musfeldt, F. Borondics, G. Klupp, K. Kamaras, C.A. Kuntscher, Rotational Dynamics in C_{70}: temperature- and pressure-dependent infrared studies, J. Phys. Chem. C 115, 3646 (2011)

[194] P.W. Stephens, L. Mihaly, P.L. Lee, R.L. Whetten, S.M. Huang, R. Kaner, F. Diederich, K. Holczer, Structure of single-phase superconducting K_3C_{60}, Nature 351, 632 (1991)

[195] R.M. Fleming, A.P. Ramirez, M.J. Rosseinsky, D.W. Murphy, R.C. Haddon, S.M. Zahurak, A.V. Makhija, Relation of structure and superconducting transition temperatures in A_3C_{60}, Nature 352, 787 (1991)

[196] O. Zhou, J.E. Fischer, N. Coustel, S. Kycia, Q. Zhu, A.R. McGhie, W.J. Romanow, J.P. McCauley, Jr., A.B. Smith III, D.E. Cox, Structure and bonding in alkali-metal-doped C_{60}, Nature 351, 462 (1991)

[197] M.J. Rosseinsky, D.W. Murphy, R.M. Fleming, R. Tycko, A.P. Ramirez, T. Siegrist, G. Dabbagh, S.E. Barrett, Structural and electronic properties of sodium-intercalated C_{60}, Nature 356, 416 (1992)

[198] T. Yildirim, O. Zhou, J.E. Fischer, R.A. Strongin, M.A. Cichy, A.B. Smith III, C.I. Lin, R. Jelinek, Intercalation of sodium heteroclusters into the C_{60} lattice, Nature 360, 568 (1992)

[199] W. Andreoni, P. Gianozzi, J.F. Armbruster, M. Knupfer, Anomalous electronic behavior of Na_xC_{60} ($x \leq 10$): Theory and Experiment Europhys. Lett. 34, 699 (1996)

[200] L. Christofolini, M. Ricco, R. De Renzi, NMR and high-resolution x-ray diffraction evidence for an alkali-metal fulleride with large interstitial clusters: $Li_{12}C_{60}$, Phys. Rev. B 59, 8343 (1999)

[201] Q. Zhu, O. Zhou, J.E. Fischer, A.R. McGhie, W.J. Romanow, R.M. Strongin, M.A. Cichy, A.B. Smith III, Unusual thermal stability of a site-ordered MC_{60} rocksalt structure (M=K, Rb, or Cs), Phys. Rev. B 47, 13948 (1993)

[202] T. Yildirim, J.E. Fischer. A.B. Harris, P.W. Stephens, D. Liu, L. Brard, R.A. Strongin, A.B. Smith III, Orientational phase transition in Na_xC_{60} ($1 \leq x \leq 3$), Phys. Rev. Lett. 71, 1383 (1993)

[203] J. Winter, H. Kuzmany, Potassium-doped fullerene K_xC_{60} with x = 0, 1, 2, 3, 4, and 6, Solid State Commun. 83, 1321 (1992)

[204] D.M. Poirier, J.H. Weaver, KC_{60} fulleride phase formation: an x-ray photoemission study, Phys. Rev. B 47, 10959 (1993)

[205] P.W. Stephens, G. Bortel, G. Faigel, M. Tezge, A. Janossy, S. Pekker, G. Oszlanyi, L. Forro, Polymeric fullerene chains in RbC_{60} and KC_{60}, Nature 370, 636 (1994)

[206] A.R. Kortan, N. Kopylov, R.M. Fleming, O. Zhou, F.A. Thiel, R.C. Haddon, K.M. Rabe, Novel A15 phase in barium-doped fullerite, Phys. Rev. B 47, 13070 (1993)

[207] A.R. Kortan, N. Kopylov, E. , A. Özdas, A.P. Ramirez, R.M. Fleming, R.C. Haddon, Strontium doped fullerite compounds, Chem. Phys. Lett. 223, 501 (1994)

[208] S. Saito, A. Oshiyama, Electronic and geometric structures of fullerenes and metallofullerenes, Jpn. J. Appl. Phys. 32, 1438 (1993)

[209] P. Zhou, A.M. Rao, K.A. Wang, J.D. Robertson, C. Eloi, M.S. Meier, S.L. Ren, X.X. Bi, P.C. Eklund, M.S. Dresselhaus, Photo-assisted structural transition and oxygen diffusion in solid C_{60} films, Appl. Phys. Lett. 60, 2871 (1992)

[210] Y. Iwasa, T. Arima, R.M. Fleming, T. Siegrist, O. Zhou, R.C. Haddon et al., New phases of C_{60} synthesized at high pressure, Science 264, 1570 (1994)

[211] M.Núez-Regueiro, L. Marques, J.L. Hodeau, O.Béthoux, M. Perroux, Polymerized fullerite structures Phys. Rev. Lett. 74, 278 (1994)

[212] K.P. Meletov, G.A. Kourouklis, Pressure- and temperature-induced transformations in crystalline polymers of C_{60}, J. Exp. Theo. Phys. 115, 706 (2012)

[213] L. Pisani, J.A. Chan, B. Montanari, N.M. Harrison, Electronic structure and magnetic properties of graphitic ribbons, Phys. Rev. B 75, 064418 (2007)

[214] M. Fujita, K. Wakabayashi, N. Nakada, K. Kusakabe, Peculiar localized state at zigzag graphite edge, J. Phys. Soc. Jpn. 65, 1920 (1996)

[215] K. Wakabayashi, Electronic and magnetic properties of nanographites, Chapter 12 in: *Carbon-Based Magnetism*, T. Makarova, F. Palacio, eds. Elsevier 2006

[216] Y.W. Son, M.L. Cohen, S.G. Louie, Energy gaps in graphene nanoribbons, Phys. Rev. Lett. 97, 216803 (2006)

[217] M.M. Ugeda, I. Brihuega, F. Guinea, J.M. Gomez-Rodriguez, Missing atom as a source of carbon magnetism, Phys. Rev. Lett. 104, 096804 (2010)

[218] P.O. Lehtinen, A.S. Foster, Y. Ma, V. Krasheninnikov, R.M. Nieminen, Irradiation-induced magnetism in graphite: a density-functional study, Phys. Rev. Lett. 93, 187202 (2004)

[219] J. Červenka, M. Katsnelson, C.F. Flipse, Room-temperature ferromagnetism in graphite driven by two-dimensional networks of point defects, Nature Physics, 5, 840 (2009)
[220] J. Chen, L. Li, W.G. Cullen, E.D. Williams, M.S. Fuhrer, Tunable Kondo effect in graphene with defects, Nature Physics 7, 535 (2011)
[221] V.L. Joly, M. Kiguchi, S. Hao et al., Observation of magnetic edge state in graphene nanoribbons, Phys. Rev. B 81, 245428 (2010)
[222] Y. Shibayama, H. Sato, T. Enoki, M. Endo, Disordered magnetism at the metal-insulator threshold in nano-graphite-based carbon materials, Phys. Rev. Lett. 84, 1744 (2000)
[223] V.L. Joly, K. Takahara, K. Takai, K. Sugihara, T. Enoki, M. Koshino, H. Tanaka, Effect of electron localization on the edge-state spins in a disordered network of nano graphene sheets, Phys. Rev. B 81, 115408 (2010)
[224] K. Tada, J. Haruyama, H.X. Yang, M. Chshiev, T. Matsui, H. Fukuyama, Ferromagnetism in hydrogenated graphene nanopore arrays, Phys. Rev. Lett. 107, 217203 (2011)
[225] J.J. Palacios, J. Fernandez-Rossier, L. Brey, Vacancy-induced magnetism in graphene and graphene ribbons, Phys. Rev. B 77, 195428 (2008)
[226] V.M. Pereira, J.M.B. Lopes dos Santos, A.H. Castro Neto, Modeling disorder in graphene, Phys. Rev. B 77, 115109 (2008)
[227] H.A. Mizes, J.S. Foster, Long-range electronic perturbations caused by defects using scanning tunneling microscopy, Science 244, 559 (1989)
[228] K.F. Kelly, N.J. Halas, Determination of α and β site defects on graphite using C_{60}-adsorbed STM tips, Surf. Sci. 416, L1085 (1998)
[229] P. Ruffieux, O. Gröning, P. Schwaller, L. Schlapbach, P. Gröning, Hydrogen atoms cause long-range electronic effects on graphite, Phys. Rev. Lett. 84, 4910 (2000)
[230] S. Fajtlowicz, P.E. John, H. Sachs, On maximum matchings and eigenvalues of benzenoid graphs, Croat. Chem. Acta 78, 195 (2005)
[231] B. Cheng, B. Liu, On the nullity of graphs, El. Jour. Lin. Alg. 16, 60 (2007)
[232] E.H. Lieb, Two theorems on the Hubbard model, Phys. Rev. Lett. 62, 1201 (1989)
[233] A.A. Abrikosov, On the magnetic properties of superconductors of the second group, Soviet Physics, JETP 5, 1774 (1957)
[234] O.V. Yazyev, L. Helm, Defect-induced magnetism in graphene, Phys. Rev. B 75, 125408 (2007)
[235] Ö.Girit, J.C. Meyer, R. Erni, M.D. Rossell, C. Kisielowski, L. Yang, C.H. Park et al., graphene at the edge: stability and dynamics, Science 323, 1705 (2009)
[236] X.Y. Cui, R.K. Zheng, Z.W. Liu, L. Li, B. Delley, C. Stampfl, S.P. Ringer, Magic magnetic nanoholes for tunable semiconducting graphene, Phys. Rev. B 84, 125410 (2011)
[237] Y. Wang, Y. Huang, Y. Song, X. Zhang, Y. Ma, J. Liang, Y. Chen, Room-temperature ferromagnetism of graphene, Nano Lett. 9, 220 (2009)
[238] M. Topsakal, E. Aktürk, H. Sevinli, S. Ciraci, First-principles approach to monitoring the band gap and magnetic state of a graphene nanoribbon via its vacancies, Phys. Rev. B 78, 235435 (2008)
[239] N.D. Mermin, H. Wagner, Absence of ferromagnetism or antiferromagnetism in one- or two-dimensional isotropic Heisenberg models, Phys. Rev. Lett. 17, 1133 (1966)
[240] O.V. Yazyev, M.I. Katsnelson, Magnetic correlations at graphene edges: basis for novel spintronics devices, Phys. Rev. Lett. 100, 047209 (2008)

[241] T. Oda, A. Pasquarello, R. Car, Fully unconstrained approach to noncollinear magnetism: application to small Fe clusters, Phys. Rev. Lett. 80, 3622 (1998)

[242] R. Gebauer, S. Baroni, Magnons in real materials from density-functional theory, Phys. Rev. B 61, R6459 (2000)

[243] D. Huertas-Hernando, F. Guinea, A. Brataas, Spin-orbit coupling in curved graphene, fullerenes, nanotubes, and nanotube caps, Phys. Rev. B 74,155426 (2006)

[244] S.V. Morozov, K.S. Novoselov, M.I. Katsnelson, F. Schedin, L.A. Ponomarenko, D. Jiang, A.K. Geim, Strong suppression of weak localization in graphene, Phys. Rev. Lett. 97, 016801 (2006)

[245] J.C. Meyer, A.K. Geim, M.I. Katsnelson,K.S. Novoselov, T.J. Booth, S. Roth, The structure of suspended graphene sheets, Nature (London) 446, 60 (2007)

[246] D. Barzola-Quiquia, P. Esquinazi, M. Rothermel, D. Spemann, T. Butz, N. Garcia, Experimental evidence for two-dimensional magnetic order in proton bombarded graphite, Phys. Rev. B 76, 161403 (R) (2007)

[247] A.C. Faria, A.T. Pires, Symmetry crossover in the one-dimensional Heisenberg ferromagnet, J. Phys. C 12, 2637 (1979)

[248] H. Terrones, R. Lv, M. Terrones, M.S. Dresselhaus, The role of defects and doping in 2D graphene sheets and 1D nanoribbons, Rep. Prog. Phys. 75, 062501 (2012)

[249] P.A. Thrower, The study of defects in graphite by transmission electron microscopy, Chem. Phys. Carbon 5, 217 (1969)

[250] M.E. Fisher, Magnetism in one-dimensional systems – the Heisenberg model for infinite spin, Am. J. Phys. 32, 343 (1964)

[251] T. Enoki, K. Takai, Unconventional magnetic properties of nanographite, Chapter 17 in: T. Makarova, F. Palacio, eds., *Carbon-Based Magnetism*, Elsevier 2006

[252] A.J. Stone, D.J. Wales, Theoretical studies of icosahedral C_{60} and some related species, Chem. Phys. Lett. 128, 501 (1986)

[253] A.R. Botello-Mndez, E. Cruz-Silva, F. Lopez-Uras, B.G. Sumpter, V. Meunier, M. Terrones, H. Terrones, Spin polarized conductance in hybrid graphene nanoribbons using 5-7 defects, ACS Nano 3, 3606 (2009)

[254] S. Okada, K. Nakada, K. Kuwabara, K. Daigoku, T. Kawai, Ferromagnetic spin ordering on carbon nanotubes with topological line defects, Phys. Rev. B 74, 121412 (2006)

[255] J. Lahiri, Y. Lin, P. Bozkurt, I.I. Oleynik, M. Batzill, An extended defect in graphene as a metallic wire, Nature Nanotech. 5, 326 (2010)

[256] X. Lin, J. Ni, Half-metallicity in graphene nanoribbons with topological line defects, Phys. Rev. B 84, 075461 (2011)

[257] M. Kan, J. Zhou, Q. Sun, Q. Wang, Y. Kawazoe, P. Jena, Tuning magnetic properties of graphene nanoribbons with topological line defects: from antiferromagnetic to ferromagnetic, Phys. Rev. B 85, 155450 (2012)

[258] P. Simonis, C. Goffaux, P.A. Thiry, L.P. Biro, P. Lambin, V. Meunier, STM study of a grain boundary in graphite, Surf. Sci. 511, 319 (2002)

[259] L. Kou, C. Tang, W. Guo, C. Chen, Tunable magnetism in strained graphene with topological line defects, ACS Nano 5, 1012 (2011)

[260] J. Wu, F. Hagelberg, Magnetism in finite-sized single-walled carbon nanotubes of the zigzag type, Phys. Rev. B 79, 115436 (2009)

[261] J. Wu, F. Hagelberg, Impact of tube curvature on the ground-state magnetism of axially confined single-walled carbon nanotubes of the zigzag type, Chem. Phys. Chem. 14, 1696 (2013)

[262] J. Wu, W. Nolting, Influence of uncorrelated overlayers on the magnetism in thin itinerant-electron films, Phys. Rev. B 60, 12226 (1999)

[263] A. Mananes, F. Duque, A. Ayuela, M.J. Lopez, J.A. Alonso, Half-metallic finite zigzag single-walled carbon nanotubes from first principles, Phys. Rev. B 78, 035432 (2009)

[264] A.J. Du, Y. Chen, G.Q. Lu, S.C. Smith, Half metallicity in finite-length zigzag single-walled carbon nanotube: a first-principle prediction, Appl. Phys. Lett. 93, 073101 (2008)

[265] S.S.P. Parkin, N. More, K.P. Roche, Oscillations in exchange coupling and magnetoresistance in metallic superlattice structures: Co/Ru, Co/Cr, and Fe/Cr, Phys. Rev. Lett. 64, 2304 (1990)

[266] D.M. Edwards, J. Mathon, R.B. Muniz, M.S. Phan, Oscillations of the exchange in magnetic multilayers as an analog of de Haas-van-Alphen effect, Phys. Rev. Lett. 67, 493 (1991)

[267] P. Bruno, C. Chappert, Oscillatory coupling between ferromagnetic layers separated by a nonmagnetic metal spacer, Phys. Rev. Lett. 67, 1602 (1991)

[268] M.S. Ferreira, J. dAlbuquerque e Castro, D.M. Edwards, J. Mathon, Fundamental oscillation periods of the interlayer exchange coupling beyond the RKKY approximation, J. Phys.: Condens. Matter 8, 11259 (1996)

[269] N. Park, M. Yoon, S. Berber, J. Ihm, E. Osawa, D. Tomanek, Magnetism in all-carbon nanostructures with negative gaussian curvature, Phys. Rev. Lett. 91, 237204 (2003)

[270] M. Menon, D. Srivastava, carbon nanotube T junctions: nanoscale metal-semiconductor-metal contact devices, Phys. Rev. Lett. 79, 4453 (1997)

[271] A.N. Andriotis, M. Menon, D. Srivastava, L. Chernozatonskii, Rectification properties of carbon nanotube Y-junctions, Phys. Rev. Lett. 87, 066802 (2002)

[272] J. Wu, Anahita Ayasoufi, J. Leszczynski, F. Hagelberg, Geometric, magnetic, and adsorption properties of cross-linking carbon nanotubes: a computational study, J. Phys. Chem. C 117, 3646 (2013)

[273] D. Qian, G.J. Wagner, W.K. Liu, M.-F. Yu, R.D. Ruoff, Mechanics of carbon nanotubes, Appl. Mech. Rev. 55, 495 (2002)

[274] A.N. Andriotis, M. Menon, R.M. Sheetz, E. Richter, McConnel model for the magnetism of C_{60}-based polymers, chapter 21 in: *Carbon-Based Magnetism*, T. Makarova, F. Palacio, eds., Elsevier 2006

[275] O. Chauvet, G. Oszlnyi, L. Forro, P.W. Stephens, M. Tegze, G. Faigel, and A. Jnossy, Quasi-one-dimensional electronic structure in orthorhombic RbC_{60}, Phys. Rev. Lett. 72, 2721 (1994)

[276] T.L. Makarova, B. Sundqvist, R.Höhne, P. Esquinazi, Y. Kopelevich, P. Scharff, V.A. Davydov, L.S. Kashevarova, A.V. Rakhmanina, Magnetic carbon, Nature 413, 716 (2001)

[277] R.Höhne, P. Esquinazi, Can carbon be ferromagnetic?, Adv. Mater. 14, 753 (2002)

[278] R.A. Wood, M.H. Lewis, M.R. Lees, S.M. Bennington, M.G. Cain, N. Kitamura, Ferromagnetic fullerene, J. Phys.: Condens. Matter 14, L385 (2002)

[279] V.N. Narozhnyi, K-H.Müller, D. Eckert, A. Teresiak, L. Dunsch, V.A. Davydov, L.S. Kashevarova, A.V. Rakhmanina, Ferromagnetic carbon with enhanced Curie temperature, Physica B 329, 1217 (2003)

[280] M. Mito, T. Kawae, K. Takeda, S. Takagi, Y. Matsushita, H. Deguchi, J.M. Rawson, F. Palacio, Pressure-induced enhancement of the transition temperature of a genuine organic weak-ferromagnet up to 65 K, Polyhedron 20,1509 (2001)

[281] J.A. Chan, B. Montanari, N.M. Harrison, Ferromagnetism in defective polymerised C_{60}, Chapter 23 in: T. Makarova, F. Palacio, eds., *Carbon-Based Magnetism*, Elsevier 2006

[282] J. Ribas-Ariño, J.J. Nuvoa, The origin of the magnetic moments in compressed crsytals of polymeric C_{60}, Angew. Chem. Int. Ed. 43, 577 (2004)

[283] D. Spemann, K.-H. Han, R.Höhne, T. Makarova, P. Esquinazi, T. Butz, Evidence for intrinsic weak ferromagnetism in a C_{60} polymer by PIXE and MFM, Nucl. Inst. Meth. B 210, 531 (2003)

[284] D.W. Boukhvalov, P.F. Karimov, E.Z. Kurmaev, T. Hamilton, A. Moewes, L.D. Finkelstein, M.I. Katsnelson, V.A. Davydov, A.V. Rakhmanina, T.L. Makarova, Y. Kopelevich, S. Chiuzbaian, M. Neumann, Testing the magnetism of polymerized fullerene, Phys. Rev. B 69, 1154 25 (2004)

[285] T.L. Makarova, B. Sundqvist, R.Höhne, P. Esquinazi, Y. Kopelevich, P. Scharff, V.A. Davydov, L.S. Kashevarova, A.V. Rakhmanina, Magnetic carbon, Nature 440, 707 (2006)

[286] A. Kumar, D.K. Avasthi, J.C. Pivin, A. Tripathi, F. Singh, ferromagnetism induced by heavy-ion irradiation in fullerene films, Phys. Rev. B 74, 153409 (2006)

[287] A. Kumar, D. Kumar Avasthi, J.C. Pivin, Defect induced intrinsic ferromagnetism in fullerene thin films, Appl. Phys. Express 1, 125002 (2008)

[288] F. Kuemmeth, *Spin States and Spin-Orbit Coupling in Nanostructures*, Dissertation, Cornell 2008

[289] *http://physics.nist.gov/PhysicsRefData/DFTdata/*

[290] S. Konschuh, M. Gmitra, J. Fabian, Tight-binding theory of the spin-orbit coupling in graphene, Phys. Rev. B 82, 245412 (2010)

[291] H. Min, J.E. Hill, N.A. Sinitsyn, B.R. Sahu, L. Kleinman, A.H. MacDonald, Intrinsic and Rashba spin-orbit interactions in graphene sheets, Phys. Rev. B 74, 165310 (2006)

[292] Y. Yao, F. Ye, X.L. Qi, S.C. Zhang, Z. Fang, Spin-orbit gap of graphene: First-principles calculations, Phys. Rev. B 75, 041401 (2007)

[293] J.C. Slater, G.F. Koster, Wave functions for impurity levels, Phys. Rev. 94, 1498 (1954)

[294] D.A. Papaconstantopoulos, M.J. Mehl, The Slater-Koster tight-binding method: a computationally efficient and accurate approach, J. Phys.: Condens. Matter 15 R413 (2003)

[295] M. Gmitra, S. Konschuh, C. Ertler, C. Ambrosch-Draxl, J. Fabian, Band structure topologies of graphene: spin-orbit coupling effects from first principles, Phys. Rev. B 80, 235431 (2009)

[296] J.C. Boettger, S.B. Trickey, First-principles calculation of the spin-orbit splitting in graphene, Phys. Rev. B 75, 121402 (2007)

[297] D.C. Elias, R.V. Gorbachev, A. S. Mayorov, S. V. Morozov, A. A. Zhukov, P. Blake, L.A. Ponomarenko, I.V. Grigorieva, K.S. Novoselov, F. Guinea, A.K. Geim, Dirac cones reshaped by interaction effects in suspended graphene, Nature Materials 7, 701 (2011)

[298] A. Avsar, T.Y. Yang, S. Bae, J. Balakrishnan, F. Volmer, M. Jaiswal, Z. Yi, S.R. Ali, G.Güntherodt, B.H. Hong, B. Beschoten, B. Özyilmaz, Toward wafer scale fabrication of graphene based spin valve devices, Nano Lett. 11, 2363 (2011)

[299] A.H. Castro Neto, N.M.R. Peres, Impurity-induced spin-orbit coupling in graphene, Phys. Rev. Lett. 103, 126804(2009)

[300] S. Jo, D.K. Chi, D. Jeong, H.J. Lee, S. Kettermann, Spin relaxation properties in graphene due to its linear dispersion, Phys. Rev. B 84 075453 (2011)

[301] E.D. Minot, Y. Yaish, V. Sazonova, P.L. McEuen, Determination of electron orbital magnetic moments in carbon nanotubes, Nature 428, 536 (2004)

[302] W. Liang, M. Bockrath, H. Park, Electronic shell filling in metallic single-walled carbon nanotubes, Phys. Rev. Lett. 88, 126801 (2002)

[303] P. Jarillo-Herrero, J. Kong, H.S.J.van der Zant, C. Dekker, L.P. Kouwenhoven, S. De Franceschi, Electronic transport spectroscopy of carbon nanotubes in a magnetic field, Phys. Rev. Lett. 94, 156802 (2005)

[304] T. Ando, Spin-orbit interaction in carbon nanotubes, J. Phys. Soc. Jpn. 69, 1757 (2000)

[305] W. Izumida, K. Sato, R. Saito, Spin-orbit interaction in single wall carbon nanotubes: symmetry adapted tight-binding calculation and effective model analysis, J. Phys. Soc. Jpn. 78, 074707 (2009)

[306] J.-S. Jeong, H.-W. Lee, Curvature-enhanced spin-orbit coupling in a carbon nano, Phys. Rev. B 80, 075409 (2009)

[307] S.H. Jhang, M. Marganska, Y. Skourski, D. Preusche, B. Witkamp, M. Grifoni, H. van der Zant, J. Wosnitza, C. Strunk, Spin-orbit interaction in chiral carbon nanotubes probed in pulsed magnetic fields, Phys. Rev. B 82, 041404 (2010)

[308] G.A. Steele, F. Pei, E.A. Laird, J.M. Jol, H.B. Meerwaldt, L.p.Kouwenhoven, Large spin-orbit coupling in carbon nanotubes, Nat. Commun. 4, 1573 (2013)

[309] S. Maekawa, S.O. Valenzuela, E. Saitoh, T. Kimura, *Spin Current*, Oxford University Press, Oxford 2012

[310] J. Fabian, A. Matos-Abiague, C. Ertler, P. Stano, I.Žutiċ, Semiconductor spintronics, Acta Phys. Slov. 57, 565 (2007)

[311] I.Žutiċ, J. Fabian, S. Das Sarma, Spintronics: Fundamentals and applications, Rev. Mod. Phys. 76, 323 (2004)

[312] A. Fert, I.A. Campbell, Two-current conduction in nickel, Phys. Rev. Lett. 21, 1190 (1968)

[313] A. Fert, I.A. Campbell, Transport properties of ferromagnetic transition metals, J. Phys. 32, C1-46 (1971)

[314] A. Fert, I.A. Campbell, Electrical resistivity of ferromagnetic nickel and iron based alloys, J. Phys. F: Met. Phys. 6, 849 (1976)

[315] B.K. Nicolić, L.P. Zârbo, S. Souma, *The Oxford Handbook on Nanoscience and Technology: Frontiersand Advances*, Vol. 1, Chapter 24, A.V. Narlikar, Y.Y. Fu, eds., Oxford University Press, Oxford 2010

[316] M.N. Baibich, J.M. Broto, A. Fert, F. Nguyen Van Dau, F. Petroff, P. Etienne, G. Creuzet, A. Friederich, J. Chazelas, Phys. Rev. Lett. 61, 2472 (1988)

[317] J. Bass, W.P. Pratt, Current-perpendicular (CPP) magnetoresistance in magnetic metallic multilayers, J. Mag. Mag. Mat. 200, 274 (1999)

[318] G.A. Prinz, Magnetoelectronics, Science 282, 1660 (1998)

[319] S. Datta, B. Das, Electronic analog of the electro-optic modulator, Appl. Phys. Lett. 56, 665 (1990)

[320] Y.A. Bychkov, E.I. Rashba, Oscillatory effects and the magnetic susceptibility of carriers in inversion layers, J. Phys. C, 17, 6039 (1984)

[321] H.C. Koo, J.H. Kwon, J. Eom, J. Chang, S.H. Han, M. Johnson, Control of spin precession in a spin-injected field effect transistor. Science 325, 1515 (2009).

[322] M.E. Flatté, Z.G. Yu, E. Johnston-Halperin, D.D. Awschalom, Theory of semiconductor magnetic bipolar transistors, Appl. Phys. Lett. 82, 4740 (2003)

[323] R.A.de Groot, F.M. Mueller, P.J. van Engen, K.H.J. Buschow, New class of materials: half-metallic ferromagnets, Phys. Rev. Lett. 50, 2024 (1983)

[324] M. Springborg, B. Sang, M.L. Persson, Electronic and structural properties of two semi-Heusler alloys: NbIrSn and NbIrSb, J. Phys.: Condens. Matt. 11 6169-6178 (1999)

[325] J.H. Park, E. Vescovo, H.J. Kim, C. Kwon, R. Ramesh, T. Venkatesan, Direct evidence for a half-metallic ferromagnet, Nature 392, 794-796 (1998)
[326] I. Galanakis, P. Mavropoulos, Zinc-blende compounds of transition elements with N, P, As, Sb, S, Se and Te as half-metallic systems, Phys. Rev. B 67, 104417 (2003)
[327] W.H. Rippard, R.A. Buhrmann, Spin-dependent hot-electron transport in Co/Cu thin films, Phys. Rev. Lett. 88, 046805 (2000)
[328] C. Cacho, Y. Lassailly, H. Drouhin, G. Lampel, J. Peretti, Spin filtering of free electrons by magnetic multilayers: towards an efficient self-calibrated spin multimeter, Phys. Rev. Lett. 88, 066601 (2002)
[329] S. van Dijken, X. Jiang, S.S.P. Parkin, Spin-dependent hot electron transport in $Ni_{81}Fe_{19}$ and $Co_{84}Fe_{16}$ films on GaAs(001), Phys. Rev. B 66, 094417 (2002)
[330] P.S. Anil Kumar, R. Jansen, O.M.J.van't Erve, R. Vlutters, P.de Haan, J.C. Lodder, Low-field magnetocurrent above 200 per cent in a spin-valve transistor at room temperature, Jour. Magn. Magn. Mater. 214, L1-L6 (2000)
[331] I.Žutiċ, J. Fabian, S. Das Sarma, Spin injection through the depletion layer: a theory of spin-polarized $p-n$ junctions and solar cells, Phys. Rev. B 64, 121201 (2001)
[332] A. Einwanger, *Spin-Injection into GaAs using ferromagnetic (Ga, Mn)As contacts*, Dissertation, Regensburg 2012
[333] J. Fabian, I.Žutiċ, The standard model of spin injection, in : *From GMR to Quantum Information*, S. Blügel, ed., Forschungszentrum Jülich 2009.
[334] R.J. Elliott, Theory of the effect of spin-orbit coupling on magnetic resonance in some semiconductors, Phys. Rev. 96, 266-279 (1954)
[335] P. Boross, B. Dora, A. Kiss, F. Simon, A unified theory of spin-relaxation due to spin-orbit coupling in metals and semiconductors, Sci. Rep. 3, 3233 (2013)
[336] E.Y. Tsymbal, I. Žutiċ, *Handbook of spin transport and magnetism*, CRC Press 2012
[337] Y. Yafet, g-factors and spin-lattice relaxation, in: *Solid State Physics*, Vol. 14, Academic Press, New York 1963
[338] *www.physik.uni-regensburg.de/forschung/fabian/pages/mainframes/lecturenotes/lecturenotes_files/Spin-orbit-coupling-in-solids.pdf*
[339] J.S. Vrentas, C.M. Vrentas, *Diffusion and Mass Transfer*, CRC Press 2012
[340] G.L. Bir, A.G. Arononv, G.E. Pikus, Sov. Phys. JETP 33,1053 (1976)
[341] P.H. Song, K.W. Kim, Spin relaxation of conduction electrons in bulk III-V semiconductors, Phys. Rev. B 66, 035207 (2002)
[342] L. Cywiński, W.M. Witzel, S. Das Sarma, Electron dephasing due to hyperfine intercations with a nuclear spin bath, Phys. Rev. Lett. 102, 057601 (2009)
[343] M. Seck, M. Potemski, W. Wyder, High-field spin resonance of weakly bound electrons in GaAs, Phys. Rev. B 56, 7422-742 (1997)
[344] M. Wojtaszek, I.J. Vera-Marun, E. Whiteway, M. Hilke, B.J.van Wees, Absence of hyperfine effects in ^{13}C-graphene spin-valve devices, Phys. Rev. B 89, 035417 (2014)
[345] F. Bloch, Nuclear induction, Phys. Rev. 70, 460-474 (1946)
[346] H.C. Torrey, Bloch equations with diffusion terms, Phys. Rev. 104, 563-565 (1956)
[347] K.I. Bolotin, K.J. Sikes, Z. Jing, M. Klima, G. Fudenberg, J. Hone, P. Kim, H.L. Stormer, Ultrahigh electron mobility in suspended graphene, Solid State Commun., 146, 351 (2008)

[348] L. Liao, Y.C. Lin, M. Bao, R. Cheng, J. Bai, Y. Liu, Y. Qu, K.L. Wang, Y. Huang, X. Duan, High-speed graphene transistors with a self-aligned nanowire gate, Nature 467, 305 (2010)
[349] Y.M. Lin, A. Valdes-Garcia, S.J. Han, D.B. Farmer, I. Meric, Y. Sun, Y. Wu, C. Dimitrakopoulos, A. Grill, P. Avouris, K.A. Jenkins, Wafer-scale graphene integrated circuit, Science 332, 1294 (2011)
[350] W. Han, R.K. Kawakami, M. Gmitra, J. Fabian, Graphene spintronics, Nature Nanotech. 9, 794 (2014)
[351] W.Y. Kim, K.S. Kim, Prediction of very large values of magnetoresistance in a graphene nanoribbon device, Nature Nanotech. 3, 408 (2008)
[352] Y.W. Son, M.L. Cohen, S.G. Louie, Half-metallic graphene nanoribbons, Nature 444, 347 (2006)
[353] S. Yuasa, T. Nagahama, A. Fukushima, Y. Suzuki, K. Ando, Giant room-temperature magnetoresistance in single-crystal Fe/MgO/Fe magnetic tunnel junctions. Nature Mater. 3, 868 (2004)
[354] Th.Maassen, *Electron Spin Transport in Graphene-Based Devices*, Dissertation, Groningen 2013
[355] G. Schmidt, D. Ferrand, L.W. Molenkamp, A.T. Filip, B.J.van Wees, Fundamental obstacle for electrical spin injection from a ferromagnetic metal into a diffusive semiconductor, Phys. Rev. B 62, R4790 (2000)
[356] A. Fert, H. Jaffrès, Conditions for efficient spin injection from a ferromagnetic metal into a semiconductor, Phys. Rev. B 64, 184420 (2001)
[357] W. Han, R.K. Kawakami, Spin relaxation in single-layer and bilayer graphene, Phys. Rev. Lett 107, 047207 (2011)
[358] F.J. Jedema, *Electrical Spin Injection in Metallic Mesoscopic Spin Valves*, Dissertation, Rijksuniversiteit Groningen 2002
[359] H. Ochoa, A.H. Castro Neto, F. Guinea, Elliott–Yafet mechanism in graphene, Phys. Rev. Lett. 108, 206808 (2012)
[360] D. Pesin, A.H. MacDonald, Spintronics and pseudo spintronics in graphene and topological insulators, Nature Materials 11, 409 (2012)
[361] G. Chen, *Nanoscale Energy Transport and Conversion: A Parallel Treatment of Electrons, Molecules, Phonons and Photons* Oxford University Press 2005
[362] E.W. Hill, A.K. Geim, K. Novoselov, F. Schedin, P. Blake, Graphene spin valve devices, IEEE Trans. Magn. 42, 2694 (2006)
[363] S. Cho, Y.F. Chen, M.S. Fuhrer, Gate-tunable graphene spin valve, Appl. Phys. Lett. 91, 123105 (2007)
[364] M. Nishioka, A.M. Goldman, Spin transport through multilayer graphene, Appl. Phys. Lett. 90, 252505 (2007)
[365] M. Ohishi, M. Shiraishi, R. Nouchi, T. Nozaki, T. Shinjo, Yoshishige Suzuki Spin injection into a graphene thin film at room temperature, Jpn. J. Appl. Phys. 46, L605L607 (2007)
[366] W.H. Wang, K. Pi, Y. Li, Y.F. Chiang, P. Wei, J. Shi, R.K. Kawakami, Magneto-transport properties of mesoscopic graphite spin valves. Phys. Rev. B 77, 020402 (2008)
[367] C.Józsa, T. Maassen, M. Popinciuc, P.J. Zomer, A. Veligura, H.T. Jonkman, B.J. van Wees, Linear scaling between momentum and spin scattering in graphene. Phys. Rev. B 80, 241403(R) (2009)
[368] W. Han, K.M. McCreary, K. Pi, W.H. Wang, Y. Li, H. Wen, J.R. Chen, R.K. Kawakami, Spin transport and relaxation in graphene, J. Magn. Magn. Mater. 324, 369 (2012)

[369] W. Han, J.R. Chen, D. Wang, K.M. McCreary, H. Wen, A.G. Swartz, J. Shi, R.K. Kawakami, Spin relaxation in single-layer graphene with tunable mobility, Nano Lett. 12, 3443 (2012)
[370] N. Tombros, S. Tanabe, A. Veligura, C.Józsa, M. Popinciuc, H.T. Jonkman, B.J.van Wees, Anisotropic spin relaxation in graphene. Phys. Rev. Lett. 101, 046601 (2008)
[371] L.F. Han, Y.G. Zhu, X.H. Zhang, P.H. Tan, H.Q. Ni, Z.C. Niu, Temperature and electron density dependence of spin relaxation in GaAs/AlGaAs quantum well, Nanoscale Res. Lett. 6, 84 (2011)
[372] T.Y. Yang, J. Balakrishnan, F. Volmer, A. Avsar, M. Jaiswal, J. Samm, S.R. Ali, A. Pachoud, M. Zeng, M. Popinciuc, G. Güntherodt, B. Beschoten, B. Özyilmaz, Observation of long spin-relaxation Times in bilayer graphene at room temperature, Phys. Rev. Lett. 107, 047206 (2011)
[373] M.J. Biercuck, S. Garaj, N. Mason, J.M. Chow, C.M. Marcus, Gate-defined quantum dots on carbon nanotubes, Nano Lett. 5, 1267 (2005)
[374] S. Sapmaz, C. Meyer, P. Beliczynski, P. Jarillo-Herrero, L.P. Kouwenhoven, Excited state spectroscopy in carbon nanotube double quantum dots. Nano Lett. 6, 1350, (2006)
[375] M.R. GrÅNaber, W.A. Coish, C. Hoffmann, M. Weiss, J. Furer, S. Oberholzer, D. Loss, C. Schönenberger, Molecular states in carbon nanotube double quantum dots, Phys. Rev. B 74, 075427, (2006)
[376] S. Pecker, F. Kuemmeth, A. Secchi, M. Rontani, D.C. Ralph, P.L. McEuen, S. Ilani, Observation and spectroscopy of a two-electron Wigner molecule in an ultraclean carbon nanotube, Nature Physics 9, 576 (2013)
[377] W.G.van der Wiel, S. De Franceschi, J.M. Elzerman, T. Fujisawa, S. Tarucha, L.P. Kouwenhoven, Electron transport through double quantum dots, Rev. Mod. Phys. 75, 1 (2003)
[378] K. Ono, D.G. Austing, Y. Tokura, S. Tarucha, Current rectification by Pauli exclusion in a weakly coupled double quantum dot system, Science 297, 1313 (2002)
[379] H.O.H. Churchill, A.J. Bestwick, J.W. Harlow, F. Kuemmeth, D. Marcos, C.H. Stwertka, S.K. Watson, C.M. Marcus, Electronnuclear interaction in ^{13}C nanotube double quantum dots, Nature Physics 5, 321 (2009)
[380] A.P'alyi, G. Burkard, Hyperfine-induced valley mixing and the spin-valley blockade in carbon-based quantum dots, Phys. Rev. B 80, 201404 (2009)
[381] A.P'alyi, G. Burkard, Spin-valley blockade in carbon nanotube double quantum dots, Phys. Rev. B 80, 255424 (2010)
[382] P.W. Anderson, Absence of diffusion in certain random lattices, Phys. Rev. 109, 1492 (1958)
[383] N.F. Mott, Metal-insulator transition, Rev. Mod. Phys. 40, 677 (1977)
[384] B. Kramer, A. MacKinnon, Localization: theory and experiment, Rep. Prog. Phys. 56, 1469 (1993)
[385] P. Henseler, *Interplay of Anderson Localization and Strong Interactions in Disordered Systems*, Dissertation, Bonn 2009 (*http://hss.ulb.uni-bonn.de/2010/2040/2040.pdf*)
[386] D. Vollhardt, P.Wölfle, Self-consistent theory of Anderson localization, in: W. Hanke, Y.V. Kopaev, eds., *Modern Problems in Condensed Matter Sciences*, Vol.32, North Holland 1992
[387] B.L. Altshuler, A. Aronov, G. Arkady, B.Z. Spivak, Boris Z., The Aharonov–Bohm effect in disordered conductors, JETP Letters 33, 94 (1981)

[388] A. Bachtold, C. Strunk, J.-P. Salvetat, J.-M. Bonard, L. Forró, C. Nussbaumer, L. Schöneberger, Aharonov–Bohm oscillations in carbon nanotubes, Nature 397, 673 (1999)

[389] L. Langer, L. Stockman, J.P. Heremans, V. Bayot, C.H. Olk, C. van Haesendonck, Y. Bruynseraede, J.-P. Issi, Electrical resistance of a carbon nanotube bundle, J. Mater. Res. 9, 927 (1994)

[390] L. Langer, V. Bayot, E. Grivei, J.-P. Issi, J.P. Heremans, C.H. Olk, L. Stockman, C. Van Haesendonck, Y. Bruynseraede, Quantum transport in a multiwalled carbon nanotube, Phys. Rev. Lett. 76, 479 (1996)

[391] H. Suzuura, T. Ando, Crossover from symplectic to orthogonal class in a two-dimensional honeycomb lattice, Phys. Rev. Lett. 89, 266603 (2002)

[392] F.V. Tikhonenko, A.A. Kozikov, A.K. Savchenko, R.V. Gorbachev, Transition between electron localization and antilocalization in graphene, 103, 226801 (2009)

[393] T. Ando, T. Nakanishi, R. Saito, Berry's phase and absence of back scattering in carbon nanotubes, Jour. Phys. Soc. Jpn. 67, 2857 (1998)

[394] P.L. McEuen, M. Bockrath, D.H. Cobden, Y.-G. Yoon, S.G. Louie, Disorder, pseudospins, and backscattering in carbon nanotubes, Phys. Rev. Lett. 83, 5098 (1999)

[395] E. McCann, K. Kechedzhi, V.I. Falko, H. Suzuura, T. Ando, B.L. Altshuler, Weak-localization magnetoresistance and valley symmetry in graphene, Phys. Rev. Lett. 97,146805 (2006)

[396] M. Biercuck, S. Ilani, C.M. Marcus, M.L. McEuen, Electrical transport in single-wall carbon nanotubes, in: *Carbon Nanotubes*, A. Jorio, G. Dresselhaus, M.S. Dresselhaus, eds., Topics Appl. Physics 111, 455, Springer 2008

[397] J. Moser, H. Tao, S. Roche, F. Alzina, C.M. Sotomayor Torres, A. Bachtold, Magnetotransport in disordered graphene exposed to ozone: from weak to strong localization, Phys. Rev. Lett. 81, 205445 (2010)

[398] H. Suzuura, T. Ando, Phonons and electron-phonon scattering in carbon nanotubes, Phys. Rev. B 65, 235412 (2002)

[399] K. Sasaki, Y. Kawazoe, R. Saito, Local energy gap in deformed carbon nanotubes, Prog. Theo. Phys., 113, 463 (2005)

[400] K.von Klitzing, G. Dorda, M. Pepper, New method for high-accuracy determination of the fine-structure constant based on quantized Hall resistance, Phys. Rev. Lett. 45, 494 (1980)

[401] *http://oer.physics.manchester.ac.uk/AQM2/Notes/Notes-4.4.html*

[402] H. Aoki, Computer-simulation of 2-dimensional disordered electron-systems in strong magnetic-fields, Jour. Phys. C 10, 14, 2583 (1977)

[403] J. Oswald, M. Oswald, Circuit type simulations of magneto-transport in the quantum Hall effect regime, J. Phys.: Condens. Matter 18, R101 (2006)

[404] E. Fradkin, Field theories of condensed matter physics, 2nd ed., Cambridge University Press 2013

[405] *www.nanowerk.com/news/newsid=20771.php*

[406] *www.sp.phy.cam.ac.uk/research/fundamentals-of-low-dimensional-semiconductor-systems/lowD*

[407] C.L. Kane, M.P.A. Fisher, Edge state transport, in: *Perspectives in the Quantum Hall Effect*, S. Das Sarma and A. Pinczuk, eds., Wiley 1997

[408] P. Phillips, *Advanced Solid State Physics*, Westview Press 2003

[409] Y. Zhang, Y.-W. Tan, H.L. Stormer, P. Kim, Experimental observation of the quantum Hall effect and Berrys phase in graphene, Nature 438, 201 (2005)

[410] K.S. Novoselov, E. McCann, S.V. Morozov, V.I. Fal'ko, M.I. Katsnelson, U. Zeitler, D. Jiang. F. Schedin, A.K. Geim, Unconventional quantum Hall effect and Berrys phase of 2π in bilayer graphene, Nature Phys. 2, 177 (2006)
[411] L.W. Schubnikov, W.J.de Haas, Proceedings of the Royal Netherlands Academy of Arts and Science 33, 130 (1930)
[412] D.C. Tsui, H.L. Störmer, A.C. Gossard, Two-dimensional magnetotransport in the extreme quantum limit, Phys. Rev. Lett. 42, 1559 (1982)
[413] V.M. Apalkov, T. Chakraborty, Fractional quantum Hall states of Dirac electrons in graphene, Phys. Rev. Lett. 97, 126801 (2006)
[414] K.I. Bolotin, F. Ghahari, M.D. Shulman, H.L. Störmer, P. Kim, Observation of the fractional quantum Hall effect in graphene, Nature 462, 196 (2009)
[415] E. Hall, On a new action of the magnet on electric currents, American Jour. of Math. 2, 287 (1879)
[416] J. Maciejko, T.L. Hughes, S.-C. Zhang, The quantum spin Hall effect, Ann. Rev. Cond. Mat. Phys. 2, 31 (2011)
[417] J. Inoue, H. Ohno, Taking the Hall effect for a spin, Science 309, 2004 (2005)
[418] F.D.M. Haldane, Model for a quantum Hall effect without Landau levels: condensed-matter realization of the parity anomaly, Phys. Rev. B 61, 2015 (1988)
[419] *www.phys.ufl.edu/ pjh/teaching/phz7427/7427notes/ch6.pdf*
[420] A. Bohm, A. Mostafazadeh, H. Koizumi, Q. Niu, J. Zwanziger, *The Geometric Phase in Quantum Systems*, Springer 2003
[421] M.Z. Hasan, C.L. Kane, Colloquium: topological insulators, Rev. Mod. Phys. 82, 3046 (2010)
[422] C.L. Kane, E.J. Mele, Z_2 topological order and the quantum spin Hall effect, Phys. Rev. Lett. 95, 146802 (2005)
[423] H. Funasaka, K. Sugiyama, K. Yamamoto, T. Takahashi, Magnetic properties of rare-earth metallofullerenes, J. Phys. Chem. 99, 1826 (1995)
[424] H. Funasaka, K. Sugiyama, K. Yamamoto, T. Takahashi, Magnetic properties of Gd@C82 metallofullerene, Chem. Phys. Lett. 232, 273 (1995)
[425] H.J. Huang, S.H. Yang, X.X. Zhang, Magnetic Behavior of Pure Endohedral Metallofullerene Ho@C_{82}: A Comparison with Gd@C_{82}, J. Phys. Chem. B 103, 5928 (1999)
[426] H.J. Huang, S.H. Yang, X.X. Zhang, Magnetic Properties of Heavy Rare-Earth Metallofullerenes M@C_{82} (M = Gd, Tb, Dy, Ho, and Er), J. Phys. Chem. B 104, 1473 (2000)
[427] K. Furukawa, S. Okubo, H. Kato, H. Shinohara, T. Kato, J. Phys. Chem. A 107, 10933 (2003)
[428] R.J. Van Zee, S. Li, W. Weltner, Jr., Gd_2: the highest spin diatomic molecule?, J. Chem. Phys. 100, 4010 (1994)
[429] M.B. Knickelbein, Experimental observation of superparamagnetism in manganese clusters, Phys. Rev. Lett. 86, 5255 (2001)
[430] M. Wolf, K.H.Müller, D. Eckert, Y. Skourski, P. Georgi, R. Marczak, M. Krause, L. Dunsch, Magnetic moments in Ho_3N@C_{80} and Tb_3N@C_{80}, J. Magn. Magn. Mater. 290, 290 (2005).
[431] T. Akasaka, S. Nagase, *Endofullerenes: A New Family of Carbon Clusters*, Kluwer Academic Publishers 2002
[432] J. Wu, F. Hagelberg, Computational study on C_80 enclosing mixed trimetallic nitride clusters of the form $Gd_xM_{3-x}N$ (M = Sc, Sm, Lu), J. Phys. Chem. C 112, 5770 (2008)

[433] Dan Liu, F. Hagelberg, S.S. Park, Charge transfer and electron backdonation in metallofullerenes encapsulating NSc_3, Chem. Phys. 330, 380 (2006)

[434] A.A. Popov, L. Dunsch, Hindered cluster rotation and ^{45}Sc hyperfine splitting constant in distonoid anion radical $Sc_3^{-3}N@C_{80}^{-}$, and spatial spin-charge separation as a general principle for anions of endohedral fullerenes with metal-localized lowest unoccupied molecular orbitals, J. Am. Chem. Soc. 130, 17726 (2008)

[435] L. Chen, E.E. Carpenter, C.S. Hellberg, H.C. Dorn, M. Shultz, W. Wernsdorfer, I. Chiorescu, Spin transition in $Gd_3N@C_{80}$, detected by low-temperature on-chip SQUID technique, J. Appl. Phys. 109, 07B101 (2011)

[436] B.Náfrádi, A. Antal, A. Pásztor, L. Forró, L.F. Kiss, T. Fehéer, E. Kováts, S. Pekker, A.Jánossy, Molecular and spin dynamics in the paramagnetic endohedral fullerene $Gd_3N@C_{80}$, Jour. Phys. Chem. Lett. 3, 3291 (2012)

[437] M. Qian, S.V. Ong, S.N. Khanna, M.B. Knickelbein, Magnetic endohedral metallofullerenes with floppy interiors, Phys. Rev. B 75, 104424 (2007)

[438] S. Yang, A. Popov, M. Kalbac, L. Dunsch, The isomers of gadolinium scandium nitride cluster fullerenes $Gd_xSc_{3-x}N@C_{80}$ and their influence on cluster structure, Chemistry- A Europ. Jour. 14, 2084 (2008)

[439] J. Lu, R.F. Sabirianov, W. Ning Mei, Y. Gao, C. Duan, X. Zeng, Structural and magnetic properties of $Gd_3N@C_{80}$, Jour. Phys. Chem. B, 110, 23637 (2006)

[440] R. Westerström, J. Dreiser, C. Piamonteze, M. Muntwiler, S. Weyeneth, H. Brune, S. Rusponi, F. Nolting, A. Popov, S. Yang, L. Dunsch, and T. Greber, An endohedral single-molecule magnet with long relaxation times: $DySc_2N@C_{80}$, J. Am. Chem. Soc. 134, 9840 (2012)

[441] R. Westerström, J. Dreiser, C. Piamonteze, M. Muntwiler, S. Weyeneth, K. Krämer, S.-X. Liu, S. Decurtins, A. Popov, S. Yang, L. Dunsch, T. Greber, Tunneling, remanence, and frustration in dysprosium-based endohedral single-molecule magnets, Phys. Rev. B 89, 060406(R) (2014)

[442] R. Westerström, A.-C. Uldry, R. Stania, J. Dreiser, C. Piamonteze, M. Muntwiler, F. Matsui, S. Rusponi, H. Brune, S. Yang, A. Popov, B.Büchner, B. Delley, T. Greber, Surface aligned magnetic moments and hysteresis of an endohedral single-molecule magnet on a metal, Phys. Rev. Lett. 114, 087201 (2015)

[443] K. Lips, M. Waiblinger, B. Pietzak, A. Weidinger, Atomic nitrogen encapsulated in fullerenes: realization of a chemical Faraday cage, Phys. Stat. Sol. 177, 81 (2000)

[444] W. Harneit, C. Meyer, A. Weidinger, D. Suter, J. Twamley, Architectures for a spin quantum computer based on endohedral fullerenes, Phys. Stat. Sol. 233, 453 (2002)

[445] C.J. Nuttall, Y. Inada, K. Nagai, Y. Iwasa, Low-temperature bistability in the magnetic properties of solvent-including lanthanide metallofullerene crystals, Phys. Rev. B 62, 8592 (2000)

[446] M. Inakuma, H. Kato, A. Taninaka, H. Shinohara, T. Enoki, Magnetic anisotropy of cerium endohedral metallofullerenes, Phys. Chem. B 107, 6965 (2003)

[447] B. Diggs, A. Zhou, C. Silva, S. Kirkpatrick, N.T. Nuhfer, M.E. McHenry, D. Petasis, S.A. Majetich, B. Brunett, J.O. Artman, S.W. Staley, Magnetic properties of carbon-coated rare-earth carbide nanocrystallites produced by a carbon arc method, J. Appl. Phys. 75, 5879 (1994)

[448] A. Tiwari, G. Dantelle, K. Porfyrakis, A. Watt, A. Ardavan, G. Andrew, D. Briggs, Magnetic properties of $ErSc_2@C_{80}$, $Er_2Sc@C_{80}$, and $Er_3@C_{80}$ fullerenes, Chem. Phys. Lett. 466, 155 (2008)

[449] T.I. Smirnova, A.I. Smirnov, T.G. Chadwick, K.L. Walker, Characterization of magnetic and electronic properties of trimetallic nitride endohedral fullerenes by SQUID magnetometry and electron paramagnetic resonance, Chem. Phys. Lett. 453, 233 (2008)

[450] T. Zuo, M.M. Olmstead, C.M. Beavers, A.L. Balch, G. Wang, G.T. Yee, C. Shu, L. Xu, B. Elliott, L. Echegoyen, J.C. Duchamp, H.C. Dorn, Preparation and structural characterization of the I_h and the D_{5h} isomers of the endohedral fullerenes $Tm_3N@C_{80}$: icosahedral C_{80} cage encapsulation of a trimetallic nitride magnetic cluster with three uncoupled Tm^{3+} ions, inorg. Chem. 47, 5234 (2008)

[451] J. Ishikawaa, T. Miyaharab, Y. Hiratoa, H. Ishiia, T. Kodamac, K. Kikuchic, T. Nakamurad, K. Kodamad, D. Asakurae, T. Koidee, MCD study on $Ce@C_{82}$ and $Ce_2@C_{80}$ in the soft-x-ray region, Jour. Electron Spectrosc. Relat. Phenom. 184, 284 (2011)

[452] R. Kitaura, H. Okimoto, H. Shinohara, T. Nakamura, H. Osawa, Magnetism of the endohedral metallofullerenes $M@C_{82}$ (M=Gd,Dy) and the corresponding nanoscale peapods: synchrotron soft x-ray magnetic circular dichroism and density-functional theory calculations, Phys. Rev. B 76, 172409 (2007)

[453] C. De Nadai, A. Mirone, S.S. Dhesi, P. Bencok, N.B. Brookes, I. Marenne, P. Rudolf, N. Tagmatarchis, H. Shinohara, T.J.S. Dennis, Local magnetism in rare-earth metals encapsulated in fullerenes, Phys. Rev. B 69, 184421 (2004)

[454] F. Bondino, C. Cepek, N. Tagmatarchis, M. Prato, H. Shinohara, A. Goldoni, Element-specific probe of the magnetic and electronic properties of dy incar-fullerenes, Jour. Phys. Chem. B 110, 7289 (2006)

[455] H. Okimoto, R. Kitaura, T. Nakamura, Y. Ito, Y. Kitamura, T. Akachi, D. Ogawa, N. Imazu, Y. Kato, Y. Asada, T. Sugai, H. Osawa, T. Matsushita, T. Muro, H. Shinohara, Element-specific magnetic properties of di-erbium $Er_2@C_{82}$ and $Er_2C_2@C_{82}$ metallofullerenes: a synchrotron soft x-ray magnetic circular dichroism study, Jour. Phys. Chem. C 112, 6103 (2008).

[456] A. Weidinger, M. Waiblinger, B. Pietzak, T. Almeida Murphy, Atomic nitrogen in C_{60}:$N@C_{60}$, J. Appl. Phys. A 66, 287 (1998)

[457] S.C. Benjamin, A. Ardavan, G.A.D. Briggs, D.A. Britz, D. Gunlycke, J. Jefferson, M.A.G. Jones, D.F. Leigh, B.W. Lovett, A.N. Khlobystov, S.A. Lyon, J.J.L. Morton, K. Porfyrakis, M.R. Sambrook, A.M. Tyryshkin, Towards a fullerene-based quantum computer, J. Phys.: Condens. Matter 18, S867 (2006)

[458] J.J.L. Morton, A.M. Tyryshkin, A. Ardavan, K. Porfyrakis, S.A. Lyon, G.A.D. Briggs, High fidelity single qubit operations using pulsed electron paramagnetic resonance, Phys. Rev. Lett. 95, 200501 (2005)

[459] T. Fujisawal, D.G. Austing, Y. Tokura, Y. Hirayama, S. Tarucha, Allowed and forbidden transitions in artificial hydrogen and helium atoms, Nature 419, 278 (2002)

[460] F. Jelezko, T. Gaebel, I. Popa, A. Gruber, J. Wrachtrup, Observation of coherent oscillations in a single electron spin, Phys. Rev. Lett. 92, 076401 (2004)

[461] Y. Nakamura, Yu.A. Pashkin, J.S. Tsai, Coherent control of macroscopic quantum states in a single-Cooper-pair box, Nature 398, 786 (1999)

[462] D. Suter, K. Lim, Scalable architecture for spin-based quantum computers with a single type of gate, Phys. Rev. A 65, 052309 (2002)

[463] R.M. Brown, A.M. Tyryshkin, K. Porfyrakis, E.M. Gauger, B.W. Lovett, A. Ardavan, S.A. Lyon, G.A.D. Briggs, J.J.L. Morton, Coherent state transfer between an electron and nuclear spin in $^{15}N@C_{60}$, Phys. Rev. Lett. 106, 110504 (2011)

[464] W. Harneit, Fullerene-based electron-spin quantum computer, Phys. Rev. A 65, 032322 (2002)

[465] S.C. Benjamin, Schemes for parallel quantum computation without local control of qubits, Phys. Rev. A (R) 61, 020301 (2000)

[466] J. Twamley, Quantum-cellular-automata quantum computing with endohedral fullerenes, Phys. Rev. A 67, 052318 (2003)

[467] T. Hovland, M. Sc.Thesis, *Quantum Computation with Global Pulses*, National University of Ireland, Maynooth 2001, *http://planck.thphys.may.ie/jtwamley/jtwamley.htm*

[468] S.W.D. Bailey, C.J. Lambert, The electronic transport properties of N@C_{60}@(n,m) carbon nanotube peapods, Physica E 40, 99 (2007)

[469] B.W. Smith, M. Monthioux, D.E. Luzzi, Encapsulated C_{60} in carbon nanotubes, Nature 396, 323 (1998)

[470] F. Simon, H. Kuzmany, Náfrádi, T. Fehér, L. Forró, F.Fülöp, A.Jánossy, L. Korecz, A. Rockenbauer, F. Hauke, A. Hirsch, Magnetic fullerenes inside single-wall carbon nanotubes, Phys. Rev. Lett. 97, 136801 (2006)

[471] R.C. Haddon, A.F. Hebard, M.J. Rosseinsky, D.W. Murphy, S.J. Duclos, K.B. Lyons, B. Miller, J.M. Rosamilia, R.M. Fleming, A.R. Kortan, S.H. Glarum, A. V. Makhija, A.J. Muller, R.H. Eick, S.M. Zahurak, R. Tycko, G. Dabbagh, F.A. Thiel, Conducting films of C_{60} and C_{70} by alkali-metal doping, Nature, 350, 320 (1991)

[472] K. Kamarás, G. Klupp, Metallicity in fullerides, Dalton Transactions 43, 7366 (2014)

[473] H. Jahn, E. Teller, Stability of degenerate electronic states in polyatomic molecules, Phys. Rev. 49, 874 (1936)

[474] I.B. Bersuker, *The Jahn–Teller Effect*, Cambridge University Press 2006

[475] M.C.M O'Brian, C.C. Chancey, The Jahn–Teller effect: an introduction and current review, Am. J. Phys. 61, 688 (1993)

[476] M. Lannoo, G.A. Baraff, M. Schlüter, D. Tomanek, Jahn–Teller effect for the negatively charged C_{60} molecule: analogy with the silicon vacancy, Phys. Rev. B 44, 12106 (1991)

[477] N. Koga, K. Morokuma, *Ab initio* MO study of the C_{60} anion radical: the Jahn–Teller distortion and electronic structure, Chem. Phys. Lett. 191, 196 (1994)

[478] I.D. Hands, J.L. Dunn, C.A. Bates, Visualization of static Jahn–Teller effects in the fullerene anion C_{60}^{-}, Phys. Rev. B 82, 155425 (2010)

[479] A. Ceulemans, L.F. Chibotaru, F. Cimpoesu, Intramolecular charge disproportionation and the band structure of A_3C_{60} superconductors, Phys. Rev. Lett. 78, 3725 (1997)

[480] L. Forró, L. Mihály, Electronic properties of doped fullerenes, Rep. Prog. Phys. 64, 649 (2001)

[481] M. Knupfer, M. Merkel, M.S. Golden, J. Fink, O. Gunnarsson, V.P. Antropov, Satellites in the photoemission spectrum of A_3C_{60} (A=K and Rb), Phys. Rev. B 47, 13944(R) (1993)

[482] R. Macovez, M.R.C. Hunt, J. Shan, A. Goldoni, T. Pichler, M. Pedio, P. Moras, C. Castellarin-Cudia, J. Schiessling, L. Venema, P. Rudolf, Metal-to-insulator transition in thin-film polymeric AC_{60}, New Jour. Phys. 11, 023035 (2009)

[483] C. Coulon, A. Pnicaud, R. Clrac, R. Moret, P. Launois, Metal-insulator and structural phase transition observed by ESR spectroscopy and x-ray diffraction in K@C_{60}, Phys. Rev. Lett. 86, 4346 (2001)

[484] H. Alloul, V. Brouet, E. Lafontaine, L. Malier, L. Forró, ^{13}C magic-angle-spinning NMR study of the electronic properties of the AC_{60} polymers (A = K,Rb,Cs), Phys. Rev. Lett. 76, 2922 (1996)
[485] Y.J. Uemura, K. Kojima, G.M. Luke, W.D. Wu, G. Oszlányi, O. Chauvet, L. Forró, Static magnetic order in the one-dimensional conductor RbC_{60}, Phys. Rev. B 52, R6991 (1995)
[486] A.F. Hebard, M.J. Rosseinsky, R.C. Haddon, D.W. Murphy, S.H. Glarum, T.T.M. Palstra, A.P. Ramirez, A.R. Kortan, Superconductivity at 18 K in potassium-doped C_{60}, Nature, 350, 600 (1991)
[487] O. Gunnarson, Superconductivity in fullerides, Rev. Mod. Phys. 69, 575 (1997)
[488] A.Y. Ganin, Y. Takabayashi, Y.Z. Khimyak, S. Margadonna, A. Tamai, M.J. Rosseinsky, K. Prassides, Bulk superconductivity at 38 K in a molccular system, Nature Mater. 7, 367 (2008)
[489] Y. Iwasa, T. Takenobu, Superconductivity, Mott-Hubbard states, and molecular orbital order in intercalated fullerides, J. Phys.: Condens. Matter 15, R495 (2003)
[490] A.Y. Ganin, Y. Takabayashi, P. Jeglic, D. Arĉon, Anton Potoĉnik, P.J. Baker, Y. Ohishi, M.T. McDonald, M.D. Tzirakis, A. McLennan, G.R. Darling, M. Takata, M.J. Rosseinsky, K. Prassides, Polymorphism control of superconductivity and magnetism in Cs_3C_{60} close to the Mott transition, Nature 466, 221 (2010)
[491] O. Gunnarsson, E. Koch, R.M. Martin, Mott transition in degenerate Hubbard models: application to doped fullerenes, Phys. Rev. B 54, R11026 (1996)
[492] J.P. Lu, Metal-insulator transitions in degenerate Hubbard models and A_xC_{60}, Phys. Rev. B 49, 5687 (1994)
[493] K. Ishii, T. Watanuki, Akihiko Fujiwara, H. Suematsu, Y. Iwasa, H. Shimoda, T. Mitani, H. Nakao, Y. Fujii, Y. Murakami, H. Kawada, Structural phase transition in the ammoniated alkali C_{60} compound $NH_3K_3C_{60}$, Phys. Rev. B 59, 3956 (1999)
[494] H. Tou, Y. Maniwa, Y. Iwasa, H. Shimoda, T. Mitani, NMR evidence for Mott-Hubbard localization in $(NH_3)K_3C_{60}$, Phys. Rev. B 62, R775 (2000)
[495] S. Margadonna, K. Prassides, H. Shimoda, T. Takenobu, Y. Iwasa, Orientational ordering of C_{60} in the antiferromagnetic $NH_3K_3C_{60}$ phase, Phys. Rev. B 64, 132414 (2001)
[496] J. Arvanitidis, K. Papagelisa, T. Takenobub, I. Margiolakia, K. Brigattia, K. Prassidesa, Y. Iwasab, A. Lappas, Antiferromagnetic ordering in the expanded $NH_3Rb_3C_{60}$ fulleride, Physica B 326, 572 (2003)
[497] M. Tinkham, *Introduction to superconductivity*, 2nd ed., Dover Publications 2004
[498] M.J. Rosseinsky, D.W. Murphy, R.M. Fleming, O. Zhou, Intercalation of ammonia into K_3C_{60}, Nature 364, 425 (1993)
[499] K. Prassides, S. Margadonna, D. Arcon, A. Lappas, H. Shimoda, Y. Iwasa, Magnetic ordering in the ammoniated fulleride $ND_3K_3C_{60}$, J. Am. Chem. Soc. 121, 11227 (1999)
[500] P.M. Allemand, K.C. Khemani, A. Koch, F. Wudl, K. Holczer, S. Donovan, G. Grüner, J.D. Thompson, Organic molecular soft ferromagnetism in a fullerene C_{60}, Science 253, 301 (1991)
[501] M. Fujiwara, T. Kambe, K. Oshima, Structural differences in two polymorphs of tetra-*kis*-(dimethylamino)-ethylene-C_{60}: an x-ray diffraction study, Phys. Rev. B 71, 174424 (2005)
[502] T. Kambe, K. Oshima, Dynamically fluctuating electric dipole moments in fullerene-based magnets, Sci. Rep. 4, 6419 (2014)

[503] T. Kambe, K. Kajiyoshi, M. Fujiwara, K. Oshima, Antiferromagnetic ordering driven by the molecular orbital order of C_{60} in α'-tetra-kis-(Dimethylamino)-ethylene-C_{60}, Phys. Rev. Lett. 99, 177205 (2007)
[504] T. Kambe, Y. Nogami, K. Oshima, Annealing effects on the magnetic and structural properties of single-crystal TDAE-C_{60}, Phys. Rev. B 61, R862 (2000)
[505] H. Yamaoka, T. Kambe, T. Sato, Y. Ishida, M. Matsunami, R. Eguchi, Y. Senba, H. Ohashi, Electronic state of an organic molecular magnet: soft x-ray spectroscopy study of α-TDAE-C_{60} single crystal, Phys. Rev. B 84, 161404 (R) (2011)
[506] T. Kambe, Y. Nogami, K. Oshima, Annealing effects on the magnetic and structural properties of single-crystal TDAE-C_{60}, Phys. Rev. B 61, R862 (2000)
[507] D. Arcon, J. Dolinsek, R. Blinc, ^{13}C NMR of the organic ferromagnet TDAE-C_{60}, Phys. Rev. B 53, 9137 (1996)
[508] T. Kawamoto, A theoretical model for ferromagnetism of TDAE-C_{60}, Solid State Commun. 101, 231 (1997).
[509] K. Mizoguchi, M. Machino, H. Sakamoto,1 T. Kawamoto, M. Tokumoto, A. Omerzu, D. Mihailovic, Pressure effect in TDAE-C_{60} ferromagnet: mechanism and polymerization, Phys. Rev. B 63, 140417 (2001)
[510] K. Mizoguchi, M. Takei, M. Machino, H. Sakamoto, M. Tokumoto, T. Kawamoto, A. Omerzu, D. Mihailovic, Magnetism of α- and β-TDAE-C_{60}, Jour. Magn. Magn. Mater., 272–276, E215 (2004)
[511] W.Śliva, Cycloaddition reactions of fullerenes, Fullerene Sci. Technol. 3, 243 (1995)
[512] S. Garaj, T. Kambe, L. Forró, A. Sienkiewicz, M. Fujiwara, K. Oshima, Polymer phase of the tetrakis(dimethylamino)ethylene-C_{60} organic ferromagnet, Phys. Rev. B 68, 144430 (2003)
[513] A. Omerzu, D. Mijatovic, D. Mihailovic, Revealing of the re-entrant spin-glass phase in TDAE-C 60 with linear and non-linear AC susceptibility measurements, Synth. Metals 121, 1155 (2001)
[514] A. Omerzu, M. Tokumoto, B. Tadić, D. Mihailovic, Critical exponents at the ferromagnetic transition in Tetrakis(dimethylamino)ethylene-C_{60} (TDAE-C_{60}), Phys. Rev. Lett. 87, 177205 (2001)
[515] D.C. Elias, R.R. Nair, T.M. Mohiuddin, S.V. Morozov, P. Blake, M.P. Halsall, A.C. Ferrari, D.W. Boukhvalov, M.I. Katsnelson, A.K. Geim, K.S. Novoselov, Control of graphene's properties by reversible hydrogenation: evidence for graphane, Science 323, 5914 (2009)
[516] Y. Wang, X. Xu, J. Lu, M. Lin, Q. Bao, B. Özyilmaz, K.P. Loh, Toward high throughput interconvertible graphane-to-graphene growth and patterning, ACS Nano 4, 6146 (2010).
[517] C. Zhou, S. Chen, J. Lou, J. Wang, Q. Yang, C. Liu, D. Huang, T. Zhou, Graphenes cousin: the present and future of graphane, Nanoscale Res. Lett. 9, 26 (2014)
[518] M.H.F. Sluiter, Y. Kawazoe, Cluster expansion method for adsorption: Application to hydrogen chemisorption on graphene, Phys. Rev. B 68, 085410 (2003)
[519] J.O. Sofo, A.S. Chaudhari, G.D. Barber, Graphane: a two-dimensional hydrocarbon, Phys. Rev. B 75, 153401 (2003)
[520] Metal-insulator transition from graphene to graphane, Nanomater. Nanotechnol. 3, Art. 10 (2013)
[521] S. Lebégue, M. Klintenberg, O. Eriksson, M.I. Katsnelson, Accurate electronic band gap of pure and functionalized graphane from GW calculations, Phys. Rev. B 79, 245117 (2009)

[522] M.S. Hybertsen, S.G. Louie, First-principles theory of quasiparticles: calculation of band gaps in semiconductors and insulators, Phys. Rev. Lett. 55, 1418 (1985)

[523] C. Lin, Y. Feng, Y. Xiao, M. Dürr, X. Huang, X. Xu, R. Zhao, E. Wang, X.-Z. Li, Z. Hu, Direct observation of ordered configurations of hydrogen adatoms on graphene, Nanolett. 15, 903 (2015)

[524] Ž. Šljivančanin, M. Andersen, L. Hornekr, B. Hammer, Structure and stability of small H clusters on graphene, Phys. Rev. B 83, 205426 (2011)

[525] J. Zhou, Q. Wang, Q. Sun, X.S. Chen, Y. Kawazoe, P. Jena, Ferromagnetism in semihydrogenated graphene sheet, Nanolett. 9, 3867 (2009)

[526] J. Berashevich, T. Chakraborty, Sustained ferromagnetism induced by H-vacancies in graphane, Nanotech. 21, 355201 (2010)

[527] Ž. Šljivančanin, R. Balog, L. Hornekr, Magnetism in graphene induced by hydrogen adsorbates, Chem. Phys. Lett. 541, 70 (2012)

[528] D. Haberer, C.E. Giusca, Y. Wang, H. Sachdev, A.V. Fedorov, M. Farjam, S.A. Jafari, D.V. Vyalikh, D. Usachov, X. Liu, U. Treske, M. Grobosch, O. Vilkov, V.K. Adamchuk, S. Irle, S.R.P. Silva, M. Knupfer, B.Büchner, A. Grüneis, Evidence for a new two-dimensional C4H-type polymer based on hydrogenated graphene, Adv. Mater. 23, 4497 (2011)

[529] M.L. Ng, R. Balog, L. Hornekr, A.B. Preobrajenski, N.A. Vinogradov, N.Mårtensson, K. Schulte, Controlling hydrogenation of graphene on transition metals, J. Phys. Chem. C 131, 18559 (2010)

[530] W. Zhao, J. Gebhardt, Florian Späth, K. Gotterbarm, C. Gleichweit, H.-P. Steinrück, A.Görling, C. Papp, Reversible hydrogenation of graphene on Ni(111)-Synthesis of graphone, Chemistry- Europ. Jour. 21, 3347 (2015)

[531] K.M. McCreary, A.G. Swartz, W. Han, J. Fabian, R.K. Kawakami, Magnetic moment formation in graphene detected by scattering of pure spin currents, Phys. Rev. Lett. 109, 186604 (2012)

[532] K.A. Worsley, P. Ramesh, S.K. Mandal, S. Niyogi, M.E. Itkis, R.C. Haddon, Soluble graphene derived from graphite fluoride, Chem. Phys. Lett. 445, 51 (2007)

[533] K. Shoda, H. Kohno, Y. Kobayashi, D. Takagi, S. Takeda, Feasibility study for sidewall fluorination of SWCNTs in CF_4 plasma, J. Appl. Phys. 104, 113529 (2008)

[534] X. Hong, S.H. Cheng, C. Herding, J. Zhu, Colossal negative magnetoresistance in dilute fluorinated graphene, Phys. Rev. B 83, 085410 (2011)

[535] R.R. Nair, W. Ren, R. Jali, I. Riaz, V.G. Kravets, L. Britnell, P. Blake, F. Schedin, A.S. Mayorov, S. Yuan, M.I. Katsnelson, H.-M. Cheng, W. Strupinski, L.G. Bulusheva, A.V. Okotrub, I.V. Grigorieva, A.N. Grigorenko, K.S. Novoselov, A.K. Geim, Fluorographene: a two-dimensional counterpart of Teflon, Small 6, 2877 (2010)

[536] H.Şahin, M. Topsakal, S. Ciraci, Structures of fluorinated graphene and their signatures, Phys. Rev. B 83, 115432 (2011)

[537] S. Osuna, M. Torrent-Sucarrat, M. Solá, P. Geerlings, C.P. Ewels, G. Van Lier, Reaction mechanisms for graphene and carbon nanotube fluorination, Jour. Phys. Chem. C 114,3340 (2010)

[538] T.G. Rappoport, B. Uchoa, A.H. Castro Neto, Magnetism and magnetotransport in disordered graphene, Phys. Rev. B 80, 245408 (2009)

[539] D.W. Boukhvalov, M.I. Katsnelson, Enhancement of chemical activity in corrugated graphene, Jour. Phys. Chem. C 113, 14176 (2009)

[540] J.O. Sofo, A.M. Suarez, G. Usaj, P.S. Cornaglia, A.D. Hernandez-Nieves, C.A. Balseiro, Electrical control of the chemical bonding of fluorine on graphene, Phys. Rev. B 83, 081411 (2011)

[541] H.J. Kim, J.H. Cho, Fluorine-induced local magnetic moment in graphene: a hybrid DFT study, Phys. Rev. B 87, 174435 (2013)

[542] A.V. Krukau, O.A. Vydrov, A.F. Izmaylov, G.E. Scuseria, Influence of the exchange screening parameter on the performance of screened hybrid functionals, Jour. Chem. Phys. 125, 224106 (2006)

[543] X. Hong, K. Zou, B. Wang, S.-H. Cheng, J. Zhu, Evidence for spin-flip scattering and local moments in dilute fluorinated graphene, Phys. Rev. Lett. 108, 226602 (2012)

[544] F.V. Tikhonenko, D.W. Horsell, R.V. Gorbachev, A.K. Savchenko, Weak localization in graphene flakes, Phys. Rev. Lett. 100, 056802 (2008)

[545] D.-K. Ki, D. Jeong, J.-H. Choi, H.-J. Lee, K.-S. Park, Inelastic scattering in a monolayer graphene sheet: a weak-localization study, Phys. Rev. B 78, 125409 (2008).

[546] M. Gmitra, D. Kochan, J. Fabian, Spin-orbit coupling in hydrogenated graphene, Phys. Rev. Lett. 110, 246602 (2013)

[547] S. Irmer, T. Frank, S. Putz, M. Gmitra, D. Kochan, J. Fabian, Spin-orbit coupling in fluorinated graphene, Phys. Rev. B 91, 115141 (2015)

[548] A. Ferreira, T.G. Rappoport, M.A. Cazalilla, A.H. Castro Neto, Extrinsic spin Hall effect induced by resonant skew scattering in graphene, Phys. Rev. Lett. 112, 066601 (2014)

[549] J. Balakrishnan, G.K. Koon, M. Jaiswal, A.H. Castro Neto, B.Özyilmas, Colossal enhancement of spin-orbit coupling in weakly hydrogenated graphene, Nature Physics 9, 284 (2013)

[550] D.A. Abanin, A.V. Shytov, L.S. Levitov, B.I. Halperin, Nonlocal charge transport mediated by spin diffusion in the spin Hall effect regime, Phys. Rev. B 79, 035304 (2009)

[551] Y.-J. Yu, Y. Zhao, S. Ryu, L.E. Brus, K.S. Kim, P. Kim, Tuning the graphene work function by electric field effect, Nano Lett. 9, 3430 (2009)

[552] S. Das Sarma, S. Adam, E.H. Hwang, E. Rossi, Electronic transport in two-dimensional graphene, Rev. Mod. Phys. 83, 407 (2011)

[553] J.D. Bergeson, S.J. Etzkorn, M.B. Murphey, L. Qu, J. Yang, L. Dai, A.J. Epstein, Iron nanoparticle driven spin-valve behavior in aligned carbon nanotube arrays, Appl. Phys. Lett. 93, 172505 (2008)

[554] O.V. Khavryuchenko, G.H. Peslherbe, F. Hagelberg, Spin filter circuit design based on a finite single-walled carbon nanotube of the zigzag type, J. Phys. Chem. C 119, 3740 (2015)

[555] A.T. Costa, D.F. Kirwan, M.S. Ferreira, Indirect exchange coupling between magnetic adatoms in carbon nanotubes, Phys. Rev. B 72, 085402 (2005)

[556] D.F. Kirwan, V.M. de Menezes, C.G. Rocha, A.T. Costa, R.B. Muniz, S.B. Fagand, M.S. Ferreira, Enhanced spin-valve effect in magnetically doped carbon nanotubes, Carbon 47, 2528 (2009)

[557] S. Saremi, RKKY in half-filled bipartite lattices: graphene as an example, Phys. Rev. B 76, 184430 (2007)

[558] S. Power, M.S. Ferreira, Indirect exchange and Ruderman–Kittel–Kasuya–Yosida (RKKY) interactions in magnetically-doped graphene, Crystal 3, 49 (2013)

[559] Z. Zanolli, J.C. Charlier, Single nanotube sensing using carbon nanotubes decorated with magnetic clusters, ACS Nano, 6, 10786 (2012)

[560] Y.K. Chen, A. Chu, J. Cook, M.L.H. Green, P.J.F. Harris, R. Heesom, M. Humphries, J. Sloan, S.C. Tsang, J.F.C. Turner, Synthesis of carbon nanotubes

containing metal oxides and metals of the d-block and f-block transition metals and related studies, Jour. Mater. Chem. 7, 545 (1997)

[561] F.C. Dillon, A. Bajpai, A. Koós, S. Downes, Z. Aslam, N. Grobert, Tuning the magnetic properties of iron-filled carbon nanotubes, Carbon 50, 3674 (2012)

[562] A. Winkler, T. Mühl, S. Menzel, R. Kozhuharova-Koseva, S. Hampel, A. Leonhardt, B.Büchner, Magnetic force microscopy sensors using iron-filled carbon nanotubes, J. Appl. Phys. 99, 104905 (2006)

[563] I.Mönch, A. Leonhardt, A. Meye, S. Hampcl, R. Kozhuharova-Koseva, D. Elefant, M.P. Wirth, B.Büchner, Synthesis and characteristics of Fe-filled multi-walled carbon nanotubes for biomedical application, Jour. Phys.: Conf. Ser. 61, 820 (2007)

[564] S.L. Prischepa, A.L. Danilyuk, A.L. Prudnikava, I.V. Komissarov, V.A. Labunov, K.I. Yanushkevich, F. Le Normand, Magnetic properties of nanocomposites based on magnetically functionalized carbon nanotubes, in: *Nano Magnetism*, J.M. Gonzalez Estevez, eds., One Central Press (OCP) 2014

[565] D. Qian, E.C. Dickey, R. Andrews, D. Jacques, Growth mechanisms and kinetics of CVD-derived carbon nanotubes, Center for Applied Energy Research, Univ. of Kentucky, *www.caer.uky.edu /carbon/posters/qian.pdf*

[566] U. Ritter, P. Scharff, G.E. Grechnev, V.A. Desnenko, A.V. Fedorchenko, A.S. Panfilov, Y.I. Prylutskyy, Y.A. Kolesnichenko, Structure and magnetic properties of multi-walled carbon nanotubes modified with cobalt, Carbon 49, 4443 (2011)

[567] L. Grover, Quantum computers can search arbitrarily large databases by a single query, Phys. Rev. Lett. 79, 4709 (1997)

[568] O. Boykin, T. Mor, V. Roychowdhury, F. Vatan, Algorithms on ensemble quantum computers, Nat. Computing 9, 329 (2010)

[569] A.V. Rode, S.T. Hyde, E.G. Gamaly, R.G. Elliman, D.R. McKenzie, S. Bulcock, Structural analysis of a carbon foam formed by high pulse-rate laser ablation, Appl. Phys. A: Mater. Sci. Process. A 69, S755 (1999)

[570] A.V. Rode, E.G. Gamaly, B. Luther-Davies, Formation of cluster-assembled carbon nano-foam by high repetition-rate laser ablation, Appl. Phys. A: Mater. Sci. Process. A70, 135 (2000)

[571] P.J.F. Harris, Fullerene-related structure of commercial glassy carbons, Phil. Mag. 84, 3159 (2004)

[572] D. Arčon, Z. Jagličič, A. Zorko, A.V. Rode, A.G. Christy, N.R. Madsen, E.G. Gamaly, B. Luther-Davis, Origin of magnetic moments in carbon nanofoam, Phys. Rev. B 74, 014438 (2006)

[573] A.V. Rode, E.G. Gamaly, A.G. Christy, J.G. Fitz Gerald, S.T. Hyde, R.G. Elliman, B. Luther-Davies A.I. Veinger, J. Androulakis, J. Giapintzakis, Unconventional magnetism in all-carbon nanofoam, Phys. Rev. B 70, 054407 (2004)

[574] A.V. Rode, R.G. Elliman, E.G. Gamaly, A.I. Veinger, A.G. Christy, S.T. Hyde, B. Luther-Davies, Electronic and magnetic properties of carbon nanofoam produced by high-repetition-rate laser ablation, Appl. Surf. Sci. 197, 644 (2002)

[575] H.A. Schwarz, *Gesammelte Mathemtische Abhandlungen*, Vols. 1 and 2, Springer, Berlin 1890

[576] L. Colombo, A. Fasolino. *Computer-Based Modeling of Novel Carbon Systems and Their Properties: Beyond Nanotubes*, Springer Science & Business Media 2010

[577] T. Lenosky, X. Gonze, M. Teter, V. Elser, Energetics of negatively curved graphitic carbon, Nature 355, 333 (1992)

[578] D. Vanderbilt, J. Tersoff, Negative-curvature fullerene analog of C_{60}, Phys. Rev. Lett. 68, 511 (1992)

[579] P.K. Chu, L. Li, Characterization of amorphous and nanocrystalline carbon films, Mat. Chem. Phys. 96, 253 (2006)

[580] W. Assman, H. Huber, C. Steinhausen, M. Dobler, H. Glückler, A. Weidinger, Elastic recoil detection analysis with heavy atoms, Nucl. Inst. Meth. B89, 131 (1994), 131 (1994)

[581] A.V. Rode, A.G. Christy, N.R. Madsen, E.G. Gamaly, S.T. Hyde, B. Luther-Davies, Positive magnetisation in carbon nanostructures, Curr. Appl. Phys. 6, 549 (2006)

[582] C. Mathioudakis, P.C. Kelires, Atomistic simulations of low-density nanoporous materials: carbon nanofoams, Phys. Rev. B 87, 195408 (2013)

[583] J. Tersoff, Empirical interatomic potential for carbon, with applications to amorphous carbon, Phys. Rev. Lett. 61, 2879 (1988)

[584] A.A. Deshmukh, S.D. Mhlanga, N.J. Coville, Carbon spheres, Mat. Sci. Eng. R 70, 1 (2010)

[585] S. Iijima, Direct observation of the tetrahedral bonding in graphitized carbon black by high-resolution electron microscopy, J. Cryst. Growth 50, 675 (1980)

[586] D. Ugarte, Curling and closure of graphitic networks under electron-beam irradiation, Nature 359, 707 (1992)

[587] V.L. Kuznetsov, A.L. Chuvilin, Y.V. Butenko, I.VMalkov, V.M. Titov, Onion-like carbon from ultra-disperse diamond. Chem. Phys. Lett. 222, 343 (1994)

[588] L Hawelek, A. Brodka, S. Tomita, J.C. Dore, V. Honkimaki, A. Burian, Transformation of nano-diamonds to carbon nano-onions studied by x-ray diffraction and molecular dynamics, Diamond Rel. Mat. 20, 1333 (2011)

[589] J.K. McDonough, Y. Gogotsi, Carbon onions: synthesis and electrochemical applications, The Electrochemic. Soc. Interface, Fall 2013, 61 (2013)

[590] C.P. Chen, T.H. Chang, T.F. Wang, Synthesis of magnetic nano-composite particles, J. Ceram. Int. 28, 925 (2002)

[591] G.H. Lee, S.H. Huh, J.W. Jeong, H.C. Ri, Excellent magnetic properties of fullerene encapsulated ferromagnetic nanoclusters, J. Mag. Mag. Mat. 246, 404 (2002)

[592] A.-Y. Ge, B.-S. Xu, X.-M. Wang, T.B. Lee, P.D. Han, X.G. Liu, Study on electromagnetic property of nano onion-like fullerenes, J. Acta Physico-Chimica Sinica, 22, 203 (2006)

[593] R.S. Ruoff, D.C. Lorents, B. Chan, R. Malhotra, S. Subramoney, Single crystal metals encapsulated in carbon nanoparticles, Science 259, 346 (1993)

[594] S.A. Majetich, J.O. Artman, M.E. McHenry, N.T. Nuhfer, S.W. Staley, Preparation and properties of carbon-coated magnetic nanocrystallites, Phys. Rev. B 48, 16845 (1993)

[595] M.E. McHenry, S.A. Majetich, J.O. Artman, M. DeGraef, S.W. Staley, Superparamagnetism in carbon-coated Co particles produced by the Kratschmer carbon arc process, Phys. Rev. B 49, 11358 (1994)

[596] J. Goldstein, *Scanning Electron Microscopy and X-ray Microanalysis*, 3rd ed., Springer 2003

[597] V.P. Dravid, J.J. Host, M.H. Teng, B. Elliott, J. Hwang, D.L. Johnson, T.O. Mason, J.R. Weertman, Controlled-size nanocapsules, Nature 374, 602 (1995)

[598] Y. Lu, Z. Zhu, Z. Liu, Carbon-encapsulated Fe nanoparticles from detonation-induced pyrolysis of ferrocene, Carbon 43, 369 (2005)

[599] T. Hayashi, S. Hirono, M. Tomita, S. Ummemura, Magnetic thin films of cobalt nanocrystals encapsulated in graphite-like carbon, Nature 381, 772 (1996)

[600] Z.H. Wang, C.J. Choi, B.K. Kim, J.C. Kim, Z.D. Zhang, Characterization and magnetic properties of carbon-coated cobalt nanocapsules synthesized by the chemical vapor condensation process, Carbon 41, 1751 (2003)

[601] R.P. Chaudhary, S.K. Mohanty, A.R. Koymen, New method for synthesis of Pt nanoparticles embedded in a carbon matrix. J. Nanosci. Nanotechnol. 11 10396 (2011)
[602] R.P. Chaudhary, S.K. Mohanty, A.R. Koymen, Novel method for synthesis of Fe core and C shell magnetic nanoparticles, Carbon 79, 67 (2014)
[603] V.N. Mochalin, O. Shenderova, D. Ho, Y. Gogotsi, The properties and applications of nanodiamonds, Nature Nanotech. 7, 11 (2012)
[604] G.W. Yang, J.B. Wang, Q.X. Liu, Preparation of nano-crystalline diamonds using pulsed laser induced reactive quenching, J. Phys. Condens. Mat. 10, 79237927 (1998)
[605] M. Frenklach, W. Howard, D. Huang, J. Yuan, K.E. Spear, R. Koba. Induced nucleation of diamond powder. Appl. Phys. Lett. 59, 546 (1991)
[606] T.L. Daulton, M.A. Kirk, R.S. Lewis, L.E. Rehn, Production of nanodiamonds by high-energy ion irradiation of graphite at room temperature, Nucl. Instrum. Meth. B 175, 12 (2001)
[607] F. Banhart, P.M. Ajayan, Carbon onions as nanoscopic pressure cells for diamond formation, Nature 382, 433 (1996)
[608] É.M. Galimov, A.M. Kudin, V.N. Skorobogatskii, V.G. Plotnichenko, O.L. Bondarev, B.G. Zarubin, V.V. Strazdovskii, A.S. Aronin, A.V. Fisenko, I.V. Bykov, A.Y. Barinov, Experimental corroboration of the synthesis of diamond in the cavitation process, Dokl. Phys. 49, 150 (2004)
[609] J.E. Dahl, S.G. Liu, R.M.K. Carlson, Isolation and structure of higher diamondoids, nanometer-sized diamond molecules, Science 299, 96 (2003)
[610] J.Y. Raty, G. Galli, Ultradispersity of diamond at the nanoscale, Nature Mater. 2, 792795 (2003)
[611] O.E. Andersson, B.L.V. Prasad, H. Sato, T. Enoki, Y. Hishiyama, Y. Kaburagi, M. Yoshikawa, S. Bandow, Structure and electronic properties of graphite nanoparticles, Phys. Rev. B 58, 16387 (1998)
[612] B.L.V. Prasad, H. Sato, T. Enoki, Y. Hishiyama, Y. Kaburagi, A.M. Rao, P.C. Eklund, K. Oshida, M. Endo, Heat-treatment effect on the nanosized graphite p-electron system during diamond to graphite conversion, Phys. Rev. B 62, 11209 (2000)
[613] M. Mezard, G. Raris, M.R. Virasoro, *Spin Glass Theory and Beyond*, World Scientific, New Jersey 1987
[614] T. Enoki, M. Suzuki, M. Endo, *Graphite Intercalation Compounds and Applications*, Oxford University Press, New York 2003
[615] K. Takai, S. Eto, M. Inaguma, T. Enoki, H. Ogata, M. Tokita, J. Watanabe, Magnetic potassium clusters in a nanographite host system, Phys. Rev. Lett. 98, 017203 (2007)
[616] T. Kyotani, T. Nagai, S. Inoue, A. Tomita, Formation of new type of porous carbon by carbonization in zeolite nanochannels, Chem. Mater. 9, 609 (1997)
[617] K. Takai, T. Suzuki, T. Enoki, H. Nishihara, T. Kyotani, Structure and magnetic properties of curved graphene networks and the effects of bromine and potassium adsorption, Phys. Rev. B 81, 205420 (2010)
[618] K. Takai, T. Suzuki, H. Nishihara, T. Kyotani, T. Enoki, Magnetic properties of host-guest material using network of curved nanocarbon sheet, Jour. Phys. Chem. Sol. 73, 1436 (2012)
[619] Y. Nozue, T. Kodaira, T. Goto, Ferromagnetism of potassium clusters incorporated into zeolite LTA, Phys. Rev. Lett. 68, 3789 (1992)

[620] Y. Nozue, T. Kodaira, S. Ohwashi, T. Goto, O. Terasaki, Ferromagnetic properties of potassium clusters incorporated into zeolite LTA, Phys. Rev. B 48, 12253 (1993)
[621] G.M. Canfield, M. Bizimis, S.E. Latturner, Sodalite ion exchange in polyethylene oxide oligomer solvents, Jour. Mat. Chem. 17, 4530 (2007)
[622] K. Takai, H. Kumagai, H. Sato, T. Enoki, Bromine-adsorption-induced change in the electronic and magnetic properties of nanographite network systems, Phys. Rev. B 73, 035435 (2006)
[623] K. Nakamura, T. Koretsune, R. Arita, *Ab initio* derivation of the low-energy model for alkali-cluster-loaded sodalites, Phys. Rev. B 80, 174420 (2009)
[624] K. Murata, H. Ushijima, Effect of pyrolysis temperature on the magnetic properties of the carbon materials prepared from trialkylboranes, J. Appl. Phys. 79, 978 (1996)
[625] T. Saito, D. Nishio-Hamane, S. Yoshii, T. Nojima, Ferromagnetic carbon materials obtained from polyacrylonitrile, J. Appl. Phys. 98, 052506 (2011)
[626] A.A. Ovchinnikov, V.N. Spector, Organic ferromagnets. New results, Synth. Met. 27, B615 (1988)
[627] H. Araki, R. Matsuoka, K. Yoshino, Ferromagnetic behavior of pyrolyzed o,m,p-phenylenediamine and triazine derivatives, Solid State Comm. 79, 443 (1991)
[628] K. Tanaka, M. Kobashi, H. Sanekafa, A. Takata, T. Yamabe, S. Mizogami, K. Kawabata, Y. Yamauchi, Peculiar magnetic property of pyrolytic carbon prepared from adamantane, J. Appl. Phys. 71, 836 (1992)
[629] T. Saito, Magnetic properties of carbon materials prepared from polyvinyl chloride, J. Appl. Phys. 105, 013902 (2009)
[630] T. Makarova, Ferromagnetic carbonaceous compounds, in *Carbon-Based Magnetism*, Elsevier Science 2006
[631] K. Kostarelos, The long and short of carbon nanotube toxicity, Nature Biotech 26, 774 (2008)
[632] Y. Liu, Y. Zhao, B. Sun, C. Chen, Understanding the toxicity of carbon nanotubes, Accts Chem. Res. 46, 702 (2013)
[633] D. Yang, F. Yang, J. Hu, J. Long, C. Wang, D. Fu, Q. Ni, Hydrophilic multi-walled carbon nanotubes decorated with magnetite nanoparticles as lymphatic targeted drug delivery vehicles, Chem. Commun. 29, 4447 (2009)
[634] V. Georgakilas, M. Otyepka, A.B. Bourlinos, V. Chandra, N. Kim, K.C. Kemp, P. Hobza, R. Zboril, K.S. Kim, Functionalization of graphene: covalent and non-covalent approaches, derivatives and applications, Chem. Rev. 112, 6156 (2012)
[635] I.R. Vlahov, C.P. Leamon, Engineering folate-drug conjugates to target cancer: from chemistry to clinic, Bioconj. Chem. 23, 1357 (2012)
[636] R. Li, R. Wu, L. Zhao, Z. Hu, S. Guo, X. Pan, H. Zou, Folate and iron difunctionalized multiwall carbon nanotubes as dual-targeted drug nanocarrier to cancer cells, Carbon 49, 1797 (2011)
[637] S. Boncel, A.P. Herman, K.Z. Walczak, Magnetic carbon nanostructures in medicine, J. Mater. Chem. 22, 31 (2012)
[638] X. Yang, X. Zhang, Y. Ma, Y. Huang, Y. Wang, Y. Chen, Superparamagnetic graphene oxide Fe_3O_4 nanoparticles hybrid for controlled targeted drug carriers, Jour. Mater. Chem. 19, 2710 (2009)
[639] R. Weissleder, P.F. Hahn, D.D. Stark, E. Rummeny, S. Saini, J. Wittenberg, J.T. Ferrucci, MR imaging of splenic metastases: ferrite-enhanced detection in rats, AJR Am. J. Roentgenol. 149, 723 (1987)

[640] for instance: M.T. Vlaardingerbroek, J.A. den Boer, *Magnetic Resonance Imaging*, 2nd ed., Springer, Berlin 1999

[641] H.Günther, *NMR Spectroscopy: Basic Principles, Concepts and Applications in Chemistry*, 3rd ed., VCH Wiley 2013

[642] T.N. Narayanan, A.P. Reena Mary, M.M. Shaijumon, L. Ci, P.M. Ajayan, M.R. Anantharaman, On the synthesis and magnetic properties of multiwall carbon nanotube superparamagnetic iron oxide nanoparticle nanocomposites, Nanotechnology 20, 055607 (2009)

[643] E. Amstad, M. Textor, E. Reimhult, Stabilization and functionalization of iron oxide nanoparticles for biomedical applications, Nanoscale 3, 2819 (2011)

[644] K.J. Murphy, J.A. Brunberg, R.H. Cohan, Adverse reactions to gadolinium contrast media: a review of 36 cases, Am. Jour. Roentgenology 167, 847 (1996)

[645] H.S. Thomsen, S.K. Morcos and P. Dawson, Is there a causal relation between the administration of gadolinium-based contrast media and the development of nephrogenic systemic fibrosis?, Clin. Radiology 61, 905 (2006)

[646] L. Wilson, Medical applications of fullerenes and metallofullerenes, Elec. Chem. Soc. Interface, Winter 1999, 24 (1999)

[647] M. Mikawa, H. Kato, M. Okumura, M. Narazaki, Y. Kanazawa, N. Miwa, H. Shinohara, Paramagnetic water-soluble metallofullerenes having the highest relaxivity for MRI contrast agents, Bioconjugate Chem. 12, 510 (2001)

[648] B. Sitharaman, K.R. Kissell, K.B. Hartman, L.A. Tran, A. Baikalov, I. Rusakova, Y. Sun, H.A. Khant, S.J. Ludtke, W. Chiu, S. Laus, E. Toth, L. Helm, A.E. Merbach, L.J. Wilson, Superparamagnetic gadonanotubes are high-performance MRI contrast agents, Chem. Commun. 3915 (2005)

[649] K.H. Mueller, L. Dunsch, D. Eckert, M. Wolf, A. Bartl, Magnetic properties of endohedral fullerene complexes, Synthetic Materials 103, 2417 (1999)

[650] E. Toth, R.D. Bolskar, A. Borel, G. Gonzalez, L. Helm, A.E. Merbach, B. Sitharaman, L.J. Wilson, Water soluble gadofullerenes toward high-relaxivity, pH responsive MRI contrast agents, J. Am. Chem. Soc. 127, 799 (2005)

[651] S. Stevenson, R.R. Stephen, T.M. Amos, V.R. Cadorette, J.E. Reid, J.P. Phillips, Synthesis and purification of a metallic nitride fullerene bisadduct: exploring the reactivity of $Gd_3@C_{80}$, J. Am. Chem. Soc. 127, 76 (2005)

[652] P.P. Fatouros, F.D. Corwin, Z.J. Chen, W.C. Broaddus, J.L. Tatum, B. Kettenmann, Z. Ge, H.W. Gibson, J.L. Russ, A.P. Leonard, J.C. Duchamp, H.C. Dorn, In vitro and in vivo imaging studies of a new metallofullerene nanoparticle, radiology 240, 756 (2006)

[653] J. Zhang, P.P. Fatouros, C. Shu, J. Reid, L.S. Owens, T. Cai, H.W. Gibson, G.L. Long, F.D. Corwin, Z.J. Chen, H.C. Dorn, High relaxivity trimetallic nitride (Gd_3N) metallofullerene MRI contrast agents with optimized functionality, Bioconjug Chem. 21, 610 (2010)

[654] E.-Y. Zhang, C.-Y. Shu, L. Feng, C.-R. Wang, Preparation and characterization of two new water-soluble endohedral metallofullerenes as magnetic resonance imaging contrast agents, J. Phys. Chem. B 111, 14223 (2007)

[655] K. Braun, L. Dunsch, R. Pipkorn, M. Bock, T. Baeuerle, S. Yang, W. Waldeck, M. Wiessler, Gain of a 500-fold sensitivity on an intravital MR contrast agent based on an endohedral gadolinium-cluster-fullerene-conjugate: a new chance in cancer diagnostics, Int. Jour. Med. Sci. 7, 136 (2010)

[656] A. Svitova, K. Braun, A.A. Popov, L. Dunsch, A platform for specific delivery of lanthanide scandium mixed-metal cluster fullerenes into target cells, Chem. Open 1, 207 (2012)

[657] H.L. Fillmore, M.D. Shultz, S.C. Henderson, P. Cooper, W.C. Broaddus, Z.J. Chen, C.Y. Shu, J. Zhang, J. Ge, H.C. Dorn, F. Corwin, J.I. Hirsch, J. Wilson, P.P. Fatouros, Conjugation of functionalized gadolinium metallofullerenes with IL-13 peptides for targeting and imaging glial tumors, Nanomedicine 6, 449 (2011)

[658] K.B. Hartman, S. Laus, R.D. Bolskar, R. Muthupillai, L. Helm, E. Toth, A.E. Merbach, L.J. Wilson, Gadonanotubes as Ultrasensitive pH-smart probes for magnetic resonance imaging, Nanolett. 8, 415 (2008)

[659] A. Bianco, K. Kostarelos, C.D. Partidos, M. Prato, Biomedical applications of functionalised carbon nanotubes, Chem. Commun. 571 (2005)

[660] D. Pantarotto, J. Briand, M. Prato, A. Bianco, Translocation of bioactive peptides across cell membranes by carbon nanotubes, Chem. Commun. 16 (2004)

[661] C.M. Sayes, F. Liang, J.L. Hudson, J. Mendez, W. Guo, J.M. Beach, V.C. Moore, C.D. Doyle, J.L. West, W.E. Billups, K.D. Ausman, V.L. Colvin, Functionalization density dependence of single-walled carbon nanotubes cytotoxicity in vitro, Toxicol. Lett. 161, 135 (2006)

[662] R. Singh, D. Pantarotto, L. Lacerda, G. Pastorin, C. Klumpp, M. Prato, A. Bianco, K. Kostarelos, Tissue biodistribution and blood clearance rates of intravenously administered carbon nanotube radiotracers, Proc. Natl. Acad. Sci. USA 103, 3357 (2006)

[663] J.M. Ashcroft, K.B. Hartman, Y. Mackeyev, C. Hofmann, S. Pheasant, L.B. Alemany, L.J. Wilson, Functionalization of individual ultra-short single-walled carbon nanotubes, Nanotechnology 17, 5033 (2006)

[664] A. Hashimoto, H. Yorimitsu, K. Ajima, K. Suenaga, H. Isobe, J. Miyawaki, M. Yudasaka, S. Iijima, E. Nakamura, Selective deposition of a gadoliunium (III) cluster in a hole opening of single-wall carbon nanohorn, Proc. Natl. Acad. Sci. USA 101, 8527 (2004)

[665] V.V. Mody, A. Cox, S. Shah, A. Singh, W. Bevins, H. Parihar, Magnetic nanoparticle drug delivery systems for targeting tumor, Appl. Nanosci. 4, 385 (2014)

[666] J.H. Park, G. Saravanakumar, K. Kim, I.C. Kwon, Targeted delivery of low molecular drugs using chitosan and its derivatives, Adv. Drug Deliv. Rev. 62, 28 (2010)

[667] Y. Hirota, Y. Akiyama, Y. Izumi, S. Nishijima, Fundamental study for development magnetic drug delivery system, Physica C 469, 1853 (2009)

[668] A. Al Faraj, A.P. Shaik, A.S. Shaik1, Magnetic single-walled carbon nanotubes as efficient drug delivery nanocarriers in breast cancer murine model: noninvasive monitoring using diffusion-weighted magnetic resonance imaging as sensitive imaging biomarker, Int. Jour. Nanomed. 10, 157 (2015)

[669] C. Tripisciano, K. Kraemer, A. Taylor, E. Borowiak-Palen, Single-wall carbon nanotubes based anticancer drug delivery system, Chem. Phys. Lett. 478, 200 (2009)

[670] S.P. Sherlock, S.M. Tabakman, L. Xie, H. Dai, Photothermally enhanced drug delivery by ultrasmall multifunctional FeCo/graphitic shell nanocrystals, ACS Nano 5, 1505 (2011)

[671] W.S. Seo, J.H. Lee, X.M. Sun, Y. Suzuki, D. Mann, Z. Liu, M. Terashima, P.C. Yang, M.V. McConnell, D.G. Nishimura, et al., FeCo/graphitic-shell nanocrystals as advanced magnetic-resonance-imaging and near-infrared agents, Nature Mater. 5, 971 (2006)

[672] K. Kataoka, T. Matsumoto, M. Yokoyama, T. Okano, Y. Sakurai, S. Fukushima, K. Okamoto, G.S. Kwon, Doxorubicin-loaded poly(ethylene glycol)-poly(beta-benzyl-Laspartate)copolymer micelles: their pharmaceutical characteristics and biological significance. Jour. Controlled Release 64, 143 (2000)

[673] K. Yang, S. Zhang, G. Zhang, X. Sun, S.-T. Lee, Z. Liu, Graphene in mice: ultra-high in vivo tumor uptake and efficient photothermal therapy, Nano. Lett. 10, 3318 (2010)

[674] J.T. Robinson, K. Welsher, S.M. Tabakman, S.P. Sherlock, H. Wang, R. Luong, H. Dai, High performance in vivo near-IR ($> 1\mu$m) imaging and photothermal cancer therapy with carbon nanotubes, Nano Res. 3, 779 (2010)

[675] F.H. Chen, L.M. Zhang, Q.T. Chen, Y. Zhang, Z.J. Zhang, Synthesis of a novel magnetic drug delivery system composed of doxorubicin-conjugated Fe_3O_4 nanoparticle cores and a PEG-functionalized porous silica shell, Chem. Commun. 46, 8633 (2010)

[676] R. Li, R. Wu, L. Zhao, M. Wu, L. Yang, H. Zou, P-glycoprotein antibody functionalized carbon nanotube overcomes the multidrug resistance of human leukemia cells, ACS Nano 4, 1399 (2010)

[677] X. Huang, C.S. Brazel, On the importance and mechanisms of burst release in matrix-controlled drug delivery systems, Jour. Controlled Release 73, 121 (2001)

[678] A. Jordan, R. Scholz, P. Wust, H.Fähling, R. Felix, Magnetic fluid hyperthermia (MFH): cancer treatment with AC magnetic field induced excitation of biocompatible superparamagnetic nanoparticles, Jour. Magn. Magn. Mater. 201, 413 (1999)

[679] R.K. Gilchrist, R. Medal, W.D. Shorey, R.C. Hanselman, J.C. Parrott, C.B. Taylor, Selective inductive heating of lymph nodes. Ann. Surgery 146, 596 (1957)

[680] A. Jordan, P. Wust, H.Fähling, W. John, A. Hinz, R. Felix, Inductive heating of ferrimagnetic particles and magnetic fluids: physical evaluation of their potential for hyperthermia, Int. Jour. Hyperthermia 9, 51 (1993)

[681] M. Johannsen, U. Gneveckow, L. Eckelt, A. Feissner, N. Waldörfner, R. Scholz, S. Deger, P. Wust, S.A. Loening, A. Jordan, Clinical hyperthermia of prostate cancer using magnetic nanoparticles: presentation of a new interstitial technique, Int. J. Hyperthermia 21, 637 (2005)

[682] E. Rosensweig, Heating magnetic fluid with alternating magnetic field, Jour. Magn. Magn. Mater. 252, 370 (2002)

[683] Y. Krupskaya, C. Mahn, A. Parameswaran, A. Taylor, K. Krämer, S. Hampel, A. Leonhardt, M. Ritschel, B.Büchner, R. Klingeler, Magnetic study of iron-containing carbon nanotubes: feasibility for magnetic hyperthermia, Jour. Magn. Magn. Mater. 321, 4067 (2009)

[684] A. Rycerz, J. Tworzydlo, C.W.J. Beenakker, Valley filter and valley valve in graphene, Nature Physics 3,172 (2007)

[685] J. Isberg, M. Gabrysch, J. Hammersberg, S. Majdi, K.K. Kovi, D.J. Twitchen, Generation, transport and detection of valley-polarized electrons in diamond, Nature Mat. 12, 760 (2013)

Index